Shells of the Philippines

Shells of the Philippines

F.J. SPRINGSTEEN & F.M. LEOBRERA

Conceived & Inspired by

CARLOS B. LEOBRERA

Editors & Publishers: CARLOS & FELY LEOBRERA
Copy Editor: TERESITA E. MORENO
Scientific Researcher: F. J. SPRINGSTEEN
Book Designers: OÑATE & PEREZ
Design Assistant: EMIL GONZALES
Artists: JUN GULAPA, ERIC SOCRATES,
ARTHUR RACELIS, BOBBY ROMERO
Photographer: NEAL OSHIMA
Production Assistant: JANE JOHNSON
Indexer: GLORIA F. PINEDA

Published 1986 by:
CARFEL SEASHELL MUSEUM
1786 A. Mabini St., Malate
M. Manila, Philippines

ISBN: 971-91029-0-x

Printed by Kyodo Printing Co., Inc.

This book is dedicated to —

my lovely wife Roseline, and my parents, Foster & Peggy.

my dear husband Charlie, my mother — Felisa Vda. de Moreno and my precious — Monette, Alan & Albert.

Foreword

Of all the natural regions of the world, the richest and most diverse marine life is found in the area known as the Indo-Polynesian province. This is a vast constellation of islands stretching across the equator from Malaysia to Polynesia. Near its centre lie the Philippines, which are themselves believed to harbour more species of molluscs than any other comparable area in the world. This is largely due to the incredible number and complexity of microhabitats available among its estimated 7,100 islands, to the potential for isolation offered thereby, and to the accelerated rate of speciation normally associated with tropical environment.

Despite (or perhaps because of) the richness of the Philippine molluscan fauna, our knowledge has grown very slowly. Indeed, much of the published information is based on collections made between 1836 and 1840 by the famous Hugh Cuming, who was the first collector to visit the archipelago. His material, acquired by both dredging and shore-collecting, was studied by the foremost conchologists of the day. Although subsequent collectors such as Semper, Quadras and Elera contributed much new data, the only catalogue for the region ever published — Hidalgo's *Catalogo de los Molluscos testaceos de las Islas Filipinas, Jolo y Marianas* (1904-1905) — still drew heavily on the sometimes unreliable records of Cuming. Even this superficial listing of Philippine marine molluscs totalled no less than 3,121 species.

In the eight decades that have followed Hidalgo's catalogue, enormous collections of Philippine shells have been made, both by scientific institutes and by amateurs, and many new species have been discovered. Since the 1970's, gill and tangle nets used by commercial fishermen in deep waters off the central and southern Philippines have brought to light a veritable treasure-trove of exciting shells. Localities such as Punta Engaño, Bohol and Davao have become magical names among shell collectors. From these localities have come unprecedented numbers of fabulous rarities such as *Cypraea valentia, C. leucodon, Pterynotus leobbeckei* and *Conus excelsus,* together with a host of new species such as the exquisite *Angaria vicdani, Pterynotus miyokoae* and *Latiaxis fruticosus.*

No identification guide to even the common Philippine molluscs exists, and the available information is scattered through hundreds of publications. Now, at long last, such a handbook is available. Conceived by the Leobreras and researched and written by James Springsteen, this sumptuous book should go a long way towards satisfying the demand. Jim Springsteen is a dedicated research worker with an extensive knowledge of Indo-Pacific molluscs. It is to his credit that **Shells of the Philippines** can boast the most up-to-date nomenclature and taxonomy possible. Carlos (Charlie) and Fely Leobrera of Carfel Seashell Museum will need no introduction to collectors. Their years of practical experience on Philippine shells and their sage advice have done much to make the project a success.

Shells of the Philippines covers nearly 1700 species, all illustrated in full colour. Most of the species likely to be found in the area are treated, plus many rarities, among which are a number figured in no other generally available work. The less popular and often neglected families such as the Turridae and Cerithiidae are well covered, and even land and freshwater molluscas are not overlooked.

I believe that this book will prove a boon to amateurs and professionals alike, and will undoubtedly stimulate further interest in this vast and complex fauna.

RICHARD N. KILBURN, Ph. D.
Curator of Molluscs Natal Museum South Africa

Contents

Preface

Shells of the Philippines is not, nor does it profess to be, a complete regional guide to all the molluscs occurring in Philipppine waters. Such an undertaking would require many years of work and research, and was never envisaged as the principal object of the book. Rather, this guide was designed to meet the needs of the ever-increasing number of collectors who require an identification manual for the species most likely to be encountered on a trip to the country. A broad selection of most molluscan groups have been covered, including a small section on Land Snails and Fresh Water Gastropods. Nearly all of the commonly encountered molluscan families have been touched on, some briefly, others in greater detail. Many species, some endemic, others unnamed, have never before been illustrated in any current identification guide, and are now, for the first time, figured and described.

The authors have refrained from a lengthy introduction in order to devote as much space as possible to the figuring and description of as many molluscan species as possible. Such topics as the biology and classification of molluscs, the cleaning of shells and the arrangement of a collection have been deliberately omitted, as it was felt that information on these topics is freely available in specialized literature.

Careful attention has been paid to taxonomic accuracy within the guide. At the start of each family section, a full classification is given, followed by a brief but concise outline of the salient features characterizing the family concerned. Within the APPENDIX section, a fuller account of the families, subfamilies, genera and subgenera used within the text is annotated, together with author and date.

The correct identification of molluscs is often the cause of much confusion and controversy among collectors and professional malacologists alike. The sciences of malacology and taxonomy are both extremely dynamic and in a state of constant flux, with new ideas and concepts presenting themselves everyday. It is little wonder that the collector often despairs of ever being able to correctly identify his shells. For this reason, the authors sought the help of various people, all acknowledged experts in particular groups, and the names annotated within this guide are considered by us to be the most accurate at the time of publication. As with any branch of science, there are the inevitable problem areas where disagreement occurs, even among experts. Many such problem species are illustrated here, and the name appended to them was arrived at only after considerable research had been undertaken, including the examination of type figures and descriptions, several of which have been incorporated within the text in order to justify the name utilized. Naturally, such type figures are subject to a variety of interpretation by different people, and this invariably accounts for the somewhat lengthy synonymy often associated with some species. The authors have deliberately refrained from annotating a lengthy synonymy within the text, as such a list would serve no useful purpose and can readily be obtained in monographs devoted to these problem species. In some cases however, some synonyms are included, particularly when a species has been known by a more popular name for some time and had been changed for taxonomic reasons.

The degree of morphological variation within a species population group is well documented. Frequently, what appears to be a distinct species is in fact no more than an ecological variation of an extremely polymorphic taxon. These apparently genetically isolated populations only appear as such because the geographical areas separating them have not been thoroughly investigated. Many taxa occurring in the Philippines often appear morphologically distinct from their counterparts in other geographical localities. The illustrated species and their corresponding descriptions are based on the form most frequently encountered within the confines of the archipelago, but should not be construed as an absolute guide to the morphology and colour of that taxon, especially if it is known to have a fairly extensive geographical distribution.

Conchological descriptions, of necessity, require the implementation of a specialized terminology, many terms of which may be unfamiliar to the collector. Where possible, the authors have kept descriptions simple and concise, but where any specialized terminology has been employed, a glossary can be found in the Appendix.

For a variety of reasons, some practical, others technical, no strict taxonomic order has been followed in presenting all the families covered, although some attempt has been made to assemble together closely related groups. Similarly, within a particular family, no attempt has been made to assemble the various species in any generic or subgeneric order, although similar looking species, and those within a particular size limit have, where possible, been grouped together. Although the majority of species illustrated within this guide have distributions outside the Philippines, the data presented only covers distribution within the archipelago proper.

The terms small, medium and large as used within the text of the guide are only meant to be considered in a comparative sense for members of the same family group, and not as an absolute criterion for all families discussed.

In several sections throughout the guide, the term *form* has been employed, especially in the section devoted to the Olividae. This terminology has no taxonomic validity, and has only been used to facilitate collectors in distinguishing different morphological and colour variations in an extremely variable taxon.

Strictly speaking, convention dictates that shells are correctly illustrated in a vertical position, with the anterior portion pointing upwards, and the posterior end pointing downwards. Based on this criterion, all gastropod species illustrated within this guide are depicted "upside-down" in a conventional sense. Apart from being aesthetically pleasing to the eye, it is an accepted practice among both amateur and professional malacologists alike, and the authors can see no valid reason for departing from it. In the case of bivalves, because of their distinct morphology, and in order to illustrate their salient characteristics, valves are displayed with their anterior and posterior portions along a horizontal axis, and the dorsal and ventral sections along a vertical axis. The Tridacnidae present a specialized case and collectors are referred to the appropriate section to understand the reason for this.

Shells of the Philippines is the end result of a lot of hard work by many world renowned malacological experts who unselfishly gave of their time and expertise in order to provide the authors with the groundwork to assemble a clear, concise identification guide to most of the Philippine shells likely to be encountered by a collector on a visit to the country. The task has not been easy and continually fraught with many difficulties and frustrations, but if we have succeeded in clarifying even a few of the problems and misconceptions surrounding the Philippine molluscan fauna, and have been instrumental in collectors gaining a better appreciation of their shells, then our task has been worthwhile.

F. J. Springsteen & F. M. Leobrera

Introduction

Philippine Shell Collecting: A Personal Recollection

The shells of the Philippines have been treasured by famous museums and leading collectors all over the world ever since conchology became a firmly entrenched hobby in the 17th century. What shell collector has not heard the famous story of the well-known British collector, Hugh Cuming "fainting with delight" on a Bohol reef after finding two specimens of *Conus gloriamaris.* This is surely the most well-remembered incident in the annals of shell collecting.

However, it is usually the European or American collectors who have visited the Philippines on famous international expeditions that become recorded in the literature. Much less well known is the fact that there have always been shell collectors within the Philippines, and that the preeminence of Philippine specimens in museums and collections all over the world can be directly attributed in a large part to this dedicated group. It seems appropriate, with the publication in the Philippines of this first comprehensive book on Philippine shells, that the contribution of these individuals be documented. Since I have been assigned the task of historian on this auspicious occasion, this history of recent Philippine conchologists will take the form of a personal memoir.

As a novice shell collector still in high school, enthusiastic and ignorant, I had the greatest pleasure of meeting Mr. Pedro de Mesa, the towering figure in Philippine shell collecting from the pre-war days. I met him in his last years, but his clear dedication to the scientific study of shells, as well as his gentlemanly way of giving advice to my untutored and naive mind made a deep impression on me. Another collector to make a lasting impression was Mr. Evaristo Zambo of Cebu. Mr. Zambo was in the shell business in the early fifties, the forerunner of the significant industry that Philippine shells are today. In many ways, as a shell collector, Mr. Zambo was and is unique: he knew the habits of living molluscs more than any other collector in the Philippines; he knew where to find them.

The vast majority of shell collectors in the Philippines have been attracted to the hobby simply from the sense of wonder, a sense of admiration for these remarkable natural objects. That was my own initial motivation. As a young collector starting out, I saw some of the noteworthy collections of the 50's. The Franco and Quisumbing collections in Cebu, the Sarenas collection in Davao, and the Bautista, Guerrero and Leonor collections in Manila.

A key event in the development of conchology in the Philippines was the formation of the Association of Philippine Shell Collectors (APSCOL). The organizational work was done by Eufrosinio Taningco, and under the early leadership of Paz L. Bautista, the interests of shell collectors became more focused and organized. Perhaps the key activity that really changed the world of Philippine conchology were the exhibits held by APSCOL, starting from the mid-fifties through the sixties. These were superbly organized by Fernando Dayrit; everyone who was at these exhibits can remember the enormous crowds that they drew, greatly spreading interest in Philippine shells as never before. It was a very pleasant group of collectors that participated in the APSCOL; among the most active were Gloria Guerrero and Mercedes Peregrino (who put together what was surely the most beautiful collection in the Philippines, a dazzling visual display), Vicente P. Tagle, Juliana Leonor, Donald and Victor Dan, Wendel Lim, Leonardo Naval, Fely and Charlie Leobrera, Soledad Lasquety, Amparo T. Liwanag, Consuelo Lumanlan, Manuel Montilla, Larry Gotuaco, Mary Payne, Victoria Javier and many many others.

What continuous excitement there was and has been in Philippine shell collecting! I can remember a few highlights from the last 30 years: a day after a typhoon when Gloria Guerrero and Mercedes Peregrino came to see me bringing in a box the first *Conus gloriamaris* found in the 20th century; my excitement as a 15-year old student preparing a very special shell collection for the Emperor of Japan, to be given as gift by the President of the Philippines; and finally, like fallen idols, one-by-one, the great rarities of shell collecting found in Philippine waters, became more and more available to the shell-collecting world: *Cypraea valentina; Cypraea guttata; Cypraea leucodon; Conus dusaveli; Conus excelsus; Conus cervus;* and equally exciting, the emergence of all the species never seen before. In the past 10 years, a remarkable number of new species have been named for Philippine shell collectors.

In the last three decades, Philippine shell collecting has changed in a myriad of ways. When I first went to Cebu and Zamboanga in the early 1950's, only Evaristo Zambo, Roque Cantos and Joe Co dealt in shells at all; today there are perhaps over 80 dealers. It is big business which not inconsequentially, has led to a rather commercially important shellcraft industry. Shells now support a significant population.

It is perhaps fitting to close this history with a tribute to the real collectors, the fishermen and divers, many of them names we will not know, who have learned when and where and how to look for shells. In the last decade, my initial feelings of admiration and aesthetic pleasure for the molluscan shell have undergone a gradual transformation into a serious scientific interest in the molecular biology and chemistry of the fish-eating cones. When I have a question about the biology of *Conus,* and I find that neither in books, nor from museum curators nor from learned biologists at academic institutions am I able to get the answers I seek, it is to this shell collector out in the field that I turn. Let us hope that the vast accumulation of knowledge of the Philippine field collector will begin to be recorded. There is much to learn about Philippine molluscs, and with the publication of this book, perhaps a new phase in the history of Philippine conchology has begun.

BALDOMERO M. OLIVERA, JR., Ph. D.
Prof. of Biology
University of Utah
Salt Lake, Utah, U.S.A.

[The author, known as "Toto" to his shell-collecting friends, has been an enthusiastic shell collector since he was nine years old. For many years he has been involved in a collaborative project on Conus research with Lourdes J. Cruz, Chairperson of the Biochemistry Department, University of the Philippines.]

Geography

The Philippine archipelago is composed of a vast network of some 7,107 islands, with a total land area of around 115,830 sq. miles, and a coastline which stretches for some 10,850 statute miles. The islands extend from Sibuto Island (4°40′ north latitude), to Y'Ami Island on the southern side of the Bashi Channel (21°5′ north latitude); and from 116°55′ to 126°35′ east longitude. The archipelago lies approximately midway between Taiwan in the north, and Borneo in the South, with the Western Pacific Ocean and the South China Sea bordering its western limits.

The coastline is extremely rugged and is dotted with many natural harbors and land-locked straits. The country is fairly mountainous, with the Cordillera, Sierra Madre and Caraballo mountain ranges of Central Luzon being the largest and most prominent. Many volcanoes dot the Philippine landscape, 10 of which are still active, and the country is frequently subjected to earth tremors. The origin of these special geographical features will be discussed in a later section.

Surrounding the various island groups of the archipelago are extensive coral reef systems, estimated to cover some 44,097 sq. kms. (Wells, 1982)[1], and made up of over 400 described species, making it one of the richest reef systems in the world.

Geographically, the islands of the Philippines are generally divided into three areas:

1. *The Luzon Island Group:* This includes Luzon Island (the largest of the islands - 40,814 sq. miles), the Batanes Islands to the far north, the islands of Romblon, Mindoro (3,794 sq. miles), Marinduque and Masbate, which lie within the Sibuyan Sea, Catanduanes which lies off the coast of Bicol, Corregidor Island just outside Manila Bay, and Palawan (4,500 sq. miles), which lies in the far southwest corner of the archipelago, between the South China Sea and the Sulu Sea.
2. *The Visayan Island Group:* These are the islands located within the center of the archipelago and consist of Samar (5,124 sq. miles), Leyte (2,799 sq. miles), Cebu (1,695 sq. miles), Bohol (1,534 sq. miles), Negros (4,903 sq. miles) and Panay (4,448 sq. miles).
3. *The Mindanao Island Group:* This is the most southerly group and is made up principally of Mindanao (the second largest island - 36,906 sq. miles), together with numerous smaller islands and island groups, including the Sulu Archipelago and Camiguin and Basilan Islands.

CLIMATE

The Philippine archipelago enjoys a typically tropical climate, with abundant rainfall, mild temperatures and a relatively high humidity. There are six recorded weather systems which control weather within the archipelago.

1. *Trade Wind effects:* This includes the northeast monsoons which occur between October and June, and the southwest monsoons which occur from July to September.
2. *The Inter-Tropical Convergent Zone* (ITCZ): This is essentially a zone formed from the meeting of northern and southern trade wind systems. It is a movable zone which in essence "follows" the sun. Between the autumnal equinox in September and the vernal equinox in March, the system shifts from North to South. A reversal of direction from South to North occurs between the months of April and August.
3. *Cyclonic effects:* These are large areas of intense low pressure which have high velocity winds. Known locally as "typhoons", these weather disturbances occur mainly between 7° and 10° North latitude, and an average of 18 to 20 occur annually in the Philippines.
4. *Local Disturbances:* The main factor at work here is a phenomenon termed "Orographic Lifting", and depends mainly on the prevailing trade winds and also the topography of the land affected. It appears to occur mainly between monsoon seasons. Warm, moisture-laden trade winds on meeting mountainous terrain are forced upwards. The decrease in air pressure allows air expansion, and the resultant cooling

[1] Wells, Susan M., 1982: *Marine Conservation in the Philippines and Papua New Guinea with special emphasis on the ornamental coral and shell trade.* (Report of the Conservation Monitoring Centre, Cambridge, England).

MAP OF THE PHILIPPINES
BABUYAN IS.
Ilocos
Cagayan
Nueva Vizcaya
LUZON
PACIFIC OCEAN
Polilio
Quezon City
Manila
Bataan
MANILA BAY
Batangas
Camarines Sur
CATANDUANES
MARINDUQUE
Albay
TAYABAS BAY
MINDORO
SORSOGON
Romblon
MASBATE
SAMAR
Culion
SOUTH CHINA SEA
VISAYAS
Antique
CUYO IS.
ILOILO
PUNTA ENGAÑO
LEYTE
CEBU
PALAWAN
NEGROS
BOHOL
Puerto Princesa
Dumaguete
Surigao
SULU SEA
Zamboanga
MINDANAO
Davao
Cotabato
Jolo
CELEBES SEA
SULU
SULU ARCHIPELAGO

results in condensation in the form of rain. This most notably occurs on the eastern side of this archipelago.

5. *Frontal Systems:* These are randomly occurring features of the Philippine weather picture and essentially represent the boundary between warm and cold air masses. They usually bring rain and thunderstorms.
6. *The Easterly Wave:* This is a series of wave-like perturbations moving east to west, and caused by the westward flowing current systems. The effect of these current systems is discussed in more detail in a later section.

The inter-action of all these systems results in three pronounced seasons: The Wet Season (June to November), the Cool, Dry Season (November to February), and the Hot Dry Season (March to May).

A Brief Geological and Biological History of the Philippines

The oldest geological rocks presently known from the Philippines are the cherts and associated schists, which date back to the Jurrasic period of the Mid-Mesozoic era. At that time, the Philippines was still a part of a larger land mass, and towards the end of the Jurrasic, initial uplifting and mountain formation is believed to have occurred. The actual formation of the archipelago is generally believed to have started during the middle Miocene, as marine sediments from earlier ages have not yet been recognized. At this time, large areas were lowered below the sea, and in the early Pliocene, extensive coral beds are believed to have been laid down.

Hydrographic and geological evidence indicates that prior to this land submersion, there existed four well-marked land bridges. The oldest was the connection between Northern Luzon and Taiwan, which itself was probably further connected to the Asian land mass. This is thought to have disappeared during Oligocene or Miocene times, probably as a result of violent earth movements along a well defined major fault line. A second land bridge occurred between Borneo and Palawan, and is believed to have disappeared during the late Pleistocene Epoch. The Sulu Archipelago once formed a stepping stone bridge between Southwest Mindanao and Borneo, and a fourth bridge is known to have occurred between the Celebes and the Sarangani Peninsula off Southern Mindanao.

From an evolutionary point of view, the molluscan fauna of the archipelago has remained virtually unchanged since middle Miocene times, as an examination of numerous Marine Pleistocene terraces throughout the islands has revealed no extinct species. As temperature change is a major factor in promoting molluscan evolution, it can be inferred that the Philippine climate has changed very little since the Pleistocene epoch. Indeed, there is considerable evidence which indicates that the archipelago was effectively buffered against the direct chill effect of the Pleistocene glaciation. However, indirect effects are known to have had a pronounced effect on the marine environment.

During the early stages of glaciation in the Northern Hemisphere, large volumes of water were removed from the tropical oceans with the result that many of the shallow seas within and around the archipelago were laid bare, and many larger island masses became interconnected with numerous land bridges. The opening of many new land portals no doubt facilitated the dissemination of many terrestrial groups, but in the case of marine molluscs, the opposite was true. With the lowering of the water level, the size of many marine habitats became smaller and more restricted, creating an increased biological pressure, which in essence necessitated rapid biological and environmental adjustment. This biological pressure was further augmented by the fact that during this time, many island masses were uplifted, as evidenced by the number of well-marked step terraces across many islands, especially Cebu. This uplifting was undoubtedly due to massive earth movements, which are still in evidence today. In addition, paleographic studies indicate that many of the deep areas around the archipelago were formed during this era.

With the advent of deglaciation in the late Pleistocene, the seas began to return to their original levels, effectively cutting the previously formed land bridges, but restoring the continuity of the marine environment. However, the presence of extremely deep areas around the islands provided effective physical barriers to molluscs, and undoubtedly antagonized any migration into wider habitats they had previously enjoyed. Other barriers no doubt also were responsible in part for antagonizing any molluscan migratory tendency. These factors will be discussed in more detail in the next section.

THE PHILIPPINES AS AN INSULATED BIOTA

A consideration of the geological history of the Philippines provides some insight into the physical processes which were responsible for the geographical separation of the archipelago from its parent land masses, into an isolated biota. In terms of insulation, the islands present a somewhat complex picture, and a brief account of the factors involved is necessary in order to understand the unique features characteristic of the Philippine molluscan fauna. From the ensuing discussion, it will be obvious that the Philippine Archipelago cannot be considered a typically insulated biota in terms of the characteristics which distinguish well-documented insular biotas, such as the Indo-Pacific island groups of Fiji and Hawaii.

1. *Geographical Isolation* This is perhaps the most important factor which characterizes truly insular biotas. Two consequences of this isolation are (a) a high degree of endemism and (b) a high gastropod:bivalve species ratio. In a typically isolated biota such as the Hawaiian Islands, endemicity has been estimated at 20% on a species and subspecies level. However, for the Philippines, a different picture emerges. Based on the 1,500 or so species described within this guide (excluding land snails and fresh water gastropods), only around 55 were considered to be endemic (approx. 4%). Similarly, the gastropod:bivalve species ratio has been quoted at 67:33 (approx. 2:1) for the Philippines[2] but is in the region of 5 or 6:1 for the Hawaiian Islands.[3] Such figures are not consistent with insular effects seen in geographically isolated biotas, and are not entirely unexpected, as the archipelago is geographically very close to the Asian land mass in the north, and the Malaysian land mass in the south.

   Presumably, the species most affected by these deep sea areas would be the ones which had been forced to partition into selective habitats, the boundaries of which had been critically defined. A case in point might well be illustrated by the Volutid species *Cymbiola (Aulica) aulica* (Sowerby I, 1825). This species is known to come only from a restricted area in the Sulu Archipelago, and although extremely close to Borneo, there have been no substantiated reports of the species occurring in that area. Similarly, no findings in waters further west have been as yet recorded.[4] This restricted distribution might be explainable in terms of its bathymetric isolation, due in part to the Sulu Sea Deep in the west, and the deep Sibutu passage lying to the south, midway between Tawi-Tawi and the Sibutu Islands. An extension of this idea might offer a plausible explanation to account for the occurrence of other species in discrete geographical localities within the Philippines (i.e. *Conus boholensis* and *Distorsio kurzi* are known only from Bohol, which is surrounded by extremely deep water).

   The direct effect of bathymetric isolation on the degree of insulation is difficult to assess, but indirect effects do play an important contributory feature in discussing the third type of isolation which is unique to the Philippines as an insulated biota

2. *Bathymetric Isolation:* Although being surrounded by areas of extremely deep water would probably have minimal insular effects on geographically isolated biotas, its occurrence around island groups close to major continental land masses would be of far greater importance. As previously discussed, the formation of many of the deep bathymetric zones around the archipelago during Pleistocene times no doubt constituted a physical barrier to subsequent radiation of molluscan species after the glacial melt. Among the more important of these deep sea

[2] E. Alison Kay, 1979: *Hawaiian Marine Shells*
[3] Figures based on an actual species count as annotated in *Hawaiian Marine Shells*

[4] Recently a few distinct forms of this species have been found in Palawan.

areas are the Philippine Trough (extending northwards from Samar), the Mindanao Deep (off the east coast of Mindanao, and interconnecting with the Philippine Trough), the Sulu Sea Deep, the Mindoro Deep and the Palawan Deep. There are numerous other deep sea areas in and around the archipelago, all of which lie on very active fault lines. These fault lines mark the boundaries of colliding tectonic plates, and represent areas of extremely violent earth movement. Such movement is responsible for the earthquake and volcanic activity still in evidence throughout the islands. When two tectonic plates come into collision, one plate may be forced under the other and eventually resorbed into the lower mantle of the earth. This process is termed subduction, and the deep troughs so common throughout the archipelago mark the graves of subducting plate edges.

3. *Climatic Isolation:* Paleogeographical evidence has shown that the Philippine marine climate has remained relatively stable since the Tertiary. Despite its large size in relation to other isolated biotas, the Philippine Archipelago shows remarkably little variation in ocean temperatures, ranging from about 29°C in the southern areas (approx. 4° North latitude) to around 26°C in the far north (approx. 19° North latitude), a change of only 3°C over 15° of latitude. However, just north of Taiwan, there is a sharp decrease in ocean temperature to around 14° or 15°C, over a very short distance, which no doubt marks the end of the zone of temperature insulation. This insular effect is more pronounced in waters on the eastern side of the archipelago, and extends outwards to a considerable distance. However, in waters to the west of the archipelago, this effect is less pronounced, as is evidenced by somewhat lower temperatures for points on the same latitude. The reason for this stability in ocean temperatures in the waters surrounding the Philippines is tied up with a complex interaction between ocean currents, trade winds, and land mass deflection.

The surface waters of the Central Pacific are blown by the Northeast Trade winds towards the Equator, in a westerly direction. On meeting the Philippine land mass, these waters are diverted northwards along the eastern border of the archipelago and mark the beginnings of the Kuro-Shio Current. This northerly deflection is probably further augmented by the extreme changes in submarine topography, marked by the deep areas off Mindanao, Samar, and Northern Luzon. The end result is a pronounced northward shifting of equatorial currents. Further west, equatorial currents are deflected into the South China Sea by the land masses of Java, Sumatra and the Malay Peninsula, but to a far lesser extent, as is evidenced by the temperature differences mentioned earlier. Such current deflections in equatorial waters have a remarkable insulating effect which is most pronounced within the Philippines. This effect is so strong in the waters east of the archipelago, that many equatorial marine species are able to live much farther north than would be possible in the Indo-Pacific at similar latitudes.

The importance of climatic insulation within the Philippines is reflected not so much in the uniqueness of the fauna, or even in the degree of endemism, but rather in the unusually high concentration and diversity of species one encounters. Although no definite figure is available, conservative estimates place the molluscan fauna of the archipelago at around 5,000 valid species with all major families fairly well represented. There is little doubt that the concentration of molluscan species in the Philippines is the highest anywhere in the Pacific, and probably, in the world. Although considerable research is still to be done, based on the currently accepted ideas relating to the geological history of the Indo-Pacific, it seems highly probable that much of the Central Pacific molluscan fauna was derived from areas around the Philippine Archipelago. An extremely high percentage of Indo-Pacific species is represented in the Philippine Molluscan fauna and, at a family level, the species count is higher than in any other Indo-West Pacific locality. This fact is strikingly demonstrated when one considers two well-known, mainly tropical gastropod families, namely, the Conidae and Cypraeidae. The Philippines has over 170 recorded *Conus* species, but this number drops dramatically to 60 in the Fijian Islands, and again to 40 in the Hawaiian Islands. Similarly, some 75 species of Cypraeidae are recorded from the Philippines, 60 from Fiji and 30 from Hawaii. Undoubtedly, a higher number of species than is presently recorded from these island groups, radiated outwards from far western areas in earlier geological times. Presumably, the lack of climatic insulation played an important part in the viability of many species, and only those capable of adapting to the different environmental conditions survived. These conditions were much more narrowly defined than those existing in the Philippines, and the increased biological pressures imposed by such restricted factors no doubt played a role in the adaptive evolutionary processes which resulted in the high degree of endemism commonly seen in these more geographically isolated biotas. This may well be an underlying cause in explaining why some marine molluscs are found in some tropical areas, but not in others. An examination of fossil deposits in several tropical areas seems to substantiate this contention. The species *Lyria santoensis*, *"Fasciolaria" walleri* and *Conus chiangi* are all recorded as living in the Philippines but are known only from fossil deposits in the New Hebrides (Ladd, 1975), an island group which is situated in similar geographic latitudes (but South of the Equator), enjoys similar tropical conditions to the Philippines, but lacks its climatic insulation.

Molluscan Distribution Patterns

An accurate knowledge of distribution patterns can provide the malacologist with a deeper insight into the ecology of many molluscan species and more importantly, can often be the starting point in the establishment of biological and taxonomic relationships between closely allied taxa. It is somewhat unfortunate that distribution patterns of a large number of Philippine molluscs is lacking, and the little that is known is, in many cases, fragmentary. There are numerous reasons for this, only a few of which will be discussed here.

For numerous socio-economic reasons, it is frequently impractical for trained field workers to visit many areas, especially many outlying islands groups. More often than not, collection points are often ascertained, solely on the word of island fishermen, whose area of operation often encompasses an exceedingly large geographic area, making it almost impossible to determine the nature and locality of biological habitats, with any degree of certainty. Coupled with this, fishermen are often reluctant to divulge exact collection points, especially if the material has a considerable commercial value. Such vague, non-specific localities as the Visayan Sea, Palawan, Sulu Sea or South China Sea are of little value to the malacologist trying to establish an accurate distribution profile for a particular species. Such responses by the fishermen are not intended as antagonistic, but merely reflect the desire to protect one's source of livelihood from a competitor.

Another factor which can complicate matters is the fact that many fishermen rely on agents to buy and sell their shells. These agents purchase direct from the fishermen, and then usually ship them back to central distribution points such as Cebu or Zamboanga. At this stage, it is almost impossible to ascertain the exact origin of most of the shells, with the result that the locality appended to many of these specimens is frequently the same as the distribution point.

Despite these somewhat formidable obstacles, much has been gleaned over the years, and slowly but surely, a clearer picture of the distribution characteristics and the biological habitats of many molluscs is starting to emerge.

A discussion of distribution patterns within the Philippine Archipelago at anything above a species or a subspecies level is ill-advised. Nevertheless, certain curious anomalies do warrant consideration, especially at a family level. Within the Cypraeidae, around 75% of the more commonly encountered species appear concentrated in the waters of Eastern Samar. Similarly, with the exception of 4 or 5 of fairly uncommon species, most members of the family Olividae appear to exhibit their highest concentration in the Sulu Sea. In the family Pleurotomariidae, all species recorded from the

Philippines appear to occur only in the deep waters of Southern Mindanao. Placing any exact connotation on these observations would be foolhardy, but speaking in a fairly broad sense, there appear to be environmental factors at work which convey a distinct biological advantage to these families, as evidenced by the high species concentration in the areas mentioned.

At species or subspecies level, distribution patterns need to be considered on an individual basis, although broadly speaking, two types can be observed. The first is essentially a general distribution pattern where the species involved occur in a fairly continuous mode throughout the archipelago. Mainly restricted to shallow intertidal levels, or to areas just below the low tide mark, the majority of these species are fairly common, and represent successful adaption to an environment which is subjected to a number of physical fluctuations, such as changes in water level, temperature and salinity. Examples of this group are represented by some members of the Muricidae (*Drupa morum, Thais savignyi*), the Angariidae (*Angaria delphinus*), the Patellidae (*Cellana testudinaria, Patelloida saccharina*), and an assemblage of species belonging to other families such as the Conidae, the Mitridae, and the Nassaridae.

The second type of distribution pattern observed within the archipelago is a more discontinuous one, composed of species which may fall into one of three categories. The first is represented by those species whose distribution is discontinuous and restricted to a few sporadic localities. Examples in this group are *Conus gloriamaris* (from Corregidor, Cebu, Bohol and Negros), *Cypraea valentia* (from Cebu and Bohol), and *Chicoreus annandalei* (from Cebu and Bohol). The second group are those species whose distribution is restricted to a single area, with a narrow but fairly well defined range. Two examples in this group are *Cymbiola aulica* (from the Sulu Archipelago), and *Angaria melanacantha* (only from a well defined area in Palawan). The third group is represented by those species whose distribution is restricted to a single, fairly isolated locality, with no appreciable geographical range extension. Two examples of this are *Conus pseudokimioi* (only from a small area around Siquijor Island), and *Distorsio constricta habei* (so far only recorded from a narrow area off Panglao, Bohol).

Although caution should be exercised in drawing too many conclusions regarding those species with a conspicuously discontinuous distribution, several facts are worthy of consideration, even if only in the most general terms. Firstly, an unusually large proportion of species in this second group appear to occur in fairly deep water, and presumably this reflects (at least in part) an adaption to a much narrower range of ecological conditions under which they can survive. Deep water habitats are less affected by environmental changes, and no doubt this is responsible for a more successful partitioning into selective niches. Such selective partitioning of necessity restricts species dissemination, and not surprisingly, many species within this group are either endemic to the archipelago, or are considered uncommon to rare.

Distribution patterns outside the Philippines are often curious, to say the least. Frequently, a species with a fairly wide Indo-West Pacific distribution is found to be absent from the Philippines. A case in point is the common species *Bufonaria rana* (BURSIDAE), which is known from the Indo-Pacific, New Guinea, Taiwan, China and Japan, but apparently not seen in the Philippines, where it is replaced by the superficially similar, endemic *Bufonaria subgranosa*.

Another curious distribution pattern is offered by those species occurring in the Philippines, which were originally described from geographically distant localities, and which have no documented occurrence from intermediate areas. Several examples are known including *Chicoreus venustulus* (from the Marquesa Islands), *Vexillum salisburyi* and *Bursa latitudo wolfei* (from Hawaii), *Tectarius grandinatus* (from South East Polynesia) and *Turridrupa prestoni* (from the Andaman Islands). Such erratic distribution patterns most likely reflect a poor knowledge of the fauna from intermediate areas, and, as more detailed collecting is done, these large gaps will probably be filled.

As well as access to differing geographic areas, an accurate determination of distribution patterns depends very heavily on adequate sampling being taken in the area under investigation. Such sampling will in turn be dependent on the way in which the material is collected. Within the Philippines, a large percentage of material is acquired by fishermen by the use of gill nets. These finely meshed nets are lowered into moderately deep water (up to 200 meters), and the material acquired is usually a rather "hit and miss" proposition. In reality, the large volume of material presently originating from Philippine waters is not so much a reflection of the efficiency of this means of shell collecting, but rather of the incredibly large number of nets which are often used in an area. Clearly, the degree of sampling in any area is far from adequate, and water depths often constitute a limiting factor on the type of material obtained. In addition, many of the smaller species are undoubtedly missed, as they would tend to pass through the net mesh. With the ever increasing interest in Philippine molluscan fauna, a number of expeditions have been mounted by overseas institutions, the most notable of which (in recent years) have been the Murostrom I Expedition (in 1976), and the Murostrom II (in 1980), under the supervision of the French Museum of Natural History. Although primarily concerned with Crustacea, these expeditions, by means of advanced trawling techniques, were able to perform more adequate sampling of the ocean floor, and more importantly, could work at much greater depths. In addition to many new species being uncovered, new locality records were established, one of the most notable being the discovery of *Thatcheria mirabilis* (TURRIDAE) in deep water off Mindoro. This species was originally considered restricted to Japanese waters, but access to its unique habitat, coupled with adequate sampling techniques, demonstrated that this taxon had a much wider range than had previously been suspected. The discovery of *Abyssochrysos melvilli*, a representative of a primitive group of snails which was common in the Paleozoic and Mesozoic eras, offered further evidence that the Philippine marine climate has changed very little since early geological times.

It is unfortunate that the prevailing economic climate of the Philippines has, at the present time, prohibited such well-equipped expeditions, which would undoubtedly solve many questions regarding the molluscan fauna of the Philippines. However, a somewhat indirect method of studying distribution patterns is available to the serious malacologist, and involves the common Carrier Shell species *Xenophora pallidula* Reeve. This remarkable species is relatively common in waters around the archipelago and has the peculiar habit of cementing shells and marine debris to its outer surface by means of a "nacreous glue". Because the species has a limited migratory range (it is extremely timid and slow-moving), an analysis of attached shells, especially the smaller "micro species", has in recent years provided some very interesting facts regarding distribution patterns of species not yet recorded within the Philippines. Two such examples are provided by the species *Microdrilla sagamiensis* (TURRIDAE) and *Homalopoma nocturnum* (TURBINIDAE). Both species were originally described from and believed restricted to Sagami Bay (Japan), but their occurrence in Philippine waters has been verified by their attachment to specimens of *Xenophora pallidula* taken in local waters and clearly demonstrate a much wider distribution than had previously been suspected. Many other smaller species have had their existence in the Philippines confirmed by this method, particularly some smaller members of the Turridae.

Two gastropod species occurring in the Philippines are worthy of special mention. The first is *Crucibulum spinosum*. Members of the genus *Crucibulum* are generally confined to the Western Atlantic and Tropical Eastern Pacific, although the species has been reported from Hawaii (possibly introduced). Although known to have a circumglobal distribution (Kay, 1979), its occurrence in Philippine waters is curious, considering the great distances involved. A second noteworthy species, *Carinodrilla quadrilirata* is recorded from the Philippines, and represents the first known record of the genus in the Western Pacific (Maes in corresp.). Unfortunately, very little can be inferred from these two discoveries until more intensive collecting is done in intermediate areas. ***

Acknowledgments

This book would not have come into existence were it not for the help of a number of people who gave so generously of their time and expertise. Many of these people are acknowledged experts in their fields, and it is to these people that we are especially indebted.

Mrs. V. Orr-Maes (Academy of Natural Sciences, Philadelphia, U.S.A.) - Turridae; Dr. E. H. Vokes (Tulane University, Louisiana, U.S.A.) - Muricidae; Mr. A. D'Attilio (Natural History Museum, San Diego, California, U.S.A.) - Coralliophilidae & Muricidae; W.O. Cernohorsky (Auckland Institute and Museum, Auckland, New Zealand) - Mitridae and Costellariidae; Dr. A. G. Beu (D.S.I.R., Lower Hutt, New Zealand) - Cymatiidae, Bursidae and Tonnidae; Fernando Dayrit (Quezon City, Philippines) - Land Snails; A. J. da Motta (Bangkok, Thailand) - Conidae.

We are grateful to our friend, Dr. Baldomero Olivera, Jr. (University of Utah, Salt Lake, Utah, U.S.A.) for writing a brief history on shell collecting in the Philippines and for his encouragement and support during the writing of our first shell book.

Although most of the specimens illustrated in this guide were taken from the Leobrera collection, several slides as well as other rare shells were generously loaned to us by friends.

To the following people, we extend our gratitude for supplying slides, specimens, personal observations and general encouragement:

Victor Dan (Manila, Philippines) for supplying many of the rarer specimens illustrated within this guide, and also sharing with us his detailed knowledge of the Philippine molluscan fauna, gained from his many years as a collector and dealer; Roger and Samia Martin (Cebu, Philippines) for slides and ecological information on the Conidae; T. C. Lan (Taiwan) for slides of some rare species; Jaime Cabrera (National Museum, Manila, Philippines) for allowing the authors access to museum material particularly many of the Bivalvia illustrated within the guide; Dr. Dieter Rockel (W. Germany) for photographs and additional information on the Conidae; Phillip W. Clover (Glen Ellen, California) for slides on Marginellidae; Manfred Blocher (W. Germany) for slides on rare Cypraea; Gloria P. Guerrero for allowing us to use some of her rare species for photography; Corazon Alvina for her valuable technical advise; and Nilo Valdez of Kyodo Printing Inc. (Manila) for preparatory and actual press production counselling. We also wish to thank our many friends who encouraged us to undertake this project and all those people who, in one way or another, gave their time, and shared their ideas in the production of this book.

SPECIAL ACKNOWLEDGMENT: We are deeply indebted to Dr. R. N. Kilburn (Natal Museum, South Africa). Dr. Kilburn has played an active part in this project since its inception, as well as providing identifications to and information on members of lesser known family groups (especially the Bivalvia). He was both mentor and advisor in the many problem areas encountered during the preparation of this guide. We are especially grateful to him for meticulously explaining the subtle differences between many morphologically similar taxa, and for providing numerous type figures and descriptions, some of which have been incorporated within the book.

Shells of the Philippines

Class Gastropoda
Subclass Prosobranchia
Order Archaegastropoda
Superfamily Pleurotomariacea
Family PLEUROTOMARIIDAE

A very small family of Gastropods, the members of which are the most primitive within the group. The shells are large, trochiform, may or may not be umbilicate, possess a characteristic slit band on all the whorls, and a slit on the outer margin of the aperture. The surface is finely granular, the interior nacreous, and the operculum horny, circular and multi-spiralled. All species live in very deep water and are quite rare. The Philippines contains the highest species concentration for the family, with 1 or 2 recent discoveries possibly representing new undescribed taxa. The family has only recently been discovered (over the last 7 years) within the archipelago and represents a new locality record for this primitive group. *Pleurotomaria hirasei* Pilsbry, 1903 was found very recently in Bohol, in deep water. Compared to the Japanese species, the specimens found here are lighter, smaller in size, and lighter in colour.

1. Perotrochus (Entemnotrochus) rumphii (Schepman, 1879)
128mm x 172mm; Taiwan.
Shell large, trochiform, width exceeding the height, sutures impressed, whorls angled with a slit band; sculptured with numerous close-set arcuate axial growth striae forming a finely granular surface with overriding spiral lirae; base rounded with a deep umbilicus; slit long, occupying almost half the body whorl; operculum much larger than the aperture which is nacreous within; colour yellowish white with fine, red axial streaks; reaches 195mm in diameter; recently confirmed from the Philippines, but specimens are virtually unobtainable at this time. They are morphologically identical to Taiwanese specimens (here illustrated).
DISTRIBUTION: Palawan and Southern Mindanao.

2. Perotrochus (Perotrochus) vicdani (Kosuge, 1980) (Paratype)
43.3mm x 45.8mm; Panglao, Bohol.
Shell small, trochiform, spire tall and sharply pointed, sutures weakly impressed, whorls angulate from a slit band which terminates in a short slit at the apertural margin; whorls finely beaded, base flat with fine, granulose spiral lirae, increasing in size towards the umbilicus which is shallowly excavated and calloused; shell creamy yellow with reddish brown axial streaks, base pale cream; reaches 44.5 mm x 49.5mm (Holotype dimensions).
DISTRIBUTION: Known only from Bohol.

3. Perotrochus (Perotrochus) teramachii Kuroda, 1955
103.7mm x 120.5mm; Japan.
Shell large, thin, broadly conical, with weakly impressed sutures and a finely granular surface formed from intersecting axial and spiral lirae; a smooth slit band on the middle of each whorl and a relatively long slit extending from the outer lip margin; the whorls are angulated by the slit band, and the umbilicus is narrowly open; colour pale orange-brown with darker axial streaking and banding; reaches up to 130mm in diameter. Recently reported from the Philippines, but specimens are currently unavailable. They appear to be morphologically indistinguishable from Japanese specimens (here illustrated).
DISTRIBUTION: So far only from Bohol.

Perotrochus (Mikadotrochus) salmiana (Rolle, 1889)
Balut Is., Southern Mindanao (Photo: V. Dan)

Perotrochus (Mikadotrochus) salmiana (Rolle, 1889)
85.0mm x 75.0mm; Balut Is., Southern Mindanao.
Shell large, solid, distinctly trochiform; sutures impressed, whorls strongly convex on the upper half, almost straight-sided on the lower half; a short but prominent slit at the apertural margin, continuing as a slit band and dividing the upper and lower area of each whorl; whorls sculptured with numerous granulose spiral lirae and weak axial threads; base with numerous fine spiral lirae; umbilicus calloused, slightly open and with a slight basal projection; colour pale orange-yellow with reddish brown streaks and blotches, a presutural row of brown quadrate spots, a pale coloured base, and a nacreous silver aperture; largest size seen illustrated here.
DISTRIBUTION: Southern Mindanao.

PLATE NO. 1 1. *Perotrochus rumphii* Schepman 2. *Perotrochus vicdani* (Kosuge) 3. *Perotrochus teramachii* (Kuroda)

PLATE 1

Class Gastropoda
Subclass Prosobranchia
Order Archaegastropoda
Superfamily Pleurotomariacea
Family HALIOTIDAE

A moderately large family of herbivorous, shallow-water gastropods which attach themselves to their rocky substrate via a large muscular foot. Shells are moderately large, ovate, have a depressed spire, a nacreous interior, a flattened columellar ridge, and may be smooth or have a corded or ribbed surface. There is a row of holes along the upper margin. Few Philippine species. On a commercial basis, the shells of this family are used in shellcraft and the animal is preserved and shipped abroad as a food commodity.

1. Haliotis (Haliotis) asinina Linné, 1758
83.6mm, 77.0mm; Lingayen Gulf, Pangasinan.
Shell medium sized, rather long, curved, width approximately half the shell length; sculptured with 3 or 4 ribs along the columellar ridge and numerous radial folds; the last 6 or 7 holes are open; brown-coloured with cream-coloured triangular patches, interior iridescent; reaches 120mm.
DISTRIBUTION: Throughout the Philippines.

2. Haliotis (Sanhaliotis) planata Sowerby, 1833
38.8mm, 41.7mm; Samar.
Shell small, ovate, apex rather depressed midway between the center and peripheral margin; the columellar margin has 4 to 6 granulose cords and numerous perforations, the last 4 or 5 being open; surface weakly tubercular, crossed by radial cords and weaker concentric striae; greenish brown with white mottling and stripes, interior nacreous silver; reaches 45mm.
DISTRIBUTION: Throughout the Philippines.

3. Haliotis (Ovinotis) ovina Gmelin, 1791
57.5mm, 56.4mm; Cebu.
Shell medium sized, ovate, spire sub-central and moderately elevated; surface extremely tubercular, with a strongly corded columellar ridge and a strongly elevated row of perforations, the last 5 or 6 of which are open; surface with coarse radial cords and weak concentric lirae; greenish to reddish brown with 5 or 6 white radial stripes, interior nacreous silver; reaches 70mm.
DISTRIBUTION: Throughout the Philippines.

4. Haliotis (Schismotis) glabra Gmelin, 1791
47.5mm, 44.7mm; Palawan.
Shell medium sized, elongate-ovate with a sunken spire; surface relatively smooth and sculptured with concentric grooves and overriding axial striae; 5 to 6 open, slightly raised perforations; columellar ridge flattened, partially obscuring the umbilical opening; external coloration greenish brown with white streaks and blotches; interior nacreous silver; reaches 50mm.
DISTRIBUTION: Throughout the Philippines

Class Gastropoda
Subclass Prosobranchia
Order Heterogastropoda
Suborder Ptenoglossa
Superfamily Architectonicacea
Family ARCHITECTONICIDAE

A comparatively small family of gastropods inhabiting sandy, shallow (occasionally deep) water in tropical and subtropical areas. The shells are discoidal, with a flat spire, a large body whorl, and a prominently open umbilicus which is bordered by a nodular keel. The family is fairly well represented in the Philippine molluscan faunal list, but only the larger and more commonly encountered species are illustrated here.

5. Architectonica (Architectonica) perspectiva (Linné, 1758)
33.0mm x 68.2mm, 34.2mm x 62.3mm; Manila Bay.
Shell medium to large sized, regularly coiled, depressed conic with well incised sutures; whorls with 2 deep grooves, one below the suture, the other above the bicingulate peripheral keel which is obscured by the sutures in earlier whorls, a further 2 grooves on the base bordering the umbilicus; surface sculptured with oblique grooves, forming elongated segments; umbilicus wide, bordered by elongate nodules; aperture sub-rhomboidal; externally greyish brown in colour, with black and white subsutural bands and 2 rows of black and white spots on the periphery and base; reaches 70mm; typical specimens from Manila Bay are usually around 20mm in diameter and have continuous subsutural banding; the specimen illustrated here is unusually large with the subsutural colour band broken up into spots on the last whorl.
DISTRIBUTION: Throughout the Philippines.

The following 2 species are poorly understood and recent work has indicated that the types of both *maxima* and *trochlearis* are different to conventional interpretations (fide Kilburn). The descriptions presented here are based on currently accepted ideas but may necessarily change as more research is done.

6. Architectonica (Architectonica) trochlearis (Hinds, 1844)
27.5mm x 55.4mm, 28.4mm x 56.0mm; Samar.
Extremely close to *A. maxima* (fig. 7), but generally

PLATE NO. 2 1. *Haliotis asinina* Linné 2. *Haliotis planata* Sowerby 3. *Haliotis ovina* Gmelin 4. *Haliotis glabra* Gmelin 5. *Architectonica perspectiva* (Linné) 6. *Architectonica trochlearis* (Hinds) 7. *Architectonica maxima* (Philippi) 8. *Architectonica fuliginosum* (Hinds) 9. *Architectonica modesta* (Philippi) 10. *Architectonica picta* (Philippi) 11. *Discotectonica acutissima* (Sowerby) 12. *Heliacus variegatus* (Gmelin) 13. *Heliacus dorsuosus* (Hinds) 14. *Philippia oxytropis* A. Adams 15. *Philippia radiata* (Röding)

PLATE 2

has a taller spire and only 2 grooves on the body whorl in contrast to the 3 seen in *maxima;* the brown and yellow sutural spotting appears to be larger and more intense in *A. trochlearis,* and the umbilical ridges are dark orange-brown, whereas in *maxima* they tend to be off-white; reaches 70mm in diameter.
DISTRIBUTION: Throughout the Philippines but in limited quantity.

7. Architectonica (Architectonica) maxima (Philippi, 1848)
32.9mm x 65.7mm, 32.5mm x 65.1mm; Palawan.
Shell large with a depressed spire and the whorls with 4 rows of gemmate cords which become obsolete on the body whorl; sutures deeply impressed, base convex with a deep peripheral groove and a granulose cord, bordered by 2 deep clefts next to a deep, sharply crenulate umbilicus; aperture subquadrate; colour greyish brown with white and brown bands above and below the sutures and 2 rows of brown spots at the peripheral keel and on the base; reaches 70mm in diameter; inhabits deep water.
DISTRIBUTION: Throughout the Philippines but in limited quantity.

8. Architectonica (Architectonica) fuliginosum (Hinds, 1844)
15.2mm x 30.9mm, 14.0mm x 29.5mm; Cebu.
Close to *A. perspectiva* of which it may only be a form; the brown band below the subsutural groove is replaced by long brown axial streaks. The oblique grooves which normally become obsolete on the body whorl persist in the 2 specimens illustrated here.
DISTRIBUTION: Manila Bay and Visayan Sea.

9. Architectonica (Architectonica) modesta (Philippi, 1848)
14.2mm x 27.4mm, 12.3mm x 25.3mm; Samar.
Very close to *A. perspectiva,* but differs in the following ways: the granulose subsutural cord is totally white, not brown and white as in *A. perspectiva;* the brown band below this cord has a sulcus below it, not seen in *A. perspectiva,* and the umbilicus is pure white, not brown as in *A. perspectiva;* reaches 30mm in diameter.
DISTRIBUTION: Manila Bay, Visayan Sea and Samar.

10. Architectonica (Architectonica) picta (Philippi, 1849)
27.5mm x 44.0mm, 21.0mm x 44.5mm; Palawan.
Shell medium sized, distinctly conoidal; surface sculpturing close to *A. maxima,* but the axial grooving is weaker, becoming obsolete on the last 2 whorls, and the spaces between the peripheral and sutural grooves have very finely incised spiral lines; colour brownish grey with rows of brown and white spots, a rather pale base, and white umbilical teeth with some brown spotting; reaches 45mm in diameter.
DISTRIBUTION: Palawan, Visayan Sea and Manila Bay.

11. Discotectonica acutissima (Sowerby, 1914)
11.0mm x 33.5mm; Leyte.
Shell small, discoidal, flat-spired and sharply keeled; sutures adpressed and the whorls sculptured with fine spiral striae and obsolete axial scratches; base moderately convex, flattened at the periphery and sculptured with axial and spiral threads; umbilicus wide with a finely crenulate margin, bordered by 2 or 3 weaker grooves; greyish white with pale yellow axial streaks; reaches 40mm in diameter. Normally taken from *Xenophora* species.
DISTRIBUTION: Mindoro and Visayan Sea.

12. Heliacus (Heliacus) variegatus (Gmelin, 1791)
9.5mm x 17.1mm, 9.5mm x 13.8mm, 9.0mm x 10.2mm; Sulu Archipelago.
Shell small, rather variable in sculpturing, being tall-spired or compressed; 4 spiral cords per whorl, segmented by oblique axial grooves, periphery well rounded with an additional 4 cords on the base, umbilicus narrow, umbilical cord divided into 2 segments by a deep groove; cream coloured with creamy white-and-brown mottling; reaches 20mm in diameter.
DISTRIBUTION: Sporadically distributed throughout the Philippines.

13. Heliacus (Torinista) dorsuosus (Hinds, 1844)
6.2mm x 12.2mm; Cebu.
Shell small, discoidal and flat-spired; whorls with 4 wide spiral cords, segmented by numerous axial grooves, 2 larger nodulose cords on the peripheral keel, separated by a deep groove; base with 4 finely grooved spiral cords; umbilicus narrow, deep and finely margined; colour red-brown with white and brown spots on the keel and sutures, umbilical channel white; reaches 15mm in diameter.
DISTRIBUTION: Visayan Sea and Camotes Sea.

14. Philippia (Psilaxis) oxytropis A. Adams, 1854
13.2mm x 18.5mm; Cebu.
Shell small, triangularly ovate, sutures impressed, spire whorls strongly convex, body whorl with a sharply angulate keel; whorls smooth except for 2 supra-sutural grooves and a group of 4 or 5 closely-set grooves at the body whorl periphery; base with obsolete axial and spiral threads, umbilicus with a dentate margin, surrounded by granulose spiral cords; shell golden brown with white tent patches at the sutures, umbilicus white with a brown-margined peri-umbilical cord; reaches 20mm in diameter.
DISTRIBUTION: Visayan Sea and Camotes Sea.

15. Philippia (Psilaxis) radiata (Röding, 1798)
6.5mm x 11.2mm, 6.1mm x 11.0mm; Cebu.
Shell small, discoidal, low-spired, relatively smooth, sutures weakly incised; whorls sculptured with 1 or 2 weak cords at the base; umbilicus narrow, bordered by a finely crenulate umbilical cord and a deep groove adjacent; colour deep brown with a white band at the sutures and on the body whorl keel, base greenish brown, umbilical channel white; reaches 25mm in diameter.
DISTRIBUTION: Visayan Sea and Camotes Sea.

Class Gastropoda
Subclass Prosobranchia
Order Archaegastropoda
Superfamily Patellacea
Family PATELLIDAE

A comparatively large family of herbivorous gastropods which live attached to rocks in the intertidal zone of most tropical and temperate seas. The shells are small to medium sized, cap or saucer shaped, have a conical or depressed spire, may be smooth or sculptured with radial and spiral cords, and may have an iridescent *(Cellana spp.)* or porcelaneous *(Patella spp.)* interior. Few recorded Philippine species.

1. Patella (Scutellastra) flexuosa flexuosa Quoy & Gaimard, 1834
29.2mm, 27.8mm; Palawan.
Shell small to medium sized, elongate-ovate, flat to moderately conical, sculptured with 8 to 9 rounded folds and numerous weaker, scabrous interstitial folds, all projecting at the margin; interior white with a yellowish orange spatula, exterior dull white with brown spots; reaches 40mm in length.
DISTRIBUTION: Throughout the Philippines.

2. Cellana rota (Gmelin, 1791)
Syn: "*C. radiata* (Born)" of Powell (1973) (non Born, 1778) (=*capensis* Gmelin, 1791).
An extremely variable taxon which is still very poorly understood. There are several recognizable "forms" two of which are considered here.
2a. typical form. Shell medium sized, broadly ovate, with a sub-central apex; sculpture regular, with numerous equi-sized radial riblets; colour off-white, with 11 to 12 black, doubled axial streaks, and 1 or 2 brown lines between; aperture nacreous silver, spatula brown to orange; reaches around 35mm in length; 27.8mm, 29.5mm; Samar.
2b. form *eneagona* Reeve, 1854: This form is characterized by its more elongate shape, coarser primary radial folds, irregular shell outline, and dark brown spatula; considered a subspecies but is sympatric with normal *rota;* 36.3mm, 18.5mm, 28.0mm; Samar.
DISTRIBUTION: Throughout the Philippines.

3. Cellana testudinaria (Linné, 1758)
73.2mm, 78.5mm; Palawan.
Shell large, broadly ovate with an elevated apex at the anterior third of the shell; sculptured with weak axial riblets and numerous concentric furrows; external colour dark brown with off-white chevrons and streaks, interior nacreous silver-blue; reaches 90mm in length.
DISTRIBUTION: Throughout the Philippines.

4. Cellana grata (Gould, 1859)
32.8mm; Palawan.
Shell medium sized, with a narrowly ovate outline and a moderately tall, anteriorly displaced spire; sculptured with numerous scaly radial ribs which are frequently eroded smooth, as is the spire; external colouration greyish yellow with broad, intermittent reddish radial bands, internally a dark brown spatula with peripheral brown maculations; reaches 56mm in length.
DISTRIBUTION: Throughout the Philippines but in limited quantity.

Class Gastropoda
Subclass Prosobranchia
Order Archaegastropoda
Superfamily Patellacea
Family ACMAEIDAE

Another large family of herbivorous gastropods which are almost morphologically identical to members of the Patellidae but can generally be separated by the possession of a porcelaneous, not iridescent interior, a dark coloured peripheral margin, and various anatomical characteristics. Few recorded Philippine species.

5. Patelloida striata (Quoy & Gaimard, 1834)
28.5mm, 30.8mm, 28.8mm; Samar.
Shell medium sized, lightweight, spire depressed, frequently eroded and situated at the anterior third of the shell; sculptured with numerous radial striae; exterior brownish black with white striae, interior bluish, spatula brown, margin purple-brown; reaches 35mm in length.
DISTRIBUTION: Throughout the Philippines.

6. Patelloida saccharina (Linné, 1758)
27.3mm, 26.2mm, 29.2mm; Mindoro.
A small to medium sized shell with a tall, subcentral, frequently eroded apex and sculptured with 7 radiating ribs and weaker interstitial riblets; colour greyish white with darker interstitial banding; interior white with a bluish brown speckled border and a centrally located brown-coloured spatula; reaches 35mm in length.
DISTRIBUTION: Throughout the Philippines.

7. Patelloida saccharinoides Habe & Kosuge, 1966
44.3mm, 34.2mm; Samar.
Shell large, elongate-ovate, with a flat, eroded spire, and sculptured with up to 12 large radial ribs, all projecting well beyond the margin; exterior colouration deep blue, the major ribs whitish, interior bluish white with a continuous bluish black margin and a small, ill-defined brown-and-purple blotched spatula; reaches 50mm; considered a form of *P. saccharina* by some, but is consistently different and no intergrades have been observed.
DISTRIBUTION: Throughout the Philippines.

Class Gastropoda
Subclass Prosobranchia
Order Archaegastropoda
Superfamily Fissurellacea
Family FISSURELLIDAE

A moderately large family of herbivorous, shallow water gastropods living attached to rocks and coral. The shells are mainly conical with a tall, perforate spire, a fine to coarsely sculptured surface and a smooth or crenulate peripheral margin. Few recorded Philippine species.

8. Scutus unquis (Linné, 1758)
24.4mm; Samar.
Shell small to medium sized, with a depressed posteriorly displaced, slightly curved apex, a wide, rounded posterior margin, and a narrower, indented anterior margin; sculptured with numerous fine concentric growth striae; exterior creamy white, interior porcelaneous white; reaches 40mm in length.
DISTRIBUTION: Throughout the Philippines in limited quantity.

9. Diodora clathrata (Reeve, 1849)
12.2mm x 24.5mm; Mindoro.
Shell medium sized, elongate-ovate, with a tall, anteriorly displaced apex bearing an ovate anal perforation; surface sculptured with major and minor radial ribs, decussate from overriding spiral cords, basal margin serrate; externally pale cream with green staining, internally porcelaneous white; reaches 40mm in length.
DISTRIBUTION: Throughout the Philippines in limited quantity.

10. Hemitoma (Montfortia) panhi (Quoy & Gaimard, 1834)
8.5mm x 13.9mm; Samar.
Shell small, elongate-ovate, with a tall central apex, anterior slope concave, posterior slope convex; sculptured with several heavy axial ribs, posterior one strongest, weaker interstitial riblets and overriding spiral cords producing a cancellate surface; apex non-perforate, marginal slit short, margins crenulate; off-white with a greenish tinge; reaches 15mm in length.
DISTRIBUTION: Throughout the Philippines in limited quantity.

11. Diodora sp.
Shell small, elongate-ovate, with a tall, perforate, anteriorly displaced apex; sculptured with up to 15 primary ribs, 2 to 3 interstitial riblets, and overriding spiral cords producing a finely cancellate surface, margin crenulate from projecting primary ribs; off-white with darker mottling; reaches 15mm in length; unable to locate a name for this species at this time.
DISTRIBUTION: Probably throughout the Philippines.

Class Gastropoda
Subsclass Prosobranchia
Order Mesogastropoda
Suborder Taenioglossa
Superfamily Littoriniacea
Family LITTORINIDAE

A large group of small, herbivorous, intertidal-dwelling gastropods with a worldwide distribution. The shells are generally small, lightweight, ovate or elongate, have rounded whorls, a pointed spire, may be smooth or sculptured and have a corneous operculum. Not strongly represented in the Philippines. A few of the more common species are illustrated here.

12. Littorina (Littorinopsis) scabra scabra (Linné, 1758)
22.0mm; Palawan.
Shell medium sized, thin, turbinate, with rounded whorls; sculptured with spiral striae which become raised on the body whorl forming spiral cords; body whorl periphery carinate, with a large round aperture, a thin outer lip and a basally dentate columella; colour variable, normally pale cream with brown axial lines, aperture pale brown; reaches 45mm in height.
DISTRIBUTION: Sporadically found throughout the Philippines.

13. Littorina (Littoraria) coccinea (Gmelin, 1791)
14.6mm; Palawan.
Shell small, conic-turbinate, with strongly convex spire whorls and a large, peripherally keeled body whorl; sculptured with faint spiral threads and obsolete axial growth striae; aperture ovate, outer lip thick, columella arcuate and pinched anteriorly; colour light purple with white subsutural banding which is also present on the peripheral keel, aperture pinkish orange; reaches 25mm in height.
DISTRIBUTION: Sporadically found throughout the Philippines.

14. Tectarius (Tectarius) coronatus Valenciennes, 1832
36.0mm; Palawan.
A medium sized conoidal shell with a tall spire, weakly channeled sutures, and sculptured with spiral rows of granulose cords, the subsutural row largest, base with 7 weakly granular spiral cords; aperture

---

PLATE NO. 3 1. *Patella flexuosa flexuosa* Quoy & Gaimard 2a. *Cellana rota* (Gmelin) 2b. Form *eneagona* Reeve 3. *Cellana testudinaria* (Linné) 4. *Cellana grata* (Gould) 5. *Patelloida striata* (Quoy & Gaimard) 6. *Patelloida saccharina* (Linné) 7. *Patelloida saccharinoides* Habe & Kosuge 8. *Scutus unquis* (Linné) 9. *Diodora clathrata* (Reeve) 10. *Hemitoma panhi* (Quoy & Gaimard) 11. *Diodora sp.* 12. *Littorina scabra scabra* (Linné) 13. *Littorina coccinea* (Gmelin) 14. *Tectarius coronatus* Valenciennes 15. *Tectarius pagodus* (Linné) 16. *Tectarius grandinatus* (Gmelin) 17. *Tectarius coronatus* Valenciennes 18. *Echininus cumingi spinulosus* (Philippi) 19. *Liotina depressa* (Reeve)

PLATE 3
1
2a
2b
3
4
5
6
7
8
9
10
11
12
13
14
15
16
17
18
19

ovate and lirate within, columella smooth with an anterior denticle; pinkish orange with a black subsutural band; reaches 40mm in height.
DISTRIBUTION: Throughout the Philippines.

15. Tectarius (Tectarius) pagodus (Linné, 1758)
58.5mm; Samar.
Shell large, heavy and pagoda form; sculptured with weak axial folds and nodulose spiral cords, a row of carinate spines forming on the shoulder angle of the spire whorls, and 2 on the body whorl; aperture large, ovate, lirate within, columella with an anterior denticle, base with 9 to 10 granulose spiral cords; early whorls bluish grey becoming pinkish orange, aperture white; reaches 60mm in height.
DISTRIBUTION: Throughout the Philippines in limited quantity.

16. Tectarius (Tectarius) grandinatus (Gmelin, 1791)
38.5mm; Mindoro.
Like a large heavy *T. coronatus*, but more conoidal, with 4 rows of irregularly spaced, granulose spiral cords and interstitial threads; base with 5 to 6 granulose spiral cords, outermost strongest; aperture ovate, cream coloured and lirate within, columella smooth with an anterior denticle; greyish pink with a darker subsutural band; reaches 40mm in height. Originally thought to be endemic to Southeastern Polynesia, this Philippine record represents a considerable westward extension of its range.
DISTRIBUTION: Throughout the Philippines.

17. Tectarius (Tectarius) coronatus Valenciennes, 1832
31.5mm; Samar.
An unusual form of the species originally thought to be another taxon entirely, but now known to be one of several ecomorphs of *T. coronatus.* It is more conoidal than the specimen shown in fig. 14, with brighter colouration and lacking the heavy surface sculpturing; the interstitial sculpture is also noticeably absent.
DISTRIBUTION: Throughout the Philippines.

18. Echininus (Tectarius) cumingi spinulosus (Philippi, 1847)
17.4mm; Punta Engaño, Cebu.
Shell small, conoidal, with well incised sutures, and sculptured with 3 rows of spines per whorl, a weak subsutural row, another stronger row on the sharply angled shoulder, and a third row at the whorl base; interstitial structure of finely granular cords, base with fine spiral lirae; aperture ovate, smooth within, columella arcuate with a slightly open umbilicus; pale pink in colour with a white aperture and spines; reaches 20mm.
DISTRIBUTION: Cebu and Bohol.

Class Gastropoda
Subclass Prosobranchia
Order Archaegastropoda
Superfamily Trochacea
Family CYCLOSTREMATIDAE

A small group of deep-water gastropods, common in most world seas. The shells are small, top-shaped, with a short spire and strongly keeled whorls. The surface may be corded, nodulose or spinose and frequently has a somewhat cancellate or pitted appearance. Shells are deeply umbilicate, the aperture circular, nacreous silver within, and the operculum small and multispiralled with small calcareous granules. Few Philippine species presently recorded.

19. Liotina depressa (Reeve, 1843)
18.0mm x 23.4mm, 15.4mm x 18.7mm; Samar.
Shell small with a depressed spire and a sharply angled presutural ramp; sculptured with 5 to 6 large, scabrous axial ribs and spiral cords giving the surface a cancellate or pitted appearance; aperture circular, outer lip thick, smooth within, umbilicus deep; colour off-white; reaches 20mm in length; close to *L. peronii* (Kiener, 1839) which has 8 to 10 axial ribs.
DISTRIBUTION: Samar, Cebu, Bohol and Palawan.

Trochidae — Plate 4

Class Gastropoda
Subclass Prosobranchia
Order Archaegastropoda
Superfamily Trochacea
Family TROCHIDAE

A large, worldwide group of herbivorous gastropods living in seaweed or on weed-covered rocks in the shallow waters of most temperate and tropical seas. Most species are intertidal, and the shells range from small to large, are usually conoidal or turbinate, may have a smooth or corded surface, a rounded aperture with a nacreous, occasionally iridescent interior, and a large rounded, multispiralled horny operculum. Also included are members of the Stomatellidae, a small group of oval, somewhat flattened shells with a depressed spire, a large nacreous aperture and a somewhat trochiform shape. The Trochidae are well represented in the Philippines, with a significant number of species being below 10mm in size. Many of these smaller species (not illustrated here) remain unidentified, and our coverage is restricted to the more commonly encountered members of this family.

PLATE NO. 4 1. *Tectus pyramis* (Born) 2. *Tectus conus* (Gmelin) 3. *Trochus hanleyanus* Reeve 4. *Trochus sacellum rota* Dunker in Kira 5. *Trochus stellatus* Gmelin 6. *Trochus maculatus* Linné 7. *Tectus triserialis* (Lamarck) 8. *Trochus niloticus* Linne 9. *Tectus fenestratus* (Gmelin) 10. *Monodonta labio* (Linné) 11. *Euchelus atratus* (Gmelin) 12. *Monilea callifera* (Lamarck) 13. *Monodonta canalifera* Lamarck 14. *Chrysostoma paradoxum* (Born) 15. *Clanculus atropurpureus* (Gould) 16. *Pseudostomatella papyracea* (Gmelin) 17. *Stomatia phymotis* Helbling 18. *Pseudostomatella rubra* (Lamarck) 19. *Calliostoma ticaonica* (A. Adams) 20. *Ethminolia stearnsii* (Pilsbry) 21. *Umbonium vestiarum* (Linné)

PLATE 4

1. Tectus pyramis (Born, 1778)
62.7mm x 64.0mm; Samar.
Shell large, tall-spired, with straight-sided whorls which are sculptured with obsolete spiral striae and weak tubercles at the sutures; base flat, white, non-umbilicate but with a wide umbilical glaze; columella occasionally notched, outer lip lirate within and with 4 marginal teeth; grey to brown with green mottling; reaches 70mm in height.
DISTRIBUTION: Throughout the Philippines.

2. Tectus conus (Gmelin, 1791)
65.4mm x 58.3mm; Samar.
Shell large, heavy, with a tall spire, weakly impressed sutures, and straight-sided whorls which bear several rows of encircling granulose lirae; base rounded, columella oblique, curved and excavated behind; aperture quadrate, outer lip smooth within; colour pinkish white with deep red maculations; reaches 80mm in height.
DISTRIBUTION: Throughout the Philippines.

3. Trochus hanleyanus Reeve, 1842
47.5mm x 40.5mm; Cebu.
Shell medium sized, solid, with flat whorls and sculptured with spiral rows of irregularly shaped granules; base concave, umbilicate, columella oblique, with 4 to 5 teeth; umbilical area wide and concentrically striated; colour variable ranging from white to pink, with fine axial red lines, base off-white with similar markings, umbilical area white; reaches 65mm in height.
DISTRIBUTION: Throughout the Philippines.

4. Trochus sacellum rota Dunker, in Kira, 1962
19.9mm x 24.6mm; Palawan.
Shell small, low conoidal in shape with weakly channeled sutures, and sculptured with numerous spiral rows of rounded granules, those below the suture largest and forming a weak carina; base weakly convex, sculptured with 8 spiral rows of small nodules; umbilicus open and concentrically lirate, aperture quadrate, lirate within and dentate on the columellar margin; deep red with white maculations; seldom exceeds 25mm in diameter.
DISTRIBUTION: Throughout the Philippines.

5. Trochus stellatus Gmelin, 1791
30.5mm x 34.1mm; Cebu.
Shell medium sized, solid, with weakly channeled sutures, a weakly convex spire whorl profile, and encircled with granulose spiral cords, the subsutural row bearing oblique, axially elongate nodules; base slightly convex with 8 rows of finely granulose spiral cords; umbilicate with a spirally lirate umbilical area; aperture ovate, lirate within, columella bluntly toothed; pale rose in colour, often with a green tinge and white axial flammules, aperture a nacreous off-white; reaches 40mm in diameter.
DISTRIBUTION: Throughout the Philippines.

6. Trochus maculatus Linné, 1758
48.5mm x 47.9mm, 42.0mm x 38.5mm; Samar.
Shell medium sized, heavy and extremely variable; whorls heavy and sculptured with rows of irregularly sized granules, frequently stepped on the body whorl; base flat, with an open, spirally corded umbilical area, 8 to 9 rows of finely beaded spiral cords, and a quadrate aperture which is lirate within and has several rounded columellar denticles; shell off-white with darker axial streaks and maculations, base off-white with fine red axial streaks; reaches 70mm in height.
DISTRIBUTION: Throughout the Philippines.

7. Tectus triserialis (Lamarck, 1822)
46.0mm x 30.5mm; Cebu.
Shell medium sized with a tall spire of 11 to 12 whorls, impressed sutures, and encircled with regular rows of granulose spiral cords, strongest on the shoulder; base flat with fine spiral striae; pseudo-umbilicate, with a somewhat twisted columella; aperture elongate and weakly lirate within; green to brown with a white base and aperture; reaches 60mm in height.
DISTRIBUTION: Throughout the Philippines in limited quantity.

8. Trochus niloticus Linné, 1767
103.5mm x 106.8mm; Samar.
A large, heavy, conical shell with 10 to 12 straight-sided whorls and slightly impressed sutures; early whorls with nodulose spiral cords, later becoming smooth; body whorl centrally concave, peripherally swollen; base slightly concave, columella smooth, terminating in a single ridge-like tooth; creamy white with deep red axial flammules, those on the base pink-coloured; reaches 150mm in height.
DISTRIBUTION: Throughout the Philippines.

9. Tectus fenestratus (Gmelin, 1791)
30.0mm x 31.9mm; Palawan.
A medium sized, tall-spired shell with 10 to 12 whorls, weakly tuberculate and with deeply impressed sutures; sculptured with oblique, axially elongate nodules on the lower half of each whorl; base flat, spirally striated and non-umbilicate, columella with a single excavated fold, aperture lirate and silvery within; grey to brown in colour with darker mottling; reaches 70mm in height.
DISTRIBUTION: Palawan and Visayan Sea.

10. Monodonta labio (Linné, 1758)
36.8mm x 32.6mm; Samar.
Shell medium sized, ovate, with weakly impressed sutures and strongly convex whorls which are sculptured with several rows of coarse, nodulose cords; columella monodentate, outer lip margin iridescent, inner area white and lirate; shell deep purple to green with lighter coloured nodules; reaches 38mm in height.
DISTRIBUTION: Throughout the Philippines.

11. Euchelus atratus (Gmelin, 1791)
28.3mm x 37.6mm; Batangas.
Normally a small shell, but reaches around 30mm in height in the Philippines; shell globose, sutures deeply channeled, whorls encircled with rows of finely beaded cords and a single interstitial thread; aperture ovate, umbilicus open, outer lip thick, lirate within, columella with a single basal denticle; colour greyish purple with a white columella and aperture.
DISTRIBUTION: Throughout the Philippines.

12. Monilea callifera (Lamarck, 1822)
19.2mm x 20.0mm; Samar.
Shell small, solid, conoidal, sculptured with strong, irregularly sized spiral cords and overriding axial striae forming a weakly granulose surface; aperture ovate, lirate within, umbilicus open and covered by a callus, columella with a single tooth; shell colour dark brownish purple with white axial blotches, aperture iridescent green within, columella white; reaches 35mm in height.
DISTRIBUTION: Throughout the Philippines.

13. Monodonta canalifera Lamarck, 1816
20.2mm x 21.1mm; Cebu.
Shell small, with globose whorls, moderately impressed sutures, and sculptured with numerous smooth, flat cords, separated by a weak cleft; non-umbilicate, the columella with a single posteriorly excavated tooth, aperture elongate-ovate, lirate within; colour greyish green, cords maculated with black and white, apertural and columellar margins white, internally iridescent green; reaches 25mm in height.
DISTRIBUTION: Throughout the Philippines.

14. Chrysostoma paradoxum (Born, 1780)
17.5mm x 14.0mm; Samar.
A small, smooth, globular shell with 4 to 5 whorls and a small, deep channel below the sutures; aperture ovate and smooth within, columella smooth, umbilical area covered with a large orange callus pad; shell off-white to cream with darker axial bands, aperture and columellar callus orange; reaches 25mm in height.
DISTRIBUTION: Sporadically found throughout the Philippines.

15. Clanculus atropurpureus (Gould, 1849)
14.0mm x 13.2mm; Mindoro.
A small squat shell with rounded whorls, well channeled sutures, and encircled with numerous rows of finely beaded lirae; base moderately convex, with 1 large and 3 small denticles within the umbilicus; columella curved and dentate, aperture iridescent green and lirate within; greyish brown with a white columella; reaches 17mm in height.
DISTRIBUTION: Throughout the Philippines.

16. Pseudostomatella (Pseudostomatella) papyracea (Gmelin, 1791)
27.4mm x 31.8mm; Palawan.
A medium sized shell with a short spire and a large, ovate body whorl; sculptured with numerous wide, flat concentric lirae; aperture elongate-ovate and smooth within, with a bright silvery pink nacre; non-umbilicate; columellar margin with an off-white border; colour greyish brown, the ribs maculated with white lines; reaches 30mm in height.
DISTRIBUTION: Sulu Sea, Palawan and Samar.

17. Stomatia phymotis Helbling, 1779
21.4mm; Cebu.
A small, rugosely sculptured shell, the spire whorls loosely coiled in a "cork screw"-like fashion, and the body whorl large, expansive, with 2 closely-set nodulose cords on the shoulder which become obsolete on the penultimate whorl; numerous oblique, axial cords on the concave presutural slope, becoming obsolete on the lower part of the body whorl; colour off-white to greyish green, with fine, red-coloured spiral striae, interior nacreous silver-pink; reaches 30mm in length.
DISTRIBUTION: Sporadically found throughout the Philippines.

18. Pseudostomatella (Stomatolina) rubra (Lamarck, 1822)
11.5mm x 15.0mm; Punta Engaño, Cebu.
Shell small, ovate, with a moderately tall spire, a large, ovate body whorl and impressed sutures; whorls bi-angulate, flat below the sutures and sculptured with 2 rows of finely tuberculate cords and weaker spiral striae, both of which are crossed by fine axial growth lines; aperture widely ovate, umbilicus weakly indented; red with white axial streaks; reaches 16mm in length.
DISTRIBUTION: Cebu, Bohol, Sulu Sea and Southern Mindanao.

19. Calliostoma ticaonica (A. Adams, 1851)
11.9mm x 10.8mm, 9.5mm x 8.6mm; Punta Engaño, Cebu.
Shell small, conoidal, with slightly convex spire whorls and corded sutures; sculptured with numerous granulose spiral cords and a weakly keeled body whorl; base convex, with fine spiral cords and overriding radial striae; aperture quadrangular, weakly lirate within, umbilicus slightly open and bordered by a strong umbilical cord; colour off-white with orange-brown blotches and white-spotted sutural cords, base creamy white, aperture internally nacreous white; reaches 15mm in height.
DISTRIBUTION: Cebu, Bohol and Southern Mindanao (deep water).

20. Ethminolia stearnsii (Pilsbry, 1895)
6.0mm x 8.4mm, 7.4mm x 10.2mm, 7.5mm x 9.8mm; Cebu.
Shell small, somewhat globose, with a moderately tall spire, impressed sutures and 5 to 6 strongly convex spire whorls; sculptured with fine spiral striae and axial growth scratches; base convex and deeply umbilicate; aperture ovate, smooth and iridescent green within; colour and pattern variable, Philippine specimens red with a black and cream-coloured subsutural band, base pale yellowish white with white

and black spiral lines; reaches around 10mm in diameter.
DISTRIBUTION: Cebu, Bohol and Southern Mindanao (deep water).

21. Umbonium (Suchium) vestiarum (Linné, 1758)
7.5mm x 11.8mm, 7.6mm x 12.0mm; Palawan.
Shell small, shiny, rather depressed, with smooth whorls occasionally with obsolete growth striae; base convex with a large, white umbilical callus; colour variable, usually white to dark brown with lighter coloured markings; reaches 20mm in diameter.
DISTRIBUTION: Palawan and Sulu Sea.

Trochidae Plate 5

1. Umbonium (Suchium) elegans (Kiener, 1838)
11.5mm x 17.5mm, 10.5mm x 17.0mm; Samar
Shell small, discoidal, with a moderately tall, straight-sided spire, and a white, strongly convex umbilical button; whorl surfaces smooth with obsolete spiral and arcuate axial scratches and a weak concave trough below the body whorl suture; aperture elongate, smooth within; whorl surfaces with black quadrangular dots in spiral lines, a black subsutural band, and a black and cream-striped periphery; reaches 18mm in diameter. This is most likely only a straight-sided form of *U. vestiarum* (Pl. 4 fig. 21).
DISTRIBUTION: Throughout the Philippines.

2. Umbonium (Suchium) costatum (Kiener, 1838)
10.5mm x 16.5mm, 10.8mm x 16.9mm; Mindoro.
Shell similar in size to *elegans*, but the surface is sculptured with fine concentric grooves (not clearly visible on the specimen illustrated); the umbilicus is smaller and less inflated and the colour, although variable, is mainly a dark greenish-grey with yellow streaks, and pink-coloured sutures; the umbilicus is dark brown to black and the aperture nacreous green; can reach 25mm in diameter.
DISTRIBUTION: Throughout the Philippines.

3. Cantharidus (Jujubinus) picturatus
(A. Adams, 1851)
15.0mm; Cebu.
Shell small, triangulate in outline with a tall, sharply acute spire; whorls sculptured with fine spiral grooves and a pronounced presutural ridge on each whorl; base more coarse and spirally striate, with axial threads in the interspaces; aperture ovate with a small umbilical tooth; pinkish red with white axial streaks, aperture white; reaches 18mm in height.
DISTRIBUTION: Cebu, Bohol and Southern Mindanao.

4. Cantharidus (Jujubinus) gilberti
(Montrouzier, in Fischer, 1878)
Syn: *Trochus giliberti* Montrouzier, in Fischer, 1880 (invalid emmendation).
16.3mm; Palawan.
Very similar to the preceding species, but slightly broader, with somewhat convex whorls; sculptured with fine spiral grooves, but lacking the presutural ridge of *C. picturatus;* base with more widely spaced grooves and fine axial scratches in the interspaces; aperture ovate, nacreous silver-blue within, umbilical tooth much reduced; greenish brown with pink and white-stained axial streaks; reaches 20mm in height.
DISTRIBUTION: Throughout the Philippines.

5. Cantharidus (Jujubinus) japonicus
(A. Adams, 1853)
14.9mm; Cebu.
Like an extremely broad form of *C. gilberti,* but the spiral grooves on the whorls are more weakly incised; the base has weaker spiral grooves, a slightly open umbilicus, and lacks the umbilical tooth; colouration slightly darker with paler axial streaks; reaches 18mm in height.
DISTRIBUTION: Cebu, Bohol and Southern Mindanao.

6. Clanculus (Mesoclanculus) denticulatus
(Gray, 1827)
8.8mm x 10.7mm; Punta Engaño, Cebu.
Shell small, low conical, with convex whorls and a distinctly flat presutural ramp; sculptured with finely beaded spiral cords and axially striate interspaces; base with larger, more widely spaced cords which are axially filleted by deep grooves, umbilicus deep; brown coloured with a pink tinted apex and a white coloured aperture and umbilical area; reaches 12mm in diameter.
DISTRIBUTION: Cebu, Bohol and Southern Mindanao.

PLATE NO. 5 1. *Umbonium elegans* (Kiener) 2. *Umbonium costatum* (Kiener) 3. *Cantharidus picturatus* (A. Adams) 4. *Cantharidus giliberti* (Montrouzier in Fischer) 5. *Cantharidus japonicus* (A. Adams) 6. *Clanculus denticulatus* (Gray) 7. *Monilea nucleus* (Philippi) 8. *Cantharidus urbanus* (Gould) 9. *Gibbula affinis cognatus* (Pilsbry) 10. *Euchelus instrictus* (Gould) 11. *Turcica concinna* A. Adams 12. *Euchelus clathratus* (A. Adams) 13. *Astele pulcherrimus* (Sowerby) 14. *Gibbula stoliczkana* (G. & H. Nevill) 15. *Lischkeia undosa* Kuroda & Kawamura 16. *Stomatella varia* (A. Adams) 17. *Phasianella solida* (Born) 18. *Phasianella modesta* (Gould) 19. *Phasianella histrio* Reeve

PLATE 5

7. Monilea nucleus (Philippi, 1849)
4.5mm x 5.0mm, 5.1mm x 5.5mm; Punta Engaño, Cebu.
Shell small, globose, with 5 to 6 concave whorls that are sculptured with numerous rows of finely beaded spiral cords; umbilicus open and deep, columella slightly excavated, finely plicate, with a prominent tooth basally; aperture sub-circular, lirate within; pale grey with brown-spotted white bands and maculations; reaches 10mm in diameter.
DISTRIBUTION: Samar, Cebu, Bohol and Southern Mindanao.

8. Cantharidus (Iwakawatrochus) urbanus (Gould, 1861)
8.3mm x 6.5mm; Punta Engaño, Cebu.
Shell small, conic with convex whorls and encircled with fine spiral cords; aperture elliptically ovate, smooth within, umbilicus narrowly open, columella arcuate and smooth; red with white axial maculations and a white aperture; reaches 10mm in height.
DISTRIBUTION: Cebu, Bohol and Southern Mindanao.

9. Gibbula (Eurytrochus) affinis cognatus (Pilsbry, 1903)
4.5mm x 6.3mm, 3.8mm x 6.4mm; Punta Engaño, Cebu.
Shell very small, discoidal, with a low spire and a sharply keeled periphery; spire whorls sculptured with finely granulose spiral cords, base with thick spiral cords which are axially filleted by deep grooves, aperture circular, lirate within, columella with fine denticles, umbilicus deep, surrounded by a highly granulose cord; black with white flecks, a black and white-spotted peripheral keel and a white spire, aperture and umbilical area; seldom exceeds 7mm in diameter.
DISTRIBUTION: Cebu, Bohol and Southern Mindanao.

10. Euchelus (Herpetopoma) instrictus (Gould, 1849)
8.0mm x 6.4mm, 7.3mm x 7.2mm; Punta Engaño, Cebu.
Shell small, elongate-ovate, with a moderately tall spire, well impressed sutures and a large body whorl; sculptured with well elevated, nodulose spiral cords and lamellate interspaces; aperture ovate, lirate within, with a large columellar tooth and a small umbilicus; cream coloured with brown axial streaks and subsutural brown spots; reaches 10mm in height.
DISTRIBUTION: Samar, Cebu, Bohol and Southern Mindanao.

11. Turcica concinna A. Adams, 1863
8.9mm x 5.7mm, 9.7mm x 5.9mm; Punta Engaño, Cebu.
Shell small, trochiform, with a tall spire, well channeled sutures and a row of deeply excavated pits above each suture line and around the sharply angled body whorl base; surface nodulose from intersecting axial and spiral cords, base with weakly nodulose spiral cords; aperture ovate, lirate within, columella smooth with a single basal denticle; cream coloured with golden-brown axial streaks; reaches 10mm in height.
DISTRIBUTION: Cebu and Bohol.

12. Euchelus (Euchelus) clathratus (A. Adams, 1853)
6.0mm x 5.2mm; Punta Engaño, Cebu.
Shell very small, somewhat globose, with well incised sutures and a large body whorl; whorls granulose from overriding axial and spiral cords, producing deeply pitted interspaces; aperture sub-circular, lirate within; shell pale cream with light brown spiral cords; reaches 8mm in height.
DISTRIBUTION: Cebu, Bohol, Palawan and Southern Mindanao.

13. Astele (Coralastele) pulcherrimus (Sowerby, 1914)
7.4mm x 6.0mm, 6.8mm x 5.5mm; Punta Engaño, Cebu.
Shell small, moderately trochiform in outline, sculptured with elevated spiral cords bearing axially compressed, somewhat echinate nodules, and oblique, axially striate, deep-pitted interspaces; 2 spiral cords on each whorl stronger than the rest, producing a bi-cingulate appearance; sutures deeply channeled, aperture sub-ovate, lirate within, columella smooth, umbilicus weakly open; pale creamy white with rows of brown spots; largest seen 10.3mm in height.
DISTRIBUTION: Cebu, Bohol, Palawan and Southern Mindanao.

14. Gibbula (Cantharidella) stoliczkana (G. & H. Nevill, 1869)
6.3mm x 5.3mm, 6.2mm x 4.8mm; Palawan.
Shell small, trochiform in outline, spire moderately tall; sculptured with numerous flat spiral cords and fine, axially striated interspaces; aperture ovate, smooth within, columella smooth, arcuate, with a weak basal tooth, umbilicus slightly open; greenish brown with white axial stripes, aperture iridescent green; reaches 12mm in height, but the above specimens are average size in the Philippines.
DISTRIBUTION: Throughout the Philippines.

15. Lischkeia (Lischkeia) undosa Kuroda & Kawamura, 1956
35.0mm x 28.2mm; Sulu Archipelago.
Shell medium sized, lightweight, with a tall spire; sculptured with a row of sharply echinate tubercles at the angulate shoulder and a second weaker row above the sutures and on the lower half of the body whorl, giving a slightly bicingulate appearance to the shell; the sutures are well impressed, bordered by a weak row of spinose nodules; rest of the shell with granulose spiral cords and weak, oblique axial striae; aperture ovate, lirate within, columella arcuate and smooth, umbilicus weakly open; creamy white in colour with a nacreous silver aperture; reaches 40mm in height; close to *L. alwinae* (Lischke), but with a taller spire and a more strongly convex base.
DISTRIBUTION: Sulu Sea.

16. Stomatella varia (A. Adams, 1850)
12.5mm, 11.0mm, 10.5mm, 9.7mm; Punta Engaño, Cebu.
Shell small, ear-shaped, elongate, sculptured with fine, radiating axial scratches and almost obsolete spiral

striae; apex slightly elevated, aperture wide, ovate, nacreous silver-pink within; extremely variable in colour and pattern, usually green, pink, or grey with lighter spots and maculations and a white band along the umbilical border; reaches 15mm in length.
DISTRIBUTION: Samar and Sulu Sea.

Class Gastropoda
Subclass Prosobranchia
Order Archaegastropoda
Superfamily Trochacea

Family PHASIANELLIDAE

A relatively small group of herbivorous gastropods inhabiting algae and sea grass in the shallow waters of most tropical and subtropical seas. The shells are small to medium sized, elongate with a smooth or, occasionally, spirally corded surface; aperture pear-shaped with a non-nacreous interior, the outer lip is thin and the operculum white, calcareous and convex in outline with a basally situated nucleus. Few Philippine species.

17. Phasianella solida (Born, 1778)
Syn: *P. variegata* Lamarck, 1822 (*non* de Roissy, 1805)
17.5mm, 15.9mm, 15.0mm; Palawan.
Shell small to medium sized, elongate-ovate, with a tall conical spire, weakly incised sutures and strongly convex whorls; surface smooth except for fine axial and spiral scratches; aperture ovate and smooth; variable in colour, usually reddish brown with white axial streaks and fine spiral lines; reaches 20mm.
DISTRIBUTION: Palawan, Samar, Visayan Sea and Sulu Sea.

18. Phasianella modesta (Gould, 1861)
11.5mm, 11.6mm; Punta Engaño, Cebu.
Shell small, elongate-ovate, with a tall, conical spire, adpressed sutures and smooth, strongly convex whorls; aperture sub-circular, smooth within; pinkish orange with white blotching and maculations; reaches 15mm.
DISTRIBUTION: Cebu, Bohol, Samar and Sulu Sea.

19. Phasianella histrio Reeve, 1862
8.2mm; Punta Engaño, Cebu.
Shell small, rather squat in outline, with smooth, strongly convex whorls and a smooth, sub-ovate aperture; light reddish brown, with brown and white spots at the sutures, brown and white-spotted spiral lines and red-bordered, white sagittal markings in the interspaces; reaches 10mm.
DISTRIBUTION: Samar, Cebu, Bohol and Sulu Sea.

Class Gastropoda
Subclass Prosobranchia
Order Archaegastropoda
Superfamily Trochacea

Family ANGARIIDAE

A small group of turbinate shells which inhabit the shallow and deep waters of most Indo-Pacific reefs. The shells are medium sized, rather solid, umbilicate, and the surface is normally sculptured with tubercles or delicate spinose processes. The aperture is circular, nacreous within, and all possess a horny operculum. Often treated as a group within the Turbinidae, they probably deserve separation on a Family level and are so treated here. The Philippine archipelago appears not only to have the highest number of *Angaria* species, but also the most spectacularly coloured and most intricately sculptured forms found anywhere in the world.

1. Angaria delphinus (Linné, 1758)
A solid, black-coloured, heavy shell, with almost flat spire whorls and a large, ovate body whorl sculptured with coarse spiral cords which may exhibit irregularly curved spines of varying lengths; aperture ovate, smooth and nacreous silver within; umbilicus wide and open; reaches around 80mm; extremely variable with several forms being recognized:

1a. form *melanacantha* (Reeve, 1842): characterized by the extremely long curved spines on the upper portion of the whorls and the densely foliate area on the lower part of the body whorl; sympatric with other forms of *delphinus* and occurring only at Palawan. No intergrades have ever been seen and it probably represents a good biological species; 68.3mm; Puerto Princesa, Palawan.
1b. typical form of *delphinus*; 61.7mm; Palawan.
1c. form *atrata* (Reeve, 1843): characterized by 2 rows of short triangular bumps or spines on the body whorl; 71.2mm; Samar.
1d. form *nodosa* (Reeve, 1846): the spire whorls are quite flattened and there are 2 spiral ridges bearing 2 rows of thick, triangular spines, larger and fewer in number than those on *atrata;* spaces between the ridges have irregular spiral cords; 53.8mm; Palawan.
1e. a transition form, midway between *nodosa* and *aculeata;* the shell is becoming smoother, the interstitial spaces developing a reddish colouration, and the spines are becoming longer and larger and developing the white colouration typical of the *aculeata* form; 77.6mm; Punta Engaño, Cebu.
1f. form *aculeata* (Reeve, 1842): a beautifully coloured form; body usually red or green and the spines are thick, slightly curved and white; the illustrated form here has black spines; 50.0mm; Panglao, Bohol.
1g. this is the typical *aculeata* form with the red body and white spines; the surface is almost smooth except

for a few scabrous spiral ridges; 51.0mm, 35.7mm, 35.5mm; Palawan.
1h. A curious variation of the *atrata* form in which the spire is sunken, the umbilicus wider and deeper than normal, and the large white spines are lacking; 23.5mm; Samar.
DISTRIBUTION: Throughout the Philippines, except the form *melanacantha* (fig. 1a) which is only found in Palawan.

2. Angria sp.
51.0mm, 41.6mm; Punta Engaño, Cebu.
A beautiful shell very similar to 1f. above, but the spines are slender and extremely long; most likely another form of *delphinus.*
DISTRIBUTION: Bohol and Cebu.

3. Angaria distorta (Linné, 1758)
51.6mm, 46.5mm; Palawan.
Possibly another form of *delphinus,* but the large, heavy, rather globose whorls of which the last 2 are always separated, the markedly flattened spire, and the surface which bears numerous small, closely-set hollow spines, seem to make it distinct; it is deeply umbilicate and never develops the long, thick spinose processes seen in most forms of *delphinus;* no specimens which might be considered intergrades have been seen; reaches around 55mm in height.
DISTRIBUTION: Palawan.

4. Angaria sphaerula (Kiener, 1838)
Shell of medium size, similar to *A. delphinus* in general shape; spire whorls not as compressed and more distinctly axially costate, and the whorls bear a row of flattened curved spines on the upper segment; the spines number 10 to 12 per whorl and may be serrated or frondose on their extremities, the lower section of the body whorl has several rows of stunted spines; the shell is usually a deep red colour and may be banded; spines may be red, green, or a combination of these two colours; largest specimen seen around 90mm in diameter.
4a. typical form; 84.2mm; Punta Engaño, Cebu.
4b. juvenile specimens in which the spire is generally sunken; the spines are blade-like on their proximal end, and the margins are coarsely serrated; 29.5mm, 31.5mm; Punta Engaño, Cebu.
4c.-4g. various specimens illustrating the great variation in colour and spinal development; sizes (including spines); 43.4mm, 33.0mm, 42.8mm, 51.9mm and 61.6mm, respectively; all from Punta Engaño, Cebu.
DISTRIBUTION: Cebu.

5. Angaria vicdani Kosuge, 1980
55.5mm, 69.2mm; Davao.
A medium sized, thick, rather solid shell of some 5 to 6 whorls, the first 3 of which are flattened; the remaining whorls loosely convolute with adpressed sutures; body whorl roundly shouldered, more so than seen in *A. sphaerula,* and bearing 8 to 10 rounded axial nodes only between the shoulder and the suture, the nodes being crossed by several spiral lirae; on the shoulder of each whorl are long, sinuous, rather thin spinose processes, usually branched at their extremities from which arise tertiary spinelets, conspicuously red coloured against the green coloured spines; on the body whorl below the shoulder spines is another row of shorter, non-branched spines, followed by several concentric rows of very short spines which are slightly curved; umbilicus open, aperture circular, nacreous silver within; operculum circular with concentric lamellae and a subcentral nucleus; colour pinkish orange, primary and secondary spinal processes green, tertiary spinelets pink; a concentric white band around the umbilicus; reaches 70mm. An endemic species named after Victor Dan, a prominent local collector.
DISTRIBUTION: Davao (deep water).

Turbinidae Plate 7

Class Gastropoda
Subclass Prosobranchia
Order Archaegastropoda
Superfamily Trochacea
Family TURBINIDAE

A relatively large group of herbivorous gastropods living in coral or seaweed in the shallow intertidal waters of most tropical and subtropical seas. The shells are solid, turbinate or trochoidal, usually with few whorls, and may be globose or depressed. The surface is often colourful and may be smooth or sculptured with spiny or nodulose spiral cords. The aperture is nacreous within and may be rounded, elliptical or sharply angled at the periphery. Columella may be smooth or umbilicate and the operculum is calcareous, has a central or eccentric nucleus, and is often colourful. The family includes such diverse genera as *Astraea, Bolma,* and *Guildfordia.* The family is fairly well represented in the Philippines, although a few of the more common Indo-Pacific species appear to be absent from the fauna.

1. Turbo (Turbo) petholatus Linné, 1758
64.2mm x 52.5mm; Samar.
Shell large, smooth, with 5 to 6 strongly convex whorls, the last 2 slightly angulate on the upper third; sutures impressed with a concave presutural ramp; aperture ovate, outer lip thin, smooth within,

PLATE NO. 6 1a. *Angaria delphinus* (Linné) form *melanacantha* (Reeve) 1b. typical form 1c. form *atrata* (Reeve) 1d. form *nodosa* (Reeve) 1e. transition form 1f. form *aculeata* (Reeve) 1g. typical *aculeata* form 1h. an ecomorph of *atrata* form 2. *Angaria sp.* 3. *Angaria distorta* (Linné) 4a. *Angaria sphaerula* (Kiener) 4b. juvenile 4c. to 4g. various forms 5. *Angaria vicdani* Kosuge

PLATE 6

columella smooth, non-umbilicate and bordered by a ridge; colour variable, usually reddish to greenish brown, encircled with darker bands and white axial flammules; aperture white, marginally iridescent yellow-green, operculum smooth, white and circular; reaches 85mm in height.
DISTRIBUTION: Throughout the Philippines but in limited quantity.

2. Turbo (Marmarostoma) argyrostomus Linné, 1758
76.4mm x 56.2mm; Sulu Archipelago.
Shell large, heavy, with angulate whorls and deeply channeled sutures; sculptured with prominent, raised, rounded spiral cords and weaker interstitial threads both of which are crossed by fine, overriding axial striae; aperture circular, outer lip thick, occasionally imbricate, columella smooth, slightly ridged and deeply channeled; light brown with brown and green maculations; aperture silvery white, operculum granular, green to white in colour; reaches 100mm in height.
DISTRIBUTION: Throughout the Philippines.

3. Turbo (Marmarostoma) chrysostomus Linné, 1758
56.5mm x 45.7mm, 63.0mm x 50.2mm, 58.0mm x 46.8mm; Sulu Archipelago.
A heavy, solid, medium sized shell with a large inflated body whorl, a short spire, and moderately impressed sutures; early whorls moderately angulate, body whorl bi-angulate, with 2 rows of spines **(3a)**, nodules **(3b)**, or elevated cords **(3c)**; rest of the whorls covered with spiral cords which may be weakly or strongly scabrous; aperture ovate, marginally serrate, columella smooth, partially umbilicate; off-white with darker coloured stripes and an orange-yellow aperture; operculum centrally pustulose, marginally striate and green coloured; reaches 70mm in height.
DISTRIBUTION: Throughout the Philippines.

4. Turbo (Marmarostoma) crassus Wood, 1828
58.5mm x 47.0mm; Samar.
Shell large, with 8 to 9 whorls, the earlier ones strongly convex, the body whorl angulate with a carinate ridge on the shoulder; sculptured with deeply incised grooves, flat ribs, and overriding axial striae giving a finely ribbed appearance to the surface; aperture ovate, outer lip and columella anteriorly expansive; off-white with irregular green and brown markings, ribs with alternating light and dark patches, aperture nacreous silver; operculum centrally pustulose, marginally fine, and deep brown in colour; reaches 85mm in height.
DISTRIBUTION: Throughout the Philippines.

5. Turbo (Marmarostoma) bruneus (Röding, 1798)
31.5mm x 29.8mm, 39.4mm x 30.8mm; Sulu Archipelago.
Shell medium sized, with 6 to 7 rounded whorls, sculptured with rough spiral cords of varying thickness, and finely foliate axial ridges; aperture ovate, nacreous silver-white within, columella white, bordered by a narrowly open umbilicus; colour pale grey to brown with darker coloured axial streaks; operculum granular and purple-coloured; reaches 60mm in height.
DISTRIBUTION: Throughout the Philippines but in limited quantity.

6. Turbo (Turbo) marmoratus Linné, 1758
146mm x 149mm; Samar.
A large, heavy, dull green coloured shell, angled on the body whorl with 2 or 3 heavily nodulose keels; axial structure of numerous fine growth striae; aperture circular to ovate, silvery white internally; outer lip thin, posteriorly angulate, columella smooth, flattened anteriorly with a heavy fasciolar ridge; shell green with bands of alternating red and grey spots; operculum large, smooth, heavy and white coloured; reaches 200mm in height.
6a. Typical adult; 146mm × 149mm; Samar.
6b. Juvenile.
DISTRIBUTION: Throughout the Philippines.

7. Turbo (Turbo) reevei Philippi, 1847
43.8mm x 39.5mm, 39.2mm x 37.3mm, 35.5mm x 34.3mm, 39.4mm x 37.2mm, 56.3mm x 46.5mm; Samar.
Structurally identical to *petholatus,* but smaller; it lacks the yellowish orange marginal colouration of that species. The shell colouration is far more variable: green, brown, orange and yellow-green with axial ornamentation; seldom exceeds 50mm in height.
DISTRIBUTION: Throughout the Philippines but the most colourful specimens come from the Sulu Sea.

8. Guildfordia aculeatus Kosuge, 1979 (November)
Syn: *Guildfordia kurzi* Petuch, 1979 (December)
43.7mm, 39.0mm (including spines); Mactan Is., Cebu.
Shell medium sized, with a low spire; sculptured with rows of granulose spiral lirae, one of which obscures the sutures; periphery with 8 short spines, normally broken in most specimens; base flat with a peripheral groove, 4 rows of granules, and a central umbilical callus which is somewhat excavated; colour pinkish grey with a white base, a grey umbilical callus and a white nacreous aperture; reaches 45mm (including spines) in diameter.
DISTRIBUTION: Cebu.

9. Bolma (Galeoastraea) guttata (A. Adams, 1853) s.s.(?)
10.3mm x 12.0mm, 9.3mm x 10.2mm; Punta Engaño, Cebu.
Two immature specimens which are very close to the South African subspecies *bathyraphis* (Smith). Shells small, trochiform, with sharply angulate spire whorls which are sculptured with rows of granulose spiral cords and a row of short, sharp spines on the

PLATE NO. 7 1. *Turbo petholatus* (Linné) 2. *Turbo argyrostomus* Linné 3a. to 3c. *Turbo chrysostomus* Linné 4. *Turbo crassus* Wood 5. *Turbo bruneus* (Röding) 6a. *Turbo marmoratus* Linné 6b. juvenile 7. *Turbo reevei* Philippi 8. *Guildfordia aculeatus* Kosuge 9. *Bolma guttata* (A. Adams) 10a. *Astralium haemotragum* (Menke) 10b. specimen covered with coralline encrustation 11. *Guildfordia yoka* (Jousseaume) 12. *Bolma persica* (Dall) 13. *Bolma guttata millegranosa* (Kuroda & Habe, in Habe) 14. *Bolma erectospina* (Habe & Okutani) 15. *Turbo cinereus* Born 16. *Turbo stenogyrum* (Fischer)

PLATE 7

shoulders; body whorl with a row of short peripheral spines, below which are 2 weaker rows; the base is encircled with rows of finely granulose cords; aperture ovate with a smooth columella and a weak umbilical groove; colour pinkish red with darker maculations, operculum white with flattened margins. Appears to belong to the *B. guttata* complex which is presently poorly understood.
DISTRIBUTION: Cebu and Bohol.

10. Astralium haemotragum (Menke, 1829)
19.2mm x 14.5mm, 16.2mm x 15.0mm, 16.2mm x 14.7mm; Punta Engaño, Cebu.
Illustrated specimens **(10a)** are juveniles of *haemotragum*, larger specimens being somewhat scarce; shell small, trochiform, with a tall spire and adpressed sutures; sculptured with granulose spiral cords and oblique axial folds which become echinate on the sharply angled periphery; base with granular cords and a slightly thickened columellar base, aperture lirate within; magenta coloured with white banding and a purple columellar base; most recorded specimens seldom exceed 20mm in height.
10a. Typical juveniles; 19.2mm × 14.5mm, 16.2mm × 15.0mm, 16.2mm × 14.7mm; Punta Engaño, Cebu.
10b. specimen covered with a curious coralline encrustation; 15.5mm × 14.5mm; Cebu.
DISTRIBUTION: Cebu and Bohol.

11. Guildfordia yoka Jousseaume, 1888
39.0mm, 31.0mm (excluding spines); Panglao, Bohol.
Slightly larger than *G. triumphans* and of very similar appearance; it differs in its longer, more curved spines, the somewhat more irregular spiral lirae, and the lighter colouration; reaches around 45mm (excluding spines) in diameter.
DISTRIBUTION: Tayabas Bay (Quezon), Cebu and Bohol.

12. Bolma (Galeoastraea) persica (Dall, 1907)
20.5mm x 27.0mm, 14.8mm x 22.5mm (including spines); Cebu.
Shell small to medium sized, spire moderately tall, whorls straight-sided, bearing long spines; body whorl strongly keeled basally, bearing 12 to 14 spines; area above the spines with 6 to 7 weakly beaded spiral cords, area below body whorl spines and basal keel with 3 gemmate lirae; base concave with 3 granulose spiral lirae and fine growth striae; aperture ovate, columella smooth with a thin basal callus; pale yellow with red radial streaks and small spots on the beads; reaches around 25mm in height.
DISTRIBUTION: Cebu, Bohol and Balut Island (Southern Mindanao).

13. Bolma (Galeoastraea) guttata millegranosa (Kuroda & Habe, in Habe, 1958)
13.0mm x 16.3mm; Punta Engaño, Cebu.
Shell small to medium sized, with a low conical spire, and weakly convex whorls; sutures deeply excavated, obscured by a row of slightly open spines at the base of each whorl; sculptured with 6 spiral rows of beaded lirae, base with several rows of granular lirae and axial growth striae; aperture ovate, outer lip thin with a smooth columella and a thickened umbilical plug; colour pale pink with darker mottling, base off-white; reaches around 35mm in height and inhabits deep water (80 to 150 meters) on sand or gravel.
DISTRIBUTION: Cebu and Bohol.

14. Bolma (Galeoastraea) erectospinosa (Habe & Okutani, 1980)
21.3mm x 36.5mm, 25.0mm x 37.0mm, 21.3mm x 33.2mm; Bohol.
A moderately sized conical shell with a smooth, spiral protoconch, flat spire whorls and sculptured with 6 to 7 rows of sharply beaded cords between which are numerous oblique, slightly raised axial growth striae; base with some 15 sharp, triangular, semi-tubular bifurcate spines, the upper row curved posteriorly; base slightly convex with 6 to 7 beaded spiral cords and fine interstitial growth striae; aperture ovate, outer lip angulate with a row of spines, columella calloused; colour pinkish with a yellow hue, crossed by oblique, radiating purple stripes, spines with purple bands; reaches 25mm in height.
An examination of a large selection of specimens of this species, the more recently described *Bolma (Galeoastraea) erectospina* Kosuge, 1983, and *Bolma (Galeoastraea) persica* (Dall, 1907) has shown that all three intergrade. Hence, this species appears to be no more than an extreme ecomorph of *Bolma (Galeoastraea) persica*, and becomes a synonym thereof.
DISTRIBUTION: Cebu and Bohol (deep water).

15. Turbo (Lunella) cinereus Born, 1778
Syn: *Turbo porphyrites* Gmelin, 1791
23.1mm x 26.4mm, 22.9mm x 28.3mm; Palawan.
A medium sized shell with a depressed spire and a broad, squat body whorl, slightly angulate at the shoulder and crossed by several large, slightly raised spiral cords and weaker interstitial lirae; axially ornamented with numerous very fine oblique striae; aperture circular, umbilicus open, the anterior portion of the shell extended in a lobe-like structure; colour variable, usually pale yellow with white and darker maculations, some specimens green or orange; reaches 35mm in diameter.
DISTRIBUTION: Throughout the Philippines.

16. Turbo (Marmarostoma) stenogyrum Fischer, 1873
11.8mm x 10.0mm, 12.2mm x 10.5mm, 11.5mm x 10.5mm; Cebu.
Shell small, ovate with a moderately tall spire, weakly impressed sutures, and sculptured with raised, rounded, widely-spaced spiral cords, a single interstitial spiral thread, and crossed by numerous oblique axial striae; aperture ovate, columella thick, concave, umbilicus weakly open; colour dark brown with white axial streaks; reaches 20mm in height.
DISTRIBUTION: Throughout the Philippines in limited quantity.

1. Guildfordia triumphans (Philippi, 1841)
20.0mm x 44.4mm, 21.4mm x 36.8mm. (excluding spines); Tayabas Bay, Quezon.
Shell medium sized, low spired, with flattened convex whorls and weakly channeled sutures; sculptured with rows of fine, closely set gemmae, strongest below the suture and which terminate on the body whorl above a row of 9 (occasionally 8) long spines; base with a thick, smooth, somewhat depressed pad in the center, and 3 to 4 spiral rows of granules and radial striae towards the periphery; colour pale pink on top, nacreous off-white underneath, with a pink banded umbilicus and a silvery white aperture; reaches 70mm (including spines) in diameter.
DISTRIBUTION: Tayabas Bay (Quezon), Cebu and Bohol.

2. Astralium calcar (Linné, 1758)
Extremely variable as regards shape; shell medium sized, solid, rather squat, sculptured with a row of partially open spines on the shoulder of the sharply angulate whorls and continuing as oblique cords on the presutural ramp; base convex with foliated spiral cords, becoming spinose towards the periphery; aperture ovate and weakly lirate within, columella smooth and non-umbilicate; shell colour pale greenish yellow, spinal tips dark green, aperture deep yellow within, marginally silvery white, with an off-white to violet columella; reaches 35mm (excluding spines) in diameter.
2a. typical form; 20.0mm x 34.3mm, 20.0mm x 28.8mm (excluding spines); Palawan.
2b. a heavy, tall-spired, spineless form; 35.0mm x 36.0mm; Samar.
DISTRIBUTION: Throughout the Philippines.

3. Bolma (Galeoastraea) henica (Watson, 1885)
Shell medium sized, with a moderately tall spire and deeply channeled sutures; early whorls flat and smooth, with a row of granules on the sharply angled shoulder, and a row of spines on the sharply angled periphery; last 1 or 2 whorls with up to 5 spiral rows of granules between the shoulder and periphery; basal angulation with a spiral beaded cord and 3 to 4 rows of granules between it and the peripheral angulation; base with 6 to 8 spiral rows of granules; aperture ovate, outer lip thin, flared and weakly lirate within, columella smooth; colour pale pink, with silvery pink sutures, a white basal area and a nacreous silver interior; reaches around 30mm in height.
3a. adult specimen; 30.0mm x 29.9mm; Balut Is., Southern Mindanao.
3b. juvenile, with weaker sculpturing, less prominent sutures, and a lower spire; 9.0mm x 15.7mm; Punta Engaño, Cebu.
DISTRIBUTION: Cebu, Bohol and Balut Is., Southern Mindanao.

4. Astralium rhodostoma (Lamarck, 1822)
15.0mm x 19.0mm; Punta Engaño, Cebu.
Shell small to medium sized, conical or depressed, with adpressed sutures and sculptured with oblique axial ribs which terminate at the periphery as a row (sometimes 2) of partially open spines; base flat with several foliate spiral cords; outer lip thin and smooth within, columella smooth and slightly excavated, with a basal denticle; external colouration off-white to yellowish green, columella silvery or purple, aperture nacreous silver; may reach 50mm in height.
DISTRIBUTION: Cebu, Bohol and Balut Is., Southern Mindanao.

5. Astralium mactanense Habe & Okutani, 1980
21.3mm x 24.5mm; Mactan Is., Cebu.
A medium sized, flatly conic shell, the spire whorls sculptured with irregularly spaced, oblique axial ribs and deep interspaces; early whorls smooth or with weak spines, becoming stronger towards the base with every 5 to 6 spines interconnected with a flat, wavy, wing-like foliation every 90° or so; base inflated with 10 to 11 scaled spiral cords; aperture ovate and lirate within; outer lip shouldered but irregularly so from the foliation, columella arcuate with a weak callus; colour off-white with a light purple hue, base deep purple, aperture vitreous white; reaches 30mm (including spines) in width. Although a uniquely distinctive shell, recent studies have shown that it is merely an unusual structural form of the more common *Astralium heimburgi* (Dunker).
DISTRIBUTION: Cebu and Bohol.

6. Bolma (Galeoastraea) girgylla (Reeve, 1843)
44.5mm x 48.0mm, 29.3mm x 45.0mm, 23.3mm x 37.2mm (including spines); Punta Engaño, Cebu.
A medium sized shell, rounded, with bi-angulate whorls bearing 2 rows of spines, the upper row being strongest and distally foliate; presutural ramp with 7 to 8 rows of granules, the first 2 being largest; base with 10 to 12 rows of beaded lirae, enlarging towards the umbilicus; aperture ovate, smooth, silvery white within, columella thick and smooth; colour variable, usually pink to reddish orange, yellow or green, occasionally with purplish dots or streaks at the sutures; reaches 60mm in height.
DISTRIBUTION: Cebu, Bohol and Southern Mindanao.

Class Gastropoda
Subclass Prosobranchia
Order Mesogastropoda
Suborder Taenioglossa
Superfamily Cerithiacea
Family VERMETIDAE

A small family of shells with a worldwide distribution. Members are characterized by their long, tubular, irregularly-coiled whorls which may be smooth or sculptured with spiral or longitudinal ridges. The earlier whorls are smooth and regularly coiled, but towards maturity become free and irregular in their coiling. They live attached to a substrate, often in huge masses, feed on plankton or detritus, and may or may not possess an operculum.

Measurements are maximum length x tube diameter. Relatively few Philippines species recorded.

7. Serpulorbis colubrinus (Röding, 1798)
36.2mm x 21.0mm; Samar.
A conspicuous shell, immediately recognized by its black exterior colouration; the shell is medium sized, rather heavy, the initial whorls being very small but enlarging rapidly and reaching up to 20mm in diameter; shell tightly coiled with coarse longitudinal growth folds; the aperture is circular, dark bluish black and marbled with white; may reach 40mm in length.
DISTRIBUTION: Throughout the Philippines.

8. Vermetus tokyoensis Pilsbry, 1895
21.0mm x 4.8mm; Leyte.
Shell relatively small, the earlier whorls tightly coiled forming a flat base, then rapidly uncoiling in a vertical position; early whorls sculptured with fine longitudinal ribs and spiral growth folds; longitudinal ribs become obsolete on the vertical coil; colour off-white with brown longitudinal ribs; reaches 25mm x 5mm.
DISTRIBUTION: Visayan Sea and Sulu Sea.

9. Serpulorbis medusae (Pilsbry, 1891)
72.0mm x 18.5mm, 111.0mm x 20.5mm; Palawan.
A large, heavy, ponderous shell; initial whorls very small, enlarging rapidly to around 25mm in diameter; sculptured with major and minor longitudinal cords and encircled with wavy growth folds; colour yellowish white with reddish brown spiral lines; may reach 105mm in length.
DISTRIBUTION: Throughout the Philippines.

10. Serpulorbis roussaei (Vaillant, 1871)
197mm x 8.0mm; Palawan.
A medium sized shell which presents itself in large entangled masses, frequently attached to some foreign object; the shell starts as a small compact coil, and enlarges rapidly with unusually pronounced vertical growth, shell pale orange-yellow with irregular dark brown wavy axial bands; individual tube size frequently difficult to assess, but masses up to 1 metre in length have been observed. May be related to or the same as *Serpulorbis sipho* (Lamarck, 1818).
DISTRIBUTION: Palawan.

Class Gastropoda
Subclass Prosobranchia
Order Mesogastropoda
Suborder Taenioglossa
Superfamily Cerithiacea
Family SILIQUARIIDAE

A small group of mud-dwelling herbivorous gastropods which usually live attached to some substrate below tidal level. The members are characterized by their irregular, loosely coiled shells which may be smooth or sculptured with ribs or spines, and differ from the very closely related Vermetidae in the presence of a long row of holes or a linear slit. Few Philippine species presently known.

11. Siliquaria armata Kuroda, Habe & Kosuge, 1971
43.4mm x 6.3mm; Leyte.
Close to *S. anguina* (Linné) (fig. 12), but is distinguished by its purplish colouration, more compact spiral whorls, and the presence of spiral cords which intersect the longitudinal ribs forming short spines; reaches 60mm in length.
DISTRIBUTION: Visayan Sea.

12. Siliquaria anguina (Linné, 1758)
57.0mm x 9.0mm, 22.0mm x 6.5mm; Palawan.
Shell medium sized, the initial whorls smooth and loosely coiled, later whorls rapidly uncoiling; sculptured with spinose longitudinal cords, and a well defined slit which becomes a series of holes in the earlier whorls; colour off-white; may exceed 70mm in length.
DISTRIBUTION: Visayan Sea and Palawan.

Naticidae Plate 9

Class Gastropoda
Subclass Prosobranchia
Order Mesogastropoda
Superfamily Naticacea
Family NATICIDAE

A moderately large group of carnivorous gastropods, preferring a sandy habitat in the mid to upper tidal areas and found in both tropical and polar regions. The shells of this family are small to medium sized, may be smooth or sculptured with fine axial or spiral growth striae, are generally elongate to ovate with a large body whorl, have a semi-lunate aperture, no siphonal canal, a prominent funicle, and an umbilical area which may be open or closed. The opercula may be horny or chitinous. The family is fairly well represented in the Philippines with several Japonic species occasionally being found.

1. Globularia fluctuata (Sowerby, 1825)
Shell moderately large, glossy, globose, with a short spire and deeply channeled sutures; aperture widely ovate, columella heavily calloused, completely obliterating the umbilicus; greyish white with white axial chevrons and three narrow brown bands,

PLATE NO. 8 1. *Guildfordia triumphans* (Philippi) 2a. to 2b. *Astralium calcar* (Linné) 3a. to 3b. *Bolma henica* (Watson) 4. *Astralium rhodostoma* (Lamarck) 5. *Astralium mactanense* Habe & Okutani 6. *Bolma girgylla* (Reeve) 7. *Serpulorbis colubrinus* (Röding) 8. *Vermetus tokyoensis* Pilsbry 9. *Serpulorbis medusae* (Pilsbry) 10. *Serpulorbis roussaei* (Vaillant) 11. *Siliquaria armata* Kuroda, Habe & Kosuge 12. *Siliquaria anguina* (Linné)

PLATE 8

aperture white with brown banding, columellar callus white with an orange-red margin; reaches 80mm in height.
1a. typical specimens; 51.2mm x 49.2mm, 52.5mm x 46.2mm; Palawan.
1b. albinistic; 49.1mm x 45.7mm; Palawan.
DISTRIBUTION: Palawan.

2. Polinices (Polinices) cumingianus (Récluz, 1844)
49.9mm x 41.2mm, 42.6mm x 37.9mm; Samar.
Shell large, glossy, pyriform-ovate and almost smooth, with weakly channeled sutures and weak axial growth lines; aperture semi-lunate, columella heavily calloused posteriorly and joining the funicle which enters the wide, deep umbilical area obliquely; off-white with a brown sutural band, a wide band on the body whorl, a white stained umbilicus and a white aperture; reaches up to 55mm in height and 50mm in width.
DISTRIBUTION: Samar, Visayan Sea and Sulu Sea.

3. Polinices (Polinices) aurantius (Röding, 1798)
Shell of moderate size, heavy, ovate, with a short, white-coloured spire, weakly channeled sutures, a slight shoulder "hump" on the body whorl, a white semi-lunate aperture, and a thick white callus which completely obliterates the umbilicus; solid orange in colour; largest specimen seen 43mm x 33mm.
3a. small, brown-banded form; 27.9mm x 23.0mm; Samar.
3b. typical specimens; 39.9mm x 30.5mm, 35.9mm x 21.9mm; Samar.
DISTRIBUTION: Samar, Visayan Sea and Sulu Sea.

4. Polinices (Mammilla) melanostomus (Gmelin, 1791)
41.5mm x 35.5mm, 29.9mm x 23.3mm; Samar.
Shell medium sized, heavy, glossy, somewhat ovate, with a short spire, weakly impressed sutures and axial growth lines; aperture semi-lunate, umbilicus slightly covered with a somewhat expansive callus; external color grey with three occasionally discontinuous dark brown bands, umbilical area dark brown, aperture white with pale banding; largest specimen seen 48mm x 42.6mm.
DISTRIBUTION: Sporadically found throughout the Philippines.

5. Polinices (Mammilla) maurus (Lamarck, 1816)
44.3mm x 37.5mm, 37.5mm x 30.5mm; Samar.
Shell medium sized, lightweight, pyriform-ovate in outline, with a matt gloss, a short spire, weakly channeled sutures, a semi-lunate aperture and a columellar callus which is thickened posteriorly but folded over the umbilical channel anteriorly; deep brown with an internally white, marginally brown aperture, and a deep brown callus; largest specimen seen 51.4mm x 43.9mm (beached).
DISTRIBUTION: Sporadically found throughout the Philippines in limited quantity.

6. Polinices (Polinices) tumidus (Swainson, 1840)
Syn: *P. pyriformis* Récluz, 1844
39.9mm x 34.0mm, 41.5mm x 35.0mm; Cebu.
Shell medium sized, glossy, porcelaneous white; pyriform-ovate in outline with a moderately tall spire, weakly incised sutures and axial growth striae; aperture semi-lunate, a thick callus pad usually obliterating the umbilicus, but a groove normally seen in younger specimens; reaches 45mm in height.
DISTRIBUTION: Throughout the Philippines.

7. Polinices (Mammilla) sebae (Récluz, 1844)
51.9mm x 40.0mm, 41.0mm x 33.5mm; Samar.
Shell moderately large, lightweight, pyriform-ovate and somewhat dorso-ventrally compressed; spire low, sutures weakly channeled, sculptured with strong axial growth lines; aperture widely ovate, columellar callus folded over and almost obliterating the umbilical opening; colour white with 3 rows of brown spots, aperture white, callus deep brown, anterior umbilical margin lighter brown; largest specimen seen here illustrated.
DISTRIBUTION: Sporadically found throughout the Philippines.

8. Polinices (Mammilla) fibrosa (Souleyet, 1852)
39.7mm x 27.0mm, 37.9mm x 28.1mm; Palawan.
Shell medium sized, lightweight, pyriform-ovate, with a moderately tall spire, adpressed sutures and crossed by fine axial and spiral growth lines; aperture elongate-ovate, with a brown-coloured columellar callus which folds over and partially occludes the umbilical opening; off-white with 2, sometimes 3, spiral bands of brown axial streaks; reaches 55mm in height. This is the *P. mammata auct.*, not the true *P. mammatus* of Röding, 1798.
DISTRIBUTION: Sporadically found throughout the Philippines.

9. Polinices (Neverita) albumen (Linné, 1758)
33.9mm, 33.0mm; Samar.
Shell medium sized, glossy, ovate in outline and dorso-ventrally compressed; spire flat, sutures weakly impressed and the whole surface with axial growth lines; aperture semi-lunate, funicle wide and bordered by a deep umbilical channel; deep brown, with the spire whorls, aperture and umbilical area all white; reaches 60mm in length.
DISTRIBUTION: Throughout the Philippines in very limited quantity.

PLATE NO. 9 1a. *Globularia fluctuata* (Sowerby) 1b. albinistic 2. *Polinices cumingianus* (Récluz) 3a. *Polinices aurantius* (Röding) brown-banded form 3b. typical form 4. *Polinices melanostomus* (Gmelin) 5. *Polinices maurus* (Lamarck) 6. *Polinices tumidus* (Swainson) 7. *Polinices sebae* (Récluz) 8. *Polinices fibrosa* (Souleyet) 9. *Polinices albumen* (Linné) 10. *Natica alapapilionis* (Röding) 11. *Polinices flemingianus* (Récluz) 12a. *Polinices peselephanti* (Link) white form 12b. orange form 13. *Natica stellata* Hedley 14. *Polinices simiae* (Deshayes, in Deshayes & Edwards) 15. *Polinices powisianus* (Récluz) 16. *Polinices sagamiensis* Pilsbry 17. *Natica fasciata* (Röding)

PLATE 9

10. Natica (Naticarius) alapapilionis (Röding, 1798)
23.9mm x 22.2mm, 24.9mm x 22.6mm; Samar.
Shell small, squat, glossy, with a short spire, well channeled sutures, and weak radially orientated subsutural ribs; aperture broadly ovate, umbilicus widely open and partially obscured by a small funicle which enters obliquely; colour greyish brown with 4 thin, brown-spotted white bands, a white umbilical area, and a lavender-coloured, marginally white aperture; reaches 30mm in height.
DISTRIBUTION: Throughout the Philippines.

11. Polinices (Polinices) flemingianus (Récluz, 1844)
38.8mm x 34.3mm, 32.8mm x 29.5mm; Bohol.
Shell medium sized, heavy, glossy, pyriform-ovate, with a moderately tall spire, adpressed sutures and axial growth striae; aperture widely ovate, a thick callus partially filling the umbilical opening; colour normally white, Philippine specimens burnt orange with a white spire, aperture and umbilicus; reaches 45mm in height.
DISTRIBUTION: Sporadically found throughout the Philippines.

12. Polinices (Neverita) peselephanti (Link, 1807)
Shell medium sized, white to pale orange in colour, rather globose with a low spire, weakly incised sutures and fine axial growth lines; aperture semi-lunate, white within, umbilicus deep with a sharply angled margin, funicle curiously stepped, partially occluding the umbilicus; reaches 40mm in height.
12a. white form; 29.7mm x 30.5mm; Samar.
12b. orange form; 39.5mm x 35.0mm, 36.9mm x 32.7mm; Samar.
DISTRIBUTION: Throughout the Philippines.

13. Natica (Natica) stellata Hedley, 1913
35.9mm x 36.1mm, 32.8mm x 31.8mm; Samar.
Shell medium sized, ovate, rather solid and glossy, with a short, somewhat flattened spire, well channeled sutures and fine axial growth striae; aperture semi-lunate, umbilicus partially covered by a thin callus extending from the posterior columellar margin; shell deep orange with 2 discontinuous white bands; largest recorded size 45mm in height.
DISTRIBUTION: Throughout the Philippines.

14. Polinices (Mammilla) simiae (Deshayes, in Deshayes & Edwards, 1838)
36.1mm x 29.5mm; Punta Engaño, Cebu.
Shell medium sized, pyriform-ovate, with a moderately tall spire, weakly channeled sutures, and fine axial growth striae; aperture wide, semi-lunate, columellar callus folded over the umbilicus but not completely obliterating it; off-white with a subsutural and basal band of brown streaks, a central brown line, and numerous axial streaks, callus light brown, umbilical channel dark brown; reaches 40mm in height.
DISTRIBUTION: Sporadically found throughout the Philippines.

15. Polinices (Neverita) powisianus (Récluz, 1844)
45.0mm x 43.4mm, 44.5mm x 43.5mm; Samar.
Shell large, solid, glossy, pyriform-ovate, with a low spire, adpressed sutures, and fine axial and spiral growth striae; aperture widely ovate, with a posteriorly thickened callus and a small funicle which partially covers a large umbilical area; spire white with an orange-yellow subsutural band and a wide brown band covering almost all the body whorl, aperture, columella and umbilicus white; reaches 50mm in height.
DISTRIBUTION: Sporadically found throughout the Philippines.

16. Polinices (Polinices) sagamiensis Pilsbry, 1904
40.3mm x 38.5mm, 39.9mm x 39.2mm; Samar.
This species is very close to *P. powisianus* (fig. 15), but differs from it in the following ways: the shell is marginally smaller with a slightly taller spire, the umbilical opening is smaller, the funicle is larger, and the colour is generally paler, not the deep red-brown of *P. powisianus;* largest size seen 46.2mm x 43.7mm.
DISTRIBUTION: Sporadically found throughout the Philippines.

17. Natica (Natica) fasciata (Röding, 1798)
23.5mm x 22.4mm, 23.1mm x 22.0mm; Palawan.
Shell small, globose, rather solid, with a short spire, weakly channeled sutures, fine axial growth striae, a semi-lunate aperture, and a deep umbilical opening which is 75% covered with a thick callus; shell dark brown with 2 paler bands, one subsuturally, the other basally, protoconch dark purple, callus a deep shiny brown, aperture pale lavender with brown bands; reaches 30mm in height.
DISTRIBUTION: Throughout the Philippines.

Naticidae Plate 10

1. Eunaticina papilla (Gmelin, 1791)
29.6mm x 23.3mm, 30.0mm x 21.5mm; Punta Engaño, Cebu.
Shell medium sized, lightweight, pyriform-ovate, sutures deeply channeled; sculptured with numerous fine spiral lirae and overriding axial growth lines; aperture large, triangularly ovate, columellar callus folded posteriorly, forming a deep umbilical channel; off-white, aperture porcelaneous white; reaches 35mm in height.
DISTRIBUTION: Sporadically found throughout the Philippines.

2. Natica (Natica) undulata (Röding, 1798)
31.5mm x 29.5mm, 28.9mm x 30.5mm; Palawan.
Shell medium sized, lightweight, globose, immediately identified by the orange-brown axial zigzag lines over a white shell; the spire is short, the sutures moderately channeled, with strong subsutural growth striae; the umbilical area is only partially obscured posteriorly by a thin callus covering, and centrally by a medium

sized funicular pad; reaches 35mm in height.
DISTRIBUTION: Sporadically found throughout the Philippines.

3. Polinices (Glossaulax) didyma (Röding, 1798)
35.4mm x 42.0mm, 33.9mm x 41.7mm; Samar.
Shell large, ovate, somewhat compressed, with a low spire, weakly channeled sutures and axial and spiral growth striae; aperture semi-lunate, umbilicus deep with some grooving; funicle with an expansive callus partially covering the umbilical opening; colour purplish grey with a darker subsutural band and a white base, aperture deep brown, funicle orange-brown; reaches 85mm in width.
DISTRIBUTION: Sporadically found throughout the Philippines.

4. Natica (Natica) violacea Sowerby, 1825
23.0mm x 19.5mm, 22.5mm x 19.6mm; Cebu.
A small shell with a short spire and a large, globular body whorl which is sculptured with fine axial scratches; aperture semi-lunate, umbilicus open, partially occluded by a callus; white with 3 brown and white sagittal bands and numerous brown axial lines, inner lip margin and parietal callus deep purple; reaches 25mm in height.
DISTRIBUTION: Sporadically found throughout the Philippines.

5. Natica (Naticarius) onca (Röding, 1798)
22.9mm x 22.1mm, 25.9mm x 22.5mm, 23.5mm x 22.0mm; Samar.
Shell small, spire low, body whorl large and ovate; sutures well impressed with radiating growth striae; aperture semi-lunate, umbilical area large, slightly open posteriorly but closed anteriorly by a wide, flat funicle and bordered by a narrow umbilical channel; white with 4 rows of brown quadrate spots; reaches 35mm in height.
DISTRIBUTION: Sporadically found throughout the Philippines.

6. Tanea lineata (Röding, 1798)
35.2mm x 31.6mm; Punta Engaño, Cebu.
Shell medium sized, globose, with a moderately tall spire, rounded spire whorls, channeled sutures, and ornamented with close-set, orange-brown lines on a white background; aperture semi-lunate, umbilicus deep, partially occluded anteriorly by a large funicular pad; reaches 55mm in height.
DISTRIBUTION: Sporadically found throughout the Philippines.

7. Tanea euzona (Récluz, 1844)
21.7mm x 22.5mm; Samar.
Shell medium sized, glossy, very lightweight, with a globose body whorl, a low spire, moderately channeled sutures, and fine, arcuate, subsutural growth striae; aperture widely ovate, umbilicus deep, open, occluded centrally by a wide funicular pad; shell off-white, with 3 white bands marked with brown chevrons, the first and third bands being strongest, and brown axial lines between the bands; base and umbilicus white with short brown streaks; reaches 30mm in height.
DISTRIBUTION: Visayan Sea, Sulu Sea and Samar.

8. Natica (Natica) arachnoidea (Gmelin, 1791)
16.4mm x 16.0mm; Punta Engaño, Cebu.
A small globose shell with a short spire, a white, semi-lunate aperture and a moderately deep umbilical channel, partially obscured posteriorly by an expansive callus; shell yellowish with orange-brown maculations; reaches 20mm in height.
DISTRIBUTION: Cebu, Bohol and Southern Mindanao.

9. Tanea picta (Récluz, 1844)
24.4mm x 24.4mm, 19.4mm x 19.2mm; Samar.
Almost identical to *T. euzona* in size and sculpturing but differs in being a dark orange-brown shell with a strong white band marked with brown chevrons just below the sutures. There are a further 2 bands, much reduced in size, on the upper and lower third of the shell and a white area around the umbilicus; reaches 30mm in height.
DISTRIBUTION: Samar, Visayan Sea and Sulu Sea.

Class Gastropoda
Subclass Prosobranchia
Order Archaegastropoda
Superfamily Neritacea
Family NERITIDAE

A comparatively large group of tropical to sub-tropical herbivorous gastropods inhabiting a variety of marine and brackish water habitats. The shells are small, globose, with a flattened base, a semi-lunate aperture, a heavy columellar callus, a dentate inner and outer lip, and a calcareous operculum. The shell may be smooth, ribbed or pustulose. The family is well represented in the Philippines, and many of the species occurring here are among the most colourful and beautifully patterned to be found anywhere. (Measurements quoted are the maximal shell length, with the specimen placed flat on a surface).

10. Nerita (Theliostyla) albicilla Linné, 1758
Shell medium sized, with a flat spire, and sculptured with irregular-sized cords; labial lip flattened, with fine teeth on the inner margin; columellar pad pustulose, columella with a few fine teeth; aperture white with a yellow tinge on the columellar pad; reaches 35mm.
10a. An unusual red-and-black banded form; 22.5mm; Punta Engaño, Cebu.
10b. typical form; 26.0mm, 25.0mm; Batangas.
DISTRIBUTION: Throughout the Philippines.

11. Nerita (Amphinerita) polita Linné, 1758
A solid, medium-sized shell with a depressed spire, weak or strong axial striae, a heavily calloused columellar pad, and a weakly dentate columella; labial denticles very fine; colour variable with various

degrees of streaking and mottling; aperture white with an orange to yellow margin; reaches 35mm.

11a. Unusually large specimens with a white aperture; this is the most common form; 37.2mm, 33.5mm; Samar.

11b. Specimens exhibiting an orange-coloured aperture; 22.5mm, 23.5mm; Marinduque.

DISTRIBUTION: Throughout the Philippines.

12. Nerita (Theliostyla) planospira Anton, 1839

21.7mm, 19.9mm; Cebu.

A medium sized shell with the spire depressed below the level of the shell surface and the upper labial lip extended beyond the columellar pad level; surface with spiral cords and striae, labial lip weakly denticulate, columella with strong teeth, columellar pad pustulose; grey with darker maculations, aperture white; reaches 35mm.

DISTRIBUTION: Throughout the Philippines.

13. Nerita (Ritena) plicata Linné, 1758

22.8mm, 20.5mm; Samar.

Shell medium sized, solid, ovate, with strong spiral cords, deep interstitial grooves and fine, overriding axial striae; spire low, columellar pad slightly convex with a few heavy wrinkles and pustules, 4 large columellar denticles, and 5 to 7 large labial denticles, the terminal ones largest; pinkish white with a lemon-coloured spire and an orange-bordered, porcelaneous white aperture; reaches 35mm.

DISTRIBUTION: Throughout the Philippines.

14. Nerita (Ritena) costata Gmelin, 1791

23.2mm, 26.7mm; Palawan.

Very close to the previous species in shape and sculpturing but the spire is lower, the cords fewer but wider, and the interstitial grooves are narrower; apertural features the same as in *N. plicata;* shell with black ribs, paler interstitial grooves and a porcelaneous white aperture; reaches 35mm.

DISTRIBUTION: Throughout the Philippines.

15. Nerita (Theliostyla) squamulata Le Guillou, 1841

18.2mm, 19.0mm, 23.8mm; Mindoro.

A medium sized nerite, spire depressed, surface with scabrous spiral cords and coarse axial striae; columellar pad sunken, pustulose and wrinkled, with 3 to 4 denticles on the margin; labial surface with elongate denticles; colour variable, normally grey with paler maculations and a white aperture; reaches 35mm.

DISTRIBUTION: Throughout the Philippines.

16. Nerita (Theliostyla) exuvia Linné, 1758

34.6mm, 35.5mm; Batangas.

Shell medium sized, solid, with a depressed spire; sculptured with 16 to 18 raised, rounded, slightly scabrous spiral cords, separated by deep, widely spaced grooves; columellar pad finely pustulose, sunken, with a well defined margin; columella with 3 to 4 weak teeth, labial lip with numerous elongate denticles, the upper two pustulose and larger than the rest; ribs black, interspaces pale yellow with black stripes, aperture white with a yellow tinge on the columellar pad; reaches 40mm.

DISTRIBUTION: Throughout the Philippines.

17. Nerita (Ritena) undata Linné, 1758

26.5mm, 21.3mm, 30.3mm, 32.3mm; Samar.

Shell medium sized, globose, solid, with a short spire; sculptured with numerous spiral cords and weakly striated grooves; columellar pad with heavy folds, columella with 4 strong teeth, labial lip with numerous fine denticles, the upper two largest; usually dark coloured with paler axial streaks or banding; aperture white with a lemon tint in the upper apertural area; reaches 40mm.

DISTRIBUTION: Throughout the Philippines.

Neritidae Plate 11

1. Neritina (Vittina) waigiensis (Lesson, 1831)

Syn: *Theodoxus (Pictoneritina) communis* (Quoy & Gaimard, 1832)

Size range for illustrated specimens:
13.9mm — 19.8mm; Davao.

Shell small, ovate, glossy, tall-spired and smooth; columellar callus smooth and thick, columella with numerous minute teeth, outer lip smooth and thin; colour and pattern extremely variable: red, mauve, yellow or pink, with contrasting bands of axial stripes and zigzags; aperture white, columellar callus grey; reaches 20mm; a mangrove dweller.

DISTRIBUTION: Throughout the Philippines.

2. Neritina (Vittoida) variegata (Lesson, 1831)

11.5mm, 14.9mm, 10.5mm, 9.2mm, 8.8mm, 8.9mm; Palawan.

Shell small and ovate with a smooth surface, and a short spire which is nearly always eroded; aperture semi-lunate, columellar callus smooth, the margin with numerous very fine teeth and a prominent central tooth; extremely variable in colour and pattern; aperture blue-white; reaches 17mm.

DISTRIBUTION: Throughout the Philippines.

3. Clithon oualaniensis (Lesson, 1831)

Size range for illustrated specimens: 7.1mm — 8.3mm; Bohol.

Shell small and ovate with a short, frequently eroded spire; aperture semi-lunate; columella, collumellar margin and outer lip smooth; colour pale cream with a black "netted" pattern; reaches 12mm.

PLATE NO. 10 1. *Eunaticina papilla* (Gmelin) 2. *Natica undulata* (Röding) 3. *Polinices didyma* (Röding) 4. *Natica violacea* Sowerby 5. *Natica onca* (Röding) 6. *Tanea lineata* (Röding) 7. *Tanea euzona* (Récluz) 8. *Natica arachnoidea* (Gmelin) 9. *Tanea picta* (Récluz) 10a. *Nerita albicilla* Linné, an unusual red and black banded form. 10b. typical form 11a. *Nerita polita* Linné, unusually large specimen with white aperture. 11b. typical form 12. *Nerita planospira* Anton 13. *Nerita plicata* Linné 14. *Nerita costata* Gmelin 15. *Nerita squamulata* Le Guillou 16. *Nerita exuvia* Linné 17. *Nerita undata* Linné

PLATE 10

DISTRIBUTION: Throughout the Philippines.

4. Smaragdia rangiana (Récluz, 1842)
Size range for illustrated specimens: 6.8mm — 7.4mm; Palawan.
Shell small, elongate-ovate, with a short spire, a smooth, swollen columellar pad, a weakly dentate margin and a smooth outer lip; green with several spiral rows of white streaks and dots lined with black; reaches 8mm.
DISTRIBUTION: Mostly from Palawan.

5. Neritina (Provittoida) coromandeliana Sowerby, 1832
18.0mm, 17.5mm, 13.7mm, 18.7mm, 11.0mm, 12.4mm; Samar.
Shell small, ovate, glossy, with a moderately tall spire; columellar pad smooth, the margin with weak teeth, and the outer lip thin; black with widely spaced axial zigzag stripes, aperture white; reaches 20mm. Sold locally as *zigzag* Lamarck, 1822 (syn. *gagates* Lamarck) which is endemic to the Indian Ocean.
DISTRIBUTION: Throughout the Philippines.

6. Septaria (Septaria) porcellana (Linné, 1758)
16.9mm, 16.7mm; Bohol.
Shell small, elongate-ovate with a large, inflated curved apex situated on the posterior margin; interior with a small ledge under the apex; exterior black with numerous close-set, greenish yellow triangular spots, interior bluish white; reaches 20mm.
DISTRIBUTION: Throughout the Philippines.

7. Clithon corona (Linné, 1758)
Syn: *Neritina brevispina* Lamarck, 1822
16.0mm, 15.1mm; Palawan.
Shell small, ovate, with a short spire, a large body whorl, and a row of long spines on the sharply angled shoulder; surface with fine arcuate axial folds; aperture ovate with a thin outer lip, columellar margin with fine teeth and a single large denticle; black with a white aperture; reaches 20mm; lives in brackish water.
DISTRIBUTION: Sporadically found thoughout the Philippines.

8. Neritina (Dostia) violacea (Gmelin, 1791)
25.5mm, 22.6mm, 22.4mm; Samar.
Shell medium sized, elongate-ovate, with a large, inflated, curved apex situated at the posterior margin; aperture semi-lunate, columellar pad expansive, forming a broad flat ledge; columellar margin finely dentate, outer lip margin smooth and flared; pale cream with dense purple maculations, aperture deep orange; reaches around 25mm.
DISTRIBUTION: Sporadically found throughout the Philippines.

9. Nerita (Ritena) signata Macleay, in Lamarck, 1822
14.6mm, 14.0mm; Palawan.
A small shell with an elevated spire and sculptured with strong scabrous cords and oblique axial striae; columellar pad smooth or weakly wrinkled with a characteristic red blotch; columella with 4 to 5 obsolete denticles, labial lip with numerous elongated teeth; variable in colour, normally grey-white with darker maculations on the cords, aperture white; reaches 20mm.
DISTRIBUTION: Sporadically found throughout the Philippines.

10. Nerita (Ritena) helicinoides Reeve, 1855
15.5mm, 15.2mm; Bataan.
Shell small with a short, lemon-coloured spire, adpressed sutures, a large, globose body whorl encircled with fine spiral cords, a thick, weakly lirate labial lip, a weakly wrinkled columellar pad, and a bi- or tri-dentate columellar margin; body whorl off-white with red staining, aperture white with a yellow margin; reaches 20mm. Often figured as *N. helicinoides laevilabris* Pilsbry, 1895. The type locality of *helicinoides* is unknown and the use of *laevilabris* as a subspecies is of doubtful validity.
DISTRIBUTION: Sporadically found throughout the Philippines.

11. Nerita (Ritena) chamaeleon Linné, 1758
20.3mm, 22.5mm, 18.5mm, 20.4mm; Bohol.
Shell small to medium sized with an elevated spire; sculptured with fine spiral cords and axial striae; columellar pad wrinkled and pustulose, columella centrally bi-dentate and the labial lip has small, nodular denticles; colour variable, usually black with white banding or mottling, occasionally red; reaches 35mm.
DISTRIBUTION: Throughout the Philippines.

12. Neritodryas dubia (Gmelin, 1791)
18.8mm, 18.2mm; Samar.
Shell medium sized, smooth, lightweight and glossy; spire moderately tall with an eroded apex; columellar pad, inner margin and labial lip all smooth; aperture semi-lunate with a thin outer lip; yellowish white with black axial zigzags, aperture marginally reddish orange, bluish white within, a large black blotch on the columellar pad; reaches 25mm.
DISTRIBUTION: Sporadically found throughout the Philippines.

13. Neritina (Vittina) turrita (Gmelin, 1791)
23.9mm, 19.0mm; Bohol.
Shell small to medium sized, ovate, with a tall, partially eroded spire and a globose body whorl; surface smooth, black coloured with numerous fine

PLATE NO. 11 1. *Neritina waigiensis* (Lesson) 2. *Neritina variegata (Lesson)* 3. *Clithon oualaniensis* (Lesson) 4. *Smaragdia rangiana* (Récluz) 5. *Neritina coromandeliana* Sowerby 6. *Septaria porcellana* (Linné) 7. *Clithon corona* (Linné) 8. *Neritina violacea* (Gmelin) 9. *Nerita signata* Macleay, in Lamarck 10. *Nerita helicinoides Reeve* 11. *Nerita chamaeleon* Linné 12. *Neritodryas dubia* (Gmelin) 13. *Neritina turrita* (Gmelin) 14. *Neritopsis radula* (Linné)

PLATE 11
1
2
3
4
5
6
7
8
9
10
11
12
13
14

white axial stripes; columellar pad inflated and smooth, columella with numerous fine teeth, outer lip off-white, thin, smooth and widely arcuate; aperture columellar pad pinkish grey; reaches 25mm; inhabits brackish water. DISTRIBUTION: Sporadically found throughout the Philippines.

Class Gastropoda
Subclass Prosobranchia
Order Archaegastropoda
Superfamily Neritacea
Family NERITOPSIDAE

A small group of gastropods, structurally very similar to the Neritidae, and possessing a globular shell, a wide aperture and a smooth columella. One Philippine species recorded.

14. Neritopsis radula (Linné, 1758)
19.8mm, 22.6mm; Cebu.
A small globular shell with a tall spire and sculptured with numerous nodulose spiral cords and fine axial striae; aperture ovate, internally lirate, columella concave with a single central tooth; cream coloured with a white aperture; reaches 25mm. DISTRIBUTION: Sporadically found throughout the Philippines.

Turritellidae, Melanellidae — Plate 12

Class Gastropoda
Subclass Prosobranchia
Order Mesogastropoda
Suborder Taenioglossa
Superfamily Cerithiacea
Family TURRITELLIDAE

A large group of herbivorous, sand dwelling gastropods living mainly in the shallow muddy waters of most temperate and tropical regions. The shells are small to medium sized, non-umbilicate, have a long tapering spire, numerous whorls which are sculptured with heavy spiral ribs, and a rounded aperture with a thin outer lip. The family appears to be poorly represented in the Philippines.

1. Turritella terebra (Linné, 1758)
133mm; Manila Bay.
The largest *Turritella* species in the Philippines, reaching around 160mm; the shell is rather heavy with a tall, acute spire and around 25 to 30 rounded whorls that are sculptured with up to 6 spiral cords and weaker interstitial spiral threads; sutures deeply impressed, the base concave and spirally striate; colour light brown to dark brown; the form *spectrum* (Reeve, 1849) (not illustrated) has more inflated whorls, deeper sutures and reduced spiral ribbing. DISTRIBUTION: Mostly from Manila Bay, but occurs throughout the Philippines.

Class Gastropoda
Subclass Prosobranchia
Order Mesogastropoda
Suborder Gymnoglossa
Superfamily Eulimacea
Family MELANELLIDAE

A large group of small sized, tall-spired, non-umbilicate shells which have a white glossy surface, numerous slightly convex whorls, and the apex occasionally bent to one side. Many of these species are parasitic on holothurians, starfish and urchins. Although only a few species are shown here, the family undoubtedly has numerous representatives in the Philippines.

2. Melanella (Balcis) candida (Marrat, 1880)
39.1mm; Palawan.
Shell large for the family, with a tall, slightly bent spire, around 15 weakly convex whorls, well impressed sutures, and a single oblique groove running the entire shell length (indicates former apertural edges); aperture "tear-drop" shaped, obliquely angled and smooth within; columella moderately thickened and smooth; colour white; reaches 40mm. DISTRIBUTION: Sporadically throughout the Philippines.

3. Melanella (Balcis) major (Sowerby, 1834)
23.9mm; Punta Engaño, Cebu.
Shell medium sized, tall-spired, with moderately impressed sutures and around 9 to 12 weakly convex whorls; the spire is acute and flexed to the right; body whorl larger than the rest, aperture ovate, outer lip

PLATE NO. 12 1. *Turritella terebra* (Linné) 2. *Melanella candida* (Marrat) 3. *Melanella major* (Sowerby) 4. *Mucronalia gigas* Kuroda & Habe 5. *Zebina tridentata* (Michaud) 6. *Rissoina reticulata* (Sowerby) 7. *Zebina gigantea* (Deshayes) 8. *Trichotropis turrita* (Habe) 9. *Amathina tricarinata* (Linné) 10. *Opimilda cancellata* (Kuroda) 11. *Vanikoro cancellata* (Lamarck) 12. *Vanikoro distans* (Récluz) 13. *Planaxis niger* Quoy & Gaimard 14. *Planaxis sulcatus* (Born) 15. *Modulus tectum* (Gmelin) 16. *Crucibulum spinosum* (Sowerby) 17. *Cheilea equestris* (Linné) 18. *Calyptraea extinctorium* Lamarck 19. *Crepidula walshi* Reeve 20. *Hipponix foliaceus* (Quoy & Gaimard) 21. *Sabia conica* (Schumacher) 22. *Dolabella auricularia* (Lightfoot)

PLATE 12

arched forward, columella weakly calloused; colour brownish white, apex and aperture pure white; reaches 25mm.
DISTRIBUTION: Sporadically found throughout the Philippines.

4. Mucronalia gigas Kuroda & Habe, 1950
21.9mm; Punta Engaño, Cebu.
Shell medium sized, white, shiny, with a tall, straight-sided spire, a small mammillate apex, and sutures which are weakly incised; aperture elongate-ovate, columella weakly calloused with a central depression and produced basally to form a very short canal with the outer lip margin; reaches around 25mm.
DISTRIBUTION: Sporadically throughout the Philippines.

Class Gastropoda
Subclass Prosobranchia
Order Mesogastropoda
Suborder Taenioglossa
Superfamily Rissoacea
Family RISSOIDAE

A large, heterogeneous group of intertidal-dwelling gastropods found in most geographical provinces. The shells are quite small, variable in shape (fusiform-ovate to fusiform-elongate), have a small aperture, a short, broad siphonal canal, and exhibit a high degree of morphological variation, some being smooth, others variously sculptured with axial and spiral cords. The following species were originally placed in the family Rissoinidae, which has now been relegated to a subfamily group within the Rissoidae (Ponder 1967). Fairly well represented in the Philippines but seldom available to the average collector unless collected in the field. The few illustrated here were taken from shell grit.

5. Zebina (Zebina) tridentata (Michaud, 1830)
9.5mm; Punta Engaño, Cebu.
A small, fusiform-ovate shell with weakly incised sutures, a large rounded body whorl, and a small ovate aperture characteristically sculptured with 3 denticles on the anterior margin of the outer lip; colour pure milky white; reaches 12mm.
DISTRIBUTION: Sporadically found throughout the Philippines.

6. Rissoina (Zebinella) reticulata (Sowerby, 1824)
18.7mm; Punta Engaño, Cebu.
Shell large for the family, solid, dirty white in colour, with convex whorls and moderately channeled sutures; earlier whorls sculptured with intersecting axial and spiral ribs with the axial sculpturing becoming obsolete towards the body whorl and only the spiral elements persisting; aperture wide and ovate, outer lip thick and arcuate, columella smooth; reaches around 25mm.
DISTRIBUTION: Sporadically found throughout the Philippines.

7. Zebina (Morchiella) gigantea (Deshayes, 1850)
12.5mm; Punta Engaño, Cebu.
Shell large, solid, elongate, with moderately incised sutures; early whorls sculptured with coarse axial riblets which are crossed by spiral threads giving a weakly cancellate appearance; spiral lirae persisting on the last 3 whorls but becoming obsolete on the body whorl; outer lip thick, expanded, arcuate and smooth within; creamy white in colour; reaches 30mm.
DISTRIBUTION: Sporadically found throughout the Philippines.

Class Gastropoda
Subclass Prosobranchia
Order Mesogastropoda
Suborder Taenioglossa
Superfamily Calyptraeacea
Family TRICHOTROPIDAE

A moderately small group of cold water gastropods found mostly in deep water. The shells are small, strongly carinate, and have a tall or depressed spire. The body whorl is large with a strongly keeled shoulder, a wide aperture, a flaring outer lip, and a hairy periostracum. Only 1 or 2 species in the Philippines.

8. Trichotropis (Turritopis) turrita (Habe, 1962)
17.1mm; Cebu.
This name appears to be occupied by *T. turrita* Dall, 1927, from Florida, U.S.A. We can find no alternative name at this time. Shell small, whitish brown, with a tall spire, a large body whorl, a wide, flaring outer lip and a slightly channeled umbilicus; spire whorls smooth with a strongly carinate shoulder; body whorl bi-cingulate, with a weaker cord between and several below the 2 carinae; outer lip thin, bi-angulate, nacreous on the inner margin; columella smooth with a weak basal "bump"; reaches 20mm.
DISTRIBUTION: Sporadically found throughout the Philippines.

9. Amathina tricarinata (Linné, 1767)
39.7mm (L) x 25.9mm (W) x 12.2mm (H); Davao.
Shell medium sized, solid, patelliform in outline, with a moderately tall, slightly coiled apex which is directed towards the shell margin; outer surface with 3 raised, strongly carinate ribs which terminate as semitubular spines on the apertural margin; aperture extremely large, smooth and porcelaneous white within, the outer lip margin somewhat flattened; external colour off-white and covered with a yellowish brown periostracum; illustrated specimen largest seen.
DISTRIBUTION: Sporadically found throughout the Philippines in limited quantity.

Class Gastropoda
Subclass Prosobranchia
Order Heterogastropoda
Suborder Ptenoglossa
Superfamily Architectonicacea
Family MATHILDIDAE

A small group of shells closely related anatomically to the Architectonicidae. The shells are small, turreted, and sculptured with axial and spiral cords. The aperture is small, smooth within, and lacks columellar folds or a canal. Only 1 or 2 species presently recorded from the Philippines.

10. Opimilda cancellata (Kuroda, 1958)
17.8mm; Punta Engaño, Cebu.
Shell small, turreted, with well impressed sutures; whorls with a strong peripheral cord and several weaker cords above and below which are crossed by axial striae, producing a reticulate surface; aperture sub-ovate, outer lip thin and smooth within; pale brownish white in colour; may reach 20mm.
DISTRIBUTION: Sporadically found throughout the Philippines.

Class Gastropoda
Subclass Prosobranchia
Order Mesogastropoda
Suborder Taenioglossa
Superfamily Hipponicacea
Family VANIKORIDAE

A small group of intertidal molluscs living on clean or muddy sand and feeding on detritus. The shells are small with a low, strongly cancellate spire and a large globose body whorl encircled with axial and spiral striae; the aperture is large, circular and smooth, and the umbilicus may be open or closed. Few Philippine species.

11. Vanikoro cancellata (Lamarck, 1822)
10.4mm x 11.0mm; Mindoro.
Shell small, globose, with a short spire and a large, ovate body whorl; sculptured with numerous axial and spiral threads, the spiral elements strongest on the body whorl; aperture large, ovate and smooth within, umbilicus open; colour dirty white; reaches around 25mm.
DISTRIBUTION: Sporadically found throughout the Philippines.

12. Vanikoro distans (Récluz, 1844)
7.5mm x 6.8mm; Punta Engaño, Cebu.
Shell small, globose, with a moderately tall spire and a large body whorl; sculptured with around 10 widely-spaced, blade-like axial cords on the body whorl and numerous fine spiral striae in the interspaces; aperture ovate, outer lip smooth, columella oblique with a narrow but deeply channeled umbilicus; colour off-white; reaches 10mm in height.
DISTRIBUTION: Sporadically found throughout the Philippines.

Class Gastropoda
Subclass Prosobranchia
Order Mesogastropoda
Suborder Taenioglossa
Superfamily Cerithiacea
Family PLANAXIDAE

A moderately large family of small to medium-sized gastropods living in shallow water, in coral, or under rocks. The shells are fairly conical, ovate, with a large body whorl and may be smooth or spirally grooved; aperture ovate and lirate within, columella concave, non-umbilicate and normally smooth but may have a single denticle, and the anterior canal is short but distinct. Moderately well represented in the Philippines but seldom seen with any frequency by the average collector unless taken in the field.

13. Planaxis niger Quoy & Gaimard, 1834
10.9mm; Palawan.
Shell small, characteristically brown-coloured, with a tall spire and an ovate body whorl; sculptured with spiral grooves which are obsolete at the center of each whorl; aperture ovate, outer lip thick and lirate within, columella smooth and calloused with a distinct parietal denticle; apertural colouration pale bluish brown; reaches 15mm.
DISTRIBUTION: Sporadically found throughout the Philippines.

14. Planaxis sulcatus (Born, 1780)
26.7mm; Batangas.
Shell medium sized, solid, with a conical spire, a large body whorl, well impressed sutures, and encircled with numerous strong spiral cords which are crossed by fine axial striae, strongest in the interspaces; aperture elongate-ovate , strongly lirate within, columella calloused, curved, with a few weak striae and a strong parietal button posteriorly; pale brownish lavender with black spots on the ribs, aperture and columella bluish white, outer lip margin brown; reaches 30mm.
DISTRIBUTION: Throughout the the Philippines.

Class Gastropoda
Subclass Prosobranchia
Order Mesogastropoda
Suborder Taenioglossa
Superfamily Cerithiacea
Family MODULIDAE

A very small group of intertidal-dwelling gas ropods which live in sand or on seaweed. The shells are of medium size, heavy, somewhat turbinate in outline, with a depressed spire and a large, angulate body whorl; surface sculptured with axial and spiral ribs, aperture wide and circular, lirate within, and the columella has a sharp basal tooth. The umbilicus is small and narrowly channeled. Few Philippine species.

15. Modulus tectum (Gmelin, 1791)
17.9mm x 19.9mm; Samar.
Shell small, solid, with a depressed spire and a large, ovate body whorl; sculptured with oblique axial cords, those on the body whorl terminating just below the roundly angulate periphery; entire surface encircled with coarse spiral striae; aperture wide, finely lirate within, columella with a sharp basal denticle; umbilicus partially occluded by a columellar callus; creamy white with darker speckling; aperture porcelaneous white; may reach 30mm in length.
DISTRIBUTION: Sporadically found throughout the Philippines.

Class Gastropoda
Subclass Prosobranchia
Order Mesogastropoda
Suborder Taenioglossa
Superfamily Calyptraeacea
Family CREPIDULIDAE

A moderately large group of limpet-like shells which are characterized by the presence of an inner deck or septum (*Crepidula spp.*), or a cup-like projection (*Crucibulum spp.*). The shells are small to medium sized, low-spired or conical, and sculptured with coarse axial and spiral cords and occasional spines. They are non-operculate, mainly ciliary feeders, and live attached to the substrate or on other shell species. They inhabit shallow to deep water in most temperate and tropical seas. Relatively few Philippine species.

16. Crucibulum spinosum (Sowerby, 1824)
21.0mm x 36.5mm; Palawan.
Shell medium sized, solid, conical, with a sharp, anteriorly curved, posteriorly displaced apex, and sculptured with numerous irregular, coarse radial ribs spinose near the irregularly ovate margin; a large cup-like projection internally; external colour greyish white, internally deep brown, marbled with white and lavender; internal cup marginally white, deep brown within; illustrated specimen the largest seen. (The genus *Crucibulum* was originally known only from America and apparently introduced into Hawaii during the last world war. Philippine records represent a considerable westward extention of the known range.)
DISTRIBUTION: So far only from the Sulu Sea.

17. Cheilea equestris (Linné, 1758)
25.0mm x 25.5mm; Mindoro.
Shell medium sized, rather lightweight, slightly translucent and with a tall, slightly curved, anteriorly displaced apex and an extremely variable shape; sculptured with numerous radial striae and concentric folds which are slightly lamellate towards the margin; internally a long, semi-circular shelf attached to the apex; colour dirty white; seldom exceeds 25mm in diameter.
DISTRIBUTION: Sporadically found throughout the Philippines in limited quantity.

18. Calyptraea (Desmaulus) extinctorium Lamarck, 1822
7.0mm x 21.8mm., 8.5mm x 20.0mm; Samar.
Shell small, lightweight, subcircular in outline, with a moderately tall, posteriorly displaced apex, and sculptured with numerous fine concentric threads; internal septa small, thin, somewhat flattened, extending outwards in a spiral fashion; colour pale brown, internally off-white with light brown marbling; illustrated specimens representative of maximum sizes seen.
DISTRIBUTION: Sporadically found throughout the Philippines in limited quantity.

19. Crepidula (Siphopatella) walshi Reeve, 1859
4.2mm x 25.5mm (max. diam.), 3.6mm x 22.2mm (max. diam.); Bataan.
Shell small, creamy white in colour, elongate-ovate in outline, the spire characteristically flattened and displaced onto the peripheral margin; sculptured with fine concentric growth striae; internally porcelaneous white with a fairly expansive septal area originating from the apex; may reach 30mm in diameter. Usually attached to *Mytilus* or *Pinna spp.*
DISTRIBUTION: Sporadically found throughout the Philippines in limited quantity.

Class Gastropoda
Subclass Prosobranchia
Order Mesogastropoda
Suborder Taenioglossa
Superfamily Hipponicacea
Family HIPPONICIDAE

A moderately large group of shells which inhabit the intertidal regions of most worldwide tropical seas. The shells are small, cap-shaped, with a curved, sub-central apex, a smooth or axially and spirally-corded surface, and a smooth interior. Shells parasitic and although non-operculate, produce a calcareous plate. Only a few species presently known from the Philippines.

20. Hipponix (Antisabia) foliaceus (Quoy & Gaimard, 1835)
8.3mm x 17.9mm; Batangas.
Shell small, solid, with a curved, somewhat flattened spire which overhangs the margin; surface sculptured with concentric lamellose folds and weak radial ridges, forming some spinose projections towards the shell periphery; frequently eroded; aperture wide, white in colour with a weakly serrate margin; reaches around 20mm in length.
DISTRIBUTION: Sporadically found throughout the Philippines.

21. Sabia conica (Schumacher, 1817)
8.5mm x 12.2mm, 6.3mm x 11.5mm; Palawan.
Shell small, conical, with a heavily scalloped margin and a tall, conical spire which is curved and overhangs the shell margin; sculptured with heavy radial cords and some concentric lirae towards the periphery; external colour off-white with a purple border, internally porcelaneous white with a brown crenulate margin; this form is often referred to as form *acuta* (Quoy & Gaimard, 1835), and is smaller and more irregularly shaped than the normal form. Found attached to a variety of host shells.
DISTRIBUTION: Sporadically found throughout the Philippines in limited quantity.

Class Gastropoda
Subclass Opisthobranchia
Order Anaspidea
Superfamily Aplysiacea
Family APLYSIIDAE

This group is better placed with other opisthobranch groups (Bullidae etc.), but is here inserted for convenience only. Members of this family are the internal shells of the so-called "sea hares", a group of slug-like animals which have a soft body and inhabit shallow waters among seaweeds on which they feed. The shells are usually quite delicate, flattened, and somewhat curved or "hook-shaped". This family has yet to be comprehensively researched here in the Philippines, although there is little doubt that it is well represented. Only a single species is discussed here.

22. Dolabella auricularia (Lightfoot, 1786)
32.2mm (L) x 23.5mm (W); Manila Bay.
Shell medium sized with a small incurved, heavily calcareous apical area which expands outwards in a semi-circular fashion, forming a thin plate-like structure of steadily increasing width; internal area porcelaneous white with fine concentric striae, surface with a light brown periostracum; may reach 45mm in length and 38mm in width. The animal is fairly large (around 140mm), brown coloured, and lives among seaweed.
DISTRIBUTION: Sporadically found throughout the Philippines.

Plate 13 — Cerithiidae

Class Gastropoda
Subclass Prosobranchia
Order Mesogastropoda
Suborder Taenioglossa
Superfamily Cerithiacea
Family CERITHIIDAE

A large group of small to medium sized herbivorous gastropods living in temperate to tropical seas and usually found in sand or around coral in shallow water, especially in the intertidal and subtidal zones. Members of this family are multi-whorled, mostly elongate, and sculptured with smooth or nodulous spiral cords, axial ribs, and occasionally, varices. The aperture is generally ovate, frequently lirate, the columella smooth or with folds, and the siphonal canal is short and frequently recurved. The family is well represented in the Philippines.

1. Pseudovertagus (Pseudovertagus) nobilis (Reeve, 1855)
116.1mm, Albay Gulf, Southern Luzon.
Shell large, heavy, spire tall and eroded; early whorls with axial cords and spiral striae, obsolete after 4 or 5 whorls; later whorls smooth with a slightly raised presutural ramp; aperture ovate, outer lip thickened and variced; columella with an expansive, basally non-adherent callus; siphonal canal short and reflexed; yellow-brown with white maculations and bands; reaches 140mm.
DISTRIBUTION: Sporadically found throughout the Philippines in limited quantity.

2. Cerithium (Cerithium) nodulosum (Bruguiére), 1792
93.0mm; Samar.
Shell large, heavy, sculptured with a row of large nodules on each whorl and crossed by deep spiral grooves; aperture ovate, outer lip foliate and expanded over the siphonal canal; columella with anterior and posterior denticles; siphonal canal short, narrow and recurved; off-white with dark axial streaks and blotches and pale blue banding; reaches 120mm.
DISTRIBUTION: Throughout the Philippines.

3. Rhinoclavis (Rhinoclavis) fasciata (Bruguiére, 1792)
73.9mm, 51.0mm, 54.0mm, 53.8mm, 56.7mm, 54.3mm; Samar.
Shell medium sized, glossy, heavy and extremely variable; spire tall and acute, spire whorls flat-sided and sculptured with axial ribs and deeply incised spiral grooves forming subsutural nodules; axial ribs obsolete on later whorls; aperture elongate-ovate, outer lip marginally thickened, columella calloused with a medial plate, siphonal canal long and reflexed; off-white with darker bands or blotching; reaches 90mm.
DISTRIBUTION: Sporadically found throughout the Philippines.

4. Cerithium (Cerithium) citrinum Sowerby, 1855
43.4mm, 34.8mm; Punta Engaño, Cebu.
Shell medium sized, lightweight, with a distinct lemon tinge on the lower whorls and on the inner apertural margin; sutures adpressed, whorls sculptured with rounded axial ribs and numerous fine, overriding spiral threads; aperture ovate, internally lirate, with a

foliate outer lip, a smooth columella and a prominent varix on the body whorl adjacent the aperture; siphonal canal long, recurved; reaches 50mm.
DISTRIBUTION: Camotes Sea.

5. Rhinoclavis (Rhinoclavis) aspera (Linné, 1758)
42.1mm, 31.0mm, 31.2mm; Samar.
Shell medium sized, elongate, with a tall spire and adpressed sutures and sculptured with raised axial ribs and varices; each whorl with 3 rows of spines crossed by weaker spiral ribs and a strong cord at the base; aperture obliquely ovate and smooth, with a central denticle on a calloused columella; colour white with occasional darker spotting or banding; reaches 60mm.
DISTRIBUTION: Sporadically found throughout the Philippines.

6. Cerithium (Cerithium) echinatum (Lamarck, 1822)
55.6mm; Palawan.
Shell moderately large, elongate-ovate, heavy, with a large, rather inflated body whorl; sculptured with a single row of blunt spines on the early spire whorls, a weaker subsutural row developing on later whorls, and 4 to 5 on the body whorl, some weak spiral threads between the spiral rows; aperture ovate, outer lip crenulate, expanded over the anterior canal; white with purple spotting and lines; reaches 75mm.
DISTRIBUTION: Throughout the Philippines.

7. Rhinoclavis (Longicerithium) longicaudata (Adams & Reeve, 1850)
44.1mm, 43.9mm, 44.9mm; Samar.
Shell medium sized, elongate, with a tall, acute spire; sutures impressed and weakly nodular; sculptured with axial ribs, overriding beaded spiral cords and a varix adjacent an internally lirate aperture; anterior canal unusually long and reflexed; pale cream to brown, occasionally with brown axial bands; reaches 60mm.
DISTRIBUTION: Visayan Sea and Camotes Sea.

8. Cerithium (Cerithium) columna Sowerby, 1834
31.5mm, 33.8mm; Samar.
Shell medium sized, broad or narrow, with a tall, acute spire and incised sutures; sculptured with angulate axial ribs and overriding nodulose spiral cords forming slightly spinose projections, most prominent on the periphery and on the body whorl base; aperture ovate, outer lip thick, marginally crenulate and lirate within; anterior canal short and recurved; off-white to cream coloured with purple lines and streaks; reaches 85mm.
DISTRIBUTION: Sporadically found throughout the Philippines.

9. Rhinoclavis (Proclava) matukense (Watson, 1880)
43.3mm; Lubang Is., Mindoro.
Shell medium sized with a tall, acute spire and weakly impressed sutures; sculptured with weak, oblique axial folds and crossed by 3 to 4 nodulose spiral cords which become flattened on the body whorl; outer lip thin and expanded; siphonal canal short and reflexed; white varices on all whorls, strongest adjacent the aperture; white with brown cords and nodules; reaches 50mm; rarely seen live; illustrated shell taken from a specimen of *Xenophora pallidula* (Reeve) in deep water.
DISTRIBUTION: Mindoro and Visayan Sea.

10. Rhinoclavis (Rhinoclavis) sinensis (Gmelin, 1791)
54.0mm, 46.1mm; Palawan.
A medium sized, glossy, somewhat inflated shell with adpressed sutures; sculptured with a row of subsutural granules, a further 2 or 3 weaker rows below, and interstitial spiral threads; aperture elongate-ovate with a varical bump adjacent; columella calloused with a medial plate; siphonal canal long and reflexed; creamy brown with mauve spots and blotches; reaches 65mm.
DISTRIBUTION: Sporadically found throughout the Philippines.

11. Clypeomorus traillii (Sowerby, 1855)
28.7mm; Cebu.
A medium sized, stoutly fusiform shell sculptured with 3 rows of granulose spiral cords which are separated by deeply incised grooves; aperture elongate-ovate with a varical hump adjacent; outer lip thickened and smooth, columella smooth and slightly calloused, anterior canal short and reflexed; creamy white with brown axial streaks; reaches 35mm.
DISTRIBUTION: Camotes Sea and Visayan Sea.

12. Rhinoclavis (Rhinoclavis) articulata (Adams & Reeve, 1850)
49.8mm, 43.6mm; Samar.
Shell medium sized, glossy, slender, with a tall acute spire and a beaded sutural cord; sculptured with weak axial cords and thick varices which are crossed by smooth, occasionally nodulose, spiral lirae; body whorl flattened ventrally, aperture elongate-ovate and smooth within, columella with an expansive callus, siphonal canal long and distally recurved; creamy white with brown spots and blotches; reaches 55mm.
DISTRIBUTION: Sporadically found throughout the Philippines.

PLATE NO. 13 1. *Pseudovertagus nobilis* (Reeve) 2. *Cerithium nodulosum* (Bruguieré) 3. *Rhinoclavis fasciata* (Bruguieré) 4. *Cerithium citrinum* (Sowerby) 5. *Rhinoclavis aspera* (Linné) 6. *Cerithium echinatum* (Lamarck) 7. *Rhinoclavis longicaudata* (Adams & Reeve) 8. *Cerithium columna* Sowerby 9. *Rhinoclavis matukense* (Watson) 10. *Rhinoclavis sinensis* (Gmelin) 11. *Clypeomorus traillii* (Sowerby) 12. *Rhinoclavis articulata* (Adams & Reeve) 13. *Rhinoclavis vertagus* (Linné) 14. *Cerithium planum* Anton 15. *Cerithium kobelti* Dunker 16. *Rhinoclavis kochi* (Philippi) 17. *Clypeomorus coralium* (Kiener) 18. *c.f. Cerithium fusiforme* Sowerby 19. *Clypeomorus brevis* (Quoy & Gaimard) 20. *Cerithium lifuense* Melvill & Standen 21. *Pseudovertagus aluco* (Linné)

PLATE 13

13. Rhinoclavis (Rhinoclavis) vertagus (Linné, 1767)
58.3mm; Samar.
Shell medium sized, heavy and elongate, with slightly impressed sutures; sculptured with coarse axial folds and overriding spiral grooves, the folds becoming obsolete towards the base; aperture obliquely ovate, outer lip thickened, columella with a thick, slightly expanded callus and a central fold; yellow-brown with pale subsutural banding, aperture white; reaches 65mm.
DISTRIBUTION: Throughout the Philippines.

14. Cerithium (Cerithium) planum Anton, 1839
29.8mm, 25.7mm, 28.0mm; Cebu.
Shell small with tall, acute spire and a nodulose, bi-cingulate cord at the sutures; whorls sculptured with rounded axial cords, nodulose from overriding spiral lirae; body whorl with a large varix and a granulose cord at the base; aperture ovate, lirate within, outer lip marginally thickened and crenulate, siphonal canal short and recurved; off-white to brown with a darker sutural band; reaches 30mm.
DISTRIBUTION: Camotes Sea and Visayan Sea.

15. Cerithium (Cerithium) kobelti Dunker, 1877
26.7mm, 28.1mm; Cebu.
Shell small and elongate with a tall, acute spire; spire whorls strongly convex, sculptured with raised axial cords, nodulose from 3 to 4 overriding spiral lirae; sutures impressed and weakly noduled; a heavy varix on the body whorl; aperture ovate, outer lip thin, siphonal canal short and recurved; brown in colour with white spiral lirae; reaches 35mm.
DISTRIBUTION: Camotes Sea and Visayan Sea.

16. Rhinoclavis (Proclava) kochi (Philippi, 1848)
36.3mm, 35.9mm; Bohol.
Shell medium sized, glossy, with a tall, acute spire, and a beaded sutural cord; sculptured with weakly elevated varices and granulose spiral cords, between which are 2 deeply incised, brown-coloured grooves; aperture elongate-ovate, outer lip crenulate, columella with 2 folds; siphonal canal short and recurved; shell colour brown with white varices, and ornamented with brown dots and axial flamules; reaches 40mm.
DISTRIBUTION: Camotes Sea and Visayan Sea.

17. Clypeomorus coralium (Kiener, 1841)
31.2mm, 25.3mm; Cebu.
Shell small with a tall spire, adpressed sutures and weak varices; sculptured with rows of granular spiral cords, 3 on the spire whorls and 6 to 7 on the body whorl, and interstitial spiral threads; aperture semi-lunate, outer lip thickened and weakly crenulate; columella with a single posterior denticle; anterior canal short, wide and slightly recurved; black to purple-brown with a light purple aperture; reaches 40mm.
DISTRIBUTION: Camotes Sea and Visayan Sea.

18. c.f. Cerithium (Cerithium) fusiforme Sowerby, 1855
22.1mm, 18.5mm, 19.6mm, 20.6mm; Cebu.
A small species with a tall, narrow spire, and a finely reticulate, finely beaded surface formed from intersecting axial and spiral cords; aperture ovate, weakly lirate within and marginally crenulate; columella with a weak central plait, anterior canal short; colour pale tan to reddish brown; similar to *C. sordidula* but has 4, not 3 granular spiral cords; also close to a young *C. kochi*, but is smaller, has a narrower spire angle, and lacks the incised spiral lines of that species; often synonymised with *C. columna* but has a different sculpturing and appears fully mature at around 20mm. Possibly a Rhinoclavid.
DISTRIBUTION: Camotes Sea.

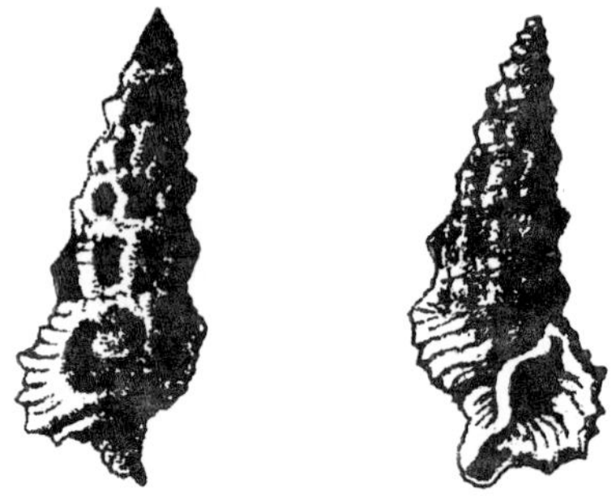

Type figure of *Cerithium fusiforme* Sowerby, 1855

19. Clypeomorus brevis (Quoy & Gaimard, 1834)
20.8mm, 20.0mm; Cebu.
Shell small and broadly fusiform, with a tall, acute spire and sculptured with axial folds, irregular varices, and overriding granulose spiral cords; aperture elongate-ovate, outer lip thick, variced and denticulate, columella smooth; pale cream with black blotching on the granules; reaches 30mm.
DISTRIBUTION: Sporadically found throughout the Philippines.

20. Cerithium (Cerithium) lifuense Melvill & Standen, 1895
28.5mm, 22.7mm; Punta Engaño, Cebu.
Shell small, lightweight, with a tall, acute spire; sculptured with a row of sharp tuberculate spines at the suture, and a smaller row below this; body whorl with an additional 2 spiral cords and a varix; yellowish white with brown streaks and white tubercles; reaches 35mm.
DISTRIBUTION: Camotes Sea and Visayan Sea.

21. Pseudovertagus (Pseudovertagus) aluco (Linné, 1758)
63.5mm, 62.3mm; Samar.
Shell medium sized, thick, heavy, fairly broad and with incised sutures; earlier whorls eroded, later whorls with a row of echinate nodules below the suture; aperture elongate-ovate and smooth within; outer lip thick and curved over the anterior canal; columellar callus non-adherent, with an expansive callus and posterior denticle; siphonal canal short, narrow and reflexed; creamy white with purple spots and axial streaks; reaches 75mm.
DISTRIBUTION: Throughout the Philippines.

1. Clypeomorus subbreviculus Oostingh, 1925
21.8mm, 21.5mm; Palawan.
A small, short shell with a large inflated body whorl; sculptured with rows of granular spiral cords and weaker interstitial threads and grooves; aperture ovate, outer lip crenulate, columella smoothly calloused with a posterior siphonal denticle; almost entirely black except for blue-white interspaces, aperture blue-white with darker streaks on the outer lip margin; reaches 25mm.
DISTRIBUTION: Throughout the Philippines.

2. Clypeomorus petrosus (Wood, 1828)
21.4mm, 17.0mm; Samar.
A small elongate-ovate shell with weakly incised sutures and sculptured with three rows of equi-sized, equi-spaced nodules, interstitial spiral threads, and a heavy varix on the body whorl dorsum; aperture ovate, with a short, oblique siphonal canal, and a thickened, crenulate outer lip margin; shell colour almost black with white interspaces and a bluish white aperture; reaches 25mm.
DISTRIBUTION: Throughout the Philippines.

3. Clypeomorus batillariaeformis Habe & Kosuge, 1966
15.5mm; Samar.
A small fusiform-elongate shell with a tall, acute spire; whorls sculptured with axial folds, irregular varices, and overriding spiral cords forming nodules; aperture elongate-ovate, outer lip thick, variced and denticulate, columella smooth; shell pale cream with black blotching on the granules; reaches 25mm.
DISTRIBUTION: Throughout the Philippines.

4. Clypeomorus humilis (Dunker, 1861)
20.4mm; Punta Engaño, Cebu.
A small elongate-ovate shell with adpressed sutures; sculptured with rounded axial ribs which become obsolete on the body whorl, and encircled with large spiral lirae and weaker interstitial threads which become slightly nodulose towards the base; aperture ovate, outer lip thick and marginally crenulate, columella calloused with a posterior denticle, siphonal canal short and recurved; black with white spiral cords, aperture bluish white with black banding; reaches 25mm.
DISTRIBUTION: Throughout the Philippines.

5. Clypeomorus zonatus (Wood, 1828)
18.6mm; Palawan.
A small broadly elongate shell, sculptured with rows of granulose spiral cords and fine spiral threads; characteristically banded with white and brown, the cords black in colour; aperture ovate, outer lip thick, siphonal canal short and recurved; reaches 30mm.
DISTRIBUTION: Throughout the Philippines.

6. Clypeomorus tuberculatus (Linné, 1758)
24.9mm; Punta Engaño, Cebu.
Very close to *C. moniliferus* in shape but slightly less elongate; the spiral cords consist of laterally-elongated, irregularly-sized nodules and weaker interstitial nodular cords; those of *moniliferus* are rounded and regular and the interstitial cords are generally lacking; off-white with diffuse grey blotching and darker coloured nodules; reaches 40mm.
DISTRIBUTION: Throughout the Philippines.

7. Clypeomorus patulum (Sowerby, 1855)
19.9mm, 18.5mm; Bohol.
Shell small, rather squat, peculiarly flattened on the ventral side of the body whorl, and possessing a distinct varical hump on the dorsum; sculptured with weak axial folds and overriding spiral cords forming nodules; aperture ovate, outer lip thickened, weakly denticulate within; colour brown with darker nodules and white interspaces; reaches 25mm.
DISTRIBUTION: Throughout the Philippines.

8. Rhinoclavis (Proclava) sordidula (Gould, 1849)
Syn: *C. pfefferi* Dunker, 1863
20.4mm; Cebu.
Shell small, with a tall, acute spire and adpressed sutures; whorls with weak varices and sculptured with three spiral cords which are crossed by axial striae, producing a reticulated, somewhat beaded surface; aperture semi-lunate, outer lip marginally crenulate, lirate within, columella weakly calloused; reddish brown with whitish granules and black interspaces; reaches 30mm.
DISTRIBUTION: Throughout the Philippines.

9. Clypeomorus bifasciatus (Sowerby, 1855)
22.3mm; Samar.
Shell small to medium sized, elongate-ovate in outline, with a tall, acute spire and a rounded body whorl; surface with three spiral rows of granulose lirae, axial cords and varices, the last of which is on the outer lip margin; aperture ovate, outer lip with a thick, rounded margin and lirate within, columella with a single posterior denticle; white with irregular brown spots and a bluish white aperture; reaches around 25mm.
DISTRIBUTION: Throughout the Philippines.

10. Cerithium (Cerithium) tenellum Sowerby, 1855
21.3mm, 18.2mm, 17.0mm; Cebu.
A small shell with an extremely granular surface formed from three rows of heavily granulose spiral cords; varices present on all whorls; aperture ovate and lirate within, outer lip thin and crenulate, siphonal canal short and obliquely angled; purple-grey in colour, black between the cords, and the aperture lined with brown streaks; reaches 25mm.
DISTRIBUTION: Throughout the Philippines.

11. Clypeomorus moniliferus (Kiener, 1841)
Shell small and squat, sculptured with three rows of rounded, regularly spaced, equi-sized granules on the spire whorls and six on the body whorl; aperture obliquely elongate-ovate, lirate within, outer lip thick and variced, siphonal canal short and recurved; colour

off-white with black nodules, aperture off-white; reaches 25mm.

11a. typical form; 21.1mm; Palawan.

11b. this species is frequently illustrated as *C. moniliferus*, but appears to be *Clypeomorus variegatus* (Quoy & Gaimard, 1834) (non Fischer, 1807); the granules are smaller, weaker and somewhat elongate, the colour pattern is of brown and white blotches or dark spots on a white shell; the aperture is white with a curious purple ring on the inner margin, and the last varix is on the dorsum, not opposite the outer lip as in *moniliferus*. Possibly an ecomorph of *C. chemnitzianus*, but probably distinct and requiring a new name; extremely variable; 20.9mm, 21mm; Palawan.

DISTRIBUTION: Throughout the Philippines.

12. Clypeomorus chemnitzianus (Pilsbry, 1901)
27.9mm, 32.8mm; Samar.
Shell medium sized, solid, broadly elongate-ovate in outline, with weakly impressed sutures and a sharply pointed spire; sculptured with irregular varices, a distinct varical hump on the body whorl dorsum, and encircled with beaded spiral cords and weaker interstitial lines; aperture ovate, outer lip thickened and lirate within, columella with a posterior denticle; white with a brown subsutural band and two on the body whorl; reaches 35mm.
DISTRIBUTION: Throughout the Philippines.

Class Gastropoda
Subclass Prosobranchia
Order Mesogastropoda
Suborder Taenioglossa
Superfamily Cerithiacea
Family POTAMIDIDAE

A comparatively small group of gastropods closely related to the Cerithiidae in form and ecological habitat, differing mainly in radula and anatomical characteristics. They are mainly tropical to subtropical in distribution, are herbivorous, have a veliger larval stage and prefer brackish, muddy water. The shells are moderately tall-spired with numerous whorls, axially and spirally sculptured, have a flared outer lip, a twisted columella, and a short, non-twisted canal. The operculum is thin and multispiraled with a central nucleus. Few recorded Philippine species, with most of the common ones being illustrated here.

13. Batillaria multiformis (Lischke, 1869)
34.4mm; Bohol.
Shell medium sized, elongate with straight sides, sutures impressed and the body whorl slightly inflated; sculptured with a granular cord below the sutures and spiral cords on the rest of the shell; aperture semi-lunate, columella calloused, heaviest at the anterior and posterior margins; shell bluish white with black cords, a bluish columella, and a black and white-lined inner lip margin; reaches 40mm.
DISTRIBUTION: Throughout the Philippines.

14. Cerithidea cingulata (Gmelin, 1791)
22.9mm, 19.4mm, 19.9mm; Bohol.
Shell medium sized, with a tall spire and deeply incised sutures; sculptured with axial ribs and spiral grooves, forming axially aligned nodules; aperture semi-lunate, outer lip thick, arcuate and weakly crenulate; colour deep brown with presutural white bands, and a brown-lined, marginally white aperture; reaches 40mm.
DISTRIBUTION: Throughout the Philippines.

15. Cerithidea rhizophorarum A. Adams, 1855
21.2mm; Palawan.
Shell medium-sized, elongate, with an eroded spire and deeply incised sutures; whorls with 1 or 2 strong varices and sculptured with axial and spiral cords producing a reticulate appearance; aperture ovate, outer lip thick, backed by a varix, siphonal canal nearly obsolete; off-white with darker banding, aperture marginally white and brown within; reaches 30mm.
DISTRIBUTION: Throughout the Philippines.

16. Terebralia sulcata (Born, 1778)
46.7mm, 44.4mm; Cebu.
Shell medium sized, elongate-ovate with a tall, acute spire and well incised sutures; surface sculptured with a few varices and numerous axial cords which are crossed by 3 spiral grooves, forming 4 rows of rounded nodules; outer lip flaring, fusing at the columellar base to form a short tubular canal; shell pale brown in colour with darker banding, white varices, and a marginally brown, white aperture; reaches 65mm.
DISTRIBUTION: Throughout the Philippines.

17. Telescopium telescopium (Linné, 1758)
69.7mm; Bohol.
Shell large, heavy and tall, with weak sutures and flat-sided whorls which have three wide and one narrow cord; base spirally corded, columella curiously twisted; aperture sub-quadrate, outer lip thin and arcuate; dark brown with white banding and a pale brown columella; reaches 100mm; inhabits mangrove areas.
DISTRIBUTION: Throughout the Philippines.

PLATE NO. 14 1. *Clypeomorus subbreviculus* Oostingh 2. *Clypeomorus petrosus* (Wood) 3. *Clypeomorus batillariaeformis* Habe & Kosuge 4. *Clypeomorus humilis* (Dunker) 5. *Clypeomorus zonatus* (Wood) 6. *Clypeomorus tuberculatus* (Linné) 7. *Clypeomorus patulum* (Sowerby) 8. *Rhinoclavis sordidula* (Gould) 9. *Clypeomorus bifasciatus* (Sowerby) 10. *Cerithium tenellum* Sowerby 11a. *Clypeomorus moniliferus* (Kiener) 11b. *Clypeomorus variegatus* (Quoy & Gaimard) 12. *Clypeomorus chemnitzianus* (Pilsbry) 13. *Batillaria multiformis* (Lischke) 14. *Cerithidea cingulata* (Gmelin) 15. *Cerithidea rhizophorarum* A. Adams 16. *Terebralia sulcata* (Born) 17. *Telescopium telescopium* (Linné) 18. *Terebralia palustris* (Linné)

PLATE 14

18. Terebralia palustris (Linné, 1767)
110mm; Palawan.
Shell large, heavy, pyramidally ovate, sutures incised, whorls flat-sided; sculptured with oblique axial folds, irregular weak varices, and crossed by three deep grooves forming rectangular nodules; aperture semi-lunate, outer lip arcuate, pointed at the base, columella calloused; dark brown in colour, the grooves off-white, the columella light brown, and the aperture blackish brown; reaches 115mm.
DISTRIBUTION: Throughout the Philippines.

Class Gastropoda
Subclass Prosobranchia
Order Mesogastropoda
Suborder Taenioglossa
Superfamily Strombacea
Family STROMBIDAE

A relatively large group of small to large-sized herbivorous gastropods inhabiting sandy and muddy areas in shallow waters near coral reefs. The shells are thick, heavy, tall-spired, have a large body whorl, a long, narrow aperture, a thickened, flaring outer lip which bears a stromboidal notch anteriorly, and a smooth or roughly noduled surface. Sexual dimorphism is observed in some species. Besides the members of the *Strombus*, three other groups are encountered within this family:
a) *Lambis:* Characterized by long, spinose digitations.
b) *Terebellum:* A single member with a smooth, "torpedo-like" body, with little resemblance to true Strombidae.
c) *Tibia:* Characterized by a long, spinose anterior canal and a deeply grooved, drawn-out posterior canal.
The family is well represented in the Philippines (around 40 species) and many species are edible and form an important source of food for many provincial areas.

1. Tibia (Tibia) fusus Linné, 1758
Shell medium sized, extremely long, with a tall, slender, sharply acute spire; spire whorls convex, sutures impressed; sculptured with axial and spiral striae on earlier whorls, becoming obsolete anteriorly; anterior canal long, thin and sinuous, posterior canal short and recurved; outer lip with 5 thick, digital processes (occasionally 6); shell brownish tan, aperture and columella white, purplish between the outer lip digitations; reaches 300mm.
1a. typical form (with 6, not 5 digitations); 273.7mm; Leyte.
1b. form *melanocheilus* Adams, 1854; smaller, purplish brown in colour with a darker aperture; 113mm; Turtle Islands (Sulu Sea).
DISTRIBUTION: Tayabas Bay (Quezon), Visayan Sea and Turtle Islands (Sulu Sea).

2. Tibia (Tibia) martini Marrat, 1877
Shell medium sized, lightweight, with a tall, acute spire and a relatively short, narrow anterior canal; spire whorls with impressed sutures, a presutural groove, and encircled with obsolete spiral threads; aperture large, semi-lunate, with 6 to 7 short digitations on the anterior portion of the slightly thickened, recurved outer lip; posterior canal short, open and recurved on the penultimate whorl; shell pale tan, aperture white margined, pale purple within, outer lip margin with a brown and white-spotted border; reaches 180mm.
2a. typical form; 157mm; Tayabas Bay, Quezon.
2b. A dwarf form of lighter weight, and with reduced outer lip digitations; 100mm; Leyte.
DISTRIBUTION: Tayabas Bay (Quezon) and Visayan Sea.

3. Tibia (Rimella) cancellata (Lamarck, 1816)
27.4mm, 34.8mm, 26.9mm; Tayabas Bay, Quezon.
Shell small, somewhat dorso-ventrally compressed, with a tall spire and a cancellate surface formed from strong axial cords and overriding spiral lirae; varices prominent on later whorls; aperture large, narrowly ovate, outer lip thick, curved, serrate on the anterior half and lirate on the inner margin; posterior canal thick, open and extending over two whorls; anterior canal short, open and recurved; columella heavily calloused; pale tan with lighter banding and a pale purple aperture; reaches 35mm.
DISTRIBUTION: Tayabas Bay (Quezon) and Visayan Sea.

4. Tibia (Rimella) crispata Sowerby II, 1842
25.0mm, 23.7mm, 26.6mm; Tayabas Bay, Quezon.
Similar to *T. cancellata* but smaller, with a shorter spire, a more cancellate surface, an outer lip margin which is serrate along its entire length, and a shorter posterior canal extending only to the penultimate whorl; aperture is brown and white banded rather than purple; reaches 30mm.

PLATE NO. 15 1a. *Tibia fusus* Linné 1b. form *melanocheilus* Adams 2a. *Tibia martini* Marrat 2b. a dwarf form 3. *Tibia cancellata* (Lamarck) 4. *Tibia crispata* Sowerby II 5. *Tibia powisi* Petit 6a. *Terebellum terebellum* (Linné) 6b. form *nebulosum* Röding 6c. form *punctulorum* Röding 6d. form *lineatum* Röding 7a. *Strombus mutabilis mutabilis* Swainson 7b.-7d. Various colour forms 7e. form *zebriolatus* Adams & Leloup 8. *Strombus microurceus* (Kira) 9. *Strombus fragilis* (Röding) 10. *Strombus dentatus* Linné 11. *Strombus terebellatus terebellatus* Sowerby II

PLATE 15

DISTRIBUTION: Tayabas Bay (Quezon) and Visayan Sea.

5. Tibia (Rimella) powisi Petit, 1842
60.0mm; Tayabas Bay, Quezon.
A small, rather elongate shell with a tall spire and strongly incised sutures; the first 3 or 4 spire whorls are smooth, the rest sculptured with strong spiral ribs and weaker interstitial axial striae; aperture elongate-ovate, outer lip thick and lirate within, the margin with five short, sharp digits; anterior canal short and open, posterior canal also short and recurved; pale creamy tan, darker at the sutures, with a white aperture and columella; reaches 60mm.
DISTRIBUTION: Tayabas Bay (Quezon) and Visayan Sea.

6. Terebellum (Terebellum) terebellum (Linné, 1758)
Shell medium sized, slender, cylindrical, lacking a stromboidal notch; surface smooth with obsolete growth striae; spire short and straight-sided, body whorl long and tapering, aperture narrow posteriorly, widening anteriorly then sharply truncated; columella smooth, slightly glazed; several forms; reaches 60mm.
6a. typical form; 59.1mm, 60.4mm; Sulu Archipelago.
6b. form *nebulosum* Röding, 1798: three brown spiral bands; 52.4mm; Sulu Archipelago.
6c. form *punctulorum* Röding, 1798: small red-brown spots over the surface; 45.5mm; Sulu Archipelago.
6d. form *lineatum* Röding, 1798: strong, narrow, brown-coloured zigzag lines; 47.0mm, 45.6mm, 48.2mm; Sulu Archipelago.
DISTRIBUTION: Sporadically found throughout the Philippines.

7. Strombus (Canarium) mutabilis mutabilis Swainson, 1821
Shell small, solid, quadrate-shaped, with 8 to 9 whorls, the earlier ones axially ribbed; body whorl smooth with 3 or 4 swollen shoulder nodules and some spiral threads basally; sutures weakly impressed, frequently girdled with a cord; aperture narrow, columella and outer lip heavily lirate, outer lip margin thick with a strong stromboidal notch anteriorly; colour variable, usually white with darker bands or patches, aperture white, red between the outer lip lirae, brown between those of the columella; reaches 40mm.
7a. typical form; 35.5mm; Samar.
7b.-7d. Various colour forms; 27.5mm, 24.7mm, 26.5mm; Sulu Archipelago.
7e. form *zebriolatus* Adams & Leloup, 1938: dark brown axial stripes on a yellow background; 27.6mm; Bohol.
DISTRIBUTION: Throughout the Philippines.

8. Strombus (Canarium) microurceus (Kira, 1959)
24.8mm, 20.2mm; Samar.
Shell small, solid, quadrate in outline, earlier whorls with axial riblets becoming large blunt shoulder nodules on the body whorl; sutures impressed, bordered by a strong cord; outer lip thickened, lirate within, and with a strong stromboidal notch anteriorly, columella lirate along its entire length; off-white with bands of darker mottling, and a brown spot at the base; columella with orange margins, purple between the lirae, aperture white internally, purple marginally; reaches 30mm.
DISTRIBUTION: Throughout the Philippines in limited quantity.

9. Strombus (Canarium) fragilis (Röding, 1798)
38.8mm; Sulu Archipelago.
Shell medium sized, thin, elongate-ovate, with smooth, rounded whorls, a few varices on the spire whorls and spiral grooves at the base; aperture narrow, widening anteriorly, commencing below the body whorl shoulder; outer lip thin, internally lirate, truncated basally with a weak stromboidal notch; cream-coloured with irregular orange-brown mottling, aperture purple within, columella smooth and dark purple; reaches 50mm.
DISTRIBUTION: Visayan Sea, Sulu Sea and Camotes Sea.

10. Strombus (Canarium) dentatus Linné, 1758
54.5mm, 58.2mm; Bohol.
Shell medium sized, glossy, rather heavy, with a moderately tall, acute spire; sculptured with angulate axial folds, originating below the well impressed sutures and continuing halfway down the whorls, those on the last two being strongest; outer lip thick, flaring, abruptly truncated anteriorly with 3 or 4 dentate projections; columella thickly calloused and smooth, with a few teeth posteriorly; pinkish cream with irregular orange-brown mottling, aperture and columella white, deep mauve between the lirae of the outer lip; reaches 60mm.
DISTRIBUTION: Visayan Sea, Sulu Sea and Camotes Sea.

11. Strombus (Canarium) terebellatus terebellatus Sowerby II, 1842
47.4mm, 47.7mm, 46.8mm; Samar.
Shell medium sized, smooth, rather lightweight, with a tall spire and well incised sutures; all the whorls smooth and strongly convex; aperture narrow, widening anteriorly then sharply truncated with two weak, widely spaced projections; columella smooth, weakly calloused, outer lip thin and smooth within; colour off-white with darker brown mottling, aperture marginally white, pale purple within; reaches 50mm.
DISTRIBUTION: Marinduque, Samar and Sulu Sea.

Strombidae — Plate 16

1. Lambis (Lambis) truncata sebae (Kiener, 1843)
276mm; Palawan.
Shell large, heavy, ponderous, sculptured with heavy spiral cords and knobs, and a somewhat pointed apex; outer lip thickened with six short, recurved digitations and a strong stromboidal notch; siphonal canal short,

open and recurved; cream with brown maculations, aperture white within, pink or purple towards the outer lip margin, columella pinkish within, brown on the outer margin; reaches 330mm.
DISTRIBUTION: Occurs mostly in the southern part of the archipelago.

2. Lambis (Millepes) wheelwrighti Greene, 1978
225mm; Palawan.
A problematical species, resembling *L. truncata* in most respects but generally smaller, with 8 or 9 short, curved digits and several faint lirae on the anterior and posterior edge of the outer lip, occasionally on the columella; shell brown externally, with a pink to orange interior, and a darker coloured columella and outer lip area; considered by many as a *L. millepeda x L. truncata* hybrid.
DISTRIBUTION: Visayan Sea, Palawan and Mindanao.

3. Lambis (Millepes) scorpius scorpius (Linné, 1758)
191mm; Albay Gulf, Southern Luzon.
Shell medium sized, rugose, sculptured with spiral cords and knobs, the first body whorl row strongest; characterized by 6 long, flattened, somewhat crimped digitations, the first four shortest and posteriorly curved; a small lobe at the base of the spire digitation, a long, twisted siphonal canal, and a marginally orange, internally purple aperture, lined with raised white lirae; columella orange, purple between the lirae, surface cream with brown mottling; reaches 200mm.
DISTRIBUTION: Throughout the Philippines.

4. Lambis (Lambis) crocata crocata (Link, 1807)
177mm; Panglao, Bohol.
A medium sized shell characterized by its smooth orange aperture, 6 long recurved digitations, and its long, curved siphonal canal; sculptured with fine spiral striae, a row of weak sutural nodules, and 3 rows of knobs on the body whorl, the first row the largest; columella smooth and calloused; colour orange with white bands and maculations; reaches 180mm.
DISTRIBUTION: Mostly from the Visayan Sea and Mindanao in limited quantity.

5. Lambis (Lambis) lambis (Linné, 1758)
Shell moderately large, solid and heavy; sculptured with rows of spiral cords and two rows of nodules on the body whorl, the first being the strongest; outer lip thick with 6 recurved digitations and a prominent stromboidal notch; aperture and columella salmon-pink within; anterior canal short, open and recurved; external colour pale cream with brown blotches or lines; reaches 200mm; exhibits sexual dimorphism.
5a. male form; smaller in size with smaller, more numerous dorsal knobs; lateral digitations short and hooked; colour quite intense; 171mm; Samar.
5b. female form; much larger, less brightly coloured, with larger and fewer dorsal knobs, and the lateral digitations frequently curved upwards; 205mm; Cebu.
DISTRIBUTION: Throughout the Philippines.

6. Lambis (Millepes) millepeda (Linné, 1758)
155mm; Cebu.
Shell moderately large, characterized by its 9 short digitations, the lower 3 or 5 of which are posteriorly curved, the light pinkish brown aperture with raised white lirae on both the outer lip and columella, the external sculpturing of spiral lirae and axially orientated nodules on the early whorls, and the three rows on the body whorl, first row largest; stromboidal notch large; pale cream with brown lines and blotches; reaches 160mm.
DISTRIBUTION: Throughout the Philippines.

7. Lambis sp. (*"scorpius x crocata"* hybrid?)
135mm; Bohol.
A curious specimen possessing certain characteristics of both the above mentioned species; the dorsal structure appears intermediate between *L. scorpius* and *L. crocata*, especially as regards to the size and number of knobs; the digitations are identical in size and curvature to those of *scorpius*, but lack the digital crimping of that species, and also the small lobe on the base of the spire digitation; the aperture is bright orange (c.f. *crocata*), but has the purple stained lirations seen in *scorpius*, though not as strong; until more material is available, best considered as a hybrid (c.f. *wheelwrighti*).
DISTRIBUTION: Visayan Sea, Camotes Sea and Mindanao.

8. Lambis (Harpago) chiragra chiragra (Linné, 1758)
Shell large, solid, with 5 long, thick, closed, hollow digitations, the marginal 3 curved, and a long siphonal canal recurved at right angles to the left; a strong stromboidal notch between the fourth and fifth digitation; aperture narrow, constricted posteriorly, lirate on the inner surface of the outer lip, columella finely lirate; spire whorls concave, body whorl with spiral cords and knobs; cream coloured with brown lines and maculations; aperture and columella pink (occasionally almost red), white within and marginally brown; exhibits sexual dimorphism.
8a. female form; large, pale in colouration, a light pink aperture and fused dorsal knobs; 220mm; Albay Gulf, Southern Luzon.
8b. male form; smaller in size, finer surface sculpturing, smaller and more numerous dorsal knobs, aperture intense pink with mauve tinted lirae; 123mm; Samar.
DISTRIBUTION: Throughout the Philippines.

9. Strombus (Tricornis) sinuatus Humphrey, 1786
107mm; Panglao, Bohol.
Shell large, solid, rather heavy, with 8 to 10 heavily noduled whorls, an outer lip with four "finger-like" projections, and a large projecting knob on the dorsum, together with several smaller ones; outer lip thick, recurved and somewhat flattened, with a strong stromboidal notch; golden brown with white bands on the dorsum and white reticulations ventrally; aperture mauve within, marginally white, columella orange; reaches 120mm.

DISTRIBUTION: Occurs more in the Visayan Sea, Mindanao and Camotes Sea.

10. Strombus (Tricornis) latissimus Linné, 1758
183mm; Cebu.
Shell large, massive, rotund, with a broadly flaring lip extending beyond the shell apex and recurving to form a heavily thickened outer margin which terminates in a deep stromboidal notch; spire whorls with a row of nodules at the sutures, body whorl with a single large knob on the shoulder; orange-brown with white maculation, aperture white, marginally brown, columella smooth, heavily glazed and white, becoming brown anteriorly; reaches 200mm. Gem specimens are rare.
DISTRIBUTION: Throughout the Philippines.

Strombidae Plate 17

1. Strombus (Lentigo) lentiginosus Linné, 1758
70.5mm; Samar.
Shell medium sized, heavy, sutures adpressed, whorls angulate, 9 to 10 in number, sculptured with large knobs and overriding striae; body whorl with 4 rows of nodules, first row largest; outer lip expansive, 3 "finger-like" projections posteriorly, a well indented stromboidal notch, and a peculiarly angulate lobe at the base; creamy white with brown axial streaks and dots, aperture salmon to orange-coloured within, marginally white, columella with a silvery glaze; reaches 105mm.
DISTRIBUTION: Throughout the Philippines.

2. Strombus (Euprotomus) aurisdianae Linné, 1758
63.5mm, 63.3mm; Cebu.
Shell medium sized, heavy, with a dull, rough exterior; crossed by strong axial ribs and spiral cords, forming a row of nodules at the base of each whorl, a row of heavy tubercles below the suture of the body whorl, and below this, 2 or 3 rows of lesser strength; outer lip thick, a "finger-like" projection posteriorly and a few weakly incised lirae below; siphonal canal recurved; cream coloured with grey spiral lines, aperture orange within, marginally white, columella heavily glazed; reaches 80mm.
DISTRIBUTION: Throughout the Philippines.

3. Strombus (Euprotomus) aratrum (Röding, 1798)
84.7mm; Cebu.
Very similar to *aurisdianae*, but is larger, with a taller spire, and is more heavily sculptured on the dorsum; the rows of knobs are larger, heavier and more numerous and the aperture is entirely orange. Typical *aratrum* has the aperture lined with black, a feature not seen in Philippine specimens and to which the name *chrysostomus* Kuroda, 1942, has often been applied.
DISTRIBUTION: Visayan Sea, Sulu Sea and the Camotes Sea.

4. Strombus (Euprotomus) bulla (Röding, 1798)
64.2mm, 66.0mm; Cebu.
Another species very close to *S. aurisdianae*, but can immediately be recognized by the following features: lighter in weight, the dorsal surface is smoother and glossier, the spiral striae are almost obsolete, and there is a single row of nodules on the body whorl dorsum; the "finger-like" projection is longer than that of *aurisdianae*, and the lirae of that species are missing in *S. bulla;* also, the columellar glazing of *bulla* is heavier and extends almost to the spire apex; reaches 75mm.
DISTRIBUTION: Throughout the Philippines.

5. Strombus (Laevistrombus) canarium Linné, 1758
Shell medium sized to large, heavy, globose, with a flaring lip which has a heavy silver glaze; spire moderately tall, spire whorls weakly reticulate, body whorl elongate-ovate and swollen with an inflated shoulder; outer lip thickened, columella smooth with a white glaze; colour variable, usually pale brown or reticulated with darker brown; reaches 100mm.
5a. dwarf form; 40mm; Palawan.
5b. typical form; 68.5mm; Cebu.
5c. giant form; 101mm; Albay Gulf, Southern Luzon.
DISTRIBUTION: Throughout the Philippines.

6. Strombus (Conomurex) luhuanus Linné, 1758
54.2mm; Samar.
Shell medium sized, well-shouldered and "cone-shaped"; characterized by its deeply canaliculate sutures, black columella, and finely striated, salmon-orange coloured aperture; exterior smooth and cream coloured with brown mottling in bands; reaches 80mm.
DISTRIBUTION: Throughout the Philippines.

7. Strombus (Doxander) vittatus vittatus Linné, 1758
Shell medium sized, lightweight, with a tall, turreted spire of some twelve whorls, each with a deep presutural groove and numerous axial ribs that are crossed by spiral lirae on the lower portion; a large, smooth, occasionally humped body whorl, spirally grooved towards the base; aperture lirate towards the outer lip margin which is thickened and incurved, columella smooth; brownish orange with whitish bands, aperture and columella white; reaches 90mm.
7a. typical form; 72.5mm; Manila Bay.

PLATE NO. 16. 1. *Lambis truncata sebae* (Kiener) 2. *Lambis wheelwrighti* Greene 3. *Lambis scorpius scorpius* (Linné) 4. *Lambis crocata crocata* (Link) 5a. *Lambis lambis* (Linné)-male form 5b. female form 6. *Lambis millepeda* (Linné) 7. *Lambis sp. ("scorpius x crocata"* hybrid?) 8a. *Lambis chiragra chiragra* (Linné)-female form 8b. male form 9. *Strombus sinuatus* Humphrey 10. *Strombus latissimus* (Linné)

PLATE 16

7b. high spired form *australis* Schroeter, 1805; 82.0mm; Manila Bay.
DISTRIBUTION: Sporadically found throughout the Philippines.

8. Strombus (Dolomena) variabilis Swainson, 1820
Shell medium sized, solid, lightweight, with 9 to 10 noduled, well-shouldered whorls; body whorl smooth ventrally with an elongate ridge at the periphery and some smaller nodules on the dorsal surface shoulder; outer lip expansive, marginally thickened and glazed, posteriorly notched; colour cream with orange-brown zigzags in 3 or 4 bands; aperture white and smooth, columella smooth, white, with or without a brown blotch; reaches 60mm. Specimens with a brown blotch on the columella are by far the most commonly encountered form in the Philippines.
8a. typical form with brown columellar blotch; 45.7mm; Samar.
8b. typical form lacking columellar blotch; 45.2mm; Palawan.
8c. colour form *athenius* Duclos, 1844: strong dark banding; 46.1mm; Cebu.
DISTRIBUTION: Throughout the Philippines.

9. Strombus (Dolomena) minimus Linné, 1771
38.7mm, 34.4mm; Tayabas Bay, Quezon.
A small, solid shell, immediately identified by its nodulose, somewhat angulate spire whorls, the wide, internally lirate, bright yellow aperture, the thick, open posterior canal, the expansive, marginally thickened, slightly incurved outer lip, and the thick white columellar callus which spreads basally; outer surface orange-brown with two white bands; reaches 40mm.
DISTRIBUTION: Sporadically found throughout the Philippines.

10. Strombus (Dolomena) plicatus pulchellus Reeve, 1851
35.9mm; Tayabas Bay, Quezon.
Shell small, lightweight, with a tall spire; spire whorls well-shouldered, the first few smooth and the rest with fine axial ribs and varices; body whorl with a few elongate shoulder nodules, outer lip expansive, marginally thin, upturned, and strongly lirate within; columella plicate but smooth centrally; pale cream with brown blotching and bands; columella brown, aperture pale cream, purple between the lirae; reaches 50mm.
DISTRIBUTION: Tayabas Bay (Quezon), Visayan Sea and Camotes Sea.

11. Strombus (Lentigo) pipus (Röding, 1798)
64.5mm; Jolo, Sulu Archipelago.
Shell medium sized, roundly quadrate, with a moderately tall spire; early whorls axially and spirally striated, later whorls with nodules on the roundly angulate shoulder; body whorl with 4 rows of nodules the shoulder row strongest; aperture lirate within, outer lip thickened, incurved, indented posteriorly and slightly undulate along its length; columella smooth, white and highly glazed; shell creamy white with brown axial stripes and bands, 5 to 6 band pairs on the outer lip margin; reaches 70mm.
DISTRIBUTION: Mostly from Mindanao.

12. Strombus (Canarium) urceus urceus Linné, 1758
Shell medium sized, with a tall spire and impressed sutures; spire whorls stepped with nodulose axial cords and varices; body whorl with some weakly incised basal grooves and 6 to 8 shoulder knobs, the largest being on the dorsum; aperture lirate, outer lip thick, stromboidal notch strong; columella plicate, smooth centrally; anterior canal short and recurved; colour variable, usually a pale base colour with darker bands and maculations; aperture pale yellow, black, or a combination of both; anterior canal black inside and out; reaches 65mm.
12a. typical form; 54.3mm; Palawan.
12b. deep water form; less heavily sculptured, very pale colouration, no black staining in the aperture or anterior canal; 53.9mm; Bohol.
12c. form *ustulatus* Schumacher, 1817; black mouth and columella; 41.4mm; Sulu Archipelago.
DISTRIBUTION: Throughout the Philippines.

13. Strombus (Dolomena) marginatus septimus Duclos, in Chenu, 1844
Shell medium sized, broadly elongate; spire whorls sharply angulate with axial ribs and varices which become nodulose on the shoulder of later whorls; early whorls spirally striated, body whorl smooth with the lirae persisting on the back of the outer lip margin; outer lip flared and lirate, the posterior canal extending over the first suture; columella plicate but smooth centrally, anterior canal recurved; brown-coloured with four "arrow-shaped" white bands, aperture and columella white; reaches 50mm.
13a. typical form; 49mm, 45.7mm; Masbate.
13b. sharply keeled specimen, a characteristic of the typical species from India; 45mm; Corregidor Island, W. Luzon.
DISTRIBUTION: Sporadically found throughout the Philippines.

PLATE NO. 17 1. *Strombus lentiginosus* Linné 2. *Strombus aurisdianae* Linné 3. *Strombus aratrum* (Röding) 4. *Strombus bulla* (Röding) 5a. *Strombus canarium* Linné, dwarf form 5b. typical form 5c. giant form 6. *Strombus luhuanus* Linné 7a. *Strombus vittatus vittatus* Linné 7b. *high spired form australis* Schroeter 8a. *Strombus variabilis* Swainson-typical form with brown columellar blotch. 8b. typical form lacking columellar blotch. 8c. colour form *athenius* Duclos 9. *Strombus minimus* Linné 10. *Strombus plicatus pulchellus* Reeve 11. *Strombus pipus* (Röding) 12a. *Strombus urceus urceus* Linné 12b. deepwater form 12c. form *ustulatus* Schumacher 13a. *Strombus marginatus septimus* Duclos, in Chenu 13b. sharply keeled specimen 14. *Strombus epidromis* Linné 15. *Strombus kleckhamae boholensis* Mühlhaüsser 16. *Strombus labiossus* Wood 17a. *Strombus labiatus labiatus* Röding, heavily ribbed form 17b. smooth form 17c. colour variations 18a. *Strombus gibberulus gibbosus* (Röding) 18b. albinistic 18c. pale form with red apertural margin 19. *Strombus dilatatus* Swainson

PLATE 17

1
2
3
4
5a
5b
5c
6
7a
7b
8a
8b
8c
9
10
11
12a
12b
12c
13a
13b
14
15
16
17a
17b
17c
18a
18b
18c
19

14. Strombus (Labiostrombus) epidromis Linné, 1758
80.0mm; Bohol.
Shell medium sized, with a relatively short spire; spire whorls rounded and slightly shouldered, with fine spiral lirae, irregular varices and small axial ribs which become shoulder nodules on last two whorls; outer lip roundly expansive, thick and porcelaneous on the margin; aperture and columella both smooth with a white glaze; externally cream coloured with brown mottling; reaches 90mm.
DISTRIBUTION: Throughout the Philippines.

15. Strombus (Dolomena) kleckhamae boholensis Mühlhäusser, 1981
57.8mm, 57.0mm; Bohol.
Shell medium sized, solid, sutures adpressed, spire whorls concave, sculptured with varices, spiral striae, and numerous finely beaded axial cords; body whorl with 8 to 9 weak nodules and spiral striae on the back of the outer lip; aperture lirate within, outer lip expanded, thickened at the edge and inflected; columella lirate, anterior lirae becoming weak; colour pale cream with orange-brown banding and streaks; aperture pale cream with a yellowish tinge, irregularly black between the lirae; columella black on the posterior half, white below; reaches 70mm. It differs from the nominate subspecies which occurs in New Britain and New Guinea, in the colour and sculpturing of the columella, the number of shoulder nodes and the surface colouration.
DISTRIBUTION: Bohol.

16. Strombus (Canarium) labiosus Wood, 1828
37.3mm, 37.8mm; Cebu.
Shell medium sized, glossy and tall-spired; spire whorls convex, angled at the shoulder, sculptured with axial cords and overriding spiral striae; body whorl rotund, with a row of peripherally situated nodes, and crossed with thick spiral threads; aperture lirate within, outer lip flaring and "U-shaped"; columella concave, calloused, but smooth centrally; colour off-white, with traces of brown, aperture white, darker internally, columella white; reaches 50mm.
DISTRIBUTION: Tayabas Bay, (Quezon), Visayan Sea and Camotes Sea.

17. Strombus (Canarium) labiatus labiatus Röding, 1798
Shell small, solid, with sharply angled shoulders and axial ribs on the spire whorls which are reduced to a few large shoulder knobs on the body whorl; base with numerous raised spiral cords; aperture and columella lirate, outer lip thickened, stromboidal notch deep; colour variable, usually brown to grey with paler streaks or blotches, aperture deep brown, columella orange with darker lirae; reaches 50mm.
17a. heavily ribbed form; 41.9mm; Palawan.
17b. smooth form; 43.5mm; Samar.
17c. colour variations; 35.0mm, 33.6mm; Samar.
DISTRIBUTION: Throughout the Philippines.

18. Strombus (Gibberulus) gibberulus gibbosus (Röding, 1798)
Shell medium sized, conoidal, with rounded whorls, the early ones with varical nodules; whorls frequently distorted, especially the penultimate one; aperture lirate, white or purple tinted, columella smooth, calloused, and stained with purple; outer lip vertical, posteriorly indented with a strong stromboidal notch anteriorly; white with brown bands and streaks; reaches 60mm.
18a. typical form; 54.0mm, 50.5mm; Samar.
18b. albinistic form; 49.3mm; Samar.
18c. pale form with red apertural margin; 37.2mm; Samar.
DISTRIBUTION: Throughout the Philippines.

19. Strombus (Dolomena) dilatatus Swainson, 1821
58.0mm, 57.2mm; Cebu.
Shell medium sized with a tall spire; spire whorls weakly shouldered with axial cords and spiral lirae, body whorl nodulose on the dorsum shoulder; aperture lirate, outer lip flaring, thickened and inflected, columella almost smooth; posterior canal pronounced, often sweeping obliquely over several spire whorls; white with brown banding, aperture white, lavender within, columella white; reaches 60mm.
DISTRIBUTION: Tayabas Bay (Quezon), Visayan Sea and Camotes Sea.

Epitoniidae Plate 18

Class Gastropoda
Subclass Prosobranchia
Order Heterogastropoda
Suborder Ptenoglossa
Superfamily Epitoniacea
Family EPITONIIDAE

A large group of small to medium sized shells which normally inhabit shallow intertidal waters among corals and rocks. The shells are typically white, tall spired and coiled. The surface is sculptured with narrow costae which may be flattened or thin and blade-like, and the aperture is rounded with a thick outer lip and a multi-spiraled operculum. Although the family is well represented in Philippine waters, most species are seldom seen and usually come from deep water by trawling or gill net fishing.

1. Epitonium (Epitonium) scalare (Linné, 1758)
Shell medium sized to large, tall-spired with an acute apex; whorls well rounded, separated, sculptured with thin, blade-like varices which connect at the sutures; interspaces smooth, aperture ovate, outer lip thick; greyish white with white varices and aperture; reaches 70mm.
1a. typical form; 68.5mm; Manila Bay.
1b. freak form which has developed an increased

number of axial costae on the last 2 whorls; 51.5mm; Manila Bay.
DISTRIBUTION: Manila Bay.

2. Epitonium (Elegantiscala) rugosum Kuroda & Ito, 1961
67.2mm; Punta Engaño, Cebu.
Shell medium sized, spire tall, sutures channeled; whorls convex and sharply shouldered, sculptured with numerous tall, narrow, somewhat fimbriate axial costae, fluted below the shoulder and smooth above; intercostal space with grooves travelling up the trailing edge of the costae, leading edge smooth; umbilicus obsolete, aperture circular and thick-walled; pure white; reaches 70mm.
DISTRIBUTION: Camotes Sea.

3. Cirsotrema varicosa Lamarck, 1822
46.7mm; Punta Engaño, Cebu.
Shell medium sized with a tall, acute spire, well rounded whorls, and sculptured with irregular varices between which are numerous flat, closely set, fluted axial costae; the whole surface is scabrous from intersecting axial and spiral striae, and a distinct ridge encircles the base of the body whorl; umbilicus obsolete, aperture circular; apex, varices and aperture white, rest of shell with a bluish hue; reaches 55mm.
DISTRIBUTION: Camotes Sea.

4. Epitonium (Papyriscala) lineolatum (Sowerby, 1844)
36.5mm; Tayabas Bay, Quezon.
Shell medium sized, lightweight, fragile, pyramidally-ovate with a tall spire and an acute apex; sutures deeply channeled, whorls well rounded, 10 in number and sculptured with up to 30 thin, slightly raised costae; umbilicus open, aperture ovate; white with obsolete brown bands; usually reaches 20mm. The specimen illustrated here is unusually large.
DISTRIBUTION: Manila Bay and Tayabas Bay, Quezon.

5. Epitonium (Epitonium) neglectum (Adams & Reeve, 1850)
31.4mm; Manla Bay.
Shell medium sized, somewhat squat, triangular in outline; spire tall, apex acute, sutures deeply channeled; sculptured with 15 to 16 continuous, blade-like varices hooked below the sutures and the interspaces with intersecting axial and spiral threads; umbilicus widely open; aperture and varices white, intercostal spaces brown, occasionally banded; reaches 35mm; often considered a form of *E. pallasii.*
DISTRIBUTION: Manila Bay and Tayabas Bay, Quezon.

6. Epitonium (Epitonium) pallasii (Kiener, 1838)
30.2mm, 20.7mm; Punta Engaño, Cebu.
Shell medium sized, with a tall spire and an acute apex; whorls rounded and slightly separated, sculptured with 12 to 14 blade-like costae, normally hooked at the suture; umbilicus open, aperture ovate, outer lip thick; brown-coloured with white costae and aperture; reaches 30mm.
DISTRIBUTION: Camotes Sea, Manila Bay and Tayabas Bay, Quezon.

7. Epitonium (Epitonium) pyramidale (Sowerby, 1844)
33.9mm, 34.9mm; Tayabas Bay, Quezon.
Shell medium sized, elongate, with a tall spire and 9 to 10 strongly convex, slightly separated whorls; sculptured with 10 to 12 thin axial costae, hooked at the sutures; aperture circular with a weak umbilicus; whorls pale brown, axial costae and aperture white; reaches 40mm.
DISTRIBUTION: Manila Bay and Tayabas Bay, Quezon.

8. Epitonium (Spiniscala) vigintifoliata (Masahito, Kuroda & Habe, 1971)
22.4mm; Punta Engaño, Cebu.
A small fragile, pyramidally ovate shell with a tall spire, an acute apex and strongly convex whorls which are sculptured with numerous close-set, rather fragile costae which occlude the deeply channeled sutures; the aperture is ovate and the shell is non-umbilicate; greyish white in colour and reaching around 25mm.
DISTRIBUTION: Camotes Sea and Manila Bay.

9. Epitonium (Gyroscala) perplexum (Pease, 1868)
25.8mm, 22.0mm; Tayabas Bay, Quezon.
Shell medium sized, triangular in outline, sutures weakly channeled, spire tall, apex acute; whorls well rounded, sculptured with numerous blade-like costae which are continuous with those on the whorl above; aperture ovate, and the base with a distinct ridge; white with a subsutural, intercostal brown band; reaches 30mm.
DISTRIBUTION: Manila Bay and Camotes Sea.

10. Epitonium (Epitonium) subauriculatum (Souverbie, in Souverbie & Montrouzier, 1866)
23.0mm; Punta Engaño, Cebu.
Shell small, elongate with a tall, sharply acute spire, deeply channeled sutures, 7 oblique, continuous costae and numerous fine intercostal striae; aperture ovate, umbilicus obsolete; pure white; reaches 25mm.
DISTRIBUTION: Camotes Sea.

11. Cirsotrema edgari (de Boury, 1912)
Syn: *Epitonium (Laniscala) eboreum* Lan, 1976
37.4mm; Cebu.
Close to *E. rugosum,* but generally much smaller; it has fewer whorls (9 to 10) than *E. rugosum* (12 to 13), the axial costae are smaller, thicker, more heavily fluted, and can number up to 40 on the body whorl (up to 25 in *rugosum);* the costae of *E. edgari* are not curved as in *rugosum,* and the interstitial threads are coarser and fewer in number; pure white; reaches 50mm.
DISTRIBUTION: Manila Bay and Camotes Sea.

12. Epitonium (Hyaloscala) hayashi (Habe, 1961)
19.9mm, 17.8mm; Manila Bay.
Shell small, elongate, with a tall spire, an acute apex

and deeply channeled sutures; sculptured with 10 to 12 thin, continuous varices, some thicker than others, and smooth interspaces; umbilicus obsolete, aperture circular; pale cream with 2 bands of brown spots; reaches 20mm.
DISTRIBUTION: Manila Bay and Camotes Sea.

13. c.f. Epitonium (Hirtoscala) replicatum Sowerby, 1844
15.5mm, 13.9mm; Tayabas Bay, Quezon.
Shell small, triangular-ovate with 8 loosely coiled whorls; 7 to 8 axially aligned costae per whorl, slightly spinose at the shoulder; aperture ovate, marginally flattened, umbilicus deeply channeled; white in colour; reaches around 17mm; typical *E. replicatum* has more pronounced shoulder spines.
DISTRIBUTION: Camotes Sea and Tayabas Bay, Quezon.

14. Epitonium (Viciniscala) liliputanum (A. Adams, 1855)
11.0mm, 11.2mm; Punta Engaño, Cebu.
Shell small, solid, rather ovate in outline with a tall spire, rounded spire whorls and a large body whorl with an open umbilicus; sculptured with 7 to 9 blade-like varices per whorl, hooked and sharply angled at the shoulder, joining those above; white with a brown body whorl; like a miniature *E. scalare*; reaches 15mm.
DISTRIBUTION: Camotes Sea and Manila Bay.

Class Gastropoda
Subclass Prosobranchia
Order Neogastropoda
Suborder Toxoglossa
Superfamily Conacea

Family CANCELLARIIDAE

A large group of small, herbivorous, sand-dwelling gastropods commonly found in shallow to deep water. They are worldwide in distribution and inhabit most tropical seas. Shells are solid, globose, have a large body whorl, and are sculptured with axial and spiral cords, producing a strongly cancellate surface. The aperture is triangular, lirate within, and the columella is strongly plicate. The family is fairly well represented in the Philippines.

15. Trigonostoma pellucida (Perry, 1811)
Syn: *Delphinula trigonostoma* Lamarck, 1822
39.4mm; Cebu.
Shell medium sized, solid, scalariform in outline, whorls strongly stepped with flat areas between the suture and the sharply keeled shoulder; sculptured with 11 to 12 axial ribs, echinate at the shoulder, and encircled with fine spiral lines; umbilicus deeply channeled, aperture triangulate, outer lip lirate within, columella with 2 plaits; colour off-white; reaches 45mm. Often confused with *T. antiquata* (Hinds, 1843), to which it is superficially similar, but *T. antiquata* is smaller, narrower, less scalariform, has a weaker spiral sculpturing and distinct shoulder lobes.
DISTRIBUTION: Cebu and Bohol.

16. Trigonostoma bicolor (Hinds, 1843)
16.0mm, 18.4mm, 21.7mm; Manila Bay.
Shell small with a tall, acute spire; spire whorls stepped, body whorl large, constricted basally, and sculptured with 8 oblique axial cords which cross the sharply angled shoulder where they become echinate and continue to the sutures; surface crossed by fine spiral striae; aperture triangulate, lirate within, umbilicus partly open, columella with 3 plaits; base colour white, brown on the shoulder; reaches around 22mm.
DISTRIBUTION: Manila Bay.

17. Trigonostoma crenifera (Sowerby, 1833)
22.5mm, 24.4mm; Manila Bay.
Shell small, solid, ovate, with a tall, acute spire; sculptured with 12 to 14 slightly oblique axial ribs, echinate at the sharply angled shoulder and crossed by fine spiral cords, forming small sharp spines; aperture triangular, lirate within, columella with 3 plaits, umbilicus open; white with irregular brown banding, aperture white and brown banded; reaches 25mm.
DISTRIBUTION: Manila Bay.

18. Trigonostoma scalariformis (Lamarck, 1822)
24.3mm, 27.3mm; Manila Bay.
Shell medium sized, triangularly ovate with deep sutures; whorls 6 in number, all sharply angled and sculptured with rounded axial cords, echinate at the sharply angled shoulder and crossed with numerous overriding spiral riblets; aperture triangulate, lirate within, columella with 3 plaits, umbilicus open; off-white with thin brown banding, aperture white and brown banded; reaches 30mm.
DISTRIBUTION: Manila Bay.

19. Trigonostoma textilis (Kiener, 1841)
32.2mm, 22.9mm; Sulu Archipelago.
A medium sized shell of 8 convex, strongly shouldered

PLATE NO. 18 1a. *Epitonium scalare* (Linné) 1b. freak form 2. *Epitonium rugosum* Kuroda & Ito 3. *Cirsotrema varicosa* Lamarck 4. *Epitonium lineolatum* (Sowerby) 5. *Epitonium neglectum* (Adams & Reeve) 6. *Epitonium pallasii* (Kiener) 7. *Epitonium pyramidale* (Sowerby) 8. *Epitonium vigintifoliata* (Masahito, Kuroda & Habe) 9. *Epitonium perplexum* (Pease) 10. *Epitonium subauriculatum* (Souverbie, in Souverbie & Montrouzier) 11. *Cirsotrema edgari* (de Boury) 12. *Epitonium hayashi* (Habe) 13. *c.f. Epitonium replicatum* (Sowerby) 14. *Epitonium liliputanum* (A. Adams) 15. *Trigonostoma pellucida* (Perry) 16. *Trigonostoma bicolor* (Hinds) 17. *Trigonostoma crenifera* (Sowerby) 18. *Trigonostoma scalariformis* (Lamarck) 19. *Trigonostoma textilis* (Kiener) 20. *Cancellaria elegans sinensis* (Reeve) 21. *Cancellaria mercadoi* Old 22. *Cancellaria asperella* Lamarck 23. *Trigonostoma obliquata* (Lamarck) 24. *Trigonostoma scalata* (Sowerby) 25. *Pisanella viva* (Habe & Okutani) 26. *Cancellaria oblonga* Sowerby

PLATE 18

whorls and deeply canaliculate sutures; sculptured with numerous strong axial ribs that are crossed by spiral threads forming nodules; aperture ovate, outer lip expanded, lirate within, columella with 3 plaits; reddish brown with white nodules and shoulder spines; reaches 35mm.
DISTRIBUTION: Camotes Sea and Sulu Sea.

20. Cancellaria (Merica) elegans sinensis (Reeve, 1856)
36.7mm, 36.0mm; Palawan.
Shell medium sized, elongate-ovate with a tall, acute spire and impressed sutures; sculptured with fine axial and spiral lirae, intersecting to produce a finely cancellate surface; aperture elongate, lirate within, columella with 3 plaits and some pustules anteriorly; orange-white with 3 brown bands; reaches 50mm.
DISTRIBUTION: Sporadically found throughout the Philippines.

21. Cancellaria mercadoi Old, 1968
33.4mm, 34.8mm; Tayabas Bay, Quezon.
Shell medium sized with a tall, acute spire, well channeled sutures and strongly convex whorls that are sculptured with 10 to 12 thick, wide axial cords, and crossed by fine undulating spiral lirae; aperture sub-ovate, lirate within, columella with 3 plaits; yellow coloured with brown spiral bands; reaches 35mm. Named after Mario Mercado of Zambales, a local shell dealer.
DISTRIBUTION: Tayabas Bay, Quezon.

22. Cancellaria (Merica) asperella Lamarck, 1822
28.2mm; Palawan.
Shell medium sized, ovate, with a moderately tall spire, channeled sutures, and sculptured with undulate axial and spiral lirae intersecting to produce a roughly cancellate surface; aperture ovate, lirate within, columella oblique and posteriorly pustulose with 3 plaits; cream coloured to reddish brown with banding; reaches 40mm.
DISTRIBUTION: Palawan and Sulu Sea.

23. Trigonostoma obliquata (Lamarck, 1822)
17.3mm, 17.9mm; Punta Engaño, Cebu.
Shell small, globose with deeply channeled sutures; sculptured with numerous (approximately 16) oblique axial costae, echinate at the shoulder, and crossed by fine spiral lirae, forming white nodules; aperture ovate, lirate within, with one posterior denticle and 3 columellar plaits; ground colour greyish white, ribs white and brown, aperture marginally white, pink within; reaches 20mm.
DISTRIBUTION: Manila Bay, Samar and Camotes Sea.

24. Trigonostoma scalata (Sowerby, 1832)
24.8mm, 25.3mm; Palawan.
Shell medium sized and loosely wound; body whorl large, spire whorls stepped with a flat presutural ramp; sculptured with numerous close-set axial ribs, spinose at the shoulder and crossed by spiral lirae, forming small nodules; aperture ovate, lirate within, columella with 2 or 3 oblique plaits, umbilicus weakly channeled; yellowish brown with darker banding and brown spots on the presutural ramp, aperture white; reaches 30mm.
DISTRIBUTION: Sporadically found throughout the Philippines in limited quantity.

25. Pisanella viva (Habe & Okutani, 1981)
37.2mm; Cebu.
Superficially resembling a *Colubraria* species, but radula studies indicate it belongs to the Cancellariidae. Shell medium sized, tall-spired, sutures impressed; sculptured with rounded axial folds and overriding spiral cords forming nodules; aperture ovate-quadrangular, outer lip marginally dentate and backed with a varix, columella with 2 to 3 plaits anteriorly; off-white with diffuse brown banding; reaches 55mm.
DISTRIBUTION: Camotes Sea.

26. Cancellaria (Merica) oblonga Sowerby, 1825
38.9mm, 32.3mm; Cebu.
Shell medium sized, fusiform-ovate, with a tall, acute spire, a large body whorl and impressed sutures; surface finely nodular from overriding axial and spiral lirae; aperture elongate-ovate, outer lip marginally serrate and lirate within, columella calloused with 2 to 3 oblique plaits; yellowish brown with darker banding, columella and outer lip margin orange, white within; reaches 45mm.
DISTRIBUTION: Sporadically found throughout the Philippines in limited quantity.

Xenophoridae — Plate 19

Class Gastropoda
Subclass Prosobranchia
Order Mesogastropoda
Suborder Taenioglossa
Superfamily Xenophoracea
Family XENOPHORIDAE

A small group of medium to large-sized shells, trochiform in shape, lightweight, with a concave base, and mainly Western Pacific in distribution. They all prefer a sandy or rocky substrate in deep water. True *Xenophora* attach marine debris (coral, shells, stones, etc.) to the shell by means of a nacreous "glue" and have a marginally elevated rim of such fragments. The *Stellaria* group are relatively shell-free, the marginal fragments replaced by hollow spines. The *Onustus*

PLATE NO. 19 1. *Xenophora solaris* (Linné) 2. *Xenophora indica* (Gmelin) 3. *Xenophora exuta* (Reeve) 4. *Xenophora chinensis chinensis* (Philippi) 5. *Xenophora solarioides solarioides* (Reeve) 6. *Xenophora cerea* (Reeve) 7. *Xenophora corrugata* (Reeve) 8. *Xenophora japonica* Kuroda & Habe 9a. *Xenophora pallidula* (Reeve) 9b. *Specimen with glass sponge* 10. *Xenophora mekranensis konoi* Habe

PLATE 19

group are also fragment-free, depressed and deeply umbilicate with a round, undulate peripheral "skirt". (Measurements refer to shell only, excluding attached fragments). The family is very well represented in the Philippines, with all the known species being shown here. Although found in sporadic populations, the family is believed to be fairly well distributed throughout the entire archipelago.

1. Xenophora (Stellaria) solaris (Linné, 1764)
123.2mm, 107.9mm (including spines); Tayabas Bay, Quezon.
Shell medium sized, discoidal, trochiform, with a moderately tall spire, a surface crossed by many axial and spiral lirae producing a rough surface, and a row of 18 to 20 hollow, upturned spines on the peripheral margin; the base is concave with a deep, partially occluded umbilicus, and a roughened surface formed from intersecting radial and spiral cords; aperture smooth and nacreous; colour light yellow-brown; reaches 130mm.
DISTRIBUTION: Tayabas Bay (Quezon) and Camotes Sea.

2. Xenophora (Onustus) indica (Gmelin, 1791)
52.0mm, 81.5mm; Tayabas Bay, Quezon.
Similar to *X. exuta* but is light cream in colour, extremely fragile, has a shiny surface sculptured with fine axial ridges, and the projecting "skirt" lacks the projections of *exuta;* umbilicus deeply channeled, and shell fragments occur only on the first few spire whorls; reaches 100mm.
DISTRIBUTION: Camotes Sea and Tayabas Bay, Quezon.

3. Xenophora (Onustus) exuta (Reeve, 1842)
86.5mm, 89.5mm; Tayabas Bay, Quezon.
Shell large, thin, lightweight and rather flat, with a few particles of sand at the apex; whorls with radial cords and spiral striae; body whorl with a thin projecting "skirt", slightly digitate at the periphery; base concave with an open umbilicus; colour light cinnamon; reaches 100mm.
DISTRIBUTION: Tayabas Bay (Quezon) and Lubang Is., Mindoro.

4. Xenophora (Stellaria) chinensis chinensis (Philippi, 1841)
81.1mm, 73.1mm; Bohol.
A thin, flat, rather discoidal species with moderately convex whorls and sculptured with fine, oblique radial striae; base strongly concave with an open umbilicus; surface cemented with shell fragments along the suture line and on the wide, thin peripheral keel; yellowish brown in colour; reaches around 80mm.
DISTRIBUTION: Camotes Sea and Bulan, Sorsogon.

5. Xenophora (Xenophora) solarioides solarioides (Reeve, 1845)
17.0mm, 15.5mm; Bohol.
Shell small, moderately conical with corded whorls; characterized by its deeply concave base and the sharply angulate, deep, widely open umbilicus; heavily covered with stone and shell fragments; reaches 40mm.
DISTRIBUTION: Camotes Sea.

6. Xenophora (Xenophora) cerea (Reeve, 1845)
42.0mm, 32.0mm; Bohol.
Shell medium sized, conoidal, sutures deeply incised, whorls rounded, somewhat overhanging and sculptured with heavy folds and threads; base characteristically brown-stained and heavily ridged, the umbilicus closed by a thin callus; shell surface and aperture deep brown, though the brown colouration may be lacking in some specimens; reaches 50mm.
DISTRIBUTION: Camotes Sea.

7. Xenophora (Xenophora) corrugata (Reeve, 1842)
38.0mm, 40.0mm; Bulan, Sorsogon.
Generally believed to be a form of *X. pallidula* (Reeve) but differs in the presence of distinct growth cords and brown spiral streaks on the base. This is the *corrugata auct.* (Cernohorsky: 1972; Habe: 1964). Recent work (Ponder: 1983) suggests that this is merely a form of *cerea* (fig. 6), true *corrugata* being an Indian Ocean endemic.
DISTRIBUTION: Camotes Sea and Bulan, Sorsogon.

8. Xenophora (Xenophora) japonica Kuroda & Habe, 1971
32.0mm, 34.0mm; Bohol.
Like a small *X. pallidula* (Reeve), but with a proportionately taller spire, and is more densely covered with debris, so much so that most of the shell surface is obscured; base flat, fairly smooth and rust coloured, with a closed or partly closed umbilicus; reaches 55mm.
DISTRIBUTION: Camotes Sea.

9. Xenophora (Xenophora) pallidula (Reeve, 1842)
Shell large, solid and trochiform; surface adorned with oblique radial folds and fine spiral threads; base smooth, narrowly umbilicate, with a dilated callus on the inner apertural margin; may be covered with shells, gravel or coral; shell white, occasionally with a brownish coloured base; reaches 100mm.
9a. typical specimens; 69mm, 73mm; Leyte.
9b. specimen with glass sponge attachment; 57mm (shell size)/142mm (sponge length); Leyte.
DISTRIBUTION: Leyte, Camotes Sea and Bulan, Sorsogon.

10. Xenophora (Xenophora) mekranensis konoi Habe, 1953
54.0mm, 65.0mm; Leyte.
Shell medium sized, solid, rather conical, with deep sutures and overhanging whorls which are sculptured with heavy, oblique radial folds and fine, overriding spiral threads; base characteristically corded and pustulose, umbilicus closed; often adorned with shell fragments, but more frequently with corals; yellow-white in colour; reaches around 65mm.
DISTRIBUTION: Visayan Sea and Camotes Sea.

Class Gastropoda
Subclass Prosobranchia
Order Mesogastropoda
Suborder Taenioglossa
Superfamily Cypraeacea
Family CYPRAEIDAE

A large group of gastropods which are predominantly tropical in their distribution, nocturnal in their habits, and live mainly in shallow waters among coral and rocks. Members of this family range in size from 10mm to 150mm, are ovate, pyriform or cylindrical in shape, and have a highly enamelled surface which may be smooth or pustulose and is frequently patterned with intricate, multicoloured designs. The aperture is narrow, the labial and labral surfaces are marginally denticulate, the terminals truncated or extended, and the base may be smooth or slightly convex. The taxonomy of this family at both a generic and species level is clearly in need of division. In many cases, inadequate knowledge of distribution patterns and a failure to appreciate generic variation within a population group has given rise to a large assemblage of species, subspecies and forms. At a higher level, numerous genera and subgenera have been employed by workers in an attempt to compartmentalize the family into well defined groups. At the present time, no satisfactory system has been demonstrated whereby such divisions can be justified, and to avoid compounding an already complex situation, the authors have described all the species within this guide under the single genus *Cypraea* Linné, 1758.

Of the 200 or so recent Cypraea species occurring worldwide, the Philippines has around 77 (or 39%) which is by far the highest species concentration for the family worldwide. In addition, the Philippines can boast to having the largest assortment of rare cowries including such collector's items as *Cypraea leucodon, valentia, porteri, guttata, langfordi* and *aurantium.*

Known locally as "sigay", a name encompassing most of the common species such as *moneta, annulus* and *tigris,* these highly glossed "jewels of the sea" have played an important part in Philippine culture, having been used as local currency, charms, talisman, decorations and as personal adornment. Today, they form an important source of raw material in the manufacture of numerous handicraft items. Apart from their natural beauty, it is only recently that the market values of certain species have increased due to the unprecedented popularity and demand by both shell collectors and overseas shell dealers.

1. Cypraea tigris Linné, 1758
A large, solid, heavy shell which grows to 160mm (Philippine specimens reach around 120mm); normally ovate to pyriform, dorsum inflated; white ground colour with a dense overlay of closely packed, irregularly sized black spots which are frequently encircled with a yellow-orange tinge, and which extend onto the sides; base flat and white, aperture narrow, widening anteriorly; teeth fine, short and numerous, mantle line prominent; shell variable in shape and surface colouration.
1a. typical form; 85.4mm; Eastern Samar.
1b. form *pardalis* Shaw, 1795; a pale form; 75.6mm; Sulu Archipelago.
1c. form *incana* Sullioti, 1924; an albinistic form; 76.9mm; Sulu Archipelago.
DISTRIBUTION: Throughout the Philippines.

2. Cypraea guttata Gmelin, 1791
66.3mm, 67.0mm; Cebu.
An extremely attractive, distinctly coloured shell which reaches around 70mm; shell pyriform-ovate, tips well produced, labral margin angulate; base convex, teeth prominent, extending over the base and partially up the dorsal wall; dorsal colour pale orange to light tan with white spots of varying sizes; mantle line slightly incised; teeth red with white interspaces; a long reddish brown callus on both labral and columellar surfaces; reaches 70mm.
DISTRIBUTION: Marinduque, Southern Mindanao, Cebu and Bohol.

3. Cypraea mappa Linné, 1758
A beautiful, distinctly marked shell reaching 90mm. Ovate, rather inflated, with slightly produced extremities; dorsum cream tan, overlaid with a dense brown network of longitudinal lines and spotting; mantle line large, irregularly shaped; base convex; color variable from cream, tan, brownish lavender, pink to almost red; aperture narrow, teeth fine, short, and restricted to the aperture.
3a. typical form; 77.8mm; Eastern Samar.
3b. unusual red form; side view showing colouration extending over entire shell; 78.2mm; Eastern Samar.
DISTRIBUTION: Throughout the Philippines but mainly from Samar.

4. Cypraea argus Linné, 1758
Shell large, cylindrical, cream-coloured with three broad, darker coloured bands and numerous brown rings of varying sizes and colour; aperture wide, teeth lined with brown; base with 3 or 4 large, brown, rectangular blotches; reaches 80mm.
4a. typical form; 69.3mm; Eastern Samar.
4b. what appears to be a dwarf race of the species; 48.9mm; Eastern Samar.
DISTRIBUTION: Mainly Eastern Samar, but sporadically found throughout the archipelago.

5. Cypraea ventriculus Lamarck, 1810
53.0mm; Eastern Samar.
Shell medium sized, ovate, reaching 70mm; margins with a heavy callus which extends onto the slightly depressed dorsal surface; aperture narrow, teeth short but well developed; dorsal colour reddish brown with 4 darker bands, frequently indistinct; marginal calluses chestnut brown with fine white lines; base white to brown; this species is normally found with structural

inclusions* on the dorsum, inclusion-free specimens being rare (*Inclusion — the term as applied here refers to the fine, hair-like fracture marks seen on the dorsal surface below the nacre. These are not structural flaws as such but represent a developmental peculiarity of the species).
DISTRIBUTION: Eastern Samar.

6. Cypraea vitellus Linné, 1758
71.9mm; Sulo Archipelago.
A medium sized, pyriform to pyriform-ovate shell reaching 80mm; dorsum colouration brown with numerous distinct, irregularly-sized white spots, and 2 or 3 pale bands; sides with fine white striations, base cream, aperture narrow, teeth short but strong.
DISTRIBUTION: Throughout the Philippines.

7. Cypraea talpa Linné, 1758
59.8mm, 77.5mm, 67.1mm; Eastern Samar.
A medium sized cylindrical shell which may reach 90mm; dorsum swollen, orange-yellow in colour with 4 dark brown bands; extremities slightly produced, especially the posterior labral margin; extremities, sides and base a deep chocolate brown; aperture slightly open and curved posteriorly; teeth numerous, short and brown stained with white interspaces.
DISTRIBUTION: Throughout the Philippines.

8. Cypraea mauritiana Linné, 1758
96.7mm; Samar.
A large, broad, heavy shell, with an inflated dorsum and rather depressed margins; reaches around 100mm; dorsal colour brownish black with pale tan spots; sides and base almost black, unspotted, with a bluish tinge; mantle line conspicuous; aperture narrow, teeth large and prominent, not extending over the base.
DISTRIBUTION: Throughout the Philippines but in limited quantity.

9. Cypraea testudinaria Linné, 1758
110.8mm; Samar.
A large, cylindrical shell reaching 130mm; dorsum creamy tan but overlaid with heavy black mottling and white spotting; extremities moderately produced, sides dark tan with large blackish spots; base flat, light tan in colour with a darker brown blotch on the central columellar surface; aperture narrow, widening anteriorly; teeth white, fine, numerous, short, and confined to the aperture; whole shell has minute white flecks in the enamel. Sulu Archipelago has specimens that are much darker in color and smaller in size.
DISTRIBUTION: Eastern Samar and Sulu Sea.

10. Cypraea valentia Perry, 1811
86.0mm, 83.0mm; Cebu.
A large, oval, heavily inflated shell reaching 100mm; extremities produced with folds; base convex, aperture narrow, widening anteriorly and curved posteriorly; teeth numerous but fine, restricted to the margin, those on the columellar margin smaller and more numerous than those on the labral surface; dorsal colour greenish brown with irregular brown blotches; spectacular specimens display a kaleidoscope of colours which are beyond description; sides tan with a lavenderish tinge and large, brown dots; base cream with a yellow to pinkish tinge.
DISTRIBUTION: Southern Mindanao, Cebu and Bohol.

11. Cypraea aurantium Gmelin, 1791
104mm, 94.0mm; Samar.
A large inflated shell reaching 120mm; dorsum a rich golden orange colour, the base white, and the teeth lined with orange; a most distinctive species. Only specimens coming from Eastern Samar have this unusual orange colour. Gem specimens are rare.
DISTRIBUTION: Eastern Samar, Davao and Masbate; mostly found on the eastern side of the Philippines.

Cypraeidae Plate 21

1. Cypraea hirasei Roberts, 1913
A solid, pyriform shell reaching around 60mm; extremities produced and margined; outer lip prominently extended and upswept; aperture narrow, teeth short but numerous; dorsal colour pale cream with a large, irregular reddish brown blotch and several smaller brown marks on the sides; base white, extremities brownish.
1a. juvenile, lacking characteristic blotch; 47.6mm; Lubang Is., Mindoro.
1b. typical form; 49.2mm, 46.5mm; Punta Engaño, Cebu.
DISTRIBUTION: Mindoro, Cebu, Bohol and Southern Mindanao.

2. Cypraea walkeri Sowerby, 1832
27.0mm, 30.8mm; Sulu Archipelago.
Shell medium sized, elongate-ovate, reaching 40mm; posterior labral margin produced, aperture wide with numerous fine, short teeth, those on the columellar surface enlarging anteriorly; the Philippine race (*surabajensis* Schilder, 1937), has a tan dorsum with fine brown spots, a strong, wide, continuous mid-dorsal band, and a heavily purple-stained columella and labral surface.
DISTRIBUTION: Sulu Sea.

3. Cypraea dayritiana Cate, 1963
19.2mm, 18.9mm; Coron, Palawan.
A small, ovate, moderately inflated shell reaching 25mm in length; aperture narrow, teeth few in number, rather coarse, and extending over the base;

PLATE NO. 20 1a. *Cypraea tigris* Linné 1b. form *pardalis* Shaw 1c. form *incana* Sullioti, albinistic form 2. *Cypraea guttata* Gmelin 3a. *Cypraea mappa* Linné 3b. unusual red form 4a. *Cypraea argus* Linné 4b. dwarf form 5. *Cypraea ventriculus* Lamarck 6. *Cypraea vitellus* Linné 7. *Cypraea talpa* Linné 8. *Cypraea mauritiana* Linné 9. *Cypraea testudinaria* Linné 10. *Cypraea valentia* Perry 11. *Cypraea aurantium* Gmelin

PLATE 20

dorsum greenish brown, mottled with brown dots and a dark spire spot, base white. Named after Fernando Dayrit, a prominent local collector.
DISTRIBUTION: Endemic to Northern Palawan, but a single recorded specimen from Bohol (Burgess: 1970).

4. Cypraea pallidula Gaskoin, 1849
21.8mm, 22.9mm; Samar.
Shell small, ovate to cylindrical, reaching 25mm; dorsum pale green with fine brown mottling and 3 to 4 dark bands, sides white with occasional brown spotting, base white; aperture narrow, teeth heavy and numerous, those on the columellar surface spreading halfway across the base.
DISTRIBUTION: Throughout the Philippines.

5. Cypraea moneta Linné, 1758
Shell small to medium sized, deltoidal in shape, with a slightly inflated dorsum, prominently widened margins, angulate sides, and a flat to convex base; aperture narrow, widening anteriorly, teeth short, solid and confined to the aperture; dorsum yellow with a fine, filamentous red ring, 2 to 3 pale green transverse bands and a white base; reaches 35mm.
5a. typical form; 31.5mm, 25.3mm; Palawan.
5b. a deep golden yellow specimen lacking the red ring; 23.0mm; Sulu Archipelago.
5c. an albinistic form; 22.5mm; Samar.
DISTRIBUTION: Throughout the Philippines.

6. Cypraea nucleus Linné, 1758
24.8mm, 25.5mm; Samar.
A small to medium sized shell, elongate-ovate in outline and somewhat inflated; glossy, pale yellow-brown in colour and ornamented with numerous irregularly shaped pustules which are interconnected by thin, slightly raised threads; dorsal sulcus deep, base convex, extremities produced; aperture narrow, teeth well defined and extending to the margin, intermediate teeth not reaching the aperture; reaches 30mm.
DISTRIBUTION: Throughout the Philippines, but mainly from Samar.

7. Cypraea annulus Linné, 1758
23.5mm, 21.5mm; Samar.
Shell small, solid, ovate, reaching around 25mm; dorsum bluish grey, sculptured with minute transverse hairlines and encircled with an orange ring; base concave, aperture wide with short, coarse teeth.
DISTRIBUTION: Throughout the Philippines.

8. Cypraea gracilis Gaskoin, 1849
21.5mm, 21.5mm; Eastern Samar.
A small, elongate-ovate shell, seldom exceeding 25mm; aperture narrow, widening anteriorly; labral teeth short, columellar teeth confined to the aperture; dorsum blue-grey with small brown dots and a large central blotch, sides cream with brown spots, extremities with a pair of brown spots.
DISTRIBUTION: Eastern Samar, Sulu Sea and Visayan Sea.

9. Cypraea clandestina Linné, 1767
19.0mm, 20.5mm; Samar.
Shell small, ovate to pyriform in outline, reaching around 25mm; tips prominent, aperture narrow posteriorly, widening anteriorly and with strong teeth; dorsum cream with pale white banding and fine, brown, zigzag lines, base creamy white.
DISTRIBUTION: Throughout the Philippines.

10. Cypraea helvola Linné, 1758
25.3mm, 24.5mm; Samar.
Shell medium sized, reaching around 35mm; broadly ovate and smooth, aperture narrow with strong teeth, slightly expanded on the labral surface; dorsum orange-brown with irregular brown and white spotting and a darker, peripheral dorsal ring; sides thickened and pitted, base tan to orange-brown.
DISTRIBUTION: Throughout the Philippines.

11. Cypraea limacina Lamarck, 1810
Shell small to medium sized, pyriform, reaching around 37mm; dorsum grey to purple-grey with white spots and frequently nodulose in adults; extremities orange-brown in colour, margins pitted; base convex and white; aperture narrow, labral teeth strong, extending to the margin, columellar teeth shorter; all teeth lined with brown.
11a. typical form; 31.0mm, 30.0mm; Eastern Samar.
11b. young specimen with a purple dorsum and lacking the white pustules; 25.0mm; Sulu Archipelago.
DISTRIBUTION: Throughout the Philippines.

12. Cypraea ovum Gmelin, 1791
23.2mm, 20.3mm; Palawan.
Shell medium sized, ovate to pyriform-ovate in outline, with a somewhat inflated dorsum which is finely spotted in greenish brown and occasionally brown-blotched; extremities produced, margins and base white, aperture narrow, widening anteriorly, teeth well defined, short and orange stained; reaches 40mm.
DISTRIBUTION: Throughout the Philippines in limited quantity.

PLATE NO. 21 1a. *Cypraea hirasei* Roberts, juvenile 1b. typical form 2. *Cypraea walkeri* Sowerby 3. *Cypraea dayritiana* Cate 4. *Cypraea pallidula* Gaskoin 5a. *Cypraea moneta* Linné 5b. a deep golden yellow specimen 5c. an albinistic form 6. *Cypraea nucleus* Linné 7. *Cypraea annulus* Linné 8. *Cypraea gracilis* Gaskoin 9. *Cypraea clandestina* Linné 10. *Cypraea helvola* Linné 11a. *Cypraea limacina* Lamarck 11b. juvenile 12. *Cypraea ovum* Gmelin 13. *Cypraea errones* Linné 14. *Cypraea poraria* Linné 15. *Cypraea cylindrica* Born 16. *Cypraea isabella* Linné 17. *Cypraea ziczac* Linné 18. *Cypraea quadrimaculata* Gray 19a. *Cypraea caputserpentis* Linné-unusual color form 19b. typical form 20. *Cypraea labrolineata* Gaskoin 21. *Cypraea asellus* Linné 22. *Cypraea erosa* Linné

PLATE 21

1a
1b
2
3
4
5a
5b
5c
6
7
8
9
10
11a
11b
12
13
14
15
16
17
18
19a
19b
20
21
22

13. Cypraea errones Linné, 1758
35.0mm, 28.7mm; Samar.
Shell medium sized, elongate-ovate, with moderately extended tips; aperture narrow, widening anteriorly, labral teeth short, columellar teeth weak and confined to the aperture; variable, usually with a greenish grey dorsum, an occasional central brown blotch, and 3 transverse bands; reaches 45mm.
DISTRIBUTION: Throughout the Philippines.

14. Cypraea poraria Linné, 1758
24.9mm, 23.2mm; Eastern Samar.
Shell small, oval to deltoidal-shaped with a slightly inflated dorsum, sides weakly pitted, and the base slightly convex; aperture narrow, slightly widening anteriorly, teeth coarse, those on the labral surface somewhat produced anteriorly and posteriorly; dorsal colour orange-brown with dark brown and white spots which are surrounded by violet rings, sides and base dark lavender, apertural teeth white; reaches 25mm.
DISTRIBUTION: Throughout the Philippines.

15. Cypraea cylindrica Born, 1778
32.0mm, 34.1mm; Eastern Samar.
Shell medium sized, cylindrically elongate and reaching around 45mm; tips moderately produced, base somewhat flat, aperture narrow posteriorly, widening anteriorly, with coarse labral teeth and weaker columellar teeth; dorsum bluish grey with a large central blotch and 2 spots at each tip, sides and base white and unspotted.
DISTRIBUTION: Throughout the Philippines.

16. Cypraea isabella Linné, 1758
37.9mm, 42.5mm; Palawan.
A medium sized, cylindrical shell reaching 60mm; dorsum yellow-grey with axially orientated black streaks; extremities moderately produced and spotted with orange, base white; aperture narrow, teeth short and numerous.
DISTRIBUTION: Throughout the Philippines.

17. Cypraea ziczac Linné, 1758
22.8mm, 22.8mm; Eastern Samar.
Shell small, pyriform and reaching 25mm; dorsum fawn to tan, crossed by 3 broad, white bands which are ornamented with brown sagittal markings; dorsum extremities, sides and base all orange-coloured and heavily spotted; aperture wide, labral teeth short and strong, columellar teeth almost obsolete posteriorly, becoming stronger anteriorly.
DISTRIBUTION: Throughout the Philippines.

18. Cypraea quadrimaculata Gray, 1824
27.3mm, 27.2mm; Eastern Samar.
Shell small, cylindrical to elongate-ovate and reaching 35mm; dorsum greenish brown, with fine brown speckling; extremities slightly produced with a pair of black terminal spots at each end; sides and base creamy white and unspotted; aperture narrow, teeth coarse, irregular, and produced across the base.
DISTRIBUTION: Throughout the Philippines.

19. Cypraea caputserpentis Linné, 1758
A medium sized, deltoidal, moderately domed shell which may reach 40mm; base flat with large heavy teeth; dorsum brown with a reticulate pattern of small white spots and a white blotch anteriorly and posteriorly; sides deep brown, base brownish white.
19a. an unusual burnt orange colour form; 31.6mm; Samar.
19b. typical form; 31.8mm, 27.9mm; Samar.
DISTRIBUTION: Throughout the Philippines.

20. Cypraea labrolineata Gaskoin, 1849
18.9mm, 18.5mm; Eastern Samar.
Shell small, elongate-ovate to cylindrical in shape, dorsum moderately inflated, margins thickened and pitted, base slightly convex, aperture narrow, widening anteriorly; teeth strong, labral ones extending almost to outer margin; base white with marginal brown spotting, dorsum olive green with white spots and a well-defined, light green mantle line; reaches 25mm.
DISTRIBUTION: Throughout the Philippines.

21. Cypraea asellus Linné, 1758
23.0mm, 24.5mm; Siasi, Sulu Archipelago.
A small, elongate-ovate shell, seldom exceeding 25mm; dorsum white with 3 broad black bands, sides and base white and unspotted; aperture narrow, with short, heavy teeth.
DISTRIBUTION: Throughout the Philippines.

22. Cypraea erosa Linné, 1758
37.9mm, 36.2mm; Cebu.
Shell medium sized, ovate to elongate-ovate, reaching 60mm; margins developed and produced, somewhat pitted; aperture narrow, teeth heavy, those on the labral surface produced across the base, columellar teeth restricted to the aperture; dorsum pale brown with white specks and dots and a pale green mantle line; base creamy white, sides with or without a brown blotch.
DISTRIBUTION: Throughout the Philippines.

Cypraeidae — Plate 22

1. Cypraea eglantina Duclos, 1833
69.0mm, 69.4mm; Sulu Archipelago.
A large, cylindrically ovate shell, moderately inflated, with rounded but not angulate sides, a convex base, and numerous thin, slightly produced teeth; dorsum colour brown to grey with irregular interrupted brown lines forming a reticulate pattern; dorsal-sulcus line distinct; base cream, spotted, teeth lined with brown; small spire blotch; reaches 70mm.
DISTRIBUTION: Throughout the Philippines.

2. Cypraea maculifera Schilder, 1932
52.5mm, 53.9mm; Samar.
A solid, heavy shell reaching 90mm; elongate-ovate, with prominent margins and extremities; dorsum deep chocolate brown with a reticulate network of large white spots and a prominent mantle line, sides greyish with large spots; base convex, aperture narrow with brown teeth, and a well defined brown blotch on the columellar surface.
DISTRIBUTION: Throughout the Philippines.

3. Cypraea arabica Linné, 1758
54.9mm, 57.6mm; Bohol.
Shell medium sized, solid, subcylindrical, reaching 80mm; sculptured with irregular interrupted brown lines over a pale cream-and-brown blotched dorsum, somewhat similar to that of *C. eglantina;* base flat, sides greyish with brown spots; labral teeth partially extended over the base, columellar teeth short, all teeth marked in brown.
DISTRIBUTION: Throughout the Philippines.

4. Cypraea depressa Gray, 1824
38.0mm, 34.2mm; Samar.
A medium sized, broadly ovate shell which may reach 60mm; dorsum depressed, margins well developed, base convex, teeth long and produced; dorsal colour white with an even reticulation of brown, sides grey with dense brown spotting, base greyish white, teeth reddish brown.
DISTRIBUTION: Throughout the Philippines.

5. Cypraea scurra Gmelin, 1791
48.5mm, 47.2mm; Eastern Samar.
Shell medium sized, cylindrical, reaching around 55mm; dorsum inflated, purple-grey in ground colour, overlain with a brown reticulate network, a faint mantle line, and 4 weak, barely visible brown bands; margins produced, pale orange-tan in colour with large purplish spots; base weakly convex, unspotted and orange-tan in colour; aperture narrow, widening anteriorly; teeth short, numerous and brown-stained.
DISTRIBUTION: Eastern Samar and Sulu Sea.

6. Cypraea caurica Linné, 1758
Shell medium sized, elongate to ovate in shape with well developed margins and heavy teeth; dorsum blue-white, heavily spotted with brown, and occasionally 3 pale bands may occur; sides pinkish with brown spots, base yellow to orange-brown, teeth white, pale orange-brown between; reaches 55mm.
6a. form *elongata* Perry, 1811; often considered an elongate subspecies of *caurica,* but occurs sympatrically with the normal form; 50.7mm; Sulu Archipelago.
6b. typical form; 37.2mm, 33.2mm; Samar.
DISTRIBUTION: Throughout the Philippines.

7. Cypraea miliaris Gmelin, 1791
Shell medium sized, elongate-ovate with prominent extremities; aperture narrow, dilating slightly anteriorly, with heavy, well defined teeth; dorsal colour of Philippine shells either yellow tan with white spots, or greyish green with darker spots, sides white and unspotted, pitted, and with a slight nacreous overlay which extends up from the white-coloured base; reaches 50mm.
7a. typical greyish green form; 32.2mm; Cebu.
7b. yellow-tan form; 41.5mm, 45.7mm; Manila Bay.
DISTRIBUTION: Throughout the Philippines.

8. Cypraea chinensis Gmelin, 1791
40.2mm, 37.5mm; Bohol.
Shell medium sized, ovate to elongate-ovate, margins well developed, base convex, teeth heavy, especially those on the labral wall; dorsum tan with cream reticulations, margins with violet spotting extending to the base, teeth cream with orange-stained interspaces; reaches 50mm.
DISTRIBUTION: Throughout the Philippines.

9. Cypraea teres Gmelin, 1791
43.3mm, 38.0mm; Bohol.
Shell small to medium sized, cylindrical to elongate-ovate in outline and reaching around 45mm; dorsum pale bluish grey with brown mottling that is arranged in 3 weak bands, the central one strongest; occasionally, a large central blotch is present; extremities moderately produced, sides greyish white with a few brown spots, base white; aperture moderately wide, teeth small and short, restricted to the aperture on the columellar side.
DISTRIBUTION: Throughout the Philippines.

10. Cypraea pyriformis Gray, 1824
31.2mm, 28.3mm; Manila Bay.
Shell small to medium sized, pyriform, reaching 35mm; dorsum slightly inflated, greenish white with numerous brown spots and an interrupted central brown band; extremities moderately produced with brown blotching on both the anterior and posterior ends; base convex, aperture slightly open, labral teeth short and coarse, columellar teeth brown-stained and produced on the posterior columellar surface.
DISTRIBUTION: Tayabas Bay (Quezon), Manila Bay, Cebu and Bohol.

11. Cypraea langfordi Kuroda, 1938
62.5mm, 52.0mm; Punta Engaño, Cebu.
Shell pyriform to elongate-ovate, dorsum rather swollen, orange-white in colour with dense brown mottling; sides and base a deep orange-brown colour; extremities well produced, aperture narrow with numerous fine weak teeth; reaches around 65mm.
DISTRIBUTION: Cebu, Bohol and Southern Mindanao.

12. Cypraea pulchella Swainson, 1823
41.9mm, 38.7mm; Bohol.
Shell small to medium sized, lightweight, elongate-ovate, reaching 45mm; dorsum greenish white, with numerous fine brown dots, 2 or 3 interrupted brown-coloured transverse bands and an occasional mid-dorsal brown blotch; extremities moderately produced

and brown-stained, sides with brown blotches, base cream with brown-stained teeth, those on the columellar surface extending across the base.
DISTRIBUTION: Tayabas Bay (Quezon), Manila Bay, Bohol and Cebu.

13. Cypraea carneola Linné, 1758
Shell medium sized, cylindrically ovate, inflated and lightweight, with an orange-tan dorsum crossed by 4 darker bands; base pale tan, aperture narrow, teeth marked in violet; may reach 95mm.
13a. typical form; 46.9mm, 43.1mm; Samar.
13b. form *propinqua* Garrett, 1879; a small, ovate, rather inflated form with lateral calluses; 31.2mm; Samar.
13c. form *leviathan* Schilder and Schilder, 1937; an unusually large form with heavily nodular, calloused margins; 81.0mm; Samar.
DISTRIBUTION: Throughout the Philippines.

14. Cypraea boivinii Kiener, 1843
29.0mm, 36.9mm; Cebu.
Shell medium sized, deltoidal to cylindrically elongate, seldom exceeding 35mm; dorsum colour off-white with a bluish tinge and small bluish spots, and a brown blotch on the spire; base white, aperture narrow, teeth heavy; dorsum occasionally with a mantle line.
DISTRIBUTION: Throughout the Philippines.

15. Cypraea porteri Cate, 1966
48.2mm, 46.9mm; Punta Engaño, Cebu.
Shell medium sized, heavy, globose, reaching 60mm; dorsum inflated, cream coloured with heavy brown mottling and dots; margins heavy, occasionally spotted, extremities moderately produced; base convex, orange-brown in colour, aperture narrow teeth short and coarse, those on the labral surface extending slightly across the base.
DISTRIBUTION: Cebu and Southern Mindanao.

Cypraeidae — Plate 23

1. Cypraea beckii Gaskoin, 1836
12.0mm, 12.3mm; Eastern Samar.
A small, inflated, cylindrical shell reaching 17mm; dorsum light yellow with small, dark brown spots in white rings; labral side pitted, columellar teeth reduced and fused anteriorly; shell margins brown spotted, teeth lined with brown.
DISTRIBUTION: Eastern Samar, Sulu Sea and Punta Engaño, Cebu.

2. Cypraea martini Schepman, 1907
15.3mm, 14.8mm; Sulu Archipelago.
A small, pyriform to elongate shell reaching 20mm; dorsum moderately inflated, extremities produced, especially the posterior one which is curved; right side of the dorsum sharply margined and curved upwards; aperture narrow, teeth fine and numerous, base convex; dorsal colour off-white to brown with 4 weak transverse bands, sides with fine brown spots, teeth stained with brown or purple.
DISTRIBUTION: Palawan, Zambales, Sulu Sea, and Punta Engaño, Cebu.

3. Cypraea contaminata Sowerby, 1832
14.0mm, 12.9mm; Laminusa, Sulu Archipelago.
Shell small, pyriform, seldom exceeding 15mm; aperture narrow, labral teeth extended across the base, columellar teeth somewhat shorter; posterior end of inner lip margin produced; dorsum off-white with 3 pale, transverse bands, a large central brown blotch, and numerous smaller brown dots; base white with brown spots.
DISTRIBUTION: Sulu Sea and Punta Engaño, Cebu.

4. Cypraea fimbriata Gmelin, 1791
13.0mm, 12.9mm; Samar.
Shell small, cylindrical, tips slightly produced, aperture narrow, widening anteriorly, teeth fine and numerous; dorsum grey to orange-brown with small brown blotches and 3 bands, the central one strongest; base white, sides with brown spots; violet staining on both anterior and posterior apertural openings, spire with a brown blotch; reaches 22mm.
DISTRIBUTION: Marinduque, Eastern Samar and Sulu Sea.

5. Cypraea microdon Gray, 1828
11.0mm, 10.7mm; Eastern Samar.
Shell small, pyriform to cylindrical, reaching 15mm; dorsum moderately inflated, orange-brown with 3 discontinuous, brown-coloured transverse bands and brown spotting; extremities produced with purple blotches, sides and base white; aperture uniformly narrow with numerous small teeth confined to the aperture.

PLATE NO. 22 1. *Cypraea eglantina* Duclos 2. *Cypraea maculifera* Schilder 3. *Cypraea arabica* Linné 4. *Cypraea depressa* Gray 5. *Cypraea scurra* Gmelin 6a. *Cypraea caurica* Linné, form *elongata* Perry 6b. typical form 7a. *Cypraea miliaris* Gmelin 7b. yellow form 8. *Cypraea chinensis* Gmelin 9. *Cypraea teres* Gmelin 10. *Cypraea pyriformis* Gray 11. *Cypraea langfordi* Kuroda 12. *Cypraea pulchella* Swainson 13a. *Cypraea carneola* Linné 13b. form *propinqua* Garrett 13c. form *leviathan* Schilder & Schilder 14. *Cypraea boivinii* Kiener 15. *Cypraea porteri* Cate

PLATE 22

DISTRIBUTION: Marinduque, Eastern Samar and Sulu Sea.

6. Cypraea hammondae Iredale, 1939
14.5mm, 14.3mm; Sulu Archipelago.
Shell small, reaching 20mm; similar to *C. fimbriata* Gmelin, but more pyriformly ovate; dorsum white to pale lilac with small brown spots, denser on the sides and crossed by several discontinuous, brown-coloured transverse bands; spire with a brown blotch and the terminals are lilac-coloured; aperture narrow, labral teeth strong, columellar teeth weak and confined to the aperture; Philippine race is *C. raysummersi* Schilder, 1960.
DISTRIBUTION: Eastern Samar and Sulu Sea.

7. Cypraea goodalli Sowerby, 1832
11.2mm, 10.4mm; Samar.
Shell small, elongate-ovate, reaching 20mm; tips produced, especially the posterior end of the labral surface; aperture narrow, labral teeth short, columellar teeth greatly reduced and confined to the aperture; dorsum smooth, white in colour with a central orange-brown blotch, sides and base sparsely spotted, and the terminals stained with small brown blotches. Scarce.
DISTRIBUTION: Eastern Samar.

8. Cypraea punctata Linné, 1771
13.9mm, 13.5mm; Samar.
Shell small, elongate-ovate to pyriform, reaching 25mm; dorsum white to yellowish with scattered brown to black spots; extremities moderately produced, frequently with a pair of terminal spots; base white, teeth long, thin, and stained with yellow.
DISTRIBUTION: Eastern Samar, Sulu Sea and Visayan Sea.

9. Cypraea felina Gmelin, 1791
17.8mm, 18.2mm; Eastern Samar.
Shell small, ovate to elongate-ovate, tips moderately produced, base convex, aperture narrow with moderately developed teeth; dorsum grey to bluish green with fine brown blotches and 2 to 3 obsolete transverse bands; base and teeth white, 2 brown spots on both anterior and posterior tips; reaches 30mm.
DISTRIBUTION: Eastern Samar, Sulu Sea and Visayan Sea.

10. Cypraea rabaulensis Schilder, 1964
20.8mm, 20.9mm; Mactan Is., Cebu.
Shell small, pyriform-elongate, reaching 25mm; dorsum light brown with 4 narrow, discontinuous brown-coloured bands and some spotting; extremities produced, especially the anterior one which is flared; sides and extremities with brown spots, base off-white; aperture narrow, widening anteriorly, teeth short and white in colour; seldom taken in good condition.
DISTRIBUTION: Cebu.

11. Cypraea margarita Dillwyn, 1817
17.6mm, 16.9mm; Samar.
A small, inflated, rather humped shell, reaching 20mm; extremities beaked and orange tinged; dorsum smooth, pale to dark brown, with a brown stained depression posteriorly; base slightly convex and yellowish brown in colour; aperture narrow, teeth fine, extending to the margins, central labral teeth very short.
DISTRIBUTION: Eastern Samar and Sulu Sea.

12. Cypraea bistrinotata Schilder & Schilder, 1937
22.5mm, 23.3mm; Eastern Samar.
Shell small, roundly inflated, seldom exceeding 20mm; extremities beaked, dorsum granular with a prominent dorsal sulcus; aperture narrow, lined with numerous fine teeth which extend to the margins; dorsum yellow-brown with 3 pairs of brown spots which are split by a sulcus, numerous smaller brown spots, and 2 pairs of brown spots on the base; teeth lined with brown.
DISTRIBUTION: Eastern Samar and Sulu Sea.

13. Cypraea globulus Linné, 1758
19.3mm, 16.7mm; Eastern Samar.
Shell small, globular, rather humped, reaching 25mm; extremities beaked, aperture narrow, teeth numerous and produced but not to the margins; colour dark orange-brown with numerous dark spots and 4 large blotches on the base; dorsal surface smooth, mantle line obsolete.
DISTRIBUTION: Sulu Sea and Eastern Samar.

14. Cypraea cicercula Linné, 1758
20.8mm, 19.7mm; Eastern Samar.
Shell small, ovate, well inflated, similar to *C. bistrinotata*, with beaked extremities, a dorsum which may or may not be granular, a narrow aperture, and fine teeth extending over the base; dorsum light yellow with brown spots, a prominent sulcus, and a dark blotch over the extreme posterior.
(Notatum: Schilder (1966), has claimed that the true *cicercula* was based on *bistrinotata*, and, if correct, the *cicercula* of authors must take the next available name, *lienardi* Jousseaume, 1874).
DISTRIBUTION: Eastern Samar and Sulu Sea.

PLATE NO. 23 1. *Cypraea beckii* Gaskoin 2. *Cypraea martini* Schepman 3. *Cypraea contaminata* Sowerby 4. *Cypraea fimbriata* Gmelin 5. *Cypraea microdon* Gray 6. *Cypraea hammondae* Iredale 7. *Cypraea goodalli* Sowerby 8. *Cypraea punctata* Linné 9. *Cypraea felina* Gmelin 10. *Cypraea rabaulensis* Schilder 11. *Cypraea margarita* Dillwyn 12. *Cypraea bistrinotata* Schilder & Schilder 13. *Cypraea globulus* Linné 14. *Cypraea cicercula* Linné 15. *Cypraea mariae* Schilder 16. *Cypraea ursellus* Gmelin 17. *Cypraea kieneri* Hidalgo 18. *Cypraea hirundo* Linné 19. *Cypraea lutea* Gmelin 20a. *Cypraea staphylaea* Linné, form *descripta* Iredale 20b. typical form 21. *Cypraea childreni* Gray

PLATE 23

15. Cypraea mariae Schilder, 1927
17.5mm, 16.7mm; Eastern Samar.
A small, pyriform-ovate shell reaching 20mm; dorsum inflated, off-white in colour and marked with brown-ringed, orange-yellow spots which extend over the sides onto a white base; aperture narrow, teeth fine and numerous; seldom live-taken, most specimens being washed ashore after a typhoon.
DISTRIBUTION: Eastern Samar and Sulu Sea.

16. Cypraea ursellus Gmelin, 1791
16.0mm, 17.6mm; Samar.
Shell small, elongate-ovate, reaching 20mm; extremities produced and marked with 4 black dots; dorsum pale yellowish brown with 3 greyish blue bands and sparse brown spots; sides and base white, aperture narrow, teeth numerous, strong and produced almost to the margins; distinguished from *C. hirundo* and *C. kieneri* by the teeth on the beak extremities.
DISTRIBUTION: Eastern Samar, Sulu Sea and Marinduque.

17. Cypraea kieneri Hidalgo, 1906
17.5mm, 12.0mm; Eastern Samar.
Shell small, cylindrically elongate, reaching 23mm; aperture narrow, teeth well marked; columellar teeth short anteriorly, becoming longer posteriorly; dorsum with large, blue-grey blotches usually in 3 bands and small orange-brown spots; sides white with larger brown spotting, base white.
DISTRIBUTION: Visayan Sea, Sulu Sea, Eastern Samar and Marinduque.

18. Cypraea hirundo Linné, 1758
21.1mm, 18.7mm; Samar.
Shell small, solid, cylindrically-ovate, reaching 23mm; extremeties moderately produced, aperture narrow, teeth strong, extending across the base almost to the margins; dorsum bluish white with irregular brown spots, a central brown blotch, some occasional banding, and a pair of spots on each extremity; sides white with brown spotting, base white.
DISTRIBUTION: Visayan Sea, Sulu Sea, Eastern Samar and Marinduque.

19. Cypraea lutea Gmelin, 1791
24.5mm, 22.9mm; Sulu Archipelago.
Shell small, elongate-ovate, reaching 25mm; dorsum olive-brown with weak brown spotting and 2 strong white bands dividing the shell into 3 equal segments; base convex and orange-coloured with dark brown spots; aperture narrow, labral teeth strong, columellar teeth weaker and confined to the aperture.
DISTRIBUTION: Throughout the Philippines.

20. Cypraea staphylaea Linné, 1758
Shell small, ovate to elongate-ovate, reaching 27mm; dorsum grey to purple with numerous lighter coloured pustules and a well defined mantle line; extremities moderately produced, pitted and orange-coloured; base deep grey to white, teeth brown, lined with dark brown and extending to the margins, interstitial teeth not reaching the aperture.
20a. form *descripta* Iredale, 1955: a somewhat paler, smoother, more globose form; 17.7mm; Eastern Samar.
20b. typical form; 19.6mm, 19.7mm, Bohol.
DISTRIBUTION: Sulu Sea, Visayan Sea and Eastern Samar.

21. Cypraea childreni Gray, 1825
23.5mm, 25.8mm; Samar.
Shell small to medium sized, ovate with beaked extremities, a flat base, a narrow aperture, and numerous fine teeth which extend across the base, encircle the shell, and meet on the mid-line of the dorsum where they form a sulcus; yellow-brown in colour, ribs slightly darker; reaches 25mm. So far, only known from beach specimens washed ashore after a typhoon.
DISTRIBUTION: Eastern Samar and Sulu Sea.

Cypraeidae Plate 24

1. Cypraea lynx Linné, 1758
Shell medium sized, ovate to elongate-ovate, occasionally almost cylindrical and reaching 65mm; dorsum moderately inflated, brown to purplish coloured with large black spots, smaller, lighter-coloured speckles, an orange-yellow mantle line (occasionally absent), spotted sides and a white base; aperture narrow with large, short white teeth which have red interspaces. Very variable.
1a. typical form; 41.0mm, 37.3mm; Eastern Samar.
1b. form *vanelli* Linné, 1758; a large cylindrical form; 67.2mm; Eastern Samar.
DISTRIBUTION: Throughout the Philippines.

2. Cypraea onyx Linné, 1758
Shell medium sized, ovate to pyriform-ovate with an inflated dorsum; extremities slightly produced, especially the posterior labral margin; base convex, aperture narrow, widening anteriorly; teeth short but strong; dorsum pale to dark brown with narrow, paler banding and a bluish white oval band around the periphery; sides and base dark brown to almost jet black; reaches 40mm.
2a. dwarf form; 22.7mm; known only from Bataan.
2b. typical form; 36.0mm, 32.3mm; Sulu Archipelago.

PLATE NO. 24 1a. *Cypraea lynx* Linné 1b. form *vanelli* Linné 2a. *Cypraea onyx* Linné (dwarf form) 2b. typical form 3. *Cypraea cribraria* Linné 4a. *Cypraea stolida* Linné 4b. ecomorph 5a. *Cypraea saulae* Gaskoin form *visayensis* Massilia (typical form) 5b. form *manilensis* Massilia 5c. form *siasiensis* Cate 5d. albinistic form 6. *Ovula costellata* Lamarck 7a. *Ovula ovum* (Linné) 7b. dwarf form 8a. *Calpurnus lacteus lacteus* (Lamarck) 8b. *s.s. semistriatus* (Pease) 9. *Calpurnus verrucosus* (Linné) 10. *Pseudosimnia marginata* (Sowerby I) 11. *Volva volva volva* (Linné)

PLATE 24

DISTRIBUTION: Throughout the Philippines but mainly from the Sulu Sea.

3. Cypraea cribraria Linné, 1758
35.7mm, 36.9mm; Eastern Samar.
Shell medium sized, ovate to elongate-ovate, base convex, aperture narrow, labral teeth heavy, columellar teeth smaller and less pronounced; dorsal colouration reddish brown with large white spots, sides and base white; reaches 40mm.
DISTRIBUTION: Throughout the Philippines.

4. Cypraea stolida Linné, 1758
Shell small to medium sized, reaching 40mm; ovate to elongate-ovate with a bluish grey dorsum, a central dark brown blotch, and additional side blotching; dorsum also with smaller brown spots, extending to the sides; extremities moderately produced with brown spotting; teeth prominent, produced, with slight orange-brown staining; base off-white. Variable from different localities.
4a. typical form; 33.9mm, 30.9mm; Sulu Archipelago.
4b. a more ovate form, lacking dorsal blotch and with unstained teeth; 25.3mm; Eastern Samar.
DISTRIBUTION: Throughout the Philippines but mainly from Eastern Samar and the Sulu Sea.

5. Cypraea saulae Gaskoin, 1843
The Philippine race of this species is *Cypraea saulae saulae* Gaskoin, 1843; a small pyriform to elongate shell reaching 30mm; posterior end of the labral margin produced, base slightly convex, aperture narrow, widening anteriorly, the teeth short, coarse and white coloured with orange interspaces.
5a. form *visayensis* Massilia, 1978; the largest form with an ivory dorsum, small brown dots, and 2 or more central blotches; this is the typical form; 34.9mm, 28.4mm; Cebu.
5b. form *manilensis* Massilia, 1978; a small, pear-shaped form with an ivory dorsum, small brown dots, larger central blotch(es), and dark tips; 22.1mm, 21.7mm; Corregidor Is., Bataan.
5c. form *siasiensis* Cate, 1960; ivory dorsum, small brown dots, one compact central blotch and a yellowish tint; 21.5mm, 22.0mm; Siasi, Sulu Archipelago.
5d. albinistic form; 28.0mm; Sulu Archipelago.
DISTRIBUTION: Bataan, Visayan Sea and Sulu Sea.

Class Gastropoda
Subclass Prosobranchia
Order Mesogastropoda
Suborder Taenioglossa
Superfamily Cypraeacea
Family OVULIDAE

A relatively large group of small to medium sized carnivorous gastropods. They may be cowry-shaped or spindle-shaped, have a smooth, porcelaneous surface, are usually white or pinkish in surface colouration, have large, narrow apertures which dilate anteriorly, a thick, occasionally dentate outer lip, and a smooth, rounded inner lip. Usually found in the deep waters of most temperate and tropical seas inhabiting corals and gorgonoids. Taxonomically speaking, this is an extraordinarily complex group to work with, and only a few of the many Philippine species are covered here. Recently, numerous new forms have been taken at Cebu, Davao and Bohol, and are currently under study.
Reference: Cate, C. 1973: A Systematic Revision of the Recent Cypraeid Family Ovulidae. *The Veliger*, Vol. 15 (supplement), 116pp., 251 figs.

6. Ovula costellata Lamarck, 1810
36.7mm, 41.5mm; Sulu Archipelago.
Shell moderately large, pyriform, rather inflated, with an obsolete dorsal carina, axial growth lines, and slightly raised transverse growth striae; extremities beaked, posterior canal opening to the side; outer lip thickly calloused, crenulate, columellar teeth absent; pure white with a lavender interior; reaches 50mm.
DISTRIBUTION: Tayabas Bay (Quezon), Cebu and Sulu Sea.

7. Ovula ovum (Linné, 1758)
Shell large, ovate, pure white, with an inflated dorsum and sculptured with obsolete axial and transverse growth striae; terminals extended, blunt and roundly beaked; outer lip thick, broad and crenulate across the entire surface; columellar surface smooth and rounded; interior pink to reddish brown; reaches 100mm.
7a. typical form; 98.0mm; Samar.
7b. dwarf form; 52mm; Sulu Archipelago.
DISTRIBUTION: Throughout the Philippines.

8. Calpurnus (Procalpurnus) lacteus lacteus (Lamarck, 1810)
Shell small, ovate to pyriform, somewhat inflated and sculptured with fine transverse striae; terminals short, non-extended, outer lip thick, irregularly dentate; aperture narrow, curving posteriorly, widening anteriorly; columella smooth, fossula well defined; pure white; reaches 20mm.
8a. typical form; 19.6mm, 19.5mm; Bohol.
8b. *s.s. semistriatus* (Pease, 1862); 14.2mm; Northern Luzon. This subspecies occurs in Japan, but its occurrence in Northern Philippine waters is not unexpected; differs from the nominate subspecies in its smaller size, more elongate shape, more regularly curved apertural opening, its weaker outer lip dentition, and its non-constricted, posterior apertural opening.
DISTRIBUTION: Nominate subspecies widely distributed all over the archipelago; *s.s. semistriatus* only known from Northern Luzon.

9. Calpurnus (Calpurnus) verrucosus (Linné, 1758)
28.2mm, 30.4mm; Samar.
Shell medium sized, rhomboidal, with a tall, humped carina and fine axial and transverse growth striae; terminals slightly projecting with an umbilical

"button" at both ends between the terminals; outer lip with numerous coarse teeth, columella smooth; pure white with a brown line around the umbilical knobs, and lavender-tinged terminals; reaches 35mm.
DISTRIBUTION: Throughout the Philippines.

10. Pseudosimnia (Inflatovula) marginata (Sowerby I, 1828)
20.9mm, 20.7mm, 16.9mm; Tayabas Bay, Quezon.
Shell small, pyriform-ovate with an inflated dorsum crossed by fine, slightly elevated transverse striae; outer lip thickened, rounded, irregularly dentate on the inner margin and projecting out from the shell and to the left at the posterior end; columella smooth, a solid funiculum posteriorly, and a deep fossular cleft anteriorly; white to pinkish with an encircling orange-yellow marginal ring; reaches 25mm.
DISTRIBUTION: Tayabas Bay (Quezon) and Visayan Sea.

11. Volva volva volva (Linné, 1758)
126mm, 134mm; Tayabas Bay, Quezon.
A large, centrally ovate, bulbously enlarged shell, with long, narrow, gently recurved terminals; body whorl smooth or weakly striate and pinkish grey in colour; terminals with fine spiral striae, outer lip thickened, aperture widening anteriorly, columellar surface rounded and smooth; reaches 140mm.
DISTRIBUTION: Manila Bay, Tayabas Bay (Quezon) and Visayan Sea.

Plate 25 Ovulidae

1. Phenacovolva (Phenacovolva) birostris (Linné, 1767)
45.2mm, 47.4mm; Sulu Archipelago.
Shell medium sized, elongate-ovate, glossy; terminals long in Philippine specimens, attenuate and tapering in a lanceolate fashion; dorsum smooth but spirally striate on the terminals; aperture narrow, widening anteriorly; outer lip thick, weakly crenulate and sharply constricted anteriorly, base smooth and ovate; usually pinkish lavender, with orange terminals and a pinkish orange outer lip; reaches 45mm.
DISTRIBUTION: Marinduque, Tayabas Bay (Quezon), Visayan Sea, Sulu Sea and Manila Bay.

2. Phenacovolva (Turbovula) brevirostris (Schumacher, 1817)
Syn. *Ovula sowerbyana* Weinkauff, Mart. & Chem., 1881
36.7mm, 36.0mm; Manila Bay.
Shell medium sized, elongate-ovate, solid; terminals short and straight with incised concentric lines; aperture narrow, dilating anteriorly; outer lip thickened, shouldered and weakly crenulate; colour rosy pink with a pale, centrally placed dorsal band, terminals orange; reaches 40mm.
DISTRIBUTION: Manila Bay, Marinduque, Tayabas Bay (Quezon), Visayan Sea and Sulu Sea.

3. Phenacovolva (Calcaria) tokioi Cate, 1973
46.0mm, 43.9mm; Punta Engaño, Cebu.
Shell large for the genus, elongate, narrow, posterior terminal longest and curved to the left, anterior terminal shorter and straight; terminals with fine spiral striae, dorsum smooth, columellar surface smooth, outer lip weakly crenulate and thickly calloused centrally, less so towards the extremities; aperture narrow, widening anteriorly; pinkish orange to brownish red, with a pale central band, deep orange terminals, and a white outer lip; interior orange to pink; reaches 50mm.
DISTRIBUTION: South China Sea, Camotes Sea and Sulu Sea.

4. Phenacovolva (Phenacovolva) rosea rosea (A. Adams, 1854)
32.4mm, 27.5mm; Sulu Archipelago.
Shell medium sized, elongate-ovate, terminals long and attenuate, incised with fine spiral striae; dorsum smooth, outer lip thickened, weakly crenulate and sharply angled anteriorly; columella smooth with a funicular welt posteriorly and sharply constricted anteriorly; pinkish lavender with a white central band, orange tips, and off-white margins and base; reaches 45mm.
DISTRIBUTION: Camotes Sea and Sulu Sea.

5. Phenacovolva (Calcaria) longirostrata (Sowerby I, 1828)
57.9mm, 44.0mm; Sulu Archipelago.
Shell long, slender, smooth, glossy white; terminals long, slender, sharply acute and curved to the left; outer lip thickened, edentulate and angulated anteriorly; aperture narrow, widening anteriorly; tips occasionally brown coloured; reaches 60mm.
DISTRIBUTION: Camotes Sea, Tayabas Bay (Quezon) and Sulu Sea.

6. Phenacovolva (Calcaria) wakayamaensis Cate & Azuma, 1973
22.0mm, 44.3mm; Punta Engaño, Cebu.
Close to *P. longirostrata*, but differs in its more fragile construction; the body whorl is thin, almost translucent, the terminals are longitudinally aligned (or nearly so) and not recurved, and the outer lip is extremely thin; the dorsal striations on the terminal beaks of *longirostrata* are present in the Philippine forms of *wakayamaensis*, though apparently absent in the Japanese forms; reaches 60mm.
DISTRIBUTION: Camotes Sea.

7. Prionovula pudica wilsoniana Cate, 1973
13.0mm, 14.4mm; Punta Engaño, Cebu.
Shell small, lightweight, ovate in shape; dorsum glossy with faint axial growth striae, beaks weakly produced, base rounded with a solid carinal ridge and

a funiculum posteriorly on the columellar surface; posterior beak curved left adapically, straight anteriorly; outer lip thick but flattened and marginally crenulate; aperture wide, widening anteriorly; dorsum pinkish lavender with paler banding, outer lip margin and aperture white; reaches 16mm.
DISTRIBUTION: Camotes Sea.

8. Dentiovula masaoi Cate, 1973
15.5mm; Manila Bay.
Shell small, sub-rhomboidal, body whorl swollen subcentrally forming a distinct carinal "hump"; surface covered with numerous incised spiral striae; terminals produced and open, the posterior one curved to the left adapically; funiculum pronounced, columellar wall depressed and flattened; outer lip thickened, flattened and dentate posteriorly; aperture narrow, widening anteriorly; rose-pink in colour with a white base and outer lip margin, and a yellowish dorsal band; reaches 15mm.
DISTRIBUTION: Manila Bay, Camotes Sea and Tayabas Bay (Quezon).

9. Phenacovolva (Pellasimnia) subreflexa (A. Adams & Reeve, 1848)
30.5mm; Manila Bay.
Shell small, elongated, with slightly produced, somewhat truncated terminals; dorsum smooth but becoming finely striated on the terminals; outer lip thickened, becoming slightly thinner towards the base where it is reflexed, causing an apertural dilation; columella smooth, lacking a funiculum near the anterior opening; dorsal colouration pinkish orange with a faint yellow line encircling the shell on the outer lip margin; base white, interior pale reddish brown; reaches 30mm.
DISTRIBUTION: Manila Bay, Tayabas Bay (Quezon), Camotes Sea and Sulu Sea.

10. Crenavolva (Crenavolva) rosewateri Cate, 1973
8.8mm, 8.5mm; Manila bay.
Shell small, elongate-ovate and humped centrally; dorsum finely striated and characteristically pinkish brown in colour with distinct white mottling; beaks moderately produced, widely open and orange coloured; posterior inner lip margin slightly longer than the outer, outer lip moderately thickened and weakly crenulate; aperture narrow, widening anteriorly; reaches 10mm.
DISTRIBUTION: Manila Bay, Tayabas Bay (Quezon), Camotes Sea and Sulu Sea.

11. Prosimnia semperi semperi (Weinkauff, 1881)
9.8mm; Manila Bay.
Shell small and cylindrically elongate; beaks short and blunt, dorsum rough, irregular and striated; outer lip thickened but flat, and weakly crenulate; aperture narrow, widening anteriorly, funicular area weakly dentate; greenish yellow with darker tips; reaches 12mm.
DISTRIBUTION: Zambales, Manila Bay and Camotes Sea.

12. Crenavolva (Crenavolva) striatula traillii (A. Adams, 1855)
12.9mm, 12.7mm; Punta Engaño, Cebu.
Shell small, elongate-ovate, slightly swollen centrally; terminals produced, open, thickened and flaring outwards, the left more so than the right; dorsum finely striated, aperture narrow, widening anteriorly; outer lip thickened, strongly dentate posteriorly; colour variable, from red to pinkish orange with a pale dorsal band, whitish margins and pinkish tips; reaches 15mm.
DISTRIBUTION: Camotes Sea and Sulu Sea.

13. Dentiovula sp.
9.8mm, 10.7mm, 10.3mm; Punta Engaño, Cebu.
Shell small, sub-rhomboidal, with a broad, rounded, subcentral shoulder abruptly sloping adapically, more gradually abapically; surface with numerous spiral striae; terminals produced, truncated and open, with 3 or 4 marginal teeth on the right adapical terminal; aperture narrow, constricted adapically, outer lip flat and dentate, a heavy callus on the base; dorsum pinkish purple, extremities and terminals paler, with 3 fine yellow bands across the dorsum and around the margin; reaches 12mm.
Appears related to *D. masaoi* (fig. 8).
DISTRIBUTION: Camotes Sea.

14. Sandalia rhodia (A. Adams, 1854)
9.0mm, 8.8mm; Sulu Archipelago.
Shell small, elongate-ovate with a smooth, slightly inflated dorsum; base convex with a large funiculum on the posterior columellar wall; posterior tips produced, open and spatulate, anterior tips weakly produced; outer lip thickened, weakly dentate posteriorly; aperture wide, increasing in width anteriorly; dorsum rosy pink, funiculum and outer lip margin white; reaches 18mm.
DISTRIBUTION: Sulu Sea and Camotes Sea.

15. Delonovolva formosa (A. Adams & Reeve, 1848)
9.0mm, 8.9mm; Punta Engaño, Cebu.
Shell small, elongate-ovate, well inflated centrally,

PLATE NO. 25 1. *Phenacovolva birostris* (Linné) 2. *Phenacovolva brevirostris* (Schumacher) 3. *Phenacovolva tokioi* Cate 4. *Phenacovolva rosea rosea* (A. Adams) 5. *Phenacovolva longirostrata* (Sowerby I) 6. *Phenacovolva wakayamaensis* Cate & Azuma 7. *Prionovolva pudica wilsoniana* Cate 8. *Dentiovula masaoi* Cate 9. *Phenacovolva subreflexa* (A. Adams & Reeve) 10. *Crenavolva rosewateri* Cate 11. *Prosimnia semperi semperi* (Weinkauff) 12. *Crenavolva striatula traillii* (A. Adams) 13. *Dentiovula sp.* 14. *Sandalia rhodia* (A. Adams) 15. *Delonovolva formosa* (A. Adams & Reeve) 16. *Pseudosimnia punctata* (Duclos) 17. *Testudovolva nebula* (Azuma & Cate) 18. *Pseudosimnia whitworthi* Cate 19. *Hiata depressa* (Sowerby III) 20. *Trivirostra oryza oryza* (Lamarck)

PLATE 25

tapering anteriorly and sharply rostrated posteriorly; adapical terminal spatulate, with a large funicular callus; outer lip thick, flattened, heavily crenulate posteriorly; aperture narrow, widening anteriorly; pale greenish yellow with orange tips and a paler margin; reaches 10mm.
DISTRIBUTION: Camotes Sea.

16. Pseudosimnia (Diminovula) punctata (Duclos, 1831)
9.5mm, 10.0mm; Punta Engaño, Cebu.
Shell small, bulbous and pyriform, with a finely striated dorsum; outer lip thickened and finely denticulate; base smooth, rounded, with a funicular ridge on the posterior wall forming the posterior canal; aperture narrow, curving to the left; white with 3 rows of large orange spots and a fine golden line encircling the shell margin; reaches 10mm.
DISTRIBUTION: Camotes Sea, Tayabas Bay (Quezon) and Sulu Sea.

17. Testudovolva nebula (Azuma & Cate, 1971)
8.7mm, 5.8mm; Punta Engaño, Cebu.
Shell small, ovate, beaks slightly projecting and curving to the left; terminal area spirally striate, dorsum smooth and glossy; outer lip thickened and finely crenulate, columella smooth, flattened, with a triangular funiculum posteriorly; dorsum pale lavender with 3 rows of longitudinally arranged brownish dots, tips orange, outer lip margin white; reaches 10mm.
DISTRIBUTION: Camotes Sea and Sulu Sea.

18. Pseudosimnia (Diminovula) whitworthi Cate, 1973
10.9mm, 12.0mm; Manila Bay.
Close to *P. punctata*, but differs in being more elongate, having denser and more numerous terminal striae, a flatter base, more highly developed labral teeth, and by its distinct colour and pattern which is a pale lavender base colour with 3 transverse, rose-coloured bands containing 6 diffuse, dark brown spots; reaches 12mm.
DISTRIBUTION: Manila Bay, Camotes Sea and Sulu Sea.

19. Hiata depressa (Sowerby III, 1875)
13.2mm; Sulu Archipelago.
Shell long, lanceolate in outline, with slightly extended, gradually tapering tips; dorsum spirally striated along its entire length; base flat, with the outer lip and columellar surface characteristically flattened, constricting the aperture except anteriorly where both constrict, causing an apertural dilation; pale greenish white with orange tips; reaches 20mm.
DISTRIBUTION: Camotes Sea and Sulu Sea.

Class Gastropoda
Subclass Prosobranchia
Order Mesogastropoda
Suborder Taenioglossa
Superfamily Cypraeacea
Family TRIVIIDAE

A comparatively small group of carnivorous gastropods living under rocks, in coral or seaweed, and inhabiting shallow to deep water in most geographical provinces. Closely related to the cowries in the similarity of their structural outline, their terminal processes, the appearance and position of the aperture, and in the similarity of their front and rear canals. They differ mainly in the presence of strong transverse ribbing. Few Philippine species, with only the most common one being illustrated here.

20. Trivirostra oryza oryza (Lamarck, 1811)
11.5mm, 10.8mm; Palawan.
A small, pure white, globular shell with a moderately inflated dorsum bearing a well delineated furrow and numerous smooth, fine tranverse ridges; aperture narrow, curving to the left posteriorly; both inner and outer lip surfaces dentate; reaches 14mm.
DISTRIBUTION: Throughout the Philippines.

Tonnidae Plate 26

Class Gastropoda
Subclass Prosobranchia
Order Mesogastropoda
Suborder Taenioglossa
Superfamily Tonnacea
Family TONNIDAE

A small group of carnivorous sand-dwelling gastropods which inhabit deep water and have a worldwide distribution. Shells are relatively large, thin and lightweight, have a short spire, a large globose body whorl, may be smooth or ribbed, have a smooth or denticulate outer lip and a prominent siphonal fasciole. Fairly well represented in the Philippines.

1. Tonna sulcosa (Born, 1778)
124.7mm; Manila Bay.
Shell medium sized with a tall spire, moderately channeled sutures and a body whorl which is ovate rather than globose; sculptured with numerous close-set, somewhat flattened spiral cords with an occasional thread in the interspaces; aperture wide with a denticulate outer lip, columella calloused and twisted anteriorly; white in colour with brown banding, dark purple spire whorls and a white

PLATE NO. 26 1. *Tonna sulcosa* (Born) 2. *Tonna cumingii* (Reeve) 3. *Tonna tesselata* (Lamarck) 4. *Tonna perdix* (Linné) 5. *Tonna allium* (Dillwyn) 6. *Tonna olearium* (Linné) 7. *Malea pomum* (Linné) 8a. to 8b. *Tonna dolium* (Linné) 9. *Tonna cepa* (Röding)

PLATE 26

aperture; reaches 130mm.
DISTRIBUTION: Sporadically found throughout the Philippines.

2. Tonna cumingii (Reeve, 1849)
120mm, 76.0mm; Cebu.
A moderately large shell with a tall spire, moderately channeled sutures and a large, globose body whorl which is sculptured with flat, closely-set ribs and grooved interspaces; aperture wide with a serrated outer lip margin, a heavily calloused columella and a slightly open umbilicus; shell pale tan in colour, ornamented with darker banding and white-and-brown streaks, spots or maculations; aperture dark brown within with a brown-bordered, white outer lip margin; reaches 125mm.
DISTRIBUTION: Sporadically found throughout the Philippines.

3. Tonna tessellata (Lamarck, 1816)
67.3mm; Marinduque.
Shell medium sized, solid, with a tall spire, deeply canaliculate sutures and an ovate body whorl; sculptured with 12 to 14 narrowly rounded ribs and smooth interspaces; aperture semi-lunate with a thick outer lip bearing 9 to 10 pairs of denticles; columella calloused with a narrowly open umbilicus and a recurved anterior canal; shell light brown with brown-spotted white cords, aperture brown within, columella and outer lip margin white; reaches 85mm.
DISTRIBUTION: Sporadically found throughout the Philippines in limited quantity.

4. Tonna perdix (Linné, 1758)
113mm; Palawan.
Shell large, lightweight, rather thin, with a tall spire, channeled sutures, and a large, pyriform-ovate body whorl; sculptured with flat, weakly elevated spiral cords and overriding axial growth striae; aperture wide, outer lip thin and fluted, columella with a thin glaze; shell brown with white axial streaks on the ribs and white interspaces, aperture brownish white with a darker margin; reaches 200mm.
DISTRIBUTION: Throughout the Philippines.

5. Tonna allium (Dillwyn, 1817)
86.2mm; Mindoro.
Shell medium sized, solid, with a tall spire, canaliculate sutures and a large, ovate body whorl; sculptured with narrow, rounded, widely-spaced spiral cords and fine axial growth striae in the interspaces; outer lip denticulate, parietal area with a weak glaze; colour pale to orange-brown, ribs usually darker, spire whorls dark purple; reaches 100mm.
DISTRIBUTION: Throughout the Philippines.

6. Tonna olearium (Linné, 1758)
182mm; Tayabas Bay, Quezon.
Shell large, solid, with a moderately tall spire, deeply canaliculate sutures and a large, elongate-ovate body whorl; sculptured with wide, rather flat spiral cords and interstitial spiral threads which persist over the entire shell length; whole surface crossed with fine axial growth striae; columella calloused basally, umbilicus weakly open, fasciolar cord heavy and twisted, outer lip thin and crenulate; shell dark brown, white around the sutures, spire whorls dark purple, aperture pale brownish lavender with a brown margin; reaches 230mm. Often confused with *Tonna galea* (Linné, 1758), a species presently not known to occur in the Philippines. It differs from *T. olearium* in its more ovate shape, shorter spire, shorter anterior canal, paler brown colouration, and the presence of interstitial spiral threads only on the upper half of the shell.
DISTRIBUTION: Throughout the Philippines.

7. Malea pomum (Linné, 1758)
66.4mm; Samar.
Shell medium sized, solid, with a tall spire, and sculptured with closely set spiral cords and axially striated interspaces; aperture narrow, outer lip thick, heavily dentate, columella plicate with a slight basal concavity; aperture white, shell cream with brown and white spots; reaches 80mm.
DISTRIBUTION: Throughout the Philippines.

8. Tonna dolium (Linné, 1758)
93.3mm, 95.5mm; Palawan.
Shell large, variable in structure, with a short spire, a large globose body whorl and deeply channeled sutures; sculptured with large spiral cords, rounder on the upper body whorl, flatter towards the base and variable interstitial threading; aperture white, columella with a peculiar twist and lacking any glazing; pale cream with red spots on the cords; reaches 150mm. Fig. 8b superficially resembles *T. tessellata* (fig. 3), but can easily be distinguished from that species by its larger size, lack of outer lip denticles, the peculiar columellar "twist", and the crowding of the spiral cords towards the base.
DISTRIBUTION: Throughout the Philippines.

9. Tonna cepa (Röding, 1798)
102mm; Samar.
Shell large, lightweight, with a short spire, canaliculate sutures and a large globose body whorl; sculptured with shallow spiral grooves and weak, overriding axial striae; aperture wide, outer lip thin, weakly crenulate, columella with a white glaze; shell brown with white and dark brown streaks, aperture brown; reaches 145mm.
DISTRIBUTION: Sporadically found throughout the Philippines.

Class Gastropoda
Subclass Prosobranchia
Order Mesogastropoda
Suborder Taenioglossa
Superfamily Tonnacea
Family CASSIDAE

A moderately large group of carnivorous gastropods inhabiting sandy areas in the shallow or deep waters of most world seas. The shells are medium to large sized, globose to elongate-ovate, have a short spire and a large body whorl. The surface may be smooth or sculptured with cords and nodules; the aperture may be narrow or fairly wide, the outer lip thick, frequently reflected and normally dentate; the columella may be smooth or plicate, the parietal shield is large and glossy and the fasciolar cord is large and recurved. Members of the genus *Morum* are generally small, elongate-ovate in shape, and are sculptured with intersecting axial and spiral cords forming strongly echinate nodules. The outer lip is thickened and dentate, and the parietal shield is normally brightly coloured and pustulose. The Cassidae are well represented in the Philippines, with the genus *Morum* having an unusually high representation.

1. Cassis (Cassis) cornuta (Linné, 1758)
210mm; Palawan.
A large massive shell, spire whorls almost flat, sutures adpressed and coronate, varices every 270°; body whorl large with 3 to 4 rows of tubercles, the shoulder row strongest, the rest weaker and in general spirally fused to form elevated cords; surface with fine pits from intersecting axial and spiral riblets; aperture narrow, outer lip thick, flattened and heavily dentate posteriorly; columellar shield heavy and expansive, columellar margin heavily dentate; shell greyish white with brown spotted ribs, outer lip and parietal shield pink, aperture deep orange; a sexually dimorphic species, the male form (not shown) being smaller, with fewer and larger shoulder tubercles; reaches 360mm, the largest specimens mostly coming from the Sulu Sea.
DISTRIBUTION: Throughout the Philippines.

2. Cypraecassis (Cypraecassis) rufa (Linné, 1758)
137mm; Jolo, Sulu Archipelago.
Shell large, heavy and glossy; spire whorls with channeled sutures, a presutural cord and fine axial and spiral striae; body whorl with 3 to 4 rows of rounded nodules, replaced by axial ridges anteriorly; aperture wide, outer lip thick, produced over the spire and heavily dentate, columella heavily calloused and plicate; shell colour brownish lavender with orange and brown bands, base and aperture orange, teeth white, purple between; reaches 160mm.
DISTRIBUTION: Sulu Sea in very limited quantity.

3. Phalium (Phalium) glaucum (Linné, 1758)
Shell large, globular, with a short spire; spire whorls finely decussate with spiral rows of nodules becoming restricted to the shoulder on the body whorl; body whorl malleate, finely and axially ribbed; outer lip thick, dentate, with four spines at the base, columellar shield flaring, weakly ridged, with a thin varix behind; colour greyish with a pinkish orange outer lip and columellar shield, aperture marginally pink, dark brown within; reaches 140mm.
3a. 135mm; Samar.
3b. sub-adult specimen; 67.5mm; Palawan.
3c. juvenile with heavily corded surface; 46.5mm; Palawan.
DISTRIBUTION: Sporadically found throughout the Philippines.

4. Phalium (Phalium) flammiferum (Röding, 1798)
Syn: *Phalium strigatum* (Gmelin, 1791) (non *strigatum* Müller, 1776)
69.2mm, 69.0mm; Cebu.
A cool water species, recently taken in the Philippines; shell medium sized, heavy, elongate-ovate, with a short obtuse spire; spire whorls decussate from intersecting axial and spiral striae; body whorl smooth, weakly shouldered and spirally grooved above and towards the base; varices strong, occurring every 240°; aperture wide, outer lip thick and denticulate, columellar shield expanded, irregularly lirate anteriorly; white with brown axial stripes; reaches 100mm.
DISTRIBUTION: Cebu and Sulu Sea in very limited quantity.

5. Phalium (Semicassis) angasi bulla (Habe, 1961)
50.0mm; Punta Engaño, Cebu.
Originally recorded as a sub-species of *P. glabratum* but is distinct and occurs sympatrically with that species. Its close similarity to *angasi* suggests it may be the Philippine sub-species of that shell. Shell small, lightweight, translucent and elongate-ovate in outline; spire short, spire whorls with grooves towards the base; aperture wide, outer lip thick and weakly dentate, columella basally lirate, parietal area moderately glazed with 3 to 4 weak posterior denticles; colour pale greyish white; reaches 60mm.
DISTRIBUTION: Camotes Sea and Southern Mindanao.

6. Phalium (Phalium) bandatum bandatum (Perry, 1811)
83.4mm; Samar.
Shell large, elongate-ovate with a tall stepped spire; spire whorls finely decussate with finely noduled shoulders extending onto the body whorl; shell surface weakly malleate; aperture wide, outer lip thickened and heavily dentate, columellar shield flaring, finely ridged and weakly variced behind; off-white in colour with spiral bands of brown quadrate spots; aperture marginally white with brown spots, darker within, teeth white; reaches 120mm.
DISTRIBUTION: Sporadically found throughout the Philippines.

7. Phalium (Phalium) areola (Linné, 1758)
79.0mm, 82.3mm; Samar.
Shell medium sized, solid and glossy; spire tall and acute, spire whorls weakly shouldered with axial lirae which become obsolete on the body whorl; some spiral lirae on the upper shoulder area of earlier whorls; the body whorl is smooth, elongate-ovate in outline, and the aperture is wide with a thick, heavily dentate outer lip, colum 'la with a flaring columellar shield bearing coarse folds anteriorly, varices only seen on later whorls; pale brown with brown axial streaks on the earlier whorls, becoming 4 to 5 spiral bands of brown quadrate spots on the body whorl; reaches 100mm.
DISTRIBUTION: Throughout the Philippines.

8. Phalium (Semicassis) bisulcatum bisulcatum (Schubert & Wagner, 1829)
An extremely variable shell which is medium sized, lightweight and glossy; ovate to globose in outline, with a tall spire and channeled sutures; spire whorls smooth or spirally lirate, the last 1 or 2 occasionally shouldered; body whorl smooth, or with numerous fine, occasionally granulose, spiral lirae; varices rarely present; aperture semi-lunate, outer lip thick and dentate, columellar shield flaring and wrinkled anteriorly; off-white with or without rows of brown spots, aperture white to brown; reaches 50mm.
8a. form *diuturnum* (Iredale, 1927): a form with flat densely packed spiral ribs and obsolete bands of spots; 54mm; Palawan.
8b. a smooth brown-spotted form; 38.2mm; Samar.
DISTRIBUTION: Throughout the Philippines.

9. Casmaria ponderosa ponderosa (Gmelin, 1791)
Often confused with *C. erinaceus* (see fig. 10), but is more ovate, has a taller spire, a well developed varix and spinose teeth along the whole length of the outer lip; colour pale cream with brown-spotted bands and subsutural brown blotches which are lacking in *erinaceus;* reaches around 70mm.
9a. form *nodulosa* Gmelin, 1791: a heavily plicate form; 38.5mm, 34.6mm; Cebu.
9b. smooth form; 44.0mm; Zambales.
9c. form *turgida* Reeve, 1848: smooth form with brown axial flammules; 73.7mm; Bohol.
DISTRIBUTION: Throughout the Philippines.

10. Casmaria erinaceus (Linné, 1758)
Shell small, lightweight, glossy, elongate-ovate in outline; spire whorls moderately globose, smooth and weakly channeled; body whorl smooth or with a heavily noduled shoulder; aperture wide, outer lip thickened with spinose projections towards the base, columella smooth, columellar shield thick and weakly lirate, occluding the umbilicus; tan coloured with weaker brown bands, aperture brown, columellar shield white with a brown blotch behind; reaches 75mm.
10a. typical form; 63.2mm; 49.8mm; Samar.
10b. form *vibex* Linné, 1758: a smooth form, generally lighter in weight, with a thinner outer lip and a less developed columellar callus; 57.9mm, 50.5mm; Palawan.
10c. a striped *vibex* form; 59.6mm; Samar.
10d. an unusually spotted form of the typical species; 57.2mm; Samar.
DISTRIBUTION: Throughout the Philippines.

11. Phalium (Echinophoria) carnosum (Kuroda & Habe, 1961)
59.1mm; Balut Is., Southern Mindanao.
Shell medium sized, globose, rather thin-walled, with a poorly developed columellar shield and a wide aperture; spire whorls shouldered and finely cancellate, the last 2 or 3 rows developing a row of sharply echinate tubercles on the shoulder; body whorl with 5 to 6 rows of echinate nodules on the upper two-thirds and fine interstitial spiral striae; base with dense spiral threads; columellar shield weakly developed and the columella is strongly twisted basally; pale orange-brown with white nodules and aperture; reaches 65mm (illustrated specimen is sub-adult, lacking outer lip development).
DISTRIBUTION: Camotes Sea and Southern Mindanao.

12. Galeodea echinophorella Hirase, 1936
64.7mm; Cebu.
Shell medium sized, elongate-ovate with a small, conic spire and a large body whorl; spire whorls sculptured with spiral cords and a row of shoulder nodules; body whorl similarly sculptured, but with 2 to 3 rows of nodules, the shoulder row strongest and the rest becoming obsolete; aperture semi-lunate, outer lip thickened, flattened and marginally crenulate, columellar callus expansive and smooth, siphonal canal long and recurved; pale brown with some darker brown suffusion, aperture and columellar callus white; reaches 65mm.
DISTRIBUTION: Camotes Sea and Southern Mindanao.

13. Phalium (Phalium) decussatum (Linné, 1758)
Shell medium sized, solid, dorso-ventrally compressed with an elongate-ovate outline; whorl surfaces highly

PLATE NO. 27 1. *Cassis cornuta* (Linné) 2. *Cypraecassis rufa* (Linné) 3a. *Phalium glaucum* (Linné) 3b. Sub-adult specimen 3c. juvenile with heavily corded surface 4. *Phalium flammiferum* (Röding) 5. *Phalium angasi bulla* (Habe) 6. *Phalium bandatum bandatum* (Perry) 7. *Phalium areola* (Linné) 8a. *Phalium bisulcatum bisulcatum* (Schubert & Wagner) form *diuturnum* (Iredale) 8b. a smooth, brown spotted form 9a. *Casmaria ponderosa ponderosa* (Gmelin) form *nodulosa* Gmelin 9b. smooth form 9c. form *turgida* Reeve 10a. *Casmaria erinaceus* (Linné) 10b. form *vibex* Linné 10c. a striped *vibex* form 10d. an unusually spotted form of the typical species 11. *Phalium carnosum* (Kuroda & Habe) 12. *Galeodea echinophorella* Hirase 13a. *Phalium decussatum* (Linné) 13b. form *flammeolum* Röding 14a. *Phalium coronadoi wyvillei* (Watson) 14b. a more globose form 15. *Morum exquisitum* (Adams & Reeve) 16. *Morum celinamarumai* Kosuge 17. *Morum Kurzi* Petuch 18. *Morum grande* (A. Adams) 19. *Morum joelgreenei* Emerson

PLATE 27

1
2
3a
3b
3c
4
5
6
7
8a
8b
9a
9b
9c
10a
10b
10c
10d
11
12
13a
13b
14a
14b
15
16
17
18
19

cancellate from intersecting axial and spiral grooves, and there are 6 to 7 varices, each bearing 2 small white ridges on the upper shoulder area; aperture wide, outer lip thick, marginally dentate, internally lirate, parietal glaze thin, columella heavily lirate basally; pale bluish white with brown axial streaks or spiral rows of spots, aperture deep brown, marginally white; reaches 70mm. Recently recorded from the Philippines.
13a. typical form; 67.5mm; Jolo, Sulu Archipelago.
13b. form *flammeolum* (Röding, 1798): colour form with spiral rows of brown quadrate blotches; 60.5mm; Jolo, Sulu Archipelago.
DISTRIBUTION: Sulu Sea.

14. Phalium (Echinophoria) coronadoi wyvillei (Watson, 1886)
Shell medium sized, globose, lightweight; spire moderately tall with convex-sided whorls which are sculptured with spiral threads and a row of shoulder spines; body whorl with 1 to 2 spiral rows of nodules and spiral threads on the shoulder and base, frequently obsolete centrally; aperture semi-lunate, outer lip thick and smooth, columellar shield weakly flaring and lirate; colour yellowish grey to pale tan; reaches 100mm.
14a. typical form; 92.2mm; Punta Engaño, Cebu.
14b. a more globose form, with nearly obsolete shoulder tubercles; 98.7mm; Punta Engaño, Cebu.
DISTRIBUTION: Camotes Sea and Southern Mindanao.

15. Morum (Oniscidia) exquisitum (Adams & Reeve, 1848)
38.0mm; Jolo, Sulu Archipelago.
Shell medium sized, solid, elongate-ovate; spire short with weak sutures, spire whorls coronate and spirally striate; body whorl with nodulose humps from intersecting axial and spiral cords; outer lip thick and heavily dentate, columella well delineated and calloused; off-white with darker banding, parietal shield and outer lip purple-pink; reaches around 40mm; endemic to the Philippines.
DISTRIBUTION: Sulu Sea.

16. Morum (Oniscidia) celinamarumai Kosuge, 1981
39.0mm; Bohol.
Shell medium sized, solid, elongate-ovate with a short spire and large body whorl; spire whorls rounded, finely nodulose from intersecting axial and spiral cords, body whorl shouldered with widely spaced axial ribs, sharply echinate from overriding spiral cords, interspaces with finely lamellate axial striae; aperture narrow, outer lip thick and glazed with 12 pairs of teeth extending over most of the outer margin, columella heavily glazed, pustulose and sharply delineated; pale orange-pink with brown spotting, columellar callus pale orange, pustules purplish white, outer lip pale orange with four areas of dark spotting; reaches 40mm.
DISTRIBUTION: Camotes Sea.

17. Morum (Oniscidia) kurzi Petuch, 1979
24.3mm, 23.5mm; Bohol.
Similar to *M. exquisitum* in general shape but much smaller, somewhat more coarsely sculptured, having spinose cords rather than nodulose "humps", and an orange-coloured callus and outer lip; reaches 25mm.
DISTRIBUTION: Camotes Sea.

18. Morum (Oniscidia) grande (A. Adams, 1855)
61.4mm; Tayabas Bay, Quezon.
The largest of the genus in the Philippines; shell large, short spired, rather heavy with weakly channeled sutures; surface sculptured with raised scabrous plates formed from intersecting axial lirae and scabrous spiral cords, and fine axial lamellae in the interspaces; outer lip thick, reflected and heavily dentate, columella sharply delineated with ridges and folds; white with 4 brown bands, aperture and columella white; reaches 70mm.
DISTRIBUTION: Camotes Sea, Southern Mindanao, Mindoro and Tayabas Bay, Quezon.

19. Morum (Oniscidia) joelgreenei Emerson, 1981
43.7mm, 42.4mm; Punta Engaño, Cebu.
A heavy, medium-sized shell, elongate-ovate in outline with a wide shoulder, a tall spire and cancellate spire whorls; surface covered with sharply spinose nodules formed from intersecting scabrous spiral cords and axial lirae, and weaker lamellate axial threads in the interspaces; outer lip thick and dentate on the inner margin only, columella with an expansive, pustulose, marginally raised, sharply delineated callus; pale yellowish white with darker banding, outer lip pale orange with darker blotches, columellar callus pale orange, pale lavender towards the columellar margin; reaches 45mm.
DISTRIBUTION: Camotes Sea.

Harpidae Plate 28

Class Gastropoda
Subclass Prosobranchia
Order Neogastropoda
Suborder Rachiglossa
Superfamily Volutacea
Family HARPIDAE

A small group of non-operculate gastropods belonging to the Volutacea and occurring worldwide in shallow to deep water, usually preferring a sandy substrate. The shell is glossy, has a large body whorl, strong axial ribbing, an extremely variable protoconch, an ovate to elongate-ovate aperture with a calloused columella and an obsolete siphonal canal. The shells are highly colourful and lack a periostracum; few Philippine species.

1. Harpa articularis Lamarck, 1822
89.4mm, 85.4mm; Manila Bay.
Shell large, broadly ovate, with a wide conical spire; sculptured with flat axial ribs, thinner than normal, spinose at the roundly angulate shoulder, and continuing to the suture as raised "blade-like" ribs; aperture wide and ovate, outer lip arcuate, columella glazed with a single large expansive brown blotch; shell colour pink to grey, ornamented with white spiral zigzag bands and brown sagittal marks, ribs strongly marked with chestnut lines and bands; reaches 100mm.
DISTRIBUTION: Sporadically found throughout the Philippines.

2. Harpa major Röding, 1798
89.4mm, 85.3mm; Sulu Archipelago.
Shell large, heavy, with an ovate body whorl and a broadly conical spire, the whorls of which are slightly cancellate and frequently covered with a heavy milky glaze; sculptured with 14 to 15 flattened ribs, spinose at the sharply angled shoulder, and coalescing at the sutures; fine axial striae between the ribs; aperture large and ovate, outer lip arcuate, columella with a large brown blotch, divided centrally; colour varies from pale pink to dark grey, ribs with or without black lines, and the interstitial spaces are festooned with brown or pink; reaches 100mm.
DISTRIBUTION: Mostly found in Southern Mindanao.

3. Harpa harpa Linné, 1758
Syn: *H. nobilis* Lamarck, 1816 (*non* Röding, 1798)
62.5mm, 60.8mm; Siasi, Sulu Archipelago.
Shell medium sized, solid, widely ovate with a low spire and sculptured with broad axial ribs which form spines at the roundly angulate shoulder; earlier whorls with a weak callus; aperture large, elongate-ovate, outer lip arcuate and weakly serrate anteriorly, columella glazed with 3 separate brown blotches; pink between the ribs with red blotching and axial zigzags, the ribs being marked with groups of close-set black lines; reaches 75mm.
DISTRIBUTION: Mostly found in Southern Mindanao.

4. Harpa kajiyamai Rehder, 1973
61.8mm; Sulu Archipelago.
Shell medium sized, elongate-ovate with a tall spire that is moderately calloused; body whorl large, ovate, with flattened axial ribs sharply spinose at the shoulder and reflected basally towards the siphonal fasciole; columella with a thin, transparent glaze and 2 brown spots, a large elongate upper one and a smaller basal one; colour similar to other *Harpa* species, though occasionally, a central band of reddish spots is visible; reaches 70mm.
DISTRIBUTION: Sulu Sea in very limited quantity.

5. Harpa amouretta Röding, 1798
Shell small, elongate-ovate, with a comparatively large, tall spire; sculptured with axial ribs, spinose at the shoulder and continuing to the sutures; earlier whorls with a nacreous overlay; aperture narrowly elongate with 3 widely spaced brown blotches on the columella, the central one being strongest, the other 2 frequently almost obsolete; colour pale yellow with deep brown spots and zigzag lines between the ribs and groups of closely set black lines on the ribs; reaches 60mm.
5a. typical form with the central blotch being prominent; 49.0mm, 46.5mm; Samar.
5b. form *crassa* Krauss, 1848: this form is paler, much heavier in weight, has a strong shoulder and is rather wide; 50.0mm; Samar.
DISTRIBUTION: Sporadically found throughout the Philippines.

Class Gastropoda
Subclass Prosobranchia
Order Neogastropoda
Suborder Rachiglossa
Superfamily Volutacea
Family TURBINELLIDAE

A comparatively small family of shells which may be discussed in terms of the 3 subfamily groups which comprise it.

I. Turbinellinae Swainson, 1840
The shells of this group are large and fusiform, with large columellar folds, usually at right angles to the shell axis. There are no known members of this subfamily in the Philippines, most members occurring in North and South America, and also in India.

II. Vasinae H. & A. Adams, 1854
Shells medium to large sized, heavy and have a strongly tuberculate surface. The aperture is narrow, the outer lip thin, crenulate and lirate within, and the columella has large, unequal-sized folds set at an oblique angle to the shell axis. The animal is highly carnivorous and inhabits sand or coral rubble in the shallow waters of most tropical seas. Only 3 or 4 species are found in the Philippines.

III. Ptychatractinae Stimpson, 1865
This group is represented in the Philippines by the genus *Teramachia* Kuroda, 1931. The members of this genus are a group of "*Tibia*"-like shells which inhabit extremely deep water. Originally placed in the Volutidae, members of this group are represented in the Philippines by around 6 species, 5 of which are depicted here (2 as yet being unnamed).

6. Vasum (Vasum) tubiferum (Anton, 1839)
61.0mm; Palawan.
Shell large, turbinate, strongly spined with a tall spire; body whorl with a row of long, upwardly curving, terminally flaring spines on the shoulder, followed by 4 rows of weaker spines, then a row of weaker, somewhat scabrous nodules; aperture narrow, outer lip crenulate and marginally brown, columella calloused, dark purplish brown in colour with a

brown blotch and 5 strong plaits, umbilicus deep, bordered with a thick ridge; off-white with black spines and brown reticulations; reaches 115mm.
DISTRIBUTION: Palawan.

7. Vasum (Vasum) turbinellum (Linné, 1758)
67.2mm; Samar.
A solid, heavy, medium-sized shell, characterized by its turbinate shape, the strong row of curved, slightly open, blunt spines on the shoulder, the additional row of spines on the body whorl, the white columella with 3 large plaits and 2 weaker ones between, and the black denticles on the outer lip margin; greyish white with dark brown mottling; reaches 85mm.
DISTRIBUTION: Throughout the Philippines.

8. Vasum (Vasum) ceramicum (Linné, 1758)
112mm; Palawan.
Shell large, heavily fusiform, with an elevated spire, adpressed sutures, a white glazed columella bearing three plaits, and brown coloured bifid teeth on the outer lip margin; spire whorls with a single row of blunt triangular spines, body whorl with several rows, decreasing in strength basally where a terminal row of spines is formed; white with brown spines and bands; reaches 150mm.
DISTRIBUTION: Sporadically found throughout the Philippines.

Turbinellidae Plate 29

1. Teramachia dalli (Bartsch, 1942)
133mm; Lubang Is., Mindoro.
Shell moderately large, fusiform-elongate, with a tall spire, eroded protoconch, widely channeled sutures and sculptured with numerous oblique axial ribs, cusped at the sutural margins and becoming obsolete on the last 2 whorls; aperture elongate-ovate, outer lip thickened, arcuate and constricted anteriorly, columella calloused, lacking plaits; dark purplish brown, paler towards the base, the aperture light purple with a brown margin; reaches 180mm.
DISTRIBUTION: Camotes Sea and Lubang Is., Mindoro.

2. Teramachia tibiaeformis Kuroda, 1931
116mm; Bohol.
Shell medium sized, slender, fusiform-elongate, with a tall, acute spire and a large body whorl; sutures deeply incised, spire whorls sculptured with numerous arcuate axial folds, slightly pinched-in subsuturally and obsolete on the last 2 whorls; aperture large and elongate, outer lip thick, arcuate and constricted anteriorly, columella smooth; colour purplish brown, early whorls lighter, aperture light purple; reaches 120mm.
DISTRIBUTION: Camotes Sea.

3. Teramachia smithi (Bartsch, 1942)
160mm; Lubang Is., Mindoro.
Shell large, fusiform-elongate, with a tall spire, large body whorl and deeply incised sutures; spire whorls convex, sculptured with numerous oblique, arcuate axial folds, becoming more widely spaced and then obsolete on the body whorl; fine spiral striae cover the entire surface; aperture elongate-ovate, outer lip thin and weakly notched, canal moderately long, columella smooth or with 2 weak plicae; shell greyish purple, earlier whorls lighter with some brown mottling, aperture pale purple; reaches 160mm.
DISTRIBUTION: Camotes Sea and Lubang Island, Mindoro.

4. Teramachia sp.
175mm; Balut Is., Southern Mindanao.
Similar to T. *smithi* (Bartsch), but larger and more elongate; the whorls are less convex, and the numerous closely-set, arcuate axial folds become obsolete on the third to the last whorl; the surface lacks the fine spiral striae of *T. smithi*, the aperture is more quadrangular, the outer lip less arcuate, and the columella has 2 weak but distinct plaits; colour pale purple with a subsutural white band, irregular brown mottling, paler spire whorls and a black-coloured sutural margin, aperture deep purple, columella dark brown anteriorly; appears to reach around 180mm.
DISTRIBUTION: Lubang Is. (Mindoro) and Southern Mindanao.

5. Teramachia sp.
210mm; Bohol.
Shell very large, very close to *T. smithi* (Bartsch), but the whorls are somewhat angulate and the axial folds become obsolete on the penultimate whorl, being replaced by irregular nondescript ridges; surface covered with very fine spiral striae, becoming stronger on the base of the body whorl; colour is a dirty brownish purple with orange-brown undertones; aperture deep brownish lavender; illustrated specimen the largest seen to date. Probably related to *Teramachia johnsoni* (Bartsch, 1942) form *williamsorum* Rehder, 1972.
DISTRIBUTION: Camotes Sea and Southern Mindanao.

PLATE NO. 28 1. *Harpa articularis* Lamarck 2. *Harpa major* Röding 3. *Harpa harpa* Linné 4. *Harpa kajiyamai* Rehder 5a. *Harpa amouretta* (Röding) 5b. form *crassa* Krauss 6. *Vasum tubiferum* (Anton) 7. *Vasum turbinellum* (Linné) 8. *Vasum ceramicum* (Linné)

PLATE 28

Class Gastropoda
Subclass Prosobranchia
Order Mesogastropoda
Suborder Taenioglossa
Superfamily Tonnacea
Family FICIDAE

A relatively small group of sand and coral-dwelling gastropods which inhabit shallow to deep water. The shells are moderately large, quite thin and pear-shaped, with a thin outer lip, a long, fairly wide aperture and a surface which is either corded or axially and spirally striated; only 3 or 4 species recorded from the Philippines.

6. Ficus gracilis (Sowerby, 1825)
162.6mm; Palawan.
Shell large with a slightly elevated spire, an elongate-ovate body whorl, and a surface sculptured with numerous strong spiral cords and weaker overriding axial striae; aperture wide, outer lip thin, smooth and arcuate, columella smooth and centrally concave, siphonal canal long; shell light brown with paler markings, aperture marginally white, brown within; reaches 200mm.
DISTRIBUTION: Sporadically found throughout the Philippines.

7. Ficus subintermedia (d'Orbigny, 1852)
110mm; Samar.
Shell medium sized with a depressed spire and a body whorl which is ovate rather than globose at the shoulder; sculptured with spiral threads and overriding axial striae producing a finely cancellate surface; aperture elongate, columella smooth and slightly concave centrally, siphonal canal very long; pale brown with white banding, dark brown spots and darker surface maculations; reaches 130mm.
DISTRIBUTION: Sporadically found throughout the Philippines.

8. Ficus filosa (Sowerby, 1892)
86.7mm; Palawan.
Shell medium sized, low spired, with a large body whorl inflated at the shoulder; sculptured with axial and spiral cords of equal strength, giving the surface a finely reticulate appearance; aperture wide, columella centrally concave, siphonal canal moderately long and slightly truncated; light brown in colour with every alternate spiral cord a deep orange-brown colour, aperture marginally white, brown within; reaches 100mm.
DISTRIBUTION: Sporadically found throughout the Philippines in limited quantity.

Class Gastropoda
Subclass Prosobranchia
Order Neogastropoda
Suborder Rachiglossa
Superfamily Buccinacea
Family MELONGENIDAE

A relatively small group of carnivorous gastropods, few of which inhabit the Philippines. The shells are medium to large sized and sculptured with either axial spiral cords, nodules or short spines. The canal may be short or long, the columella smooth, and all possess a thick, hairy periostracum and horny operculum.

9. Volema (Pugilina) cochlidium (Linné, 1758)
110mm; Manila Bay.
Shell large, heavy, with deeply incised sutures and angulate whorls; early whorls with spiral threads and axial cords below the shoulder; last 2 whorls coronated with a shoulder row of blunt, spinose tubercles; spiral threads absent on the upper body whorl, appearing again towards the base; aperture narrowly elongate-ovate, outer lip obsoletely dentate, columella smooth and heavily glazed with a strongly folded siphonal fasciole; orange-brown with a yellowish, marginally brown aperture; reaches 150mm; sold locally as *"Melongena pugilina"*.
DISTRIBUTION: Sporadically found throughout the Philippines.

---

10. Volema (Hemifusus) carinifera (Habe & Kosuge, 1965)
142mm; Manila Bay.
Shell large, fusiform-elongate, with a tall spire and a very long siphonal canal; whorls sharply angulate, sculptured with axial folds and spiral cords; earlier whorls with elevated, axially flattened nodules at the shoulders, the last 2 whorls with weaker axial folds, somewhat tuberculate at the shoulder; aperture narrowly elongate, outer lip slightly thickened, serrate and weakly lirate on the inner margin, columella smooth; colour burnt orange-brown with a creamy aperture; reaches 200mm.
DISTRIBUTION: Sporadically found throughout the Philippines.

11. Volema (Melongena) myristica Röding, 1798
Syn: *Volema galeodes* (Lamarck, 1822)
61.2mm, 66.3mm, 49.8mm; Sulu Archipelago.
Shell medium sized, solid, rather heavy, with a moderately tall spire and a broadly fusiform outline; sculptured with heavy axial folds, weaker interstitial threads, and encircled with thick spiral cords; earlier whorls nodulose, becoming spinose at the sutures of the later whorls, with an occasional row of spines on the body whorl base; aperture widely ovate, outer lip

PLATE NO. 29 1. *Teramachia dalli* (Bartsch) 2. *Teramachia tibiaeformis* Kuroda 3. *Teremachia smithi* (Bartsch) 4. *Teremachia sp.* 5. *Teramachia sp.* 6. *Ficus gracilis* (Sowerby) 7. *Ficus subentirmedia* (d'Orbigny) 8. *Ficus filosa* (Sowerby) 9. *Volema cochlidium* (Linné) 10. *Volema carinifera* (Habe & Kosuge) 11. *Volema myristica* Röding

PLATE 29

thick, angulate and weakly lirate on the inner margin, columella smooth, heavily glazed, umbilicus deeply channeled, bordered by a strong fasciolar cord; colour pale brown to orange, with or without brown-coloured cords, aperture and columella white or orange, occasionally with brown banding; an extremely variable species; reaches 80mm.
DISTRIBUTION: Throughout the Philippines.

Cymatiidae Plate 30

Class Gastropoda
Subclass Prosobranchia
Order Mesogastropoda
Suborder Taenioglossa
Superfamily Tonnacea
Family CYMATIIDAE

A relatively large, rather diverse group of gastropods which inhabit tropical and subtropical regions. They are carnivorous, prefer rocky or sandy substrates, live in shallow to moderately deep water and have a free-swimming larval stage. The shells are small to large, fusiform ovate with angulate whorls and sculptured with axial and spiral cords, nodules and varices. The aperture is ovate and usually highly thickened with dentate inner and outer lips. The siphonal canal may be short or long, the periostracum is usually thin and hairy, and the operculum horny and ovate in outline with a subcentral or anterior nucleus. There are 3 sub-family groupings: Cymatiinae, Ranellinae and Distorsioninae. The family is well represented in the Philippines.

1. Cymatium (Ranularia) sarcostomum (Reeve, 1844)
72.3mm; Samar.
Like a small *C. pyrum* (Linné) (fig. 7), but immediately distinguished by its shorter spire, its proportionately longer, narrower, less sinuous siphonal canal, smaller size, the centrally smooth columella, the small bifid teeth on the outer lip, and the peripherally pink aperture which is in contrast to the burnt orange of *pyrum;* reaches 80mm.
DISTRIBUTION: Throughout the Philippines in limited quantity.

2. Cymatium (Ranularia) gutturnium (Röding, 1798)
82.9 mm; Sulu Archipelago.
Similar to *C. caudatum* (fig. 6), but the sutures are not channeled and the surface sculpturing is coarser, consisting of weakly granular spiral cords and heavily noduled axial ribs; outer lip thick, dentate, columella plicate; pale cream with brown and orange axial markings especially on the ribs, aperture orange to pale red with a porcelaneous callus; reaches 90mm.
DISTRIBUTION: Throughout the Philippines in limited quantity.

3. Cymatium (Ranularia) sinense (Reeve, 1844)
85.4mm; Samar.
Shell medium sized, with a tall pointed spire, a rounded body whorl and an extremely long, open, centrally reflexed siphonal canal; sculptured with raised, bicingulate spiral cords, narrow axial ridges and numerous interstitial threads; aperture narrowly ovate, outer lip dentate, columella plicate; creamy white with yellow-brown streaks and bands; reaches 100mm.
DISTRIBUTION: Sporadically found throughout the Philippines in limited quantity.

4. Cymatium (Ranularia) exile (Reeve, 1844)
49.7mm, 50.2mm; Bulan, Sorsogon.
A comparatively small shell characterized by its short spire, large, ovate, heavily noduled body whorl, and the very long, sinuous, recurved siphonal canal; aperture yellowish, outer lip thickly dentate, columella plicate; white with a brown subsutural band and streaks on the body whorl; reaches 55mm.
DISTRIBUTION: Sporadically found throughout the Philippines in limited quantity.

5. Cymatium (Ranularia) encausticum (Reeve, 1844)
55.5mm; Cebu.
Shell medium sized, similar to *C. gutturnium,* but with a shorter spire, a more heavily nodulose surface, and a highly expansive, pink apertural callus which almost covers the previous whorl up to the penultimate varix; there are fewer outer lip denticles than in *gutturnium;* white with pink overtones and orange-brown streaks on the axial cords; reaches 75mm.
DISTRIBUTION: Sporadically found throughout the Philippines.

6. Cymatium (Ranularia) caudatum (Gmelin, 1791)
53.8mm; Bohol.
A solid, medium sized shell with a short or long siphonal canal, deeply channeled sutures (see vertical profile), and sculptured with raised, bicingulate spiral cords, noduled by narrow overriding axial cords; aperture ovately quadrangular, with a thick, dentate outer lip, and a lirate columella; white with orange-yellow zones; reaches 70mm.

PLATE NO. 30 1. *Cymatium sarcostomum* (Reeve) 2. *Cymatium gutturnium* (Röding) 3. *Cymatium sinense* (Reeve) 4. *Cymatium exile* (Reeve) 5. *Cymatium encausticum* (Reeve) 6. *Cymatium caudatum* (Gmelin) 7. *Cymatium pyrum* (Linné) 8. *Cymatium testudinarium* (A. Adams & Reeve) 9. *Cymatium grandimaculatum* (Reeve) 10. *Cymatium lotorium* (Linné) 11. *Charonia tritonis tritonis* (Linné) 12. *Cymatium muricinum* (Röding) 13. *Cymatium nicobaricum* (Röding) 14. *Cymatium pileare* (Linné) 15. *Linatella succincta* (Linné) 16. *Cymatium aquatile* (Reeve) 17. *Linatella caudata* (Gmelin) 18. *Cymatium flaveolum* (Röding) 19. *Cymatium hepaticum* Röding 20. *Cymatium rebeculum* (Linné)

PLATE 30

DISTRIBUTION: Throughout the Philippines mostly from Visayas and Mindanao, always in limited quantity.

7. Cymatium (Ranularia) pyrum (Linné, 1758)
109mm, 92.0mm; Samar.
Shell moderately large, solid, sutures weakly channeled, spire short, body whorl inflated, sculptured with varices on the later whorls, 6 to 7 axial rows of nodules between; surface with fine axial and spiral striae; aperture ovate, outer lip thick with 2 rows of denticles, columella plicate; orange-brown in colour with a red apertural margin and white varices, denticles and plicae; periostracum thin and fibrous; reaches 110mm.
DISTRIBUTION: Throughout the Philippines.

8. Cymatium (Ranularia) testudinarium (A. Adams & Reeve, 1850)
73.5mm, 86.2mm; Bulan, Sorsogon.
A solid, medium sized shell with a long, open, recurved siphonal canal, weakly incised sutures, 2 varices per whorl and 4 large intervarical nodes; surface sculptured with fine axial and spiral striae; aperture ovate, white, with a thick, heavily dentate outer lip and a dark red-brown to purple-tinged columella marked with white ridges; white with pale orange-brown banding; reaches 90mm.
DISTRIBUTION: Camotes Sea and Bulan, Sorsogon.

9. Cymatium (Lotoria) grandimaculatum (Reeve, 1844)
96.1mm; Samar.
Shell moderately large, heavy, with weakly channeled sutures, a tall spire, an inflated body whorl, and a long, slightly recurved canal; sculptured with flattened spiral cords and interstitial threads crossed by narrow axial ribs forming nodules where they intersect; aperture ovate, lirate within, columella smooth; orange brown, varices and outer lip brown and white banded, and 2 brown blotches on the upper and lower parietal wall; reaches 120mm.
DISTRIBUTION: Throughout the Philippines in limited quantity.

10. Cymatium (Lotoria) lotorium (Linné, 1758)
135mm; Cebu.
Shell large, heavily rugose, sutures adpressed, spire tall, siphonal canal long and sinuous; sculptured with heavy raised varices and 3 prominent intervarical nodes; additional spiral cords and fine axial striae cross the surface; aperture white, marginally brown, outer lip thick, bluntly dentate, columella smooth with 2 brown blotches; orange with brown and white banded varices; reaches 170mm.
DISTRIBUTION: Throughout the Philippines.

11. Charonia tritonis tritonis (Linné, 1758)
338mm; Camarines Sur.
Shell large, solid with a tall spire, rounded spire whorls and a very inflated body whorl; sculptured with wide, smooth, rounded varices (every 270°), broad, flat spiral ridges and a row of subsutural granules; aperture large, elongate-ovate, outer lip with 8 or 9 spinose teeth, columella strongly lirate with white ribs on a dark brown background; external colour light brown, ribs cream with brown chevrons, aperture orange within; reaches 450mm.
DISTRIBUTION: Throughout the Philippines.

12. Cymatium (Gutturnium) muricinum (Röding, 1798)
62.0mm, 50.2mm, 39.9mm, 31.2mm; Cebu.
Shell small, solid, with a low spire, inflated body whorl and a short, distally recurved siphonal canal; sculptured with heavily noduled spiral cords and strong axial ribs, there being 6 nodules between the varices which are spaced every 270°; aperture ovate, outer lip thick, white and marginally dentate, columella heavily glazed, lirate anteriorly; greyish-brown with white banding on the varices and body whorl, aperture interior red; reaches 65mm; plate shows specimens with differing canal development.
DISTRIBUTION: Sporadically found throughout the Philippines.

13. Cymatium (Septa) nicobaricum (Röding, 1798)
59.2mm; Samar.
Shell medium sized, solid, very heavily sculptured with strong varices and heavily nodular cords; nodules axially aligned and number 4 to 5 between varices; spire tall, sutures weakly channeled, siphonal canal short and recurved; aperture ovate, outer lip dentate, columella lirate; creamy white with brown banding on the whorls and varices, aperture orange; reaches 65mm.
DISTRIBUTION: Throughout the Philippines.

14. Cymatium (Septa) pileare (Linné, 1758)
116.6mm; Samar.
Shell large, heavy, fusiform-ovate, with adpressed sutures, a tall spire and a short, recurved siphonal canal; varices every 270°; sculptured with nodular spiral cords strongest at the shoulder, weaker interstitial threads, and fine overriding axial ridges; outer lip with long denticles extending into the aperture, columella lirate; creamy white with mauve overtones, and brown and white banding, aperture red with white denticles; reaches 140mm.
DISTRIBUTION: Throughout the Philippines.

15. Linatella (Gelagna) succincta (Linné, 1771)
Syn: *C. clandestinum* Lamarck, 1816
80.7mm; Zamboanga.
Shell medium sized, solid, with rounded whorls, a short to tall spire, moderately impressed sutures, and a short, slightly recurved siphonal canal; sculptured with raised, dark brown spiral cords and deep interspaces of a paler brown colouration; aperture white with brown marginal denticles, columella with a single white bifid denticle posteriorly and fine lirae anteriorly; reaches 85mm.
DISTRIBUTION: Sporadically found throughout the Philippines in limited quantity.

16. Cymatium (Septa) aquatile (Reeve, 1844)
117.9mm; Samar.
A large, tall-spired, elongate-fusiform shell; sutures impressed, body whorl large and rounded, siphonal canal short and recurved; sculptured with nodulose axial ribs and spiral cords and weaker interstitial threads, all crossing the varices which occur every 270°; 5 to 6 nodulose rows between varices; outer lip with 2 rows of short bifid teeth, columella lirate; colour orange-brown, varices brown and white-banded, aperture orange, denticles and lirae white; reaches 120mm.
DISTRIBUTION: Throughout the Philippines.

17. Linatella (Linatella) caudata (Gmelin, 1791)
Syn: *L. cingulata* (Lamarck, 1822); *L. cutacea* Lamarck, 1816
60.5mm; Manila Bay.
Shell moderately large, with a tall spire and an inflated body whorl; sculptured with numerous close-set spiral cords, slightly nodulose on the shoulder of later whorls; aperture wide, white within, outer lip thick with brown-stained denticles; columella white and smooth, siphonal canal broad and recurved; shell white with pale brown bands and blotches; reaches 100mm.
DISTRIBUTION: Manila Bay and Camotes Sea.

18. Cymatium (Septa) flaveolum (Röding, 1798)
55.5mm; Samar.
Shell medium sized, tall spired, sculptured with 4 rows of finely nodulose spiral cords on the spire whorls and 10 on the body whorl, all connected by fine axial threads; aperture ovate, white, with a red margin and white denticles and lirae; columella with white lirae and red interspaces; dark brown with a white apex, a wide orange band at the base of spire whorls and at the center of last whorl, and white cords on the varices; Röding's type figure (in Chemnitz) is unclear, and the name *flaveolum* is a *"nomen inquirendum"*
DISTRIBUTION: Eastern Samar and Sulu Sea.

19. Cymatium (Septa) hepaticum Röding, 1798
50.5mm; Samar.
Shell medium sized, elongate-ovate, sutures weakly channeled, body whorl rounded, canal short and recurved; sculptured with varices and overriding granulose spiral cords; aperture white with a red margin, a thick, dentate outer lip and a lirate columella; surface with reddish-brown cords and black interspaces, varices with white bands, denticles and lirae white; reaches 60mm.
DISTRIBUTION: Sporadically found throughout the Philippines, mostly from Eastern Samar and Sulu Sea.

20. Cymatium (Septa) rubeculum (Linné, 1758)
44.2mm; Eastern Samar.
Shell small, rather squat, with a tall spire, a rounded body whorl, and a short, recurved siphonal canal; sculptured with 4 rows of finely noduled spiral cords, connected with fine axial threads; aperture ovate, outer lip thick, dentate; columella lirate; deep red with narrow orange banding and white patches on the varices; aperture white, denticles red and white, columella lirae white with red interspaces; reaches 50mm.
DISTRIBUTION: Sporadically found throughout the Philippines, mostly from Eastern Samar and Sulu Sea.

Plate 31 Cymatiidae

1. Cymatium (Septa) tenuiliratum (Lischke, 1873)
42.1mm, 29.0mm, 22.5mm; Samar.
Shell medium sized, fusiform, spire tall and pointed, sutures moderately channeled, siphonal canal long, open and recurved at the end; sculptured with fine axial and spiral lirae and 4 nodules between the pairs of rather thin varices; aperture ovate, white, outer lip dentate, columella lirate; creamy white with brown banding and blotching and white bands on the varices; sculpture and colour more intense in younger specimens; variable in surface sculpturing; reaches 65mm.
DISTRIBUTION: Visayan Sea, Camotes Sea and Sulu Sea.

2. Cymatium (Septa) vespaceum (Lamarck, 1822)
44.4mm; Punta Engaño, Cebu.
Somewhat variable, normally medium sized, elongate-ovate, with a tall spire, weakly incised sutures, and a long siphonal canal recurved at the end; sculptured with varices, spiral cords and striae, all forming nodules with overriding axial ribs and threads; nodules strongest at the shoulder; aperture ovate, outer lip with 6 to 7 bifid teeth, columella lirate; yellowish grey with white and brown markings, aperture white; reaches 50mm.
DISTRIBUTION: Visayan Sea, Camotes Sea and Sulu Sea.

3. Cymatium (Septa) comptum (A. Adams, 1855)
Often confused with *C. vespaceum* but differs in the following ways; the shell is lighter in weight, nearly always brightly coloured (brown, red, and orange), lacking the greyish tints of *vespaceum*; the sutures are more deeply impressed; the varices are more prominent; the denticles inside the outer lip are finer and closer, the nodules are larger and fewer in number, the surface sculpturing is much finer and the spiral cords remain single on the varices, whereas in *vespaceum* they divide into two; reaches 40mm.
3a. typical form; 31.7mm, 30.9mm; Cebu.
3b. various colour forms; 30.0mm, 31.9mm, 25.8mm; Cebu.
DISTRIBUTION: Visayan Sea, Camotes Sea and Sulu Sea.

4. Gyrineum (Gyrineum) bituberculare (Lamarck, 1816)
Shell medium sized, rather pyramidal in outline with oblique, axially aligned varices, impressed sutures, a rounded body whorl, and a long, recurved siphonal

canal; sculptured with axial and spiral cords (both variable), forming nodules at the points of intersection; aperture ovate, outer lip thick with a raised, dentate outer margin, columella lirate; cream-coloured with brown nodules and varices, aperture white; reaches 45mm.
4a. typical form; 42.4m; Palawan.
4b. form *fuscocostata* (Dunker, 1862); a more heavily noduled form; 21.5mm; Palawan.
DISTRIBUTION: Visayan Sea, Camotes Sea and Sulu Sea.

5. Gyrineum (Gyrineum) cuspidatum (Reeve, 1844)
28.7mm, 26.1mm; Cebu.
Close to some forms of *G. bituberculare*, but appears to differ in being smaller, more pyramidally ovate, having 2 to 3 large, rounded, axially aligned intervarical nodes, a shorter siphonal canal, and is nearly always orange brown with white varices or white with yellow-orange banding; generally believed to be an extreme ecomorph of *bituberculare*, but is a good distinct taxon.
DISTRIBUTION: Visayan Sea and Camotes Sea.

6. Gyrineum (Biplex) perca (Perry, 1811)
49.1mm, 35.7mm, 36.9mm; Cebu.
Similar to *G. aculeatum*, but bigger, with a pale orange surface and white nodules and variceal tips; it is wider, the sutures are less strongly channeled, the lower whorls are more dorso-ventrally flattened, the variceal wings are less expanded and have smaller spines, and the surface is more glossy; variable in nodular sculpturing; reaches 45mm. in the Philippines. Japanese specimens often reach 100mm.
DISTRIBUTION: Tayabas Bay (Quezon) and Camotes Sea.

7. Gyrineum (Gyrineum) gyrinum (Linné, 1758)
32.9mm, 29.8mm; Samar.
Shell small, fusiform-ovate, with a tall pointed spire, a short, strongly reflexed siphonal canal, and nodular sutures; varices axially aligned; surface nodular from intersecting axial and spiral cords; aperture ovate, outer lip thick and dentate, columella weakly lirate; white to yellow with a dark reddish brown spiral band on the spire whorls and two on the body whorl; aperture white; reaches 45mm.
DISTRIBUTION: Throughout the Philippines.

8. Gyrineum (Biplex) aculeatum (Schepman, 1909)
42.0mm; Cebu.
Shell medium sized, dorso-ventrally compressed, sutures deeply channeled, varices winged and axially aligned; sculptured with axial cords and spiral threads intersecting to form nodules; aperture ovate, outer lip thick, raised and dentate, columella smooth; brownish-purple with pale brown nodules and varices, and a white aperture; reaches 50mm.
DISTRIBUTION: Tayabas Bay (Quezon) and Camotes Sea.

9. Gyrineum (Gyrineum) roseum (Reeve, 1844)
19.7mm, 18.5mm, 20.5mm; Punta Engaño, Cebu.
A small, rather short shell, immediately recognized by its intense pink colouration, paler nodules, cream aperture, and the orange and white banded varices; shell variably noduled from intersecting axial and spiral cords, strongest on the shoulder; outer lip dentate, columella lirate but smooth centrally; reaches 20mm. Considered by some to be a form of the next species, but the absence of intergrades and the sympatric occurrence of the two make this contention rather tenuous.
DISTRIBUTION: Camotes Sea and Eastern Samar.

10. Gyrineum (Gyrineum) pusillum (Broderip, 1833)
22.0mm, 19.0mm; Punta Engaño, Cebu.
Very close to the preceding species but not as wide, and has a distinctly clathrate surface; the shell is brown-and-white banded and lacks the intense pink colouration of *G. roseum*; the aperture is purple tinged, and although the apertural features are very similar, *G. pusillum* has a columella that is entirely plicate, lacking the centrally smooth area of *G. roseum*; reaches 30mm.
DISTRIBUTION: Eastern Samar, Camotes Sea and Sulu Sea.

11. Cymatium (Turritriton) labiosum (Wood, 1828)
24.5mm, 22.3mm; Cebu.
Shell small, ovate, tall spired, with a rounded body whorl, a short, recurved siphonal canal, a heavy fasciolar ridge and weakly incised sutures; whorls sharply angled, sculptured with axial and spiral cords intersecting to form nodules, strongest at the shoulder where they are bisected by a spiral groove; spiral cords bear 3 riblets; aperture ovate, outer lip heavily dentate, columella lirate, smooth centrally; cream to greyish brown with paler nodules; reaches 25mm.
DISTRIBUTION: Eastern Samar, Camotes Sea and Sulu Sea.

12. Cymatium (Septa) gemmatum (Reeve, 1844)
Shell small, beautifully sculptured, fusiform-elongate, with a tall spire, impressed sutures, and a moderately long, recurved siphonal canal; sculptured with varices every 270°, finely noduled spiral cords, overriding interstitial axial striae and 3 to 4 weak intervarical axial ribs; aperture ovate, outer lip with short bifid teeth, columella lirate; orange-brown with darker

PLATE NO. 31 1. *Cymatium tenuiliratum* (Lischke) 2. *Cymatium vespaceum* (Lamarck) 3a. *Cymatium comptum* (A. Adams) 3b. colour forms 4a. *Gyrineum bituberculare* (Lamarck) 4b. form *fuscocostata* (Dunker) 5. *Gyrineum cuspidatum* (Reeve) 6. *Gyrineum perca* (Perry) 7. *Gyrineum gyrinum* (Linné) 8. *Gyrineum aculeatum* (Schepman) 9. *Gyrineum roseum* (Reeve) 10. *Gyrineum pusillum* (Broderip) 11. *Cymatium labiosum* (Wood) 12a. *Cymatium gemmatum* (Reeve) albinistic form, unusually large 12b. typical form 13. *Cymatium mundum* (Gould) 14. *Sassia nassariformis semitorta* (Kuroda & Habe) 15. *Cymatium pfeifferianum* (Reeve)

PLATE 31

brown and white varical bands; reaches 45mm.
12a. an unusually large albinistic form; 44.4mm; Sulu Archipelago.
12b. typical Philippine form; 30.5mm, 27.5mm; Cebu.
DISTRIBUTION: Camotes Sea and Sulu Sea.

13. Cymatium (Septa) mundum (Gould, 1849)
28.2mm, 26.9mm; Bohol.
Incorrectly illustrated as *C. gemmatum* by many authors, but is white, shorter and wider, and more coarsely sculptured; the varices are thicker and the intervarical cords are more pronounced especially on the spire whorls; generally reaches a slightly larger size (35mm) than typical Philippine *gemmatum* (around 30mm).
DISTRIBUTION: Camotes Sea and Sulu Sea.

14. Sassia nassariformis semitorta (Kuroda & Habe, 1961)
45.9mm; Cebu.
Shell medium sized, fusiform-elongate, with a tall spire, impressed sutures, and a moderately long, recurved siphonal canal; sculptured with varices (every 270°) and intersecting axial and spiral cords forming axially orientated nodules; aperture white, ovate, outer lip thick with a raised dentate outer lip, columella lirate with an expansive callus; cream with brown nodules and varices; reaches 50mm.
DISTRIBUTION: Camotes Sea and Sulu Sea.

15. Cymatium (Septa) pfeifferianum (Reeve, 1844)
56.5; Samar.
Shell medium sized, spire tall, siphonal canal long and recurved at the end; whorls inflated, sculptured with prominent varices and many fine, closely-set nodules at the intersections of axial and spiral threads; outer lip lirate, lirae continuing into the aperture, columella plicate; white with brown bands and lines, aperture white; reaches 75mm.
DISTRIBUTION: Eastern Samar, Camotes Sea and Sulu Sea.

ADDENDA

A. The species *Gyrineum (Gyrineum) natador* (Röding, 1798) and *Cymatium (Septa) intermedium** (Pease, 1869) (not illustrated) have, on occasions, been reported from Philippine waters, but it is almost certain that both these taxa are absent from the Philippine faunal list.

B. During the course of preparation of the Cymatiidae, 2 species not previously covered in the section were brought to the attention of the authors. These taxa, namely *Cymatium (Ranularia) gallinago* (Reeve, 1844) and *Cymatium (Ranularia) trilineatum* (Reeve, 1844) were recorded from the Philippines by Reeve (see Conc. Icon. "Triton", Vol. 2: 1844). These 2 species represent 2 members of a complex triad of shells, the ecology of which is poorly known. Neither of these species has been seen by the authors and Reeve's cited localities must be treated with caution as they were taken from the Hugh Cuming collection which is known to contain numerous erroneous locality designations. We present here the original description of each species, together with a copy of the type figure and pictures of the B.M. (N.H.) syntypes. They appear to be ecomorphs of a single taxon.

Cymatium (Ranularia) gallinago (Reeve, 1844)
Syntype (BMNH: 1967593); Locality: "Philippines" (ex Hugh Cuming collection); size: 60mm. Description (from Reeve): "Shell abbreviately club-shaped, with 2 round solid varices; spire rather short, acuminated; whorls angulated at the upper part, armed at the angle with a row of sharp, peculiarly flatly-pinched tubercles, tubercles on the last whorl very irregular; whorls ribbed beneath, ribs crenulated, flexuously noduled under the tubercles, interstices between the ribs sculptured with raised striae; white, varices stained with bright orange-brown; columella rugosely plaited, interior of the aperture white, lip strongly denticulated within; canal rather long, turned upwards."

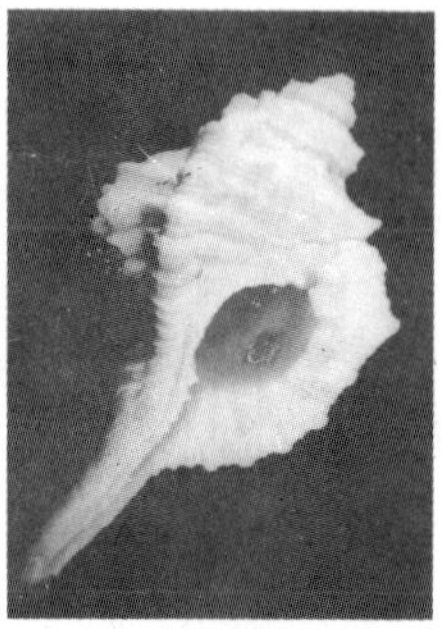

Type figure of *Triton gallinago* Reeve (from Reeve)
Syntype of *Triton gallinago* Reeve (BMNH) (Photo: A. G. Beu)

Cymatium (Ranularia) trilineatum (Reeve, 1844)
Syntype (BMNH: 1967627); Locality: "Philippines" (ex Hugh Cuming collection); size: 66mm. Description

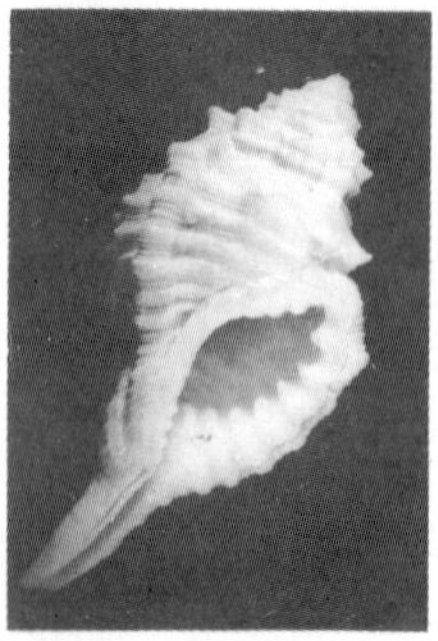

Type figure of *Triton trilineata* Reeve (from Reeve)
Syntype of *Triton trilineata* Reeve (BMNH) (Photo: A. G. Beu)

*More recently, a specimen in the author's collection from Punta Engaño, was verified as being this taxon (fide Dr. A.G. Beu).

(from Reeve); "Shell clavately-fusiform, with 3 varices; spire rather short; whorls angulated at the upper part, compressly tubercled at the angle, transversely flatly ridged, ridges very finely crenulated, interstices sculptured with three raised lines; whitish, variegated with brown; columella yellowish white, ribbed across; canal rather elongated, slightly turned upwards; lip strongly denticulately ribbed within; interior of aperture white." Most likely a form of *C. gallinago.* (fide Beu).

Plate 32 Cymatiidae

Sub Family Distorsioninae

A small but distinctive subfamily within the Cymatiidae, characterized by their distorted shape, irregular whorls, discontinuous varices, glazed columellar shield and an irregular, black chitinous operculum with a submarginal nucleus.

1. Distorsio (Distorsio) reticulata (Röding, 1798)
Shell medium sized, one of the least distorted of all the species; spire tall, body whorl large, rounded, siphonal canal long, recurved; sculptured with intersecting axial and spiral çords, clathrate on the earlier whorls, more widely reticulate on the later whorls; outer lip flattened, expanded, with 8 to 10 denticles, the first 2 bifid, the third usually largest; columella heavily denticulate, size of denticles decreasing anteriorly; columellar callus covering most of the previous whorl, with 2 parietal teeth; light to dark brown, callus pale pink to orange-brown, denticles white; largest seen 94.2mm. Extremely variable.
1a. typical form; 67.5mm; Manila Bay.
DISTRIBUTION: Manila Bay, Tayabas Bay (Quezon) and Visayan Sea.
1b. A form of *reticulata* in which the spire is somewhat more elongate than normal, the body whorl somewhat more inflated, and the surface more finely reticulate. Such specimens are usually taken in intermediate depths of between 50 and 100 meters; 71.5mm; Cebu.
DISTRIBUTION: Visayan Sea and Camotes Sea.
1c. form *ridens* (Reeve, 1844)
A deep water form found living at depths exceeding 100 meters; characterized by its more elongate shape and more finely reticulate surface; 85.0mm, 74.8mm; Punta Engaño, Cebu.
DISTRIBUTION: Camotes Sea, Tayabas Bay (Quezon) and Manila Bay.
1d. Two juvenile specimens of *reticulata,* illustrating the close morphological similarity to specimens of *D. perdistorta.* The similarity is so close in some specimens that separation can only be done by a close examination of apertural characteristics; 44.5mm, 42.9mm; Manila Bay.
DISTRIBUTION: Camotes Sea, Tayabas Bay (Quezon) and Manila Bay.
1e. form *francesae* Iredale, 1931
Another form of *reticulata* in which the shell is smaller and more squat than normal, and the surface sculpturing consists of widely spaced axial cords and spiral lirae producing a wide, latticed appearance, devoid of any pronounced nodulation; 42.8mm, 37.9mm; Manila Bay.
DISTRIBUTION: Camotes Sea, Tayabas Bay (Quezon) and Manila Bay.

2. Distorsio (Distorsio) anus (Linné, 1758)
73.5mm, 78.0mm; Palawan.
Shell large, heavy, fusiform-ovate, with a tall spire, a large body whorl, and a short, strongly recurved siphonal canal; surface with axial and spiral cords, clathrate on the spire whorls, nodulose on the body whorl; outer lip wide, flattened, strongly dentate; columella with an expansive plate-like callus covering the lower part of spire, heavily noduled and coloured pale brown with white streaks; shell white with brown banding; reaches 80mm.
DISTRIBUTION: Throughout the Philippines.

3. Distorsio (Distorsio) kurzi Petuch & Harasewych, 1980
45.0mm, 42.9mm; Bohol.
Shell medium sized, tall spired, grossly distorted, the whorls with a distinct, sharply angled "hump" on the shoulder; sculptured with intersecting axial and spiral cords, forming nodules; spiral cords wide but close, and coalescing on the shoulder; outer lip wide, flattened, with 2 rows of denticles, the third on the inner lip margin the largest; columella with 13 to 16 denticles; columellar callus wide, expanded and transparent; colour orange brown (including aperture) with darker banding and white denticles; reaches 50mm.
DISTRIBUTION: Known only from Bohol, but probably has a wider distribution, as specimens have been taken from the Andaman Islands.

4. Distorsio (Distorsio) constricta habei Lewis, 1972
49.0mm, 17.4mm; Panglao, Bohol.
A medium sized shell with a tall spire, the whorls of which appear as if axially compressed so that the last 3 have a pronounced distorted shape; surface granulose from intersecting axial and spiral cords; whorls roundly angulate with several closely spaced cords on the shoulder, separated from the first row by a pronounced gap; aperture with a deeply concave parietal wall, outer lip flatly expanded with around 8 denticles, the third being the largest; columella with 12 to 14 closely set plicae, vertically aligned; siphonal canal short, almost closed off posteriorly against the outer lip; pale brown with golden brown blotching, especially on the shoulder cords, aperture white; reaches 50mm.
DISTRIBUTION: Camotes Sea.

5. Distorsio (Distorsio) perdistorta Fulton, 1938
Shell medium sized, spire tall with some lateral distortion on last 3 whorls, body whorl rounded and swollen; surface nodulose from intersecting axial and spiral cords; distance between the first and second spiral cords greater than between lower pairs; four closely set spiral cords at the base; aperture irregularly ovate, with a deeply concave parietal wall; outer lip flattened, with 9 denticles, the third one largest; columella with around 14 closely set plicae, weakly channeled behind; anterior canal long, posteriorly constricted and recurved; columellar shield thin with a single parietal denticle; almost pure white; reaches 75mm.
5a. typical form; 50.9mm, 60.9mm; Punta Engaño, Cebu.
5b. An ecomorph of *perdistorta* in which the body whorl is lightly less ovate and the last 3 or 4 whorls show less lateral distortion; there is also some brown colouration becoming evident especially on the shoulder cords; this is a good example of character displacement towards a less distorted form similar to that seen in *D. reticulata* form *ridens* (fig. 1c); 49.0mm, 60.0mm; Bohol.
DISTRIBUTION: Camotes Sea.

6. Distorsio (Distorsio) decipiens Reeve, 1844
62.9mm, 48.8mm, 47.4mm, 36.5mm, 32.0mm, 32.5mm; Punta Engaño, Cebu.
Appears endemic to the Philippines and probably represents a replacement for *D. constricta habei* which seems to be a fringe taxon here. Shell medium sized, solid, and tall spired; whorls are roundly distorted but have a concave shoulder slope and a somewhat sharply angled shoulder on which are situated 2 closely set cords, giving the shell a distinctly "hunched" appearance; large gaps separate these shoulder cords from the first and fourth cords; aperture moderately large, irregular in contour, with a thin columellar shield and 1 or 2 parietal denticles; outer lip flattened, with 9 to 10 large denticles, the third largest and often bifid; columella with 14 to 16 plicae, size decreasing anteriorly; canal unusually long for the genus, posteriorly constricted and recurved; white with golden brown blotching, especially on the spire whorls and on the parietal surface; largest size seen 68mm. Plate illustrates a growth series demonstrating the consistency in form and colouration from the earliest stages of development. Known only from the Philippines.* Usually considered a synonym of *D. reticulata* (Röding), but is a valid taxon.
DISTRIBUTION: Camotes Sea.

A BRIEF SYNOPSIS OF THE PHILIPPINE DISTORSIO

An examination of the *Distorsio* species occurring in Philippine waters, reveals a very close relationship existing between some members, especially those occurring in deep water. These species include *D. kurzi* (fig. 3), *D. constricta habei* (fig. 4), *D. perdistorta* (figs. 5a and 5b) and *D. decipiens* (fig. 6). Indeed, the morphological similarity between these four species is so close, that they can be arranged in a sequence which very clearly demonstrates a gradual shift in shape. This phenomenon is possibly an example of character displacement. It is most strikingly obvious when one compares specimens of *D. constricta habei* with *D. decipiens. D. constricta habei* is rare, and almost certainly a fringe taxon in the Philippines, and it is possible that *D. decipiens* is a replacement for it here, judging from their extremely close morphology. The taxonomic status of *D. decipiens* is borne out by a growth series and the lack of intergrades between it and *D. constricta habei.* Another species exhibiting this phenomenon is *D. perdistorta.* There appears to be 2 forms of this species, one (fig. 5a) extremely close to *D. constricta habei,* and another form (fig. 5b), which is far less distorted and appears to occur sympatrically with the typical form.

In both the above two cases, we are observing character displacement from a more to a less distorted form. The significance of this phenomenon is, as yet, unclear, but it is easy to understand how many of these species have been confused in the past. A possible explanation is character displacement because they occur with the extremely distorted Philippine endemic *D. kurzi*** which does not occur with either *D. constricta habei* or *D. perdistorta* over most of the geographical ranges of the latter two species. So, the genetic composition of Philippine populations of strongly distorted species tends to vary away from the shape of *D. kurzi* towards less distorted shapes. This would indicate that they compete closely with *D. kurzi* for food or other environmental factors.

The development of the *ridens* form of *D. reticulata* (fig. 1c), is another example of the interesting variation of Philippine *Distorsio*. Examination of large suites of material has established that this is mainly a depth-dependant phenomenon, with the shallow water form of *reticulata* (fig. 1a), being moderately distorted, and having coarse sculpturing. Specimens from intermediate depths (fig. 1b) show the development of larger, less distorted, finely granulose forms, until one reaches deep water, with the emergence of the *ridens* form, which is less distorted, more elongate, and has a more finely granulose surface than its shallow water counterparts. It is conceivable that depth may not be the only factor at work here, but also the increasing presence of other very closely related taxa which have successfully established themselves in deeper water, and can thus exert some influence on the developmental morphology of *reticulata.* The degree of interaction between environment and the presence of a large number of other taxa in the same genus is uncertain, and one can only speculate as to which factor is primarily responsible for the unusual variation seen within members of the Philippine *Distorsio* species.

---

* Recently recorded from South Africa (fide Beu).
** *D. kurzi* has been reported from Japan, the Andaman Islands, and more recently South Africa.

PLATE NO. 32 1a. *Distorsio reticulata* (Röding) 1b. an ecomorph of *reticulata* (Röding) 1c. form *ridens* (Reeve) 1d. juvenile specimens 1e. form *francesae* Iredale 2. *Distorsio anus* (Linné) 3. *Distorsio kurzi* Petuch & Harasewych 4. *Distorsio constricta habei* Lewis 5a. *Distorsio perdistorta* Fulton 5b. an ecomorph of *perdistorta* Fulton 6. *Distorsio decipiens* Reeve (growth series)

PLATE 32

Class Gastropoda
Subclass Prosobranchia
Order Mesogastropoda
Suborder Taenioglossa
Superfamily Tonnacea
Family BURSIDAE

The Bursidae are a relatively small group of Gastropods closely related to the Cymatiidae but differing in the presence of a deep notch in the outer lip at the posterior end of the aperture. The members of this group are strongly variced, have a horny operculum, a short anterior canal, may be strongly nodular or spinose, frequently have thick coralline encrustations, and lack a periostracum. They are mainly tropical in their habitat and prefer a coral substrate. The ecology of the family is little known. The family is very well represented in the Philippines.

1. Tutufa (Tutufa) bufo (Röding, 1798)
Syn: *B. lissostoma* Smith, 1914
135mm, 103.1mm; Bohol.
Shell large, heavy, fusiform-ovate, sculptured with varices (every 240°), granular spiral cords, and a row of 5 to 6 axially compressed nodules on the shoulder of each whorl, with additional weaker rows on the body whorl; similar to *T. bubo*, but never attains its size; the columella is smooth, not wrinkled, and there is a characteristic red rim within the white aperture; reaches 140mm.
DISTRIBUTION: Throughout the Philippines in limited quantity.

2. Bursa (Bursa) lamarckii (Deshayes, 1853)
55.2mm, 56.5mm; Cebu.
This is the "unnamed, black-mouthed" species of various authors. Shell small to medium sized, rugosely sculptured, with axially aligned varices which bear previous siphonal appendages and a row of 2 or 3 large nodules between each pair of varices; aperture ovate, outer lip dentate with a raised margin, columella finely lirate, grey to pale yellow with purple blotching, outer apertural area dark brown to deep purple with orange denticles and lirae; reaches 65mm.
DISTRIBUTION: Camotes Sea and Sulu Sea in limited quantity.

3. Bursa (Bursa) leo Shikama, 1964
44.5mm, 42.0mm; Manila Bay.
A small, ovate, rather coarse shell with a short spire, weakly impressed sutures, and prominent siphonal appendages on the varices of the earlier whorls; sculptured with axially aligned varices, heavily noduled spiral cords and deep pits, especially on the body whorl; aperture ovate, lirate on the columella, irregularly wrinkled centrally; outer lip expanded, marginally crenulate, 5 to 6 pairs of bifid denticles on the inner margin, anterior canal short; greyish yellow in colour, lirae darker, aperture white; reaches 45mm.
DISTRIBUTION: Throughout the Philippines in limited quantity.

4. Tutufa (Tutufella) oyamai Habe, 1973
78.0mm; Palawan.
Shell medium sized, elongate-ovate, with a tall spire, impressed sutures, a rounded body whorl, and varices every 270°; sculptured with finely granulose major and minor spiral cords forming axially compressed nodules (strongest at the shoulders) where they cross weak axial folds; aperture ovate, outer lip expanded, crenulate and dentate on the inner margin, columella with irregular folds on a thin callus; anterior and posterior canals long; orange-brown with darker cords and varices, aperture white, light brown within; reaches 85mm.
DISTRIBUTION: Throughout the Philippines in limited quantity.

5. Bursa (Colubrellina) condita (Gmelin, 1791)
Syn: *B. candisata* Lamarck, 1822
108.8mm, 107.7mm; Sulu Archipelago.
Shell large and heavy, with a tall spire and a large body whorl; varices prominent, slightly off-set axially, and numbering 2 per whorl; sculptured with numerous fine, close-set, beaded spiral cords, some stronger than others; aperture elongate-ovate, heavily lirate within, outer lip slightly flared and variced; columella calloused, anteriorly and posteriorly lirate, centrally pustulose; yellowish white with brown-spotted cords and brown-banded varices, aperture white; reaches 110mm.
DISTRIBUTION: Eastern Samar and Sulu Sea in limited quantity.

6. Tutufa (Tutufella) rubeta (Linné, 1758)
Shell large, heavy, with a tall spire and a large, ovate body whorl; sculptured with finely nodular spiral cords and a shoulder row of 4 to 6 axially compressed, bisected nodules between the varices (which occur every 240°); aperture large, ovate, with widely open anterior and posterior canals; outer lip thick, expanded, with 2 rows of white denticles, the inner row forming short lirae in many specimens; columella with white folds on a heavily calloused parietal wall; orange-brown with darker banding, pale nodules, and a deep red aperture; reaches 110mm.
6a. typical form; 108.3mm, 103.1mm; Samar.
6b. juvenile; 43.0mm; Samar.
DISTRIBUTION: Throughout the Philippines.

7. Bufonaria (Bufonaria) crumena (Lamarck, 1816)
61.5mm, 61.2mm; Manila Bay.
Shell medium sized, heavy, somewhat ovate in

PLATE NO. 33 1. *Tutufa bufo* (Röding) 2. *Bursa lamarckii* (Deshayes) 3. *Bursa leo* Shikama 4. *Tutufa oyamai* Habe 5. *Bursa condita* (Gmelin) 6a. *Tutufa rubeta* (Linné) 6b. juvenile 7. *Bufonaria crumena* (Lamarck) 8a. *Tutufa bubo* (Linné) 8b. juvenile 9. *Bufonaria subgranosa* (Sowerby) 10. *Bufonaria gnorima* (Melvill) 11. *Bufonaria nobilis* (Reeve) 12a-12b. *Bufonaria margaritula* (Deshayes) 13. *Tutufa bubo* (Linné) form *tenuigranosa* (E. A. Smith)

PLATE 33

outline, with axially aligned and weakly spinose varices, weakly channeled sutures, a moderately tall spire, and sculptured with nodulose spiral cords, a row of spinose tubercles on the whorl shoulder and 3 rows on the body whorl; aperture ovate, with prominently open posterior and anterior canals; outer lip expanded and denticulate; columella thinly calloused with numerous plicae; pale orange-brown with white and dark brown nodules, interior of aperture white, outer lip margin and anterior columellar area orange; reaches 90mm.
DISTRIBUTION: Throughout the Philippines.

8. Tutufa (Tutufa) bubo (Linné, 1758)
Syn: *B. gigantea* Smith, 1914; *B. tenuigranosa* Smith, 1914 (see fig. 13)
Shell very large, heavy, with an inflated body whorl and tall spire; sculptured with varices every 240° and nodulose spiral cords, with a shoulder row of large tubercles on the spire whorls and 3 or 4 on the body whorl, those below the shoulder progressively weaker; aperture ovate, outer lip flared, heavily calloused and marginally dentate; columella heavily glazed with a few folds; colour white to yellow with brown blotches, aperture pale pink (white in immature specimens), orange deep inside the shell; reaches 270mm.
8a. typical form; 197mm; Bohol.
8b. juvenile, 53.0mm; Palawan.
DISTRIBUTION: Throughout the Philippines.

9. Bufonaria (Bufonaria) subgranosa (Sowerby, ex Beck MS, 1836)
Syn: *B. beckii* (Kiener, 1841)
64.8mm, 68.7mm; Manila Bay.
Shell medium sized, fusiform-ovate, dorso-ventrally compressed, with a tall spire and a large body whorl; varices axially aligned, 2 per whorl; whorls sculptured with many fine, close-set, nodulose spiral cords and a shoulder row of spinose tubercles between the varices; aperture ovate, outer lip dentate, columella lirate, siphonal canal short, open and recurved, the earlier siphonal appendages with spinose margins; pale yellow with brown blotching and banding, aperture white; reaches 100mm. This species is endemic to the Philippines and appears to replace the superficially similar *B. elegans* (Sowerby), *B. echinata* (Link), and *B. rana* (Linné), all three of which occur further West; some transition forms between *rana*, *elegans* and *subgranosa* have been observed in a few specimens taken from Palawan. Sold locally under the name *"Bursa rana"*, a species which almost certainly doesn't occur in the Philippines. Although considered endemic to the Philippines, recent information suggests a much wider distribution, with specimens exactly matching Philippine shells in their morphology having been trawled off Cape Moreton, Queensland, Australia (Fide A. G. Beu).
DISTRIBUTION: Throughout the Philippines.

10. Bufonaria (Bufonaria) gnorima (Melvill, 1918)
82.5mm, 79.7mm; Samar.
A medium sized, rather heavy shell, best described as "a tall-spired *margaritula*" or "a *nobilis* without variceal spines"; shell fusiform-ovate with rounded whorls, and sculptured with numerous fine, granulose spiral cords and a row of 5 to 6 spinose tubercles between the axially aligned varices; aperture elongate-ovate, with a heavily lirate columella, and a marginally raised, slightly flaring outer lip which is strongly plicate; light brown with white and dark brown nodules and banding, aperture white with an orange-yellow margin; reaches 85mm.
DISTRIBUTION: Throughout the Philippines.

11. Bufonaria (Bufonaria) nobilis (Reeve, 1844)
51.3mm, 54.8mm; Samar.
Shell medium sized, fusiform-ovate, with a tall spire, adpressed sutures, and sharply spinose, widely expanded, axially aligned varices; sculptured with fine, granulose spiral cords and a row of axially flattened nodules between the varices; aperture narrowly ovate, outer lip dentate with a raised margin which is slightly flaring; columella lirate, weakly so centrally; yellow-brown with brown-and-white banding, aperture white; like a tall-spired spinose *margaritula* with which it is often confused; reaches 60mm.
DISTRIBUTION: Throughout the Philippines.

12. Bufonaria (Bufonaria) margaritula (Deshayes, 1832)
Shell small to medium sized, fusiform-ovate with a short, sharp spire, weakly impressed sutures and sculptured with numerous fine, closely set granulose spiral cords and a row of 5 axially compressed shoulder nodules between each pair of varices which occur every 180°; aperture elongate-ovate, outer lip thick, dentate, with a raised, slightly flaring margin; columella lirate, weakly so centrally; yellowish brown with white and dark brown nodules and tubercles, aperture white, marginally orange; reaches 55mm; extremely variable in colour and sculpturing, some specimens being very dark brown and sculptured with an extremely high number of granulose spiral cords.
12a. A small, dark coloured, finely granulose specimen; 32.0mm; Palawan.
12b. typical form; 45 mm, 51.2 mm; Palawan.
DISTRIBUTION: Throughout the Philippines.

13. Tutufa (Tutufa) bubo (Linné, 1758) form **tenuigranosa** (E. A. Smith, 1914)
180mm, 184mm; Samar.
Once considered a distinct species, now believed to be a deep water form of *T. bubo* (Linné). It differs from typical *bubo* by its relatively more elongate shape, less coarsely noduled surface, an all-white aperture, flared inner and outer lips and a coarsely plicate columellar margin; close to *oyamai* in general apertural characteristics; largest size seen 184mm.
DISTRIBUTION: Throughout the Philippines in limited quantity.

1. Bufonaria sp.
An apparently unnamed species which has been erroneously figured by many authors as *B. rana* (Linné), a species whose presence is suspected but not yet confirmed in Philippine waters. Shell large, fusiform-ovate, somewhat dorso-ventrally compressed; spire moderately tall, sutures weakly impressed, body whorl large; sculptured with spinose, axially aligned varices, and the surface is covered with many close, finely nodulose spiral cords, a row of spinose tubercles on the shoulder of the spire whorls, and 2 to 3 on the body whorl; aperture widely elongate-ovate, with a thick, dentate outer lip bearing a unique inner lip flange, columella finely lirate; posterior and anterior canals prominent; pale brown with large areas of orange-brown suffusion, a white aperture, and orange blotching on the outer lip margin; reaches 140mm. Sold locally as *"Bursa spinosa"* [a synonym of *B. echinata* (Link)], a species which does not occur in the Philippines.
1a. Unusually large specimen; 131.8mm; Manila Bay.
1b. An unusually granular form; 64.5mm; Manila Bay.
DISTRIBUTION: Mostly found in Manila Bay and Batangas Bay; may occur all over the archipelago.

2. Bursa (Colubrellina) granularis (Röding, 1798)
A distinctive but highly variable species which has received no fewer than 10 different names. Shell solid, tall-spired, with low, axially aligned varices; whorls sculptured with weak spiral ribs bearing small to heavy axially compressed nodules (nodule size is highly variable); aperture ovate, outer lip thick with a raised margin, and 10 to 12 pairs of denticles which extend as lirae for a short distance into the aperture; columella yellowish, with darker staining between the lirae on the parietal wall; reaches 70mm.
2a. typical form; 68.9mm; Samar.
2b. form *affinis* Broderip, 1832; a form with large tubercles; 43.9mm; Palawan.
DISTRIBUTION: Throughout the Philippines.

3. Bursa sp.
60.0mm; Punta Engaño, Cebu.
Shell medium sized, solid, with a tall spire and a large, stout body whorl; 2 varices per whorl (approximately every 220°), surface sculptured with a row of nodules on the spire whorls, and 2 to 3 on the body whorl; aperture elongate-ovate, outer lip thin, highly erect with a crenulate margin, and 8 to 10 denticles on the inner margin; columella with a thin, highly expanded callus and numerous lirae along the margin; cream with pale orange-brown mottling and darker brown spots and streaks, aperture white with a dark brown blotch on the upper parietal wall; reaches 65mm. Superficially similar to *ranelloides* Reeve, but appears unnamed at this time.
DISTRIBUTION: Visayan Sea and Camotes Sea.

4. Bursa (Bursa) tuberosissima (Reeve, 1844)
44.3mm; Manila Bay.
Shell small to medium sized, ovate in outline, with adpressed sutures, axially aligned varices, and sculptured with major and minor granulose spiral cords and a shoulder row of large tubercles between the varices; aperture ovate, yellow in colour, smooth within; outer lip thick with a raised margin, dentate within, columella lirate, smooth centrally; colour pale whitish yellow with greyish nodules; reaches 45mm. Differs from *B. leo* in its paler colour and yellow aperture.
DISTRIBUTION: Throughout the Philippines.

5. Bursa (Bursa) rosa (Perry, 1811)
36.2mm; Palawan.
Shell medium sized, solid, sculptured with axial cords and overriding spiral threads, producing axially compressed nodules where they intersect on the shoulder of each whorl; varices axially aligned, 5 to 6 nodules between each pair, siphonal appendages prominent; aperture ovate, deep reddish purple within, outer lip thick and dentate, columella roughly and irregularly lirate; colour pale cream, nodules and cords darker, apertural denticles and columellar lirae white; reaches 55mm.
DISTRIBUTION: Visayan Sea, Sulu Sea, Palawan and Camotes Sea.

6. Bursa (Colubrellina) awatii Ray, 1949
Syn: *Bursa (Colubrellina) rehderi* Beu, 1978
59.0mm; Bohol.
A medium-sized, lightweight, thin-shelled species, elongate-ovate, with varices (every 270°), and sculptured with either fine or coarsely granulose spiral cords and fine axial striae; aperture white, ovate, outer lip weakly or strongly flared, smooth or weakly dentate, columella glazed and weakly lirate; anterior and posterior canals widely open; light brown with darker brown-and-white varices and nodules; reaches 75mm.
DISTRIBUTION: Visayan Sea and Camotes Sea, in limited quantity.

7. Bursa (Bursa) cruentata (Sowerby, 1835)
34.5mm; Palawan.
Shell small, solid, with concave whorls, axially aligned varices, and a large, ovate body whorl; sculptured with 2 rows of nodules on the spire whorls and 3 on the body whorl; spiral cords form projecting tubercles where they cross the varices; aperture ovate, outer lip thick, marginally expanded with 5 pairs of denticles, columella calloused and irregularly lirate; pale cream with pale violet-tinged nodules, aperture white with dark red-brown staining between the lirae of the upper parietal wall; reaches 50mm.
DISTRIBUTION: Visayan Sea, Sulu Sea and Palawan in limited quantity.

8. Bursa (Bursa) bufonia (Gmelin, 1791)
70.5mm; Cebu.
Shell medium sized, solid, heavy, fusiform-ovate, sculptured with axially aligned varices which exhibit prominent earlier siphonal appendages, thick, knobbed spiral cords, and fine spiral threads; aperture ovate, outer lip thick, with a raised, heavily dentate

outer lip margin, columella heavily lirate; colour greyish creamy white with darker mottling and a white aperture; reaches 80mm.
DISTRIBUTION: Throughout the Philippines.

9. Bursa (Bursa) rhodostoma (Sowerby, 1835)
24.3mm; Punta Engaño, Cebu.
Shell small, fusiform-ovate, with a tall pointed spire and axially aligned varices; sculptured with finely nodulose spiral cords, a row of large shoulder tubercles on the spire whorls, and 2 which may fuse axially on the body whorl. Aperture almost circular, deep reddish mauve in colour, dentate on the columella and on the marginally raised outer lip; colour pale brown with yellow tinged varices; reaches 25mm. A curious form of what appears to be this species is illustrated below. It is distinct in its pale pink aperture, and lirae on both the columella and outer lip margin which extend partly into the aperture; 18.3mm in size; Punta Engaño, Cebu.
DISTRIBUTION: Eastern Samar and Camotes Sea.

c.f. B. rhodostoma (Sowerby, 1835)
18.3mm; Punta Engaño, Cebu

10. Bursa (Colubrellina) latitudo aff. **wolfei** Beu, 1980
33.9mm; Punta Engaño, Cebu.
Shell small, lightweight, with a tall spire and a large, rather wide body whorl; varices every 240°; spire whorls sculptured with a row of axially compressed, spirally bisected, pointed nodules on the angled shoulder, and 3 rows on the body whorl, with weakly nodulose cords between the major rows; aperture ovate, outer lip reflected and lirate, columella heavily plicate; posterior canal short and open, anterior canal short, posteriorly constricted, widely open anteriorly and distally recurved; colour reddish brown with darker brown maculations; aperture white with a distinct reddish brown parietal patch; clearly a close congener of the Hawaiian subspecies, but its distinct sculpturing and size (seldom exceeds 35mm), suggest it may represent a hitherto unnamed subspecies of *latitudo*.
DISTRIBUTION: Camotes Sea.

ADDENDA

Bursa (Bursa) venustula (Reeve, 1844)
An extremely rare species, the authors being aware of only one authenticated specimen from the Philippines, presently in a private collection. The shell is somewhat intermediate between *B. rhodostoma* (Sowerby) and *B. lamarckii* (Deshayes) in size, sculpture and colouration. The shell is somewhat dorso-ventrally compressed and the nodules are smaller, more numerous, and more rounded than those of either *lamarckii* or *rhodostoma*. In comparison to the latter species, the outer lip is not flared over the varix and the aperture has a row of small distinctive nodules all around the inner margin. The colour is a pale orange brown and the aperture is a dark, purplish red-brown colour which extends only a short way into the aperture. Appears to reach around 40mm.

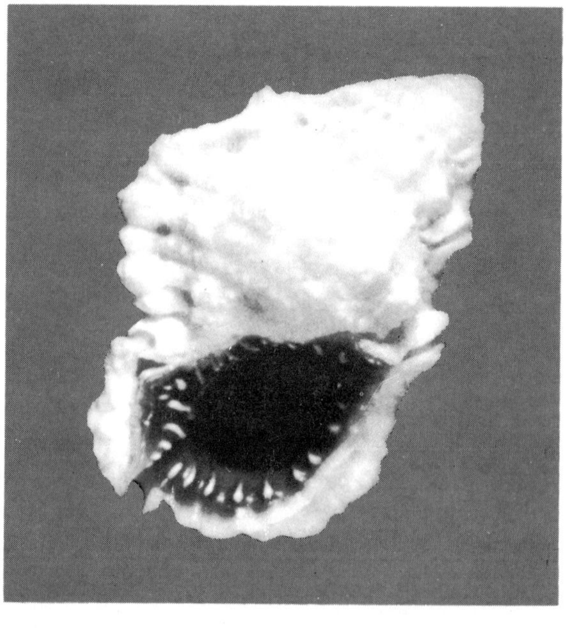

Renalla venustula Reeve, 1844 (syntype)
BM(NH) Reg. No. 196654½. 36.1 mm × 25.7 (Photo: A. G. Beu)

PLATE NO. 34 1a. *Bufonaria sp.* 1b. granular form 2a. *Bursa granularis* (Röding) 2b. form *affinis* Broderip 3. *Bursa sp.* 4. *Bursa tuberosissima* (Reeve) 5. *Bursa rosa* (Perry) 6. *Bursa awatii* Ray 7. *Bursa cruentata* (Sowerby) 8. *Bursa bufonia* (Gmelin) 9. *Bursa rhodostoma* (Sowerby) 10. *Bursa latitudo* aff. *wolfei* Beu 11. *Colubraria muricata* (Lightfoot) 12. *Colubraria soverbii* (Reeve) 13. *Colubraria nitidula* (Sowerby) 14. *Colubraria sp.* 15. *Colubraria castanea* Kuroda & Habe 16. *Colubraria sp.* 17. *Colubraria tortuosa* (Reeve) 18. *Colubraria antiquata* (Hinds) 19. *Antemetula metulina* Kuroda & Habe

1a
1b
2a
2b
3
4
5
6
7
8
9
10
11
12
13
14
15
16
17
18
19

ADDITIONAL NOTES ON THE BURSIDAE

Bursa (Bursa) rosa (Perry, 1811)
(Pl. 34, fig. 5)
This species has frequently been figured under the earlier name of *Bursa mammata* Röding, 1798. Röding based his description on figures from Martini & Chemnitz (Conchylien Cabinet, Vol. 4, Pl. 129, fig. 1240, 1241) which are illustrated here. The figures are bad and could equally apply to 1 or 2 other morphologically similar taxa. It is our opinion that the figures are unrecognizable and that *mammata* should be rejected as a *nomen dubium.*

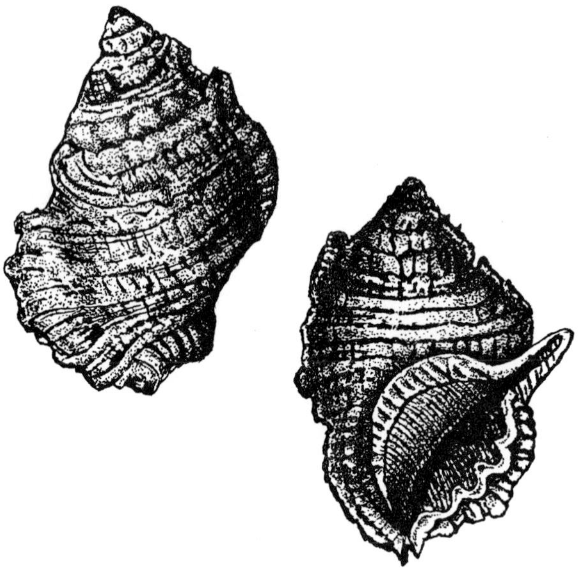

Type figures of *Bursa mammata* Röding (from Chemnitz)

Class Gastropoda
Subclass Prosobranchia
Order Neogastropoda
Superfamily Buccinacea
Family COLUBRARIIDAE

A relatively small group of intertidal, rock dwelling gastropods which inhabit most tropical seas, and appear to be suctorial in their feeding mode due to the absence of any radular ribbon. The shells vary from small to large, are tall and fusiform in outline, and possess convex whorls which are sculptured with varices and intersecting axial and spiral cords, producing a finely granulose surface. The aperture is small, ovate, with a dentate outer lip, and a thickened columellar callus which may be smooth, lirate, or finely pustulose. The siphonal canal is short, open, distally recurved, and the operculum is brown, corneous and has an anterior nucleus.

11. Colubraria (Colubraria) muricata (Lightfoot, 1786)
Syn: *Colubraria maculosa* (Gmelin, 1791)
89.3mm; Samar.
Shell large, solid, elongate-ovate, with adpressed sutures and a tall pointed spire; surface with intersecting axial and spiral cords, forming a coarsely granular surface; axial cords stronger than the spiral elements; aperture elongate-ovate, outer lip thickened, variced and marginally plicate, columella calloused, thinly glazed posteriorly, flaring anteriorly, and entirely smooth except for some weak plicae anteriorly; colour creamy white with brown blotches, frequently in bands; reaches 100mm.
DISTRIBUTION: Throughout the Philippines.

12. Colubraria (Colubraria) soverbii (Reeve, 1844)
57.6mm; Cebu.
Shell medium sized to large, with a tall spire which has the first few whorls slightly distorted; shell elongate-ovate, whorls strongly convex, sculptured with 2 varices per whorl (every 240°), and numerous arcuate, closely set axial cords which are crossed by numerous red-stained grooves, producing a finely granular surface; aperture narrowly elongate-ovate, outer lip thickened, variced and weakly lirate within; columella smooth except for a few lirae posteriorly, callus thin posteriorly, somewhat thicker anteriorly, flaring and with a scalloped free edge; colour pale orange with white and brown spotting, frequently in bands, fine red-brown spiral lines, and an orange-yellow aperture; reaches over 90mm.
DISTRIBUTION: Visayan Sea and Camotes Sea.

13. Colubraria (Colubraria) nitidula (Sowerby, 1833)
52.4mm; Sulu Archipelago.
Shell medium sized, elongate-ovate, with a tall spire and strongly convex whorls; sculptured with wide, flattened varices and numerous axial and spiral elements which intersect to produce a finely granulose surface on the earlier whorls, but become nearly obsolete on later whorls, resulting in an almost smooth surface; a few granulose spiral cords on the body whorl base; aperture narrowly ovate, outer lip thick, denticulate within, columella smooth, with a thick callus, highly expanded in the parietal area and forming a smooth, straight-sided shield basally; colour pale brown with white blotches and banding, and brown spots within the bands; reaches 55mm.
DISTRIBUTION: Visayan Sea and Sulu Sea.

14. Colubraria sp.
48.2mm; Sulu Archipelago.
Shell medium sized, with a tall, slightly distorted spire, strongly convex whorls, and sculptured with varices (2 per whorl) and numerous intersecting axial and spiral cords producing a coarsely granular surface; aperture narrowly elongate, outer lip thick, variced and lirate within, columella smooth and moderately calloused, with a few lirae basally; shell a dark grey with dark brown banding, white and brown spotted varices, a pale cream aperture, a brown spotted parietal area, and a row of reddish brown

spots along the outer lip margin; reaches 55mm.
DISTRIBUTION: Sulu Sea.

15. Colubraria (Colubraria) castanea Kuroda & Habe, 1952
Syn: *Triton (Epidromus) comptus* G. B. Sowerby III, 1874 (*non* A. Adams, 1854); *Colubraria fantomei* Garrard, 1961
69.5mm; Cebu.
An unmistakable species, unlikely to be confused with any other. Shell medium sized, solid, with a tall, relatively wide spire, well impressed sutures and strongly convex whorls which are sculptured with varices (2 per whorl, approximately every 240°) and numerous closely set axial and spiral cords, intersecting to produce a uniform, finely granular surface; aperture elongate-ovate, outer lip thick, slightly flaring, backed by a varix and lirate on the inner margin; columella smooth, with a wide callus, thin on the parietal area, and forming a thickened shield basally; pale greyish brown with wide brown banding on the lower part of each whorl and on the varices, aperture pale orange; reaches 75mm.
DISTRIBUTION: Visayan Sea and Camotes Sea in limited quantity.

16. Colubraria sp.
49.0mm; Lubang Is., Mindoro.
Shell medium sized, and probably best described as like a small *C. muricata* (Lightfoot). It differs from that species in its more finely granular surface, the more numerous outer lip denticles, and the less strongly developed columellar callus, which is irregularly lirate and pustulose, not smooth as in *muricata;* not seen to exceed 50mm.
DISTRIBUTION: Western Luzon.

17. Colubraria (Colubraria) tortuosa (Reeve, 1844)
37.4mm; Palawan.
Shell medium sized with a tall spire which has a curiously twisted axis; sutures impressed, whorls strongly convex, sculptured with intersecting axial and spiral cords of equal strength, producing a finely granulose surface; one varix per whorl, not completely axially aligned, so varices form an oblique ridge up the spire; aperture ovate, outer lip thickened, variced and lirate within; columella smooth, callus thin in the parietal area, but forming a pustulose shield anteriorly; creamy white with brown blotches, frequently in bands; reaches 65mm.
DISTRIBUTION: Eastern Samar, Visayan Sea and Sulu Sea.

18. Colubraria (Nivitriton) antiquata (Hinds, 1844)
17.8mm; Punta Engaño, Cebu.
Shell small, elongate-ovate, rather thin, with a tall spire; sculptured with 2 varices per whorl, and intersecting axial and spiral threads, producing a weakly cancellate surface; aperture narrowly elongate, outer lip angulate, weakly denticulate within; columella smooth, callus shield well developed, margin erect with a slight swelling centrally; pale brown with brown blotching; reaches 20mm. Although this species is presently placed in the Colubrariidae by most authors, more recent work by Beu & Maxwell (unpublished)suggests that it is congeneric with *Pisanella viva* (Habe & Okutani 1981),and is probably better placed in the Cancellariidae.
DISTRIBUTION: Visayan Sea and Camotes Sea.

19. Antemetula (Colubrarina) metulina Kuroda & Habe, 1971
39.0mm; Lubang Is., Mindoro.
Shell medium sized, elongate to elongate-ovate, with a moderately tall spire and a large body whorl; whorls strongly convex, sculptured with numerous intersecting axial and spiral riblets, producing a finely granulose surface; aperture narrowly elongate, outer lip thickened, somewhat marginally crenulate and lirate within; columella weakly calloused and smooth; early whorls white, later whorls pale yellowish brown with darker brown banding, aperture white; reaches 50mm.
DISTRIBUTION: Western Luzon.

ADDITIONAL NOTES ON THE COLUBRARIIDAE

Colubraria castanea Kuroda & Habe, 1952
Pl. 43, fig. 15
Colubraria fantomei Garrard, 1961 was described and illustrated in the *Journ. Malac. Soc. Aust.,* 1(5): 25, Pl. 1, fig. 5. Subsequent work by Yen demonstrated *fantomei* to be identical to the B. M. syntype of *Triton (Epidromus) comptus* G. B. Sowerby III, 1874 which was clearly illustrated in *Proc. Zool. Soc. Lon.* 1874: 598, P1. LXXII, figs. 2, 2a. (see insert*). This name is preoccupied by *Triton comptus* A. Adams, 1854, and the name *castanea* was introduced specifically as a replacement by Kuroda and Habe [*Check List and Bibliography of the Recent Marine Mollusca of Japan* (Leo W. Stack,Tokyo):48 (new name for *Triton comptus* Sowerby, 1874, not of A. Adams)]. Curiously, the shell depicted as this species by Habe (1964, Pl. 23, fig. 7) is much more slender, the whorls less ventricose, and the surface more coarsely granulose. It is very close to the specimen illustrated herein on Pl. 34, fig. 14, which at present remains unidentified.

* Type figure of *Triton comptus* G.B. Sowerby III, 1874

Class Gastropoda
Subclass Prosobranchia
Order Neogastropoda
Suborder Rachiglossa
Subfamily Muricacea
Family MURICIDAE

An extremely large group of carnivorous gastropods inhabiting a large variety of biological habitats, but generally found in mud or sand among rocks and coral. They occur in most geographical provinces and may be found in either shallow or very deep water. The shells vary from small to large and are extremely polymorphic. In shape, they may be ovate with short spires or elongated with tall spires. The whorls are generally turreted and the surface is sculptured with 3 varices per whorl (usually) and ornamented with spines, fronds or nodules. The outer lip may be dentate or crenulate, the columella is normally smooth, and the siphonal canal may be short or long and is generally partially closed. The operculum is corneous, oval or "tear-drop" in shape, and has a subcentral nucleus. The family has been divided into a number of subfamilies by different authors in an attempt to compartmentalise the multitude of differing taxa. Unfortunately in many cases, the criteria used in separating different groups is often ill-defined, with the result that a single taxon may often be assignable to more than one group. The problem will only be resolved by more detailed research and by a critical re-assessment of the features used to separate one subfamily group from another.

The Muricidae are extremely well represented in the Philippines, with members exhibiting a diversity of shape, ornamentation and colour unequaled by any other geographical locality. In particular, many of the smaller species (under 20mm) are exquisitely sculptured and their true beauty can only be appreciated when viewed under the lens. Many of these smaller shells appear to be, as yet, undescribed, and are featured here for the first time.

1. Pterynotus tripterus (Born, 1778)
63.1mm; Cebu.
Shell medium sized, fusiform, coarsely sculptured, with 8 shouldered whorls, weakly incised sutures, and a large, broad body whorl bearing 3 prominent, well developed, wing-like varices, strongly lamellose on the leading edge, and 2 to 3 intervarical nodes which are crossed by spiral cords; aperture sub-ovate, dentate on the columella and outer lip; colour dirty white with a yellowish orange aperture; reaches 60mm.
DISTRIBUTION: Sporadically found throughout the Philippines.

2. Pterynotus elongatus (Lightfoot, 1786)
82.6mm (lavender), 77.5mm (orange); Cebu.
Shell moderately large, fusiform-elongate, tall-spired, with 9 shouldered whorls and impressed sutures; sculptured with 3 varical wings and a single intervarical node on each whorl; the earlier wings are small and "hook-shaped", the later ones larger, undulate and ventrally lamellose and the surface is covered with major and minor spiral cords; aperture elongate-ovate, outer lip crenulate and dentate within, columella with 4 anterior denticles; colour white, pink, lavender or orange; reaches 90mm.
DISTRIBUTION: Sporadically found throughout the Philippines.

3. Pterynotus pinnatus (Swainson, 1822)
80.7mm; Cebu.
A beautifully exquisite fusiform-elongate shell, all white in colour, with 3 sinuous winged varices on each whorl and a single intervarical node, all crossed by numerous spiral cords; aperture ovate, outer lip margin raised and serrate, siphonal canal long and curved to the right, with a prominent siphonal fasciole projecting to the left; reaches 80mm.
DISTRIBUTION: Sporadically found throughout the Philippines.

4. Pterynotus pellucidus (Reeve, 1845)
48.2mm; Cebu.
Shell medium sized, fusiform, tall-spired, with 8 shouldered whorls, impressed sutures and a large body whorl; sculptured with 3 broad, axially aligned, wing-like varices, lamellate on the leading edge, and a single intervarical node; early varices short and spinose, the last varical wing being the most highly developed; the entire surface is crossed with finely scabrous spiral cords; outer lip erect, marginally crenulate and dentate internally, columella with 3 anterior plicae; colour white; reaches 55mm.
DISTRIBUTION: Manila Bay and Camotes Sea.

5. Pterynotus bipinnatus (Reeve, 1845)
40.5mm, 49.1mm; Cebu.
Shell medium sized, fusiform-elongate with angulate whorls, adpressed sutures and sculptured with numerous axial ribs, nodulose on the presutural ramp, and 1 or 2 foliaceous wing-like structures on the apertural varix; the surface is sculptured with axial striae and bisecting spiral cords; aperture ovate, outer lip raised, serrate and internally dentate, columella obsoletely folded, siphonal canal long and partially open; colour off-white with a violet aperture; reaches 50mm; like a small *elongatus.*
DISTRIBUTION: Sporadically found throughout the Philippines.

PLATE NO. 35 1. *Pterynotus tripterus* (Born) 2. *Pterynotus elongatus* (Lightfoot) 3. *Pterynotus pinnatus* (Swainson) 4. *Pterynotus pellucidus* (Reeve) 5. *Pterynotus bipinnatus* (Reeve) 6. *Chicoreus annandalei* (Preston) 7. *Siratus alabaster* (Reeve) 8. *Pterynotus miyokoae* Kosuge 9. *Pterynotus loebbeckei* (Kobelt) 10a.-10b. *Chicoreus saltatrix* Kuroda 11. *Chicoreus orchidiflorus* (Shikama) 12. *Chicoreus celinamarumai* (Kosuge) 13. *Pterynotus martinetana* auct. (non Röding) 14. *Typhinellus occlusus* (Garrard) 15. *Chicoreus penchinati* (Crosse) 16. *Chicoreus nobilis* Shikama 17. *Chicoreus artemis* Radwin & D'Attilio 18. *Homalocantha anatomica* (Perry) form *pele* (Pilsbry) 19. *Homalocantha zamboi* (Burch and Burch) 20a. *Homalocantha scorpio* (Linné) 20b. Albinistic form

PLATE 35

6. Chicoreus (Naquetia) annandalei (Preston, 1910)
89.2mm, 84.0mm; Cebu.
Shell large, fusiform, tall-spired, with 7 to 8 shouldered whorls and impressed sutures; body whorl large, fusoid with 3 low, webbed, expansive varices, slightly pleated on the leading surface, and 3 to 4 slender intervarical nodes, prominent at the shoulder; aperture ovate, anal sulcus deep, outer lip erect, crenulate and internally weakly lirate, columella partially adherent and smooth, siphonal canal long, straight and slightly open; pale brown with 3 brown bands, aperture pale violet within, marginally white; reaches 90mm.
DISTRIBUTION: Camotes Sea and Southern Mindanao.

7. Siratus alabaster (Reeve, 1845)
148.7mm; Bohol.
Shell large, fusiform, with 9 shouldered whorls and a large, globose body whorl; sculptured with 3 webbed varices per whorl, each of which has a long shoulder spine, 2 to 3 intervarical nodes strongly nodulose on the shoulder and encircled with obsolete lirae; aperture ovate, inner and outer lips erect and smooth, siphonal canal long and recurved, siphonal fasciole prominent; colour ivory white with orange tinted nodes; reaches 220mm.
DISTRIBUTION: Camotes Sea.

8. Pterynotus miyokoae Kosuge, 1979
55.3mm; Mactan Is., Cebu.
Shell moderately large, broadly fusiform and high-spired, with 9 to 10 whorls; each whorl with 3 varices and 2 intervarical ridges; the varices are wing-like, twisted in appearance, finely rippled and lamellate on the ventral surface, and follow the shell contour; aperture large and ovate, outer lip thick and dentate, inner lip smooth, canal long, recurved and almost closed; shell brown, encircled with darker spiral lirae and occasional white varical banding; reaches 75mm. Endemic to the Philippines.
DISTRIBUTION: Cebu and Southern Mindanao.

9. Pterynotus loebbeckei (Kobelt, 1879)
62.0mm; Bohol.
Shell moderately large with 7 to 8 weakly convex spire whorls, a sharply pointed spire and a broadly fusiform body whorl; sculptured with 3 prominently winged varices, serrated on the leading edge, 2 or 3 knobby intervarical nodes and encircled with numerous major and minor spiral cords; aperture ovate, outer lip denticulate, inner lip smooth and partially adherent, siphonal canal long, recurved and almost tubular, siphonal fasciole prominent; orange coloured, occasionally white or lavender; reaches 75mm.
DISTRIBUTION: Camotes Sea.

10. Chicoreus (Chicoreus) saltatrix Kuroda, 1964
48.5mm; Mactan Is., Cebu.
Shell medium sized, fusiform-ovate in outline, spire tall and pointed, sutures weakly incised, body whorl large, varying from ovate to elongate-ovate in outline; three axially aligned varices per whorl bearing short spines, 1 or 2 on the spire whorls, 4 on the body whorl, and 1 or 2 on the canal; 1 to 1½ intervarical nodes; in some stouter forms, there is a tendency to spinal reduction on the posterior of the last 2 whorls; surface with numerous finely scabrous spiral lirae which may be uniform in strength (10a) or with some lirae stronger than others (10b); aperture ovate, anal sulcus V-shaped, outer lip erect, crenulate and lirate within, columella smooth with 1 or 2 anterior and posterior denticles; colour variable from pale bluish white to pinkish orange, apex pink, aperture deep pinkish purple; reaches 70mm.
DISTRIBUTION: Camotes Sea.

11. Chicoreus (Chicoreus) orchidiflorus (Shikama, 1973)
38.8mm; Punta Engaño, Cebu.
A small delicate shell of 8 to 9 gently sloping whorls, sculptured with 3 varices per whorl, 1 or 2 intervarical nodes and fine spiral striae; characterized by the presence of beautifully serrated fronds, those on the earlier whorls smaller and tabulate, later ones becoming separated; aperture ovate, outer lip erect and dentate within, columella smooth and non-adherent with a parietal denticle, canal long, recurved and almost closed, siphonal fasciole pronounced; colour white to pinkish brown; reaches 40mm.
DISTRIBUTION: Camotes Sea.

12. Chicoreus (Chicoreus) celinamarumai (Kosuge, 1980) (nom. emend.)
30.8mm; Bohol.
Shell small, broadly fusiform and tall spired, with impressed sutures, moderately convex spire whorls and a large body whorl; axially sculptured with 3 well-expanded, webbed, wing-like varices that are crenulate and digitate, and possess marginal spinose projections; there are 3 intervarical costae which are crossed by major and minor cords, a marginally crenulate, internally lirate outer lip, a long, curved, almost closed canal bearing 4 foliaceous spines and a prominent siphonal fasciole; orange-coloured with darker lirae and occasional brown maculations; reaches 40mm. Close to *C. orchidiflorus*, but has finer sculpturing and a different varical-blade structure; possibly a subspecies or a local ecomorph; named after Ms. Celina S. Galindes of Cebu and Mr. M. Maruma Suzuki of Tokyo.
DISTRIBUTION: Camotes Sea and Southern Mindanao.

13. Pterynotus martinetana auct. (non Röding, 1798)
26.9mm, 25.8mm; Cebu.
Shell medium sized, broadly fusiform and tall spired with weakly impressed, undulating sutures; sculptured with 6 weakly convex whorls, a large fusiform body whorl, 4 roughly sculptured varices and a single intervarical cord; 3 groups of spiral cords (triple, single and double stranded respectively) cross the varices forming 3 spines; aperture elongate-ovate, outer lip erect, crenulate and internally dentate, columella smooth, erect and posteriorly plicate,

siphonal canal short, recurved and slightly open; colour off-white with pink varical spines and black interspaces; reaches 55mm.
DISTRIBUTION: Camotes Sea and Southern Mindanao.

14. Typhinellus occlusus (Garrard, 1963)
28.4mm; Talikod, Davao.
Shell small to medium sized, fusiform-elongate, with a tall spire and a large, broadly trigonal body whorl; sculptured with 4 thin erect varices, dorsally recurved along their free margin and forming recurved spines on the whorl shoulders; earlier siphonal canals obvious with a large prominent siphonal tube present on the shoulder of the last intervarical space; siphonal canal closed, aperture sub-ovate, and the inner and outer lips are erect and smooth; dirty white with red spots on the outer lip wall and along the closed siphonal opening; reaches 30mm.
DISTRIBUTION: Talikod, Davao.

15. Chicoreus (Chicoreus) penchinati (Crosse, 1861)
40.5mm, 30.9mm, 32.5mm, 30.5mm; Sulu Archipelago.
Shell medium sized, elongate-ovate, with a tall, sharply pointed spire, impressed sutures, 8 to 10 convex spire whorls, and a large body whorl; sculptured with 3 varices per whorl, 2 or 3 nodulose, axially aligned intervarical cords, and encircled with major and minor spiral lirae intersecting the axial cords to produce nodules and the varices to form short, thick, highly foliaceous spines; aperture ovate, outer lip erect, thick and dentate with a posterior denticle, columella smooth, canal thin and recurved; colour variable, white, orange or brown, aperture white; reaches 45mm.
DISTRIBUTION: Sulu Sea.

16. Chicoreus (Chicoreus) nobilis Shikama, 1977
55.4mm; Mactan Is., Cebu.
Shell medium sized and broadly fusiform, immediately recognized by its deep pink colour and its short, ramified, highly foliaceous spines; varices number 3 per whorl with 2 to 3 intervarical nodes; spire whorls with a single spine, the body whorl with 4 to 5, and the canal with 3, these usually being crowded towards the base; surface with numerous slightly elevated lirae and numerous finer interstitial spiral threads; aperture sub-circular, anal sulcus deep and U-shaped, outer lip raised, marginally serrate and lirate within; the intervarical nodes are generally paler than the shell as a whole, and the major spiral lirae are frequently stained with very dark pink, aperture pale pink within; can reach around 60mm.
DISTRIBUTION: Camotes Sea.

17. Chicoreus (Chicoreus) artemis Radwin & D'Attilio, 1976
55.0mm; Mactan Is., Cebu.
Shell medium sized, fusiform, with a tall, acute spire, impressed sutures and a somewhat sub-orbicular body whorl; sculptured with 3 varices per whorl, 1 or 2 intervarical nodes, and the surface is encircled with numerous major and minor cords; the varices bear short to moderately long, slightly recurved spines, 1 or 2 on the spire whorls, 5 to 6 on the body whorl and 3 on the siphonal canal; aperture ovate with a V-shaped anal sulcus, outer lip erect, marginally crenulate and internally lirate, columella smooth, erect and detached, siphonal canal long, partially open and recurved; colour pink to pinkish orange, the major spiral cords occasionally tinged a deep reddish brown, aperture white, marginally pink; reaches 60mm.
DISTRIBUTION: Camotes Sea.

18. Homalocantha anatomica (Perry, 1811) form **pele** (Pilsbry, 1918)
43.8mm, 45.9mm; Cebu.
Shell medium sized, fusiform, high-spired with deeply channeled sutures, 3 to 4 well-shouldered post nuclear whorls and a large, fusoid body whorl; sculptured with 4 varices per whorl, a single intervarical node, and crossed by major and minor spiral cords forming long, narrowly open, palmate, terminally serrated digitations, 4 on the body whorl and 3 on the canal; aperture ovate, outer lip erect and dentate, columella smooth and detached, siphonal canal long, open and distally recurved; white, rarely orange, with a white aperture; reaches 65mm. In this form, 2 of the body whorl digitations are larger than the rest.
DISTRIBUTION: Camotes Sea.

19. Homalocantha zamboi (Burch & Burch, 1960)
53.9mm, 50.5mm; Cebu.
Shell medium sized, fusiform, with 3 to 4 shouldered whorls, a large, fusoid body whorl, and 5 thick varices bearing 4 broadly palmate, terminally digitate spiny projections, the 2 medial spines largest and V-shaped; sutures impressed, "cleft-like" in appearance; aperture ovate, outer lip erect, marginally crenulate and smooth within; white with a pink aperture and pale violet between the varices; reaches 80mm; separable from *H. anatomica* by the nature of the digitations, the more acuminate shape, the possession of 2 long V-shaped spines and the pink apertural colouration. Named after a famous local collector, Evaristo Zambo of Cebu.
DISTRIBUTION: Camotes Sea.

20. Homalocantha scorpio (Linné, 1758)
Shell medium sized, tall-spired, with 5 shouldered whorls and deeply excavated sutures; body whorl large and fusoid, with 6 varices crossed by spiral cords, forming 5 equi-sized projections on the body whorl and 4 on the anterior canal; occasionally interconnecting fimbriate webbing and interstitial spinelets are seen; aperture ovate, outer lip erect, dentate and lirate within, inner lip smooth and erect, anterior canal long, open and recurved; black, brown or white with a white aperture; reaches 65mm.
20a. typical form; 42.0mm; Punta Engaño, Cebu.
20b. albinistic form; 41.2mm; Punta Engaño, Cebu.
DISTRIBUTION: Sporadically found throughout the Philippines.

1. Chicoreus (Chicomurex) superbus (G. B. Sowerby III, 1889)
Shell medium sized, solid, with well shouldered whorls and a tall, acute spire; three varices per whorl bearing spines, largest at the shoulder; 2 to 3 intervarical nodes, crossed by scabrous spiral cords; aperture large and ovate, outer lip marginally serrate, lirate within and expanded to form a webbed varix bearing spines, columella non-adherent and weakly plicate, siphonal canal long, narrow and recurved; off-white with pink mottling and brown maculations; reaches 75mm.
1a. typical form from deep water; 63.9mm; Punta Engaño, Cebu.
1b. A slightly squatter form of *superbus*, often referred to as the "Japanese *laciniatus*". It lacks the spination of *superbus*, is somewhat broader, has finer, less prominently scabrous cords, and has a single intervarical node almost entirely on the shoulder; often called *elliscrossi* Fair, 1974, and may be a good species; 56.7mm; Cebu.
DISTRIBUTION: Camotes Sea.

2. Chicoreus (Chicomurex) laciniatus (G. B. Sowerby II, 1841)
60.2mm, 62.3mm; Cebu.
Shell medium sized, heavily fusiform and tall-spired, with 7 to 8 convex whorls and deeply impressed sutures; body whorl large with 3 lamellose varices; surface sculptured with elevated lamellae formed from intersecting axial and spiral striae which become obsolete anteriorly; aperture large and ovate, outer lip erect, serrate, lirate within and bordered by a scabrous varix, canal long, nearly closed and recurved; brown to orange with darker varices, a white aperture and a purple columella; reaches 65mm.
DISTRIBUTION: Camotes Sea and Sulu Sea.

3. Chicoreus (Chicomurex) venustulus Rehder & Wilson, 1975
54.4mm, 52.2mm, 42.0mm; Cebu.
Shell medium sized with impressed sutures, an elevated spire and 8 to 9 rounded whorls, each bearing 3 varices and 2 or 3 intervarical nodes; the early whorls have short frondose spines and the surface is highly scabrous from numerous fine axial and spiral striae; aperture ovate, outer lip flared and highly lamellose, canal long, recurved and bordered with wing-like lamellae; off-white with pink and reddish brown bands; reaches 60mm; prominently banded forms sold locally as *"jickelii."*
DISTRIBUTION: Camotes Sea.

4. Siratus pliciferoides (Kuroda, 1942)
124.4mm; Cebu.
Shell large and fusiform, with a tall, acute spire, weakly channeled sutures and a large, fusoid body whorl; sculptured with 3 varices per whorl and 1 to 3 narrow intervarical nodes; encircled with spiral cords which intersect the varices to produce a single shoulder spine on the spire whorls and an additional 6 spines on the body whorl, size decreasing anteriorly; aperture ovate, outer lip dentate and lirate within, siphonal canal long, recurved and with 2 short spines; off-white to brown in colour with 3 light brown bands and a white aperture; reaches 130mm.
DISTRIBUTION: Sporadically found throughout the Philippines in limited quantity.

5. Chicoreus (Chicoreus) axicornis (Lamarck, 1822)
Shell medium sized, fusiform and tall-spired with impressed sutures; there are 9 whorls, each bearing 3 spinose varices, 2 intervarical nodes and encircled with spiral cords of varying strengths; varical spines strongest at the shoulder, then decreasing in size and curved towards the apex; aperture ovate, anal sulcus strong, outer lip erect, marginally serrate, internally lirate, and backed by a thick, spinose varix, siphonal canal long, partially closed and distally recurved; shell colour off-white to brown with a white aperture; reaches 70mm.
5a. form *kawamurai* Shikama, 1964; a smaller shell with fewer, longer, more curved spines; 57.9mm, 46.9mm; Bulan, Sorsogon.
5b. typical form; 71.4mm; Sulu Archipelago.
DISTRIBUTION: Camotes Sea, Sulu Sea and Bulan, Sorsogon.

6. Chicoreus (Naquetia) trigonulus (Lamarck, 1816)
53.5mm; Palawan.
Shell medium sized and fusiform-elongate, with a tall spire, impressed sutures, a large body whorl bearing 3 irregularly developed varices and 2 or 3 narrow intervarical costae; surface finely scabrous from intersecting axial and spiral elements; aperture elongate-ovate, outer lip erect, dentate, weakly lirate within and with a moderately webbed varix, columella smooth and adherent, canal narrow, recurved and moderately long; pale cream with brown banding and some mottling; reaches 60mm.
DISTRIBUTION: Sporadically found throughout the Philippines.

7. Chicoreus (Naquetia) triquetra (Born, 1778)
49.2mm; Punta Engaño, Cebu.
Superficially similar to *C. trigonula* but has a greater

PLATE NO. 36 1a. *Chicoreus superbus* (G. B. Sowerby III) 1b. "squatter" form 2. *Chicoreus laciniatus* (G. B. Sowerby II) 3. *Chicoreus venustulus* Rehder & Wilson 4. *Siratus pliciferoides* (Kuroda) 5a. *Chicoreus axicornis* (Lamarck) form *kawamurai* Shikama 5b. typical form 6. *Chicoreus trigonulus* (Lamarck) 7. *Chicoreus triquetra* (Born) 8a. *Chicoreus banksii* (Sowerby) 8b. deep water form 9. *Chicoreus cnissodus* (Euthyme) 10a. *Chicoreus brunneus* (Link) 10b. unusual specimen 10c. orange form 11a. *Chicoreus ramosus* (Linné) 11b. dwarf form 12. *Chicoreus rubiginosus* (Reeve) 13. *Chicoreus saulii* (Sowerby) 14. *Chicoreus rosarius* (Perry) 15a. *Chicoreus capucinus* (Lamarck) 15b. unusual specimen 16a. *Chicoreus microphyllus* (Lamarck) 16b. large form 17a. *Hexaplex cichoreum* (Gmelin) 17b. unusual colour form 17c. black form 17d. albinistic form 17e. form *depressospinosus* Dunker

PLATE 36

1a
1b
2
3
4
5a
5b
6
7
8a
8b
9
10a
10b
10c
11a
11b
12
13
14
15a
15b
16a
16b
17a
17b
17c
17d
17e

number of spire whorls, more developed intervarical nodes, the webbing on the apertural varix is far more fimbriate, the outer lip is more elevated and more serrate than in *trigonula*, the columella is finely nodulose and the colouration is much darker; reaches 70mm.
DISTRIBUTION: Sporadically found throughout the Philippines in limited quantity.

8. Chicoreus (Chicoreus) banksii (Sowerby, 1841)
Shell large, fusiform, with impressed sutures, a tall spire of 8 to 9 shouldered whorls and a large body whorl; sculptured with 3 spinose varices per whorl, 2 to 3 intervarical nodes, and crossed by scabrous spiral cords forming foliate spines on the varices, the size decreasing anteriorly, and a spinal gap towards the base; aperture small and ovate, outer lip erect, canal long, broad, partially open and distally recurved; aperture white, shell pale greyish brown with darker varices; an extremely variable species; reaches 100mm.
8a. typical form; 62.8mm; Cebu.
8b. a rare, deep water form with a highly scabrous surface and extraordinarily developed spination; 62.1mm; Punta Engaño, Cebu.
DISTRIBUTION: Camotes Sea and Sulu Sea.

9. Chicoreus (Chicoreus) cnissodus (Euthyme, 1889)
73.0mm; Cebu.
Shell medium sized, broadly fusiform, with a tall spire, deeply impressed sutures, 7 weakly shouldered whorls and a large, fusoid body whorl; sculptured with 3 foliaceous varices and 2 to 3 intervarical nodes, major and minor cords intersecting the varices to produce long, foliate spines and weaker, intercalated spinelets; aperture large and ovate, outer lip erect, marginally serrate and internally dentate; anterior canal long, curved, nearly closed; white with brown nodes and spines; reaches 75mm.
DISTRIBUTION: Manila Bay, Camotes Sea and Southern Mindanao.

10. Chicoreus (Chicoreus) brunneus (Link, 1807)
Shell large, heavy, rhomboidal and tall spired, with impressed sutures, 7 weakly angulate whorls and a large, broad body whorl; sculptured with 3 prominent varices, foliate from overriding spiral cords (12), numerous weaker spiral lirae, and a single, large intervarical node; spines large but short, the size decreasing anteriorly, with 4 on the long, broad, distally recurved canal; aperture ovate, outer lip erect and dentate; colour brownish black, pink or orange, with a pinkish orange columella; reaches 100mm.
10a. typical form; 80.1mm; Batangas.
10b. an unusual form lacking typical spination; 67.9mm; Cebu.
10c. orange form; 58.7mm; Punta Engaño, Cebu.
DISTRIBUTION: Throughout the Philippines.

11. Chicoreus (Chicoreus) ramosus (Linné, 1758)
Shell extemely large, fusiform and low spired, with 9 strongly convex whorls, deeply impressed sutures and a large, inflated body whorl; sculptured with 3 varices per whorl, each bearing short foliaceous spines formed from overriding major and minor spiral cords, and 2 to 3 intervarical nodes; aperture large and ovate with a prominent anal sulcus, outer lip coarsely dentate and finely lirate, inner lip smooth and adherent, canal long, narrow and distally recurved; white with orange sutures, a white aperture and pink apertural margin; reaches 300mm.
11a. typical form; 204mm; Cebu.
11b. a fully mature dwarf form exhibiting pronounced spinal development and curvature; 54.2mm; Punta Engaño, Cebu.
DISTRIBUTION: Throughout the Philippines.

12. Chicoreus (Chicoreus) rubiginosus (Reeve, 1845)
80.5mm, 80.9mm; Cebu.
A solid, medium-sized shell, fusiformly elongate-ovate in outline, with a tall spire and impressed sutures; each whorl is sculptured with 3 varices and a single small intervarical node; each of the varices bears 5 to 6 dorsally recurved foliaceous spines and numerous ventrally curved intercalated spinelets; aperture ovate, outer lip serrated and channeled at the shoulder spine, columella smooth; colour variable, usually brown with darker cords and spines, and a pinkish orange columella; reaches 100mm.
DISTRIBUTION: Sporadically found throughout the Philippines.

13. Chicoreus (Chicoreus) saulii (Sowerby, 1841)
110mm; Cebu.
Shell large, fusiform, tall spired, with impressed sutures, 9 convex post-nuclear whorls and a moderately sized, elongated body whorl; sculptured with 3 varices per whorl and 1 large and 2 weaker intervarical nodes; surface crossed by major and minor scabrous spiral cords forming long foliaceous spines and intercalated spinelets where they cross the varices, and number 4 on the body whorl and 3 on the canal; aperture ovate, anal sulcus prominent, outer lip erect, dentate and marginally lirate, canal long and distally recurved; orange-brown with darker lirae and a lavender apertural margin and spinal interior; reaches 130mm.
DISTRIBUTION: Sporadically found throughout the Philippines, mostly from Sulu Sea.

14. Chicoreus (Chicoreus) rosarius (Perry, 1811)
Syn: *C. palmarosae* (Lamarck, 1822)
111.4mm; Samar.
Very close to *C. saulii* in general structure and frondation but differs in several ways; *rosarius* lacks the small ventrally recurved spinelets of *saulii*, its intervarical nodes are more regular in size (also 3 in number), the outer lip denticles are weaker, and the outer columellar margin bears a row of white beads, not seen in *saulii*; *rosarius* also lacks the pink hue of *saulii* and the aperture is white with a brown margin; reaches 125mm.
DISTRIBUTION: Sporadically found throughout the Philippines.

15. Chicoreus (Naquetia) capucinus (Lamarck, 1822)
Shell medium sized, fusiform and tall spired with adpressed sutures; spire whorls are 6 in number and have convex sides; body whorl moderately large and fusiform with 2 to 3 intervarical ridges and 3 varices per whorl, fimbriate from overriding major and minor spiral cords and occasionally pleated on the last varix; aperture ovate, outer lip thick, marginally crenulate and lirate within, columella smooth and adherent, canal long, broad, and narrowly open; deep brown to black, aperture bluish white; reaches 90mm.
15a. typical form; 83.3mm; Palawan.
15b. specimen with unusually developed surface sculpturing and frondation; 63.0mm; Cebu.
DISTRIBUTION: Sporadically found throughout the Philippines.

16. Chicoreus (Chicoreus) microphyllus (Lamarck, 1816)
Shell small to large, solid, fusiform, with impressed sutures and 9 convex postnuclear whorls; body whorl large and elongate, sculptured with 3 varices per whorl which bear short spines, somewhat scabrous from overriding major and minor spiral cords (6 on the body, 3 on the canal); 1 to 3 intervarical nodes; aperture ovate, outer lip erect, serrate and the inner lip margin may be smooth or denticulate; canal moderately long, wide and recurved; off-white to chestnut brown, spiral elements darker, aperture white with an orange columella; reaches 110mm.
16a. typical *microphyllus*; 61.0mm; Jolo, Sulu Archipelago.
16b. a large form, usually figured as *C. torrefactus* (Sowerby, 1841) in most literature but considered a synonym of *microphyllus* by many authors. This is the more common form in the Philippines. The *microphyllus/torrefactus* complex is still poorly understood; 109mm; Cebu.
DISTRIBUTION: Sporadically found throughout the Philippines.

17. Hexaplex cichoreum (Gmelin, 1791)
Shell large, tall spired with 7 to 8 convex whorls, a large, globose body whorl, 6 to 8 foliaceous varices, and a single intervarical node; 5 major and minor spiral cords intersect the varices to produce long recurved spines; aperture large and ovate, outer lip with a dentate outer margin, columella smooth and partially adherent, canal long, bent and distally recurved; extremely variable in colour, usually white with 2 brown bands, dark spines, a white aperture and pink margin; reaches 150mm.
17a. typical form with unusually developed spination; 96.0mm; Palawan.
17b. colour form with orange spines; 92.0mm; Cebu.
17c. black form; 95.5mm; Cebu.
17d. albinistic form; 77.5mm; Cebu.
17e. form *depressospinosus* Dunker, 1869; an unusual form characterized by its fusiform shape, lack of spines in the earlier whorls, and the anteriorly curved body whorl spines; 81.0mm; Samar.
DISTRIBUTION: Throughout the Philippines.

Plate 37 — Muricidae

1. Haustellum haustellum (Linné, 1758)
127.3mm; Samar.
Shell large, club-shaped, low spired, with impressed, undulate sutures and a large, elongate, somewhat globose body whorl; sculptured with 3 axially aligned varices, 3 to 4 intervarical nodes which are sharply angled at the shoulder and crossed with 5 to 6 spiral cords intersecting the varices to form tubercles; aperture ovate, inductura almost totally erect and obsoletely lirate within; the siphonal canal is long, thin and almost closed, with bluntly spinose cords posteriorly; colour pinkish purple with brown banding, aperture deep pink; reaches 150mm.
DISTRIBUTION: Sporadically found throughout the Philippines.

2. Haustellum haustellum (Linné, 1758) form **kurodai** Shikama, 1964
147.8mm; Zamboanga.
Similar to typical *H. haustellum* (Linné), but the shell is much paler, the sutures more deeply channeled, the whorls more sharply angulate at the shoulders, and the tubercles are elevated, sharp, and somewhat axially fused, forming cords; the aperture is white not pink, the siphonal canal is longer in relation to the shell size, and there are a few sharp spines posteriorly; largest seen 164mm.
DISTRIBUTION: Sporadically found throughout the Philippines in limited quantity.

3. Haustellum haustellum (Linné, 1758) form **vicdani** Kosuge, 1980
117.2mm; Bulan, Sorsogon.
Close to the *kurodai* form but generally smaller and immediately recognized by the deep lavender hue, the orange-yellow aperture and the long, comparatively smooth siphonal canal; sculpturally similar to *kurodai*, but the axial fusion of the nodules is more pronounced and the sutures are far less channeled; reaches around 120mm; named after a prominent local collector, Victor Dan.
DISTRIBUTION: Bulan, Sorsogon.

4. Murex mindanoensis Sowerby, 1841
98.7mm; Tayabas Bay, Quezon.
Shell moderately large, fusiform, tall spired, with convex whorls and impressed sutures; body whorl with 3 varices, 3 to 4 weak intervarical nodes and encircled by major and minor spiral cords which form small spines and spinelets where they cross the varices; aperture ovate, outer lip erect, marginally

crenulate and lirate within, columella partially adherent and weakly dentate, siphonal canal long, partially open and distally recurved; colour white to brown, with white cords and a white aperture; reaches 200mm.
DISTRIBUTION: Sporadically found throughout the Philippines in limited quantity.

5. Murex aduncospinosus Beck, in Sowerby, 1841
Syn: *Murex trapa auct.* (non Röding, 1798)
97.9mm; Panay Island.
Shell medium sized and fusiform, with a moderately tall spire, impressed sutures, and a long, open siphonal canal which is bent and somewhat distally recurved; sculptured with 3 varices per whorl, 4 to 5 narrow intervarical costae, and encircled with strong cords crossing the varices to form a single spine on the spire whorls, 3 on the body whorl, and 5 on the canal; smaller spinelets also present; aperture ovate, outer lip erect with bifid crenulations and a prominent labial tooth anteriorly, columella partially adherent and smooth; colour brown with paler lirae and spines, aperture white, brown within; reaches 120mm.
DISTRIBUTION: Sporadically found throughout the Philippines in limited quantity.

6. Murex rectirostris Sowerby, 1841
85.8mm; Palawan.
Shell medium sized, club-shaped, with a moderately tall spire, impressed sutures, weakly angulate whorls, and a long, partially closed, distally recurved canal; sculptured with 3 varices per whorl, 3 to 4 strong intervarical costae and encircled with major and minor cords, weak above the shoulder, stronger at and below the shoulder; a single varical spine on the spire whorls and 2 on the body whorl varices; aperture ovate, outer lip erect and crenulate, columella partially adherent and weakly lirate; pale brown with 2 darker body whorl bands and deep brown spiral elements, aperture white; reaches 85mm. Sold locally as *"Murex kiiensis"*.
DISTRIBUTION: Palawan, Camotes Sea and Tayabas Bay, Quezon.

7. Murex gallinago Sowerby, 1903 form **nagaidesu** Shikama, 1970
82.5mm; Tayabas Bay, Quezon.
Shell medium sized with a tall, sharply pointed spire, deeply impressed sutures, convex spire whorls, and a long, distally recurved canal which is closed posteriorly; sculptured with 3 varices per whorl, 3 intervarical nodes, strongly nodulose at the shoulder and encircled with weak cords; whorls with 2 spines on the varices, shoulder one strongest and curved; several weaker spinelets on the base of the body whorl varices; aperture ovate, outer lip erect and weakly lirate within, columella adherent posteriorly and weakly plicate; colour off-white with dark brown spiral cords and paler banding on the body whorl and canal; aperture white; reaches 100mm; a most distinctive form.
DISTRIBUTION: Camotes Sea and Tayabas Bay, Quezon.

8. Murex sp.
74.0mm; Cebu.
Shell small to medium sized, tall spired with impressed sutures and a long, distally recurved siphonal canal which is closed posteriorly; sculptured with 3 varices per whorl, 3 to 4 intervarical costae and encircled with strong spiral cords forming a single weak, central varical spine on each varix, and 2 to 3 weaker spinelets towards the base on the body whorl varices; outer lip erect, serrate and internally lirate, columella posteriorly adherent and strongly plicate; off-white with 3 pale brown body whorl bands, spiral lirae deep brown, aperture white; reaches 80mm; appears unnamed at present.
DISTRIBUTION: Camotes Sea.

9. Murex dentifer Watson, 1883
88.5mm; Cebu.
Shell large, spire moderately tall and acute, body whorl large and fusoid, sutures impressed, canal short in relation to total shell size; sculptured with 3 large varices per whorl, 3 to 4 strong intervarical costae, and encircled by strong spiral lirae; aperture ovate, outer lip erect, marginally dentate and internally lirate, columella posteriorly adherent and irregularly plicate; the shell lacks any spination except on the early whorls; colour yellowish white with darker banding, reddish brown lirae, and a white, marginally brown-lined aperture; reaches 100mm. Erroneously sold locally under the name *"Murex hirasei"*, a species not yet recorded from the Philippines.
DISTRIBUTION: Camotes Sea.

10. Murex aduncospinosus Beck, in Sowerby, 1841 form **concinus** Reeve, 1845
62.9mm; Cebu.
A small, lightweight shell immediately recognized by its pale yellowish white colour and narrow brown threads, 1 on the spire whorls and 3 on the body whorl; these threads correspond to the short, sharp spines which arise from the 3 varices of each whorl; there are 4 to 5 irregular intervarical nodes crossed by weaker spiral lirae; the aperture is ovate, off-white and lirate within, the outer lip erect with bifid serrations, and the columella is posteriorly adherent

PLATE NO. 37 1. *Haustellum haustellum* (Linné) 2. *Haustellum haustellum* (Linné) form *kurodai* Shikama 3. *Haustellum haustellum* (Linné) form *vicdani* Kosuge 4. *Murex mindanoensis* Sowerby 5. *Murex aduncospinosus* Beck, in Sowerby 6. *Murex rectirostris* Sowerby 7. *Murex gallinago* Sowerby *form nagaidesu* Shikama 8. *Murex sp.* 9. *Murex dentifer* Watson 10. *Murex aduncospinosus* Beck, in Sowerby form *concinus* Reeve 11. *Murex sp.* 12. *Murex pecten* Lightfoot 13. *Murex troscheli* Lischke 14. *Murex nigrispinosus* Reeve 15. *Siratus pliciferoides* (Kuroda) form *propinquus* (Kuroda & Azuma) 16a. *Murex tenuirostrum* Lamarck 16b. brown-banded form

PLATE 37

and smooth; reaches 65mm. A deep water form.
DISTRIBUTION: Camotes Sea and Tayabas Bay, Quezon.

11. Murex sp.
60.0mm; Cebu.
Shell small, tall spired with impressed sutures, strongly convex whorls, and a long, narrowly open, strongly corded siphonal canal; sculptured with 3 varices per whorl and 4 to 5 intervarical costae, nodulose from regularly spaced, overriding spiral lirae; aperture ovate, outer lip erect, serrate and internally lirate, columella adherent posteriorly and irregularly plicate; off-white with one brown band on the spire whorls, and 3 on the body whorl; reaches 65mm. Close to *M. multiplicatus* Sowerby, 1875 but has somewhat different sculpturing.
DISTRIBUTION: Camotes Sea.

12. Murex pecten Lightfoot, 1786
141mm; Sorsogon.
Shell large, sub-fusiform, with a tall, acute spire, 7 to 8 convex spire whorls, deeply impressed sutures, a large, fusoid body whorl, and a long, straight, almost closed siphonal canal; sculptured with 3 varices per whorl, obsolete intervarical costae, and encircled with strong cords which form long curved spines, strongest on the shoulder and diminishing in size anteriorly; secondary and tertiary spinelets oriented at right angles to the primary spines are seen along the canal; aperture ovate, outer lip erect and strongly dentate, columella smooth and posteriorly adherent; colour pale ochre, aperture white, deep brown within; reaches 150mm.
DISTRIBUTION: Sporadically found throughout the Philippines.

13. Murex troscheli Lischke, 1868
142mm; Davao.
Shell similar to *M. pecten* but usually grows to a much larger size; the sutures are more deeply channeled, the spines are fewer in number, slightly larger, and not as curved; the aperture is entirely pure white, and the outer lip margin only weakly dentate; the colour is pale yellowish white with brown stained primary cords; it lacks the scabrous axial lamellae seen in *pecten*; reaches 200mm.
DISTRIBUTION: Sporadically found throughout the Philippines in limited quantity.

14. Murex nigrispinosus Reeve, 1845
90.5mm; Cebu.
Shell medium sized with a short, pointed spire, deeply channeled sutures, a somewhat ovate body whorl, and a long, curved, slightly open siphonal canal; sculptured with 3 varices per whorl and 3 to 4 irregular intervarical costae which are nodulose from overriding primary and secondary spiral cords; primary cords intersect the varices to form long, straight spines, 1 on the spire whorls, 3 on the body whorl and 5 to 6 on the canal; secondary spinelets occasionally seen; aperture ovate, outer lip erect and strongly crenulate, columella posteriorly adherent and smooth; bone-white with purplish-tipped spines; synonymised with *M. tribulus* but is consistently different in colour and structure; reaches 120mm.
DISTRIBUTION: Sporadically found throughout the Philippines.

15. Siratus pliciferoides (Kuroda, 1942) form **propinquus** (Kuroda and Azuma, 1961)
82.5mm; Palawan.
A peculiar form of *pliciferoides* in which the shell is more fusiform-ovate, lacks any appreciable anterior canal, and is devoid of varical spines; the spiral cords are stronger and distinctly scabrous and the shell colour is off-white with only a suggestion of pale brown banding; reaches 90mm.
DISTRIBUTION: Sporadically found throughout the Philippines in limited quantity.

16. Murex tenuirostrum Lamarck, 1822
Shell moderately large, club-shaped, with a short, acute spire, moderately channeled sutures, and a long, straight, narrowly open anterior canal; 3 varices per whorl, and 5 to 8 intervarical nodes of varying strength which are slightly nodulose from overriding spiral cords; spire whorl varices with a single spine, body whorl with 3, the dorsal shoulder spine strongest; numerous secondary spinelets also present; aperture ovate, outer lip erect and strongly crenulate, columella smooth and posteriorly adherent; brownish yellow with paler nodules, and a white aperture which is occasionally brown banded; reaches 125mm. Sold locally as *"Murex ternispina."*
16a. typical form; 95.2mm; Tayabas Bay, Quezon.
16b. brown-banded form; 80.2mm; Cebu.
DISTRIBUTION: Sporadically found throughout the Philippines.

Muricidae — Plate 38

1. Murexiella balteata (Sowerby, 1841)
26.5mm, 20.5mm, 24.0mm. 19.3mm; Punta Engaño, Cebu.
Shell medium sized, fusiform-ovate, with a tall spire and weakly impressed sutures; sculptured with narrow, elevated axial ribs (6 on the body whorl) and crossed by spiral cords forming short spines, 2 rows

PLATE NO. 38. 1. *Murexiella balteata* (Sowerby) 2. *Murexiella pelepili* (D'Attilio & Bertsch) 3. *Pterynotus barclayanus* (A. Adams) 4. *Favartia rosamiae* D'Attilio & Myers 5. *Pterynotus bibbeyi* Radwin & D'Attilio 6. *Favartia brevicula* (Sowerby) 7. *Morula parva* (Pease) 8. *Favartia sp.* 9. *Murexiella judithae* (D'Attilio & Bertsch) 10. *Murexiella martini* Shikama 11. *Attiliosa sp.* 12. *Favartia jeanae* Bertsch & D'Attilio 13. *Maculotriton serriale* (Deshayes, in Laborde & Linant) 14. *Murexiella mactanensis* Emerson & D'Attilio 15a. *Favartia dorothyae* Emerson & D'Attilio — pink form 15b. larger, paler form 16. c.f. *Morula biconica* (Blainville) 17. *Cytharamorula eximius* (Reeve)

PLATE 38
1
2
3
4
5
6
7
8
9
10
11
12
13
14
15a
15b
16
17

on the spire whorls, 7 on the body whorl, strongest on the sharply angled shoulder with the size and curvature decreasing anteriorly; aperture ovate, outer lip erect and serrate, columella smooth, canal long, narrow and recurved; colour variable, white, orange or pink with brown spines and a white to violet aperture; reaches 30mm.
DISTRIBUTION: Punta Engaño, Cebu.

2. Murexiella pelepili (D'Attilio & Bertsch, 1980)
21.8mm; Sugod, Cebu.
Shell large for the genus, fusiform-ovate, sculptured with 3 varices and 2 intervarical folds on each whorl which are all crossed by spiral cords (2 on the spire whorls, 4 to 5 on the body whorl), forming short, semi-tubular, upturned spines, strongest at the sharply angled shoulder and lamellate on the leading edges; aperture ovate, outer lip erect, serrate and smooth within, canal long, almost closed and recurved; colour pale brown; reaches 30mm.
DISTRIBUTION: Sugod, Cebu.

3. Pterynotus barclayanus (A. Adams, 1873)
23.2mm; Bohol.
Shell medium sized, elongate-ovate, with adpressed sutures and convex whorls; sculptured with irregularly spaced axial cords, overriding spiral lirae, and highly scabrous lamellae in the interspaces; aperture narrowly elongate-ovate, outer lip raised, dentate with a "wing-like" foliaceous border extending the entire body whorl length; columella smooth, canal straight, short and open; colour pinkish brown with a purple cast and a violet aperture; reaches 30mm.
DISTRIBUTION: Cebu, Bohol and Southern Mindanao.

4. Favartia rosamiae (D'Attilio & Myers, 1985)
16.2mm; Punta Engaño, Cebu.
Shell small, elongate-ovate, with deeply channeled sutures, a moderately tall spire and a large body whorl; sculptured with 4 to 5 widely spaced, somewhat elevated varices, and crossed by spiral cords, 2 on the spire whorls, 6 on the body whorl, the interspaces deeply pitted with axial lamellae; varices spinose on the sharply angled shoulder, giving the shell a somewhat "pagoda-like" appearance; outer lip erect, margin undulate, bordered with a highly expansive lamellate margin; columella smooth, canal short, narrow and recurved; off-white with a pinkish hue; reaches 20mm. Close to *F. nucula* Reeve which is described as having 6 varices.
DISTRIBUTION: Cebu and Bohol.

5. Pterynotus bibbeyi Radwin & D'Attilio, 1976
40.5mm; Bohol.
Shell medium sized, elongate-ovate, spire tall, sutures adpressed; sculptured with 4 varices and one intervarical node per whorl; encircled by a series of strong cords, 2 to 3 on the spire whorls below the sharply angled shoulder, and 7 on the body whorl (3 at the shoulder, 2 centrally and another 2 basally ; aperture elongate-ovate, outer lip erect, marginally serrate and internally dentate, columella smooth with 2 basal denticles, canal long, open and recurved; pinkish orange with a deep pink aperture; reaches 45mm.
DISTRIBUTION: Bohol, Cebu and Southern Mindanao.

6. Favartia brevicula (Sowerby, 1834)
16.2mm, 13.5mm; Punta Engaño, Cebu.
Shell medium sized, rhomboidal, with 5 to 6 shouldered whorls and a short, squat body whorl bearing 4 to 5 sharply attenuate varices which are crossed by 5 to 6 spiral cords whose strength decreases anteriorly; varical trailing edge excavated, leading edge fimbriate; aperture large and ovate, canal short and nearly closed; chalky white to grey with a white aperture; reaches 30mm.
DISTRIBUTION: Cebu and Bohol.

7. Morula parva (Pease, 1868)
7.2mm; Punta Engaño, Cebu.
Shell small, fusiform-ovate with adpressed sutures and sculptured with 2 rows of echinate nodules on the spire whorls and 5 to 6 on the body whorl; aperture narrowly elongate, outer lip with 3 to 4 denticles, the posterior one largest, columella smooth with 1 to 2 obsolete folds; off-white with every alternate nodular row black-banded; seldom exceeds 7mm. A secondary homonym for the Moruline species *Ricinula parva* Reeve, 1846 and presently appears to have no substitute name.
DISTRIBUTION: Cebu and Bohol.

8. Favartia sp.
13.9mm; Punta Engaño, Cebu.
Shell small with a tall spire, and sculptured with 6 to 7 narrow axial cords which are crossed by strong, elevated spiral lirae, 2 on the spire whorls, 4 on the body whorl; aperture ovate, outer lip margin and columella relatively smooth, siphonal canal short and nearly closed; off-white with black staining in the interstitial pits; reaches 15mm.
DISTRIBUTION: Punta Engaño, Cebu.

9. Murexiella judithae (D'Attilio & Bertsch, 1980)
17.1mm; Punta Engaño, Cebu.
Shell small, fusiform-ovate, with a moderately tall spire and a large body whorl; sculptured with 6 to 7 highly scabrous varices and crossed by strong spiral cords, 2 on the spire whorls, 6 on the body whorl; the roundly angulate shoulder of each whorl bears a row of long, open, highly branched spines with several rows of weaker spines below this; whole surface finely scabrous and spinose; aperture ovate, outer lip erect, serrate and smooth within; yellowish brown, spines off-white; reaches 25mm.
DISTRIBUTION: Cebu and Southern Mindanao.

10. Murexiella martini Shikama, 1977
25.0mm; Punta Engaño, Cebu.
Shell small, fusiform-ovate with a tall spire and 7 to 8 whorls; body whorl with 3 to 4 varices, spinose from

strong spiral cords; spines longest at the shoulders, lamellate on their leading edge, and connected by a web-like structure; shell surface entirely lamellate; aperture ovate, outer lip with 4 spinose processes, peristome entire, canal long, open and recurved; light brown with darker cords and spines, aperture white; reaches 30mm.
DISTRIBUTION: Cebu and Bohol.

11. Attiliosa sp.
11.2mm; Punta Engaño, Cebu.
Shell small to medium sized with a tall spire, adpressed sutures and a large, fusoid body whorl; shell with 6 rounded varices per whorl, each with a long, sharp shoulder spine; entire shell encircled with weak spiral cords, producing weak nodules below the level of the spines; aperture ovate, smooth within; colour greyish white; largest seen 35mm.
DISTRIBUTION: Cebu and Bohol.

12. Favartia jeanae Bertsch & D'Attilio, 1980
8.1mm; Punta Engaño, Cebu.
Shell very small, elongate-ovate with a tall spire, impressed sutures and well rounded whorls which are sculptured with numerous well spaced axial cords and crossed by numerous scabrous spiral lirae forming slightly echinate nodules at the points of intersection; aperture ovate, outer lip erect, variced and smooth within, columella short and recurved; pinkish orange, varices and aperture white; reaches 10mm.
DISTRIBUTION: Cebu and Bohol.

13. Maculotriton serriale (Deshayes, in Laborde & Linant, 1834)
11.2mm; Punta Engaño, Cebu.
Shell small, fusiform, with impressed sutures and 6 to 7 convex whorls; sculptured with 10 to 11 axial ribs, strongest on the whorl shoulder and crossed by numerous fine spiral threads; aperture narrowly elongate, outer lip variced and lirate within, canal short and recurved; off-white with or without black spots and banding; reaches 20mm.
DISTRIBUTION: Cebu and Bohol.

14. Murexiella mactanensis Emerson & D'Attilio, 1979
17.7mm; Punta Engaño, Cebu.
Shell small, fusiform-ovate, with a tall spire and impressed sutures; spire whorls with 6 varices, body whorl with 5; surface encircled with strong cords (2 on the spire whorls, 5 on the body whorl); varices lamellose and heavily spinose at the shoulder; aperture ovate, outer lip marginally crenulate, columella smooth, siphonal canal long; off-white to pink with a white aperture; reaches 25mm; readily recognized by the 5 long, prominently curved spines on the outer lip margin.
DISTRIBUTION: Cebu and Bohol.

15. Favartia dorothyae Emerson & D'Attilio, 1979
Shell small, fusiform, elongate-ovate with a tall spire and impressed sutures; sculptured with 5 axial cords on the body whorl and crossed by numerous spiral striae, producing an extremely foliate surface; aperture ovate, outer lip variced, erect, marginally serrate and internally lirate; columella smooth and posteriorly adherent; colour bright pinkish red, larger specimens pale pinkish orange; reaches 18mm.
15a. bright pink form; 10.2mm; Punta Engaño, Cebu.
15b. larger paler form; 16.3mm, 15.6mm; Punta Engaño, Cebu.
DISTRIBUTION: Cebu.

16. c. f. Morula biconica (Blainville, 1832)
13.8mm, 13.9mm; Samar.
Shell small, fusiform-ovate, sculptured with spinose axial ribs and scabrous spiral cords, 2 or 3 of which are larger and broader on the body whorl; aperture narrowly elongate with a scabrous margin, outer lip denticulate, columella irregularly plicate; colour dark brown to black, the axial cords and major spiral elements all white, aperture violet; reaches 25mm.
DISTRIBUTION: Sporadically found throughout the Philippines.

17. Cytharamorula eximius (Reeve, 1844)
13.1mm; Punta Engaño, Cebu.
Shell small, elongate with adpressed sutures, and the whorls subangulate on the presutural slope; sculptured with slender axial ribs and numerous overriding spiral threads; outer lip variced, internally denticulate, columella weakly plicate; orange to brown-coloured with darker spiral elements, aperture off-white; reaches 20mm.
DISTRIBUTION: Cebu and Bohol.

Plate 39 Muricidae

1. Drupa (Drupa) ricinus ricinus (Linné, 1758)
A medium sized shell similar to *D. morum* in apertural characteristics but smaller, and with 5 rows of somewhat longer spines; shell colour white to off-white, spines almost black, aperture white, occasionally with a peripheral orange band; reaches 35mm.
1a. white-mouthed form, often called *albolabris* Blainville, 1832; 23.0mm; Palawan.
1b. orange-spotted form, often called *arachnoides* Lamarck, 1816; 22.0mm; Cebu. (Both these forms occur with equal frequency and represent the typical species).
DISTRIBUTION: Throughout the Philippines.

2. Drupa (Drupa) morum morum Röding, 1798
33.1mm; Samar.
A medium sized globose shell with an almost obsolete spire and 4 rows of short, blunt tubercles; aperture long, narrow, and heavily toothed; colour creamy white with black tubercles, aperture deep violet; reaches 50mm.
DISTRIBUTION: Throughout the Phililippines.

3. Drupa (Drupina) grossularia Röding, 1798
24.8mm; Palawan.
A moderately small shell with a somewhat clathrate surface, a short spire, and a large body whorl which is encircled with rows of weakly rounded nodules forming 1 or 2 long, digitate processes at the apertural margin; pale brown with an orange aperture, apertural denticles and outer lip margin white; reaches 35mm.
DISTRIBUTION: Throughout the Philippines.

4. Drupa (Ricinella) clathrata clathrata (Lamarck, 1816)
26.2mm; Samar.
Shell medium sized, ovate, with a short spire and 5 rows of open tubercles interconnected axially and forming pits; aperture ovate and heavily calloused, with a dentate outer lip and 4 plicae on the anterior end of the columella; shell pale brown, outer lip margin white, teeth white with brown spots between, extending around the aperture periphery, aperture interior violet; reaches 60mm.
DISTRIBUTION: Sporadically found throughout the Philippines.

5. Drupa (Ricinella) rubusidaeus Röding, 1798
39.4mm; Palawan.
Shell large, heavy and globose, with a short spire and 5 rows of partially open spines; aperture with numerous small denticles on the outer lip and 3 or 4 plicae on the inner lip; colour yellowish, spinal tips black, outer lip margin yellow, inner area deep pinkish purple; reaches 55mm.
DISTRIBUTION: Sporadically found throughout the Philippines.

6. Attiliosa nodulifera (Sowerby, 1841)
27.9mm; Marinduque.
Shell small with a conical spire of 5 to 6 whorls bearing one row of long, open, slightly curved spines on the mid point of the earlier whorls, and 3 to 4 rows on the body whorl; varices number between 6 and 8 on the last 2 whorls; aperture ovate, outer lip denticulate, columella with 4 basal denticles; shell surface pale brown, aperture pale yellow, spines brown; reaches 32mm.
DISTRIBUTION: Sporadically found throughout the Philippines in limited quantity.

7. Spinidrupa spinosa (H. & A. Adams, 1853)
21.2mm; Punta Engaño, Cebu.
Shell small to medium sized, sculptured with 9 to 10 spinose varices, a single presutural row of spines on the spire whorls and 4 on the body whorl; surface also sculptured with numerous rows of fine spiral striae, strongest in the intervarical spaces; aperture narrowly elongate, outer lip thick with 6 denticles, columella mainly smooth with a few weak anterior folds; shell colour purple with darker varices and spinal tips, aperture also mauve with white denticles; Philippine specimens reach around 30mm.
DISTRIBUTION: Throughout the Philippines.

8. Drupella cornus (Röding, 1798)
Shell medium sized with a sharply acute spire, a single row of echinate nodules on the spire whorls, 4 on the body whorl and spiral grooves between; aperture narrow, outer lip with 6 to 7 denticles, columella with 3 to 4 weak basal plicae. Most specimens off-white with a pale coloured aperture; reaches around 30mm. Typical *cornus* occurs in the Indian Ocean and has a yellow aperture.
8a. typical Philippine form; 24.7mm; Palawan.
8b. *Drupella rugosa* (Born, 1778): often figured as a form of *D. cornus* (8a.); characterized by its rounded, more numerous nodules, and its pink to red coloured aperture bearing smaller, more numerous denticles; 27.3mm; Palawan. Syn: *D. fragum* Blainville, 1832.
DISTRIBUTION: Sporadically found throughout the Philippines.

9. Thais (Thais) savignyi (Deshayes, 1844)
Syn: *T. distinguenda* Dunker, 1866
30.8mm; Samar.
Shell medium sized, ovate-biconic with 5 to 6 whorls; sculptured with spiral ridges (frequently eroded in earlier whorls), and 4 rows of triangular denticles which decrease in size anteriorly; aperture ovate, outer lip black, with 5 paler denticles continuing as lirae into the pale blue interior; columella brownish and obsoletely folded; external colour black with white streaks and patches; reaches 65mm.
DISTRIBUTION: Sporadically found throughout the Philippines.

10. Thais sp.
30.0mm, 30.9mm; Samar.
Shell medium sized, solid, elongate-ovate in outline with adpressed sutures; sculptured with rows of axially compressed nodules, 2 on the spire whorls and 5 on the body whorl, and the entire shell is encircled with numerous interstitial scabrous cords; aperture elongate, outer lip margin brown coloured, crenulate and with 5 lirae which terminate on the outer lip as denticles; columella brown coloured with 1 or 2 obsolete folds anteriorly; external colouration deep brown with white streaks between the nodules; reaches 40mm.
DISTRIBUTION: Sporadically found throughout the Philippines.

PLATE NO. 39 1a. *Drupa ricinus ricinus* (Linné) form *albolabris* Blainville 1b. form *arachnoides* Lamarck 2. *Drupa morum morum* Röding 3. *Drupa grossularia* Röding 4. *Drupa clathrata clathrata* (Lamarck) 5. *Drupa rubusideus* Röding 6. *Attiliosa nodulifera* (Sowerby) 7. *Spinidrupa spinosa* (H. & A. Adams) 8a. *Drupella cornus* (Röding) 8b. *Drupella rugosa* (Born) 9. *Thais savignyi (Deshayes)* 10. *Thais sp.* 11. *Cronia margariticola margariticola* (Broderip) 12. *Cronia fenestrata* (Blainville) 13. *Cronia ochrostoma* (Blainville) 14. *Vexilla vexillum* (Gmelin) 15. *Cronia contracta* (Reeve) 16. *Morula uva* (Röding) 17. *Thais buccinea* Deshayes 18. *Thais echinata* (Blainville) 19. *Cronia funiculus* (Wood) 20. *Cronia fiscella* (Gmelin) 21. *Pinaxia versicolor* (Gray)

PLATE 39

11. Cronia (Cronia) margariticola margariticola (Broderip, 1833)
23.9mm, 23.9mm; Cebu.
Shell medium sized, biconic, with 7 to 8 whorls which are sculptured with 9 to 10 broad axial cords terminating as prominent shoulder nodes below the sutures; surface encircled with numerous scabrous spiral lirae; aperture elongate-ovate, outer lip with 6 to 7 marginal denticles, columella smooth, rarely with obsolete plicae; black coloured with white spiral bands, aperture mauve to brown, columella pinkish brown; reaches 40mm; extremely variable. The subspecies *crassulnata* (Hedley, 1915) is found in Northern Australia.
DISTRIBUTION: Throughout the Philippines.

12. Cronia (Muricodrupa) fenestrata (Blainville, 1832)
31.7mm; Mindoro.
Shell medium sized, biconic, solid, crossed by broad axial and spiral cords, producing rows of rounded nodules most prominent at the shoulders, and deep pits in the interspaces; outer lip with 3 to 4 denticles, columella obsoletely plicate; yellowish grey in colour with an orange-yellow aperture; reaches 35mm.
DISTRIBUTION: Sporadically found throughout the Philippines.

13. Cronia (Pascula) ochrostoma (Blainville, 1832)
18.1mm, 16.5mm; Palawan.
Shell small, squat, biconic, spire tall and sharply pointed, sutures adpressed; sculptured with 6 axial cords, nodulose at the shoulders and crossed with spiral cords intersecting to produce echinate nodules and deep brown-coloured pits in the interspaces; aperture ovate, outer lip dentate and lirate within; white with an orange aperture; reaches around 20mm.
DISTRIBUTION: Sporadically found throughout the Philippines.

14. Vexilla vexillum (Gmelin, 1791)
20.6mm, 20.9mm; Punta Engaño, Cebu.
A small, smooth, cylindrically ovate shell with a short spire and a large body whorl; aperture elongate-ovate, finely denticulate on the outer lip; columella smooth, anteriorly twisted; colour off-white to brownish yellow with darker banding; aperture cream, orange between the denticles; reaches 25mm.
DISTRIBUTION: Sporadically found throughout the Philippines in limited quantity.

15. Cronia (Ergalatax) contracta (Reeve, 1846)
28.5mm; Palawan.
Shell medium sized, solid, with 7 to 8 subangulate whorls and adpressed sutures; sculptured with around 10 axial cords and crossed by numerous scabrous spiral lirae; aperture narrowly ovate, outer lip with 7 to 8 denticles, columella calloused and weakly plicate; off-white with 2 or 3 brown bands, aperture white; reaches 37mm.
DISTRIBUTION: Sporadically found throughout the Philippines.

16. Morula (Morula) uva (Röding, 1798)
22.5mm; Samar.
Shell small, ovate, biconic, with a moderately tall spire and a large body whorl; sculptured with rows of sharply echinate tubercles, 2 on the spire whorls, 5 on the body whorl, and scabrous cords between; aperture narrow, quadrangular, outer lip with 5 denticles, the size decreasing anteriorly, columella smooth but weakly plicate anteriorly; white with lavender aperture, the tubercles frequently black; reaches 30mm.
DISTRIBUTION: Sporadically found throughout the Philippines.

17. Thais (Thais) buccinea Deshayes, 1844
49.0mm; Palawan.
Shell large, solid, elongate-ovate with adpressed sutures; sculptured with 10 to 11 broad axial folds strongest at the shoulder and crossed by numerous spiral lirae of varying strengths; aperture elongate-ovate, outer lip thick and lirate within, columella glazed with a large parietal denticle and weak basal plicae; colour dark brown with the major spiral elements paler, aperture and columella creamy orange with a brown outer lip margin; reaches 50mm.
DISTRIBUTION: Sporadically found throughout the Philippines in limited quantity.

18. Thais (Mancinella) echinata (Blainville, 1832)
27.9mm; Samar.
Shell medium sized, solid, rather ovate in shape with adpressed sutures; whorls sculptured with rows of echinate nodules, one on the spire whorls, 4 on the body whorl, and numerous interstitial spiral lirae; aperture elongate-ovate, outer lip crenulate and weakly lirate within, columella smooth, somewhat twisted, fasciolar ridge strong, bordering a pseudo-umbilicus; brownish white externally, porcelaneous white within the aperture; reaches 45mm.
DISTRIBUTION: Sporadically found throughout the Philippines.

19. Cronia (Muricodrupa) funiculus (Wood, 1828)
Syn: *Cronia (Muricodrupa) decussatus* (Reeve, 1845)
22.9mm; Punta Engaño, Cebu.
Shell small, somewhat stoutly biconic in outline, sculptured with 7 broadly scabrous axial cords, strongly echinate at the sharply angled shoulder, and encircled with 5 thick cords which intersect to produce deep, black coloured, laterally elongated pits in the interspaces; aperture elongate-ovate, outer lip heavily scabrous with 6 denticles on the inner margin, columella with 3 anterior denticles; colour off-white with a violet aperture; reaches 30mm.
DISTRIBUTION: Sporadically found throughout the Philippines in limited quantity.

20. Cronia (Muricodrupa) fiscella (Gmelin, 1791)
Syn: *Cronia (Muricodrupa) ricinuloides* (Quoy & Gaimard, 1833)
22.4mm; Punta Engaño, Cebu.
Shell small, elongate-ovate, with a moderately tall

spire, a large body whorl and sub-angulate shoulders; sculptured with 8 to 9 broad, scabrous, double-stranded axial cords, and encircled with 5 double-stranded spiral cords and weaker interstitial spiral lirae on the body whorl; interspaces weakly pitted and brown stained; aperture ovate, outer lip denticulate, columella smooth; off-white in colour with a violet aperture; reaches 30mm; superficially similar to *C. funiculus* (Wood, 1828), (fig. 19), but is more squat, has more numerous, more closely set ribs, shallower interstitial "pits", and a slightly shorter anterior canal.
DISTRIBUTION: Sporadically found throughout the Philippines in limited quantity.

21. Pinaxia versicolor (Gray, 1839)
17.5mm; Samar.
Shell small, solid, with a large body whorl, a short spire and angulate shoulders; surface crossed with spiral cords and axial striae; aperture elongate-ovate, outer lip weakly lirate within, columella smooth with a strong fold anteriorly; colour yellow with pale brown maculations and rows of black, discontinuous spiral lines; reaches 20mm.
DISTRIBUTION: Sporadically found throughout the Philippines in limited quantity.

Plate 40 — Muricidae

1. Vitularia miliaris (Gmelin, 1791)
50.5mm; Samar.
Shell medium sized, spire whorls somewhat compressed and the body whorl large; sutures deeply channeled and the surface is somewhat maleate and pitted and sculptured with 6 to 8 varices per whorl, nodulose at the shoulder; aperture elongate-ovate, denticulate on the expanded outer lip margin, columella smooth with a strong fasciolar ridge; orange-brown with brown varical spots, aperture pale orange; reaches 65mm.
DISTRIBUTION: Sporadically found throughout the Philippines.

2. Vitularia crenifer (Montrouzier in Souverbie, 1861)
21.9mm; Cebu.
Similar to *V. miliaris* but much smaller, with more compressed spire whorls and an extremely shouldered body whorl; sculptured with numerous, almost obsolete axial folds and encircled with numerous fine spiral striae producing a scaled surface; aperture ovate with a dentate outer lip and smooth columella; shell pale brown to pinkish yellow, aperture white; reaches 40mm.
DISTRIBUTION: Sporadically found throughout the Philippines.

3. Lataxiena fimbriata (Hinds, 1844)
42.0mm, 43.5mm; Punta Engaño, Cebu.
Shell medium sized, solid, tall spired with impressed sutures and a large, fusoid body whorl; whorls angulate and sculptured with numerous axial ribs and crossed by spiral cords which start below the shoulder; body whorl with additional axial lamellae and overriding scabrous spiral cords; aperture ovate, lirate within, columella smooth, umbilicus deep, fasciolar ridge corded; pale yellowish white with 2 to 3 brown bands on the body whorl, aperture orange-white; reaches 50mm. Morphologically resembling a member of the Coralliophilidae, but possessing a radula which confirms its assignment to the Muricidae (Subfamily: Trophoninae).
DISTRIBUTION: Sporadically found throughout the Philippines.

4. Thais (Thais) armigera (Link, 1807)
66.4mm; Palawan.
Shell large, heavy and biconic; spire moderately tall, body whorl large with 3 rows of heavy, blunt spines, decreasing in size anteriorly; aperture elongate-ovate, outer lip marginally dentate and lirate within; columella with 3 to 4 weak anterior cords; off-white with brown bands between the spinal rows, aperture pink, marginally white, columella white with a pink tinge; reaches 58mm.
DISTRIBUTION: Sporadically found throughout the Philippines.

5. c. f. Lataxiena fimbriata (Hinds, 1840)
40.2mm; Samar.
Shell fusiform-elongate with a moderately tall spire and impressed sutures; sculptured with fimbriate axial ribs somewhat echinate at the sharply angled shoulder, and overriding spiral cords which produce a coarsely cancellate surface; aperture ovate, outer lip serrate and lirate within, columella smooth; yellowish brown with a white aperture; reaches 60mm. Close to *fimbriata* Hinds which is a highly polymorphic taxon, but probably undescribed.
DISTRIBUTION: Sporadically found throughout the Philippines.

6. Bedeva blosvillei (Deshayes, 1832)
45.0mm; Albay Gulf, Southern Luzon.
Shell medium sized, fusiform, tall spired, with impressed, somewhat undulate sutures and a large, fusoid body whorl; whorls sharply angled, sculptured with 8 to 9 axial folds, slightly nodulose at the shoulder and crossed with major and minor spiral cords; aperture elongate-ovate with a deep posterior notch, a marginally serrate, internally lirate outer lip and a long siphonal canal; greyish yellow with irregular brown lines, an orange aperture and a brown-streaked columella; reaches 60mm.
DISTRIBUTION: Sporadically found throughout the Philippines.

7. Morula chaidea (Duclos, 1832)
21.8mm, 21.0mm; Punta Engaño, Cebu.
Shell small, fusiform-ovate, with adpressed sutures and a markedly constricted presutural area; sculptured with coarse, oblique axial cords and overriding scabrous spiral striae; aperture narrowly ovate, outer lip with 6 denticles, columella with 3 anterior folds; off-white; reaches 25mm.
DISTRIBUTION: Sporadically found throughout the Philippines.

8. Morula granulata (Duclos, 1832)
20.0mm, 17.4mm; Batangas.
Shell small, ovately biconic, sculptured with rows of large rounded nodules, 2 on the spire whorls, 4 on the body whorl and spirally striate between; aperture narrowly elongate with 4 to 5 denticles, columella with 2 or 3 weak plicae; external colour off-white, nodules black, aperture interior purple; reaches 35mm.
DISTRIBUTION: Sporadically found throughout the Philippines.

9. Morula musiva (Kiener, 1836)
23.9mm, 21.2mm; Estancia, Iloilo.
Shell small, ovately biconic, sutures adpressed, spire half the shell length; sculptured with 2 rows of nodules on the spire whorls and 4 on the body whorl; aperture narrowly ovate, outer lip thick with 4 to 6 heavy denticles, columella with 1 to 2 weak folds; colour white with black banding, nodule rows alternating red and black; usually badly eroded; reaches 25mm.
DISTRIBUTION: Sporadically found throughout the Philippines.

10. Thais (Thais) javanica (Philippi, 1848)
27.0mm; Estancia, Iloilo.
Shell small, elongate-ovate and tall spired, with a large body whorl and corded sutures; sculptured with 8 to 10 axial folds and overriding spiral cords forming axially compressed nodules, strongest at the angled shoulder; aperture elongate-ovate, outer lip marginally serrate and lirate within (6 to 8), columella smooth; colour bluish white, axial cords brown, interspaces and nodules streaked with orange, aperture bluish white with a brown margin, columella orange; reaches 30mm.
DISTRIBUTION: Sporadically found throughout the Philippines.

11. Thais (Mancinella) bufo (Lamarck, 1822)
52.0mm; Palawan.
Shell medium sized, broadly ovate, heavy, with a short spire and adpressed sutures; sculptured with numerous spiral lirae and 2 rows of bluntly rounded tubercles, one on the upper third of the body whorl, the other on the lower third; aperture large, ovate, outer lip thick and crenulate, columella posteriorly expanded forming a thick callus; brown with white lines and white patches between the tubercles, aperture cream, columella orange; reaches 75mm.
DISTRIBUTION: Sporadically found throughout the Philipines.

12. Thais (Thais) kieneri (Deshayes, 1844)
42.5mm, 37.7mm; Cebu.
Shell medium sized, solid and biconic with a tall spire, adpressed sutures and a large body whorl which is sculptured with fine spiral lines terminating as minute teeth on the outer lip; whorls sharply angled with axial cords which form a row of axially compressed nodules on the spire whorls and 4 on the body whorl, the size decreasing anteriorly; aperture ovate, outer lip dentate, columella smooth; colour dark brown, orange between the nodules, aperture and columella pale orange; reaches 60mm.
DISTRIBUTION: Sporadically found throughout the Philippines.

13. Rapana rapiformis (Born, 1778)
132mm; Estancia, Iloilo.
Shell large, heavy and bulbous, with a low spire and deeply canaliculate sutures; body whorl extremely inflated with 2 or 3 rows of tubercles, most prominent at the shoulder; lower body whorl area with prominent spiral grooves; aperture large and ovate with marginal secretions continuing as inner lip lirae; columella smooth and calloused, umbilicus deep, bordered by a heavily scabrous fasciolar cord; light brown with an orange columella and aperture; reaches 150mm.
DISTRIBUTION: Sporadically found throughout the Philippines.

14. Nassa serta (Bruguiére, 1789)
51.3mm; Samar.
Shell medium sized, fusiform-ovate in outline, with a tall spire, weakly channeled sutures and rounded whorls which are sculptured with beaded spiral cords; aperture elongate-ovate, outer lip marginally dentate, columella smooth with a single posterior denticle; reddish brown with pink or white blotches, sometimes in bands, aperture pink, outer lip margin and columella dark brown; reaches 70mm.
DISTRIBUTION: Sporadically found throughout the Philippines.

15. Nassa francolina (Bruguiére, 1789)
57.4mm; Cebu.
Very similar to *N. serta* (Bruguiére), but the spiral cords are smooth and the aperture lacks the dark

PLATE NO. 40 1. *Vitularia miliaris* (Gmelin) 2. *Vitularia crenifer* (Montrouzier in Souverbie) 3. *Lataxiena fimbriata* (Hinds) 4. *Thais armigera* (Link) 5. *c.f. Lataxiena fimbriata* (Hinds) 6. *Bedeva blosvillei* (Deshayes) 7. *Morula chaidea* (Duclos) 8. *Morula granulata* (Duclos) 9. *Morula musiva* (Kiener) 10. *Thais javanica* (Philippi) 11. *Thais bufo* (Lamarck) 12. *Thais kieneri* (Deshayes) 13. *Rapana rapiformis* (Born) 14. *Nassa serta* (Bruguiére) 15. *Nassa francolina* (Bruguiére) 16. *Thais echinulata* (Lamarck) 17. *Thais mancinella* (Linné) 18. *Thais tuberosa* (Röding) 19. *Purpura panama* (Röding) 20. *Purpura persica* (Linné)

PLATE 40

brown colouration and is usually pale pink or bluish white in colour; also reaches around 70mm.
DISTRIBUTION: Sporadically found throughout the Philippines.

16. Thais (Mancinella) echinulata (Lamarck, 1822)
45.5mm; Samar.
Shell medium sized and heavy with a short spire and adpressed sutures; body whorl large, globose, and sculptured with rows of sharply tuberculate spiral cords and weaker spiral threads; aperture elongate-ovate, outer lip crenulate, lirate within, columella smooth; off-white with an orange overtone, aperture and columella pure white with orange margins; reaches 50mm.
DISTRIBUTION: Sporadically found throughout the Philippines.

17. Thais (Mancinella) mancinella (Linné, 1758)
34.5mm; Samar.
Shell medium sized, solid and tall spired with adpressed sutures and a large, ovate body whorl; sculptured with 4 rows of blunted triangular nodules, a beaded fasciolar cord and numerous interstitial spiral striae; aperture elongate-ovate, with a deep orange glaze and red coloured lirae, terminating just short of the crenulate outer lip margin; colour pale yellowish orange with darker tubercles; reaches 55mm.
DISTRIBUTION: Throughout the Philippines.

18. Thais (Mancinella) tuberosa (Röding, 1798)
46.5mm; Cebu.
Shell medium sized, ovate-biconic in outline, with a tall spire, adpressed sutures and a large body whorl; sculptured with numerous spiral lirae, a row of blunted triangular tubercles on the spire whorls, 3 to 4 on the body whorl, and a corded fasciolar ridge; aperture elongate-ovate with red lirae within, outer lip thin and crenulate, columella heavily calloused; colour brown with white lines and spots on the tubercles, aperture cream, marginally yellow and brown, columella with brown spotting; reaches 60mm.
DISTRIBUTION: Sporadically found throughout the Philippines.

19. Purpura panama (Röding, 1798)
Syn: *Purpura rudolphi* (Lamarck, 1822)
61.4mm; Samar.
Shell medium sized, solid and fusiform-ovate, with a short spire and a large body whorl; sculptured with spiral striae, a row of low nodules on the shoulder of the spire whorls and 4 rows on the body whorl; aperture large and ovate with red lirae within; outer lip thick, marginally brown and crenulate; columella white, smooth and deeply excavated anteriorly; colour deep brown with white spots between the tubercles, aperture white; reaches 90mm.
DISTRIBUTION: Sporadically found throughout the Philippines.

20. Purpura persica (Linné, 1758)
93.3mm; Palawan.
Shell large and heavy, with adpressed sutures, a short spire and a large, ovate body whorl; early whorls weakly noduled, the body whorl almost smooth and encircled with numerous primary and secondary spiral cords; aperture large, elongate-ovate, outer lip thick and marginally crenulate, columella heavily calloused and deeply excavated anteriorly; external colour dark brown, the primary cords ornamented with black and white streaks; aperture with a dark brown margin, followed by a red-streaked orange band, then bluish white within; columella orange with a deep purple-black parietal glaze and a white-coloured basal excavation; reaches 100mm.
DISTRIBUTION: Sporadically found throughout the Philippines.

Muricidae Plate 41

1. Chicoreus (Chicoreus) crocatus auct. (non Reeve, 1845)
70.5mm; Zamboanga.
Shell large, solid and fusiform in outline; sculptured with fine, occasionally squamose spiral ridges, 3 varices per whorl, and a single large intervarical node; varices sculptured with long frondose spines, strongest at the shoulder, 2 on the spire whorls, 5 on the body whorl and 3 to 4 on the long, almost closed siphonal canal; aperture elongate-ovate, outer lip raised, sharply dentate, columella smooth, anteriorly detached; a deep orange colour with brown in the intervarical spaces, aperture white; reaches 75mm.
DISTRIBUTION: Zamboanga.

2. Homalocantha anomaliae Kosuge, 1979
40.0mm; Panglao, Bohol.
Shell medium sized with 6 to 7 shouldered whorls and impressed sutures; last 2 whorls loosely coiled, connected with heavy axial folds, body whorl large with 4 spinose varices, 3 of which have long palmate projections connected by fine fimbriate webbing, weak projections formed on the varices from overriding spiral lirae; outer lip erect, finely dentate, columella smooth and erect, siphonal canal long, partially open and distally recurved; colour off-white; reaches 50mm.

PLATE NO. 41 1. *Chicoreus crocatus* auct. (non Reeve) 2. *Homalocantha anomaliae* Kosuge 3. *Siratus vicdant* Kosuge 4. *Dermomurex neglecta* (Habe & Kosuge) 5. *Pterynotus martinetana* (Röding) 6. *Cronia sp.* 7. *Morula anaxeres* (Kiener) 8. *c.f. Cronia contracta* (Reeve) 9. *Murexiella peregrina* Olivera 10. *Nassarius sufflatus* (Gould) 11. *Nassarius concinnus* (Powys) 12. *Nassarius luridus* (Gould) 13. *Nassarius limnaeiformis* (Dunker) 14. *Nassarius livescens* (Philippi) 15. *Nassarius margaritiferus* (Dunker) 16. *Nassarius gaudiosus* (Hinds) form *zonalis* (A. Adams) (non Bruguière) 17. *Nassarius dorsatus* (Röding) 18. *Nassarius conoidalis* (Deshayes in Belanger) 19. *Nassarius pullus* (Linné) 20. *Nassarius arcularius arcularius* (Linné) 21. *Nassarius olivaceus* (Bruguière)

PLATE 41

DISTRIBUTION: Bohol and Southern Mindanao.

3. Siratus vicdani Kosuge, 1980
30.5mm; Balut Is., Southern Mindanao.
Shell small, fusiform-elongate, with 6 to 8 moderately convex whorls and impressed sutures; body whorl large with 3 varical blades bearing spinose projections and 2 intervarical nodes, strongest at the shoulder; earlier whorls crossed by 3 spiral cords forming spines on the varical blades; aperture large, ovate, outer lip with an expansive lamellate margin, columella smooth and adherent; canal long and straight; pale brown with 3 white bands on the body whorl; reaches 45mm; close to a young *S. alabaster* but differs in its spinose varical webbing and colour.
DISTRIBUTION: Southern Mindanao.

4. Dermomurex neglecta (Habe & Kosuge, 1971)
26.2mm; Bohol.
Shell small, elongate-ovate, encircled with numerous spiral threads and sculptured with 5 blade-like varices per whorl, obliquely angled and somewhat prosocline from the shoulder to the suture; aperture ovate, periostome entirely adherent and smooth, canal short, open and distally recurved, with a strong siphonal fasciole and deep umbilicus; chalky white in colour; reaches 30mm.
DISTRIBUTION: Cebu and Bohol.

5. Pterynotus martinetana (Röding, 1798)
52.7mm; Balut Is., Southern Mindanao.
Shell medium sized, fusiform-ovate, with a tall spire of 5 to 6 weakly subangulate whorls and a large, fusoid body whorl; sculptured with 5 varices on the body whorl, a single thick intervarical cord which is nodulose at the shoulder and encircled by 3 series of finely scabrous lirae, 3 on the shoulder, 2 centrally, and 2 on the canal; short spinose projections interconnected by fine webbing formed where the cords cross the varices; aperture ovate, outer lip erect, marginally serrate, internally denticulate (5 to 6); off-white with orange spines, black pit-like interspaces, and an orange-coloured aperture with white denticles; reaches 60mm.
DISTRIBUTION: Southern Mindanao.

6. Cronia sp.
20.5mm; Cebu.
Shell small, fusiform, elongate-ovate, sculptured with 9 to 10 axial cords on the body whorl, a row of sharply spinose tubercles on the sharply angled shoulder of the spire whorls and 5 rows on the body whorl; interstitial spaces with axial and spiral threads forming a highly scabrous surface; aperture elongate-ovate, outer lip marginally serrate, variced, with 7 to 8 weak denticles internally, outer lip margin constricted anteriorly, forming a short, distally recurved canal, columella with obsolete basal denticles; yellowish orange with a lavender aperture; reaches 30mm.
DISTRIBUTION: Cebu and Bohol.

7. Morula anaxeres (Kiener, 1835)
13.4mm; Punta Engaño, Cebu.
Shell small, ovate-biconic, sculptured with a row of white coloured nodules on the spire whorls, 2 on the body whorl, and a third weaker row at the base of the body whorl; aperture narrowly elongate, outer lip with 4 to 6 denticles, columella irregularly plicate; shell black with white nodules and a dirty purple-brown aperture occasionally streaked with white; reaches 30mm.
DISTRIBUTION: Sporadically found throughout the Philippines in limited quantity.

8. c. f. Cronia (Ergalatax) contracta (Reeve, 1846)
19.0mm; Palawan.
Shell small with a tall spire, adpressed sutures and a large body whorl; whorls angulate, sculptured with elevated axial cords, nodulose on the presutural ramp and crossed by spiral cords forming an additional 2 rows of black-coloured nodules on the body whorl; aperture elongate-ovate, outer lip variced, denticulate within, columella basally plicate, weakly umbilicate with a strong fasciolar cord; white with black sutural banding, black nodular cords and yellowish brown interstitial spaces, aperture pale lavender; reaches 25mm.
DISTRIBUTION: Sporadically found throughout the Philippines.

9. Murexiella peregrina Olivera, 1980
13.5mm; Punta Engaño, Cebu.
Shell small, elongate, with a tall spire and a large body whorl; sculptured with 4 elevated varices per whorl and crossed by finely fluted spiral lirae, 2 on the spire whorls and 5 to 6 on the body whorl, which form a row of long, upturned spines on the shoulders and weaker spinelets below; aperture ovate, bordered by an expansive, somewhat lamellate margin bearing 5 to 7 spines, columella smooth, siphonal canal long, narrow and strongly recurved distally; brownish white with a white aperture and reddish canal; reaches 15mm. Named after the Peregrina sisters, Mrs. Gloria P. Guerrero and the late Mercedes Peregrina, both prominent local collectors.
DISTRIBUTION: Cebu and Bohol.

ADDENDA

Talityphis bengalensis Radwin & D'Attilio, 1976
9.0mm; Off Marinduque.
Shell small, broadly biconic, with well impressed sutures, a fairly tall spire and a large body whorl; protoconch smooth, prominent, with $1\frac{1}{2}$ convex whorls; spire whorls (which number 4 to 5) well shouldered and bearing several short, open-ended tubular spines; body whorl broadly trigonal with 4 varices, 3 being thin and blade-like, the last one broad and wing-like and separated from the lower shoulder area of the previous whorl by a thin partition; aperture small, ovate, with a continuous raised

peristome, siphonal canal broad, short and sealed; entire surface smooth; colour pale tan with a pink suffusion, protoconch off-white; reaches around 10mm.
DISTRIBUTION: Presently known only from Marinduque.

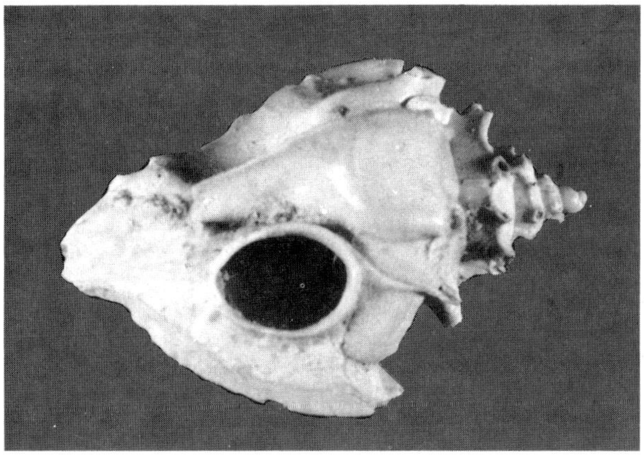

Talityphis bengalensis Radwin & D'Attilio, 1976 Marinduque

Chicoreus (Chicomurex) superbus problematicus (Lan, 1981)
77.8mm; Cebu (HOLOTYPE)
Original description: "Shell light to dark brown in colour. Entire surface covered with closely spaced dark brown interrupted spiral lines. There are two broad dark brown bands on the body whorl. The moderately high spire consists of nine whorls. Each whorl has three primary sharply pointed spines that curve slightly upwards. The sutural area has numerous tiny scales which are much more pronounced on the body whorl. The shoulder of the body whorl is slightly sloping, having more tiny scales on each varix beside the marginal spines that also recurve upwards and are aligned in spiny rows recurving backwards at the lower part of each varix. The corrugations have darker lines spirally, interrupted, possess tiny scales, stronger especially beside the varices, but weaker or obsolete on nodules or interspaces. The aperture white, glossy, oblong and oblique, serrated at the outer lip. Operculum horny, siphonal canal long and recurving backward."

Although described as a subspecies of *superbus*, the exact taxonomic status of this shell is still unresolved. Although reported from "Cebu, Bohol, the Philippines" in the original description, the shell is at present only known from Talikod Island, Davao Gulf. It has also been reported from Taiwan, but at present its actual distribution pattern is still incomplete. As far as can be ascertained at this time, it does not appear to occur sympatrically with "typical" *superbus*, and so its separation from that species may be justified. The *Chicoreus superbus* complex contains a number of closely allied shells, the inter-relationship of which is still under review. For the present time, we tentatively retain the name *problematicus* as a subspecies of *superbus*, with the provision that its status may change with the advent of future research.

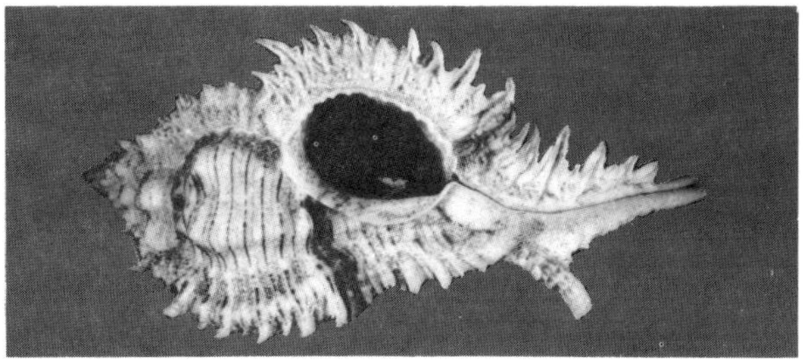

Chicoreus (Chicomurex) superbus problematicus (Lan, 1981)
77.8mm; Cebu (HOLOTYPE)

Favartia cirrosa (Hinds, 1844)
18.0mm; Sugod, Cebu.
Shell small, broadly fusiform, with a tall spire, a protoconch of 2½ smooth convex whorls and 6 sub-angulate post-nuclear whorls; body whorl broadly fusiform, with 8 strong varices, weakly lamellate on the leading edges and crossed by 6 spiral cords forming short, recurved, weakly foliate spines; siphonal canal long, open, dorsally recurved and sculptured with a bifid spiral cord; aperture widely ovate, outer lip margin slightly raised and crenulate, columella smooth, adherent posteriorly; colour light to medium amber brown, the varical spines occasionally white; reaches 20mm.
DISTRIBUTION: Presently known only from Sugod, Cebu.

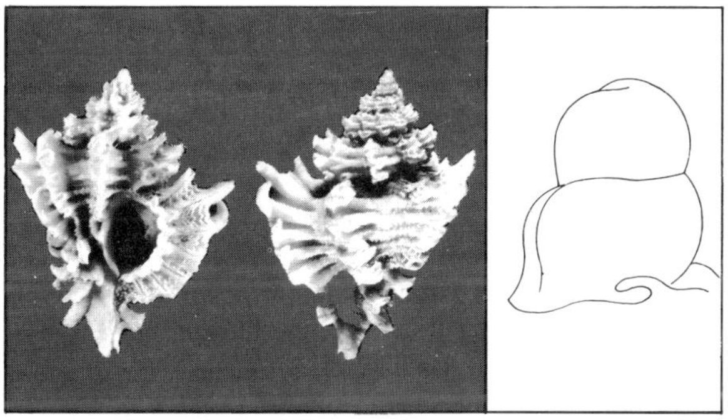

Favartia cirrosa (Hinds, 1844)
18.0mm; Sugod, Cebu (Shell: V. Dan; Photo: A. D'Attilio)
Protoconch of *F. cirrosa* (Hinds) (after D'Attilio)

ADDITIONAL NOTES ON THE MURICIDAE

Cronia (Cronia) margaritula margaritula (Broderip, 1833)/**Cronia (Muricodrupa) funiculus** (Wood, 1828)/**Cronia (Muricodrupa) fiscella** (Gmelin, 1791) (Pl. 39, figs. 11,19 and 20 respectively)
The true identity of Gmelin's *Murex fiscellus* has long been the subject of controversy among malocologists. Gmelin's type was based on figures from Chemnitz (1788: *Neues systematisches Conchylien-Cabinet.* Nurnberg 10, pl. 160, figs. 1524, 1525) and the accompanying description. A paper by Emerson & D'Attilio (1981: *Nautilus,* Vol. 95(2) contended that Chemnitz's figures were representatives of *Murex margariticola* Broderip, 1833, and that the accompanying descriptive text could equally apply to several other morphologically similar species. The status of

Cronia fiscella and *Cronia margariticola* has been well covered in a recent paper by Cernohorsky (1982), in which he traces the nomenclatural history of both species and clearly establishes them to be distinct taxa. The relationship between *Cronia funiculus* (Wood) and *Cronia fiscella* (Gmelin) however, is less clear cut. Although Cernohorsky synonymises the two, we believe the two are distinct and can be separated easily on the basis of certain morphological differences, clearly annotated in the text. Although the two might intergrade, we have seen no material to date to substantiate this contention and provisionally retain the two taxa as distinct.

Morula biconica (Blainville, 1832)
(Pl. 38, fig. 16)
A problematical species. Our figure represents the generally accepted interpretation of this species, but Blainville described it as "brown with two white zones" and pictured a much narrower shell.

Vitularia crenifer (Montrouzier, in Souverbie, 1861)
(Pl. 40, fig. 2)
The validity of this taxon is somewhat doubtful. The shell morphology and surface sculpturing appear distinct from *V. miliaris*, but may well be within the limits of variation expected for the latter species. Our specimen agrees well with the type figure and description as featured by Fair (1976, p. 34). However, no mention of the prominent posterior parietal denticle, so obvious in the type figure and omitted from the type description, has been made by more recent authors. In fact, the specimens of *crenifer* annotated by Cernohorsky (1979: p. 172, figs. 2-4), lack any suggestion of this feature. We tentatively retain the species as valid, with the provision that further work may indeed relegate it to synonymy with *miliaris*, itself a highly polymorphic species throughout its range.

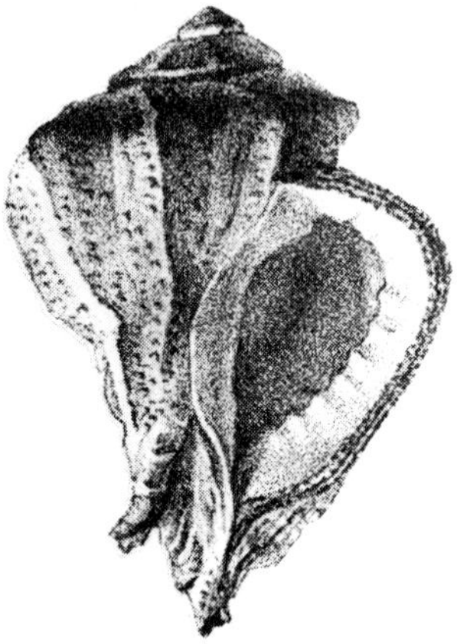

Type figure of *Vitularia cranifer* (Montrouzier)
(from Cernohorsky: 1979)

Chicoreus (Chicoreus) celinamarumai (Kosuge, 1980)
(Pl. 35, fig. 12)
On the basis of early whorl structure, this species is clearly a *Chicoreus*, not a Pterynotus. Note the spelling: *celinamarumai* not *cerinamarumi* (nom. emend.). This species appears to be very closely allied to, if not the same, as *Chicoreus orchidiflorus* Shikama, 1973 (Pl. 35, fig. 11).

Typical *orchidiflorus* is more common towards Taiwan and is well illustrated by Lan (1979, figs. 63-64). Philippine specimens differ somewhat in their varical blade structure as well as a few other points annotated in the text (see also Kosuge, 1980). Intergrades (or what appear to be) have been observed, and may well represent the connecting link between 2 end-members of a fairly polymorphic taxon. For the present we retain Kosuge's name, with the provision that future research may show the two taxa to be the conspecific or at the very least, sub-species of each other.

Pterynotus martinetana auct. (non Röding, 1798)
(Pl. 35, fig. 13)
This species is the one most frequently figured by authors as *martinetana*. It differs from true *martinetana* in a number of characteristics which are clearly outlined in the text. True *martinetana*, a somewhat rare species here in the Philippines, is well illustrated on Pl. 41, fig. 5. This species was described by Röding and was based on a poor beach shell figured in the non-binomial work of Chemnitz (Pl. 161, figs. 1536, 1537). Reeve subsequently illustrated this species under the name *fenestratus*, Dillwyn and his figure beautifully illustrated all the salient characteristics of this superb shell (1845: *Conc. Icon.*, Murex Pl. XV, figs. 58a, 58b). The type locality of Reeve's *fenestratus* (= *martinetana*) is unknown, and the *martinetana auct.* may represent a local ecomorph. The generic position of this species is still uncertain.

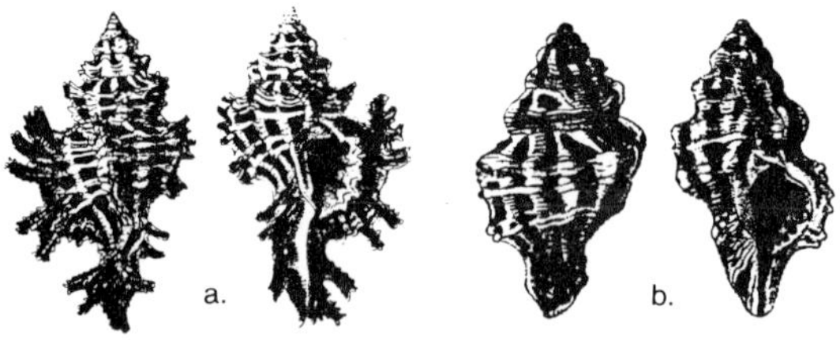

a. Type figure of *Murex fenestratus* (Dillwyn) (from Reeve)
b. Type figure of *Purpura martinetana* Röding (from Chemnitz)

Homalocantha anatomica (Perry, 1811) form **pele** (Pilsbry, 1918)
(Pl. 35, fig. 18)
Often considered a distinct species, differing from the typical form in the development of only 2 digitations. As this developmental aspect appears variable, we retain *pele* as a form only, with typical *anatomica* bearing 3 (sometimes 4) digitations.

Chicoreus microphyllus/Chicoreus torrefactus complex
(Pl. 36, fig. 16a & 16b)
An extremely difficult complex to quantify in terms of exact differences, most authors usually synonymising *torrefactus* with the earlier *microphyllus*. It seems likely that there are 2 species and collectors are given the following guide which may assist in their separation.

Chicoreus microphyllus: Generally smaller, brown and white-banded, with short, almost obsolete spination, a denticulate inner lip and 2 to 3 intervarical nodes (our fig. 16a).
Chicoreus torrefactus: Usually much larger, darker coloured, with a single intervarical node, medium sized spines and a non-denticulate inner lip (our fig. 16b). There appears to be a third form (Radwin & D'Attilio: Pl. 4, fig. 7), with 2 to 3 intervarical nodes, long spination, and a non-denticulate aperture. This agrees with the *microphyllus* of Kiener and is probably a form of *torrefactus*.

Haustellum haustellum / Haustellum kurodai / Haustellum vicdani complex
(Pl. 37, figs. 1, 2 & 3)
Both *Haustellum kurodai* Shikama, 1964 and *Haustellum vicdani* Kosuge, 1980 have been described as distinct species. Although separated by specific morphological differences, the periodic occurrence of what appear to be hybrid forms, strongly suggests that a single taxon may be involved. It is difficult to ascertain whether or not subspeciation is occurring. *H. haustellum* is widely distributed throughout the archipelago and *H. kurodai* has a somewhat sporadic distribution coming mainly from the Sulu archipelago and Zamboanga. *H. vicdani* however, appears to have a somewhat more restricted distribution, coming almost exclusively from Sorsogon (fide V. Dan). Until more accurate information as regards ecology and distribution becomes available, it is expedient to consider both *kurodai* and *vicdani* as forms of the more common *haustellum*.

Chicoreus (Chicoreus) crocatus auct. (non Reeve, 1845)
(Pl. 41, fig. 1)
Reeve's type figure *(Conc. Icon.* Murex Pl. XXXIII, sp. 168) is poor and hardly conveys the beautifully frondose nature of the shell usually portrayed as this species. Further, Reeve's description mentions " . . . a tubercle between the varices", but his figure clearly depicts a shell with 2 (possibly 3) thin axial cords between the varices. Examination of the holotype (B.M.N.H.) shows Reeve's shell to be a badly worn orange morph of *Chicoreus banksii* (Sowerby, 1841). These colour morphs, although rare, do occur in dark brown *Chicoreus spp.* and Reeve's specimen most likely comes from Northern Australia where it is of more frequent occurrence (fide Vokes). Philippine specimens of *crocatus auct.* are fairly constant in their overall morphology and are immediately recognized by the deep safron-coloured varical fronds, the single, large intervarical node (rarely some specimens show a reduction in node size and the appearance of a second, weaker axial fold), and the pale to dark brown intervarical spaces. Reeve's description and figure are conflicting, and raise the question as to whether the illustrated shell was in fact the taxon he was intending to describe. There is no doubt that *crocatus auct.* represents a valid species, but at present, it appears unnamed. It is curious that the concept of *crocatus* has become so firmly established in literature when Reeve himself stated that the shell was "a species of somewhat doubtful character".

a. Type figure of *Murex crocatus* Reeve (BMNH)
b. Type figure of *Murex crocatus* Reeve
(1845. Conc. Icon. Murex Pl. XXXIII, sp. 168)

c.f. Cronia (Ergalatax) contracta (Reeve, 1846)
(Pl. 41, fig. 8)
This appears to be a young specimen of *contracta*, differing mainly in surface and apertural colouration. It is erroneously figured as *Cronia pothuanii* (Souleyet, in Vaillant, 1852) by Habe (1964: Pl. 26, fig. 3) (note . . *phothuavii* is incorrect — error in litt.). Typical *pothuanii* is entirely different (see insert), being a more fusiform-ovate shell with a lower spire, more widely spaced axial cords and deep interstitial pits (clearly a *Muricodrupa*). It is conspecific with *Cronia (Muricodrupa) fiscella* (Gmelin, 1791) (Pl. 39, fig. 20).

Cronia (Muricodrupa) pothuanii (Souleyet in Vaillant, 1822)
(Photo: Vokes)

Pterynotus pinnatus Swainson, 1822
(Pl. 35, fig. 3)
This species is frequently figured under an earlier name, *Pterynotus alatus* (Röding, 1798). Up until recently, this name was considered to be a junior secondary homonym of *Murex alatus* Gmelin, 1791, but the subsequent transferral of Gmelin's taxon to the Turridae has precluded any such homonomy, and theoretically, made Röding's name available. However, the name cannot be applied as it has been rejected (even though erroneously) for the last 160 years. (I.C.Z.N. Code: Article 59(b):i:A junior secondary homonym rejected before 1961 is permanently rejected and cannot be restored unless the use of the existing name is contrary to existing usage).

Favartia sp.
(Pl. 38, fig. 8)
An extremely distinctive shell which appears, as yet, unnamed. It bears a close resemblance (in outline only) to *Murex mundus* Reeve, 1849, as figured by Sowerby (1879: *Thes. Conch.* Murex XXXIII, fig. 242) (see insert). However, the figure is bad and shows a somewhat spinose shell lacking the prominent double-stranded cords so obvious in our illustrated specimen.

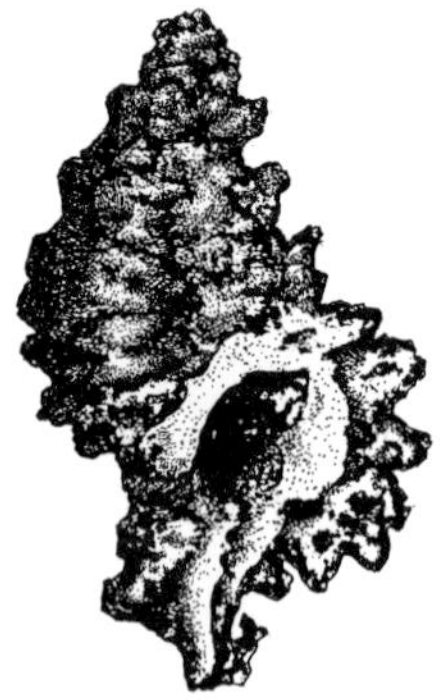

Type figure of *Murex mundus* Reeve, 1845 (from Sowerby)
Type locality "Zebu" Philippines.

Chicoreus aculeatus Lamarck, 1822 / **Chicoreus artemis** Radwin & D'Attilio, 1976
(Pl. 35, fig. 17)
Chicoreus artemis Radwin & D'Attilio, 1976 is a validly proposed taxon, introduced as a replacement for *Murex aculeatus* Lamarck, 1822 which is a secondary homonym of both *Aranea aculeata* Perry, 1811 and *Muricites aculeatus* Schlothein, 1820. The existence of homonomy between these three species has been the subject of considerable debate among various authors, but the matter was well covered by Vokes (1972, *Veliger* 15(2):102). The identity of Lamarck's taxon will probably never be known with any degree of certainty, as his original description was accompanied by neither figures nor references. Deshayes (1835-1845), in the second edition of Lamarck, merely quoted Lamarck's original description verbatim, but added references to Sowerby's Conchological Illustrations (1832-1841, fig. 63), and to Kiener (1842, p. 27, pl. 39, fig. 3). Both these references illustrate two obviously different shells. It is known that Deshayes saw the Massena collection* (which was under the custodial care of Chenu), but there is no evidence to suggest that he actually saw Lamarck's type. Reeve (1854, Murex Plate XV, fig. 60) illustrated Lamarck's taxon based on a shell from the Cuming collection, and quoted Deshayes as his reference. Both Reeve's shell and the Sowerby shell referred to by Deshayes are the same and appear to be *Chicoreus nobilis* Shikama, 1977.

The key to the identity of Lamarck's shell almost certainly lies with Kiener's figure, although it appears to be a poor, somewhat depauperate specimen, and hardly conveys the rich colour or the finely developed frondose spines so characteristic of some Philippine specimens.

Kiener's figure was appended with the acknowledgement "Collct. Lam. et Mus.", and makes it most probably the type figure, at least until the holotype is traced.

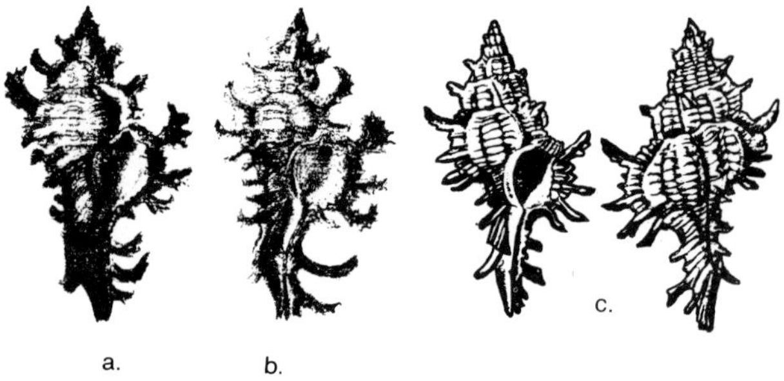

a. *Murex aculeatus* Lamarck, 1822 (from Sowerby)
b. *Murex aculeatus* Lamarck, 1822 (from Reeve)
c. *Murex aculeatus* Lamarck, 1822 (from Kiener)

*Lamarck's personal collection was acquired by Massena and subsequently came under the custodianship of Chenu. This collection was seen by many people, including Kiener and Deshayes.

Chicoreus rossiteri (Crosse, 1872) / **Chicoreus saltatrix** Kuroda, 1964
(Pl. 35, figs. 10a & 10b)
This is another problematical group which has been the subject of considerable debate among more recent authors. The concept of *rossiteri* must be based on the type figure (Crosse, 1872: *J. de Conch.* Pl. XIII, fig. 2) which clearly depicts a shell with a low spire, a distinctly rounded body whorl which lacks any posterior foliations, a lirate outer lip margin, and a distinctly crenulate upper and lower columellar wall. This species does not inhabit the Philippines and appears restricted to the central Indo-Pacific region. It is not a worn or corroded specimen as reported by some authors, living specimens exactly matching the type figure having been collected and examined (fide D'Attilio). *Chicoreus saltatrix* Kuroda, 1964 is definitely a very closely allied taxon but has a distinctly taller spire, somewhat more body whorl foliations, and lacks any discernable columellar crenulations (not mentioned in the type description). Based on the examination of large numbers of specimens (D'Attilio in corresp.), *saltatrix* is believed to be a distinct and valid taxon, and theoretically may represent a replacement for the central Indo-Pacific *rossiteri.* Philippine specimens of *saltatrix* only appear to differ in their somewhat taller spire and may represent a local morphophenotype. *Chicoreus saltatrix* and *Chicoreus artemis* have often been lumped together, but can be separated by several diagnostic features clearly alluded to in the text. The basis of the problem in this case seems to be almost identical colour patterns and generic similarities.

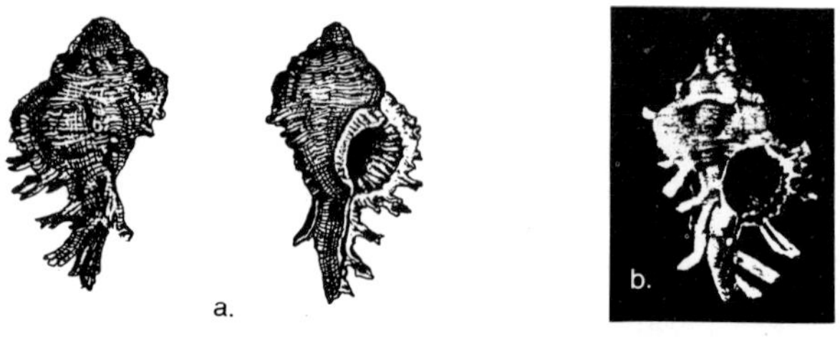

a. Type figure of *Murex rossiteri* Crosse, 1872
b. Holotype of *Chicoreus saltatrix* Kuroda, 1964

Class Gastropoda
Subclass Prosobranchia
Order Neogastropoda
Suborder Rachiglossa
Superfamily Buccinacea
Family **NASSARIIDAE**

A large group of small to medium-sized, sand-dwelling carnivorous gastropods inhabiting intertidal to deep water in most geographical provinces. The shells are ovate to elongate, may be smooth or nodulose, have a small, rounded aperture which may be dentate or lirate within, and a corneous, marginally serrated operculum. The family is fairly well represented in the Philippines.

10. Nassarius (Telasco) sufflatus (Gould, 1860)
20.8mm; Punta Engaño, Cebu.
Shell small with rather swollen whorls, the earliest of which are axially ribbed, the last 2 to 3 being smooth; aperture elongate-ovate, outer lip lirate within, columella heavily calloused anteriorly, thinner posteriorly, with a posterior parietal denticle, some basal plicae, and a smooth, excavated central area; yellowish brown in colour with subsutural purple blotching, and brown lines which are sometimes obscure but visible on the back of the outer lip, rear of siphonal canal brown-stained; reaches 25mm.
DISTRIBUTION: Throughout the Philippines.

11. Nassarius (Zeuxis) concinnus (Powys, 1835)
23.0mm; Samar.
Shell small, elongate-ovate, with a tall pointed spire, and a large rounded body whorl; sutures occluded with a row of small granules, surface finely granulose from intersecting axial cords and overriding spiral grooves; aperture elongate-ovate, outer lip with 7 to 8 short lirate denticles, columella with a few folds and a parietal denticle; pale yellow with brown banding and a white aperture; reaches 25mm.
DISTRIBUTION: Throughout the Philippines.

12. Nassarius (Telasco) luridus (Gould, 1850)
18.5mm; Palawan.
Shell small, globose, with moderately channeled sutures, early whorls with axial ribs reduced to obsolete subsutural nodules on the later whorls, base and siphonal fasciole corded; outer lip variced, lirate within, columella calloused, centrally excavated and dentate along its entire length; yellowish white with greyish banding and brown spots between the nodules, aperture dark brown, marginally white; reaches 20mm.
DISTRIBUTION: Throughout the Philippines.

13. Nassarius (Telasco) limnaeiformis (Dunker, 1847)
19.5mm; Samar.
A small, solid shell with a tall spire, moderately incised sutures, and a long body whorl; whorls convex and entirely smooth; aperture ovate, outer lip thick, variced and lirate within, columella calloused, excavated, plicate along its entire length; colour white with green marbling, brown spiral lines, and an occasional dark band centrally on the body whorl; reaches 25mm.
DISTRIBUTION: Throughout the Philippines.

14. Nassarius (Niotha) livescens (Philippi, 1849)
22.4mm; Batangas.
Shell medium sized, elongate-ovate, with a tall acute spire, deeply channeled sutures, and a row of subsutural granules which are bisected by a weak groove; surface weakly granular from intersecting axial cords and overriding spiral grooves; aperture ovate, lirate within, columella heavily calloused and irregularly plicate; pale brown with 2 to 3 yellowish brown bands, aperture purple within, marginally white; reaches 25mm; variable.
DISTRIBUTION: Throughout the Philippines.

15. Nassarius (Zeuxis) margaritiferus (Dunker, 1847)
24.7mm; Punta Engaño, Cebu.
Shell medium sized, elongate-ovate, with a tall acute spire, weakly channeled and crenulate sutures, and a smooth subsutural furrow; whorls with nodulose axial and spiral cords, the spiral cords strongest basally; aperture ovate, outer lip thick, lirate within, columella calloused and plicate; yellowish white with darker blotching, aperture brown internally, white marginally; reaches 30mm.
DISTRIBUTION: Throughout the Philippines.

16. Nassarius (Telasco) gaudiosus (Hinds, 1844) form **zonalis** (A. Adams, 1852) (non Bruguiére, 1789)
16.9mm; Palawan.
Shell small, elongate-ovate, with a tall spire, ovate body whorl, and finely nodulose sutures; surface smooth except for some basal grooves; aperture ovate, outer lip thick, lirate within, columella calloused, thin and expansive posteriorly, with a few irregular plicae; shell yellowish with brown banding, white sutures, and a brown-and-white striped aperture; reaches 25mm.
DISTRIBUTION: Throughout the Philippines.

17. Nassarius (Zeuxis) dorsatus (Röding, 1798)
36.6mm; Samar.
Shell medium sized, with a tall, acute spire, sutures deeply channeled, ledged, and finely crenulate; shell surface smooth except for some axial and spiral threads on the earlier whorls, and some spiral grooves towards the base; aperture elongate-ovate, outer lip thick, variced, lirate within and marginally serrate towards the base, columella irregularly plicate; shell colour yellowish grey, with a purple aperture; reaches 45mm.
DISTRIBUTION: Sporadically found throughout the Philippines.

18. Nassarius (Niotha) conoidalis (Deshayes, in Belanger, 1832)
26.1mm; Palawan.
Shell medium sized with a tall, acute spire and an

extremely large, globose body whorl; sculptured with numerous axial cords and overriding spiral grooves producing nodules, heaviest at the sutures; aperture elongate-ovate, outer lip marginally crenulate, lirate within, columella heavily calloused and irregularly plicate, with the body whorl structure showing through the upper parietal area; creamy white with 1 or 2 pale brown bands or markings; reaches 30mm.
DISTRIBUTION: Sporadically found throughout the Philippines.

19. Nassarius (Plicarcularia) pullus (Linné, 1758)
21.2mm; Samar.
Shell small, solid, with a moderately tall spire, weakly channeled sutures, and a prominent hump on the body whorl dorsum; sculptured with axial cords and weak interstitial lirae, the axial cords becoming obsolete to the left of the hump, base with several spiral grooves; aperture quadrangular, outer lip thick, lirate within, columella with a large expansive callus and a few denticles; colour greyish green with a thin yellow band, and brown between the axial cords; reaches 25mm.
DISTRIBUTION: Throughout the Philippines.

20. Nassarius (Nassarius) arcularius arcularius (Linné, 1758)
24.7mm; Cebu.
Shell medium sized, solid, with a short spire and a large body whorl slightly sunken behind a greatly expanded parietal callus; sculptured with heavy axial cords which become nodulose on the body whorl shoulder and obsolete on the dorsum, interstitial spaces smooth, base with spiral grooves; aperture ovate, outer lip thick and marginally crenulate, columella with some basal denticles; greenish grey with white banding; reaches 30mm.
DISTRIBUTION: Throughout the Philippines.

21. Nassarius (Zeuxis) olivaceus (Bruguiére, 1789)
35.6mm; Samar.
Shell medium sized, solid, with a tall, acute spire, sutures that are moderately channeled and ledged, and a large, ovate body whorl; sculptured with heavy axial cords becoming obsolete basally and encircled with fine striae towards the base; aperture ovate, outer lip lirate, columella heavily calloused and dentate; dark brown in colour with a thin yellow band, aperture blue-white; reaches 40mm.
DISTRIBUTION: Sporadically found throughout the Philippines.

Nassariidae — Plate 42

1. Nassarius (Alectrion) glans glans (Linné, 1758)
42.5mm; Palawan.
Shell medium sized, lightweight, with a tall, acute spire and deeply channeled sutures; early whorls nodulose from axial ribs and overriding spiral grooves, later whorls smooth; aperture widely ovate, outer lip with some basal denticles, columella moderately calloused and smooth except for a posterior denticle; pale brown with reddish brown lines, a subsutural band of brown and white spots, a creamy white aperture and a pink protoconch; reaches 45mm.
DISTRIBUTION: Sporadically found throughout the Philippines in limited quantity.

2. Nassarius (Alectrion) papillosus (Linné, 1758)
39.6mm; Laminusa, Sulu Archipelago.
Shell medium sized, elongate-ovate, with a tall, acute spire and moderately incised sutures; sculptured with spiral rows of nodules which become axially aligned on later whorls; aperture roundly ovate, outer lip with a sharply denticulate margin and lirate within, columella heavily calloused and smooth, with a posterior parietal denticle; cream with brown blotches, aperture white; reaches 45mm.
DISTRIBUTION: Sporadically found throughout the Philippines in limited quantity.

3. Nassarius (Nassarius) coronatus (Bruguiére, 1798)
33.4mm; Sulu Archipelago.
Shell medium sized, solid, rather ovate, with a glossy surface and deeply channeled sutures; earlier whorls with intersecting axial and spiral cords, later whorls smooth except for a row of large subsutural nodules and 1 or 2 cords at the base; aperture elongate-ovate and lirate within, columella smooth, with a heavy, somewhat expanded callus and a posterior parietal denticle; reddish brown with a single white band on the whorls and white subsutural nodules, aperture brown within, marginally white; reaches 35mm.
DISTRIBUTION: Throughout the Philippines.

PLATE NO. 42 1. *Nassarius glans glans* (Linné) 2. *Nassarius papillosus* (Linné) 3. *Nassarius coronatus* (Bruguiére) 4. *Nassarius distortus* (A. Adams) 5. *Nassarius albescens albescens* (Dunker) 6. *Nassarius siquijorensis* (A. Adams) 7. *Nassarius velatus* (Gould) 8. *Nassarius vitiensis* (Hombron & Jacquinot) 9. *Nassarius gruneri* (Dunker) 10. *Nassarius pauperus* (Gould) 11. *Nassarius fissilabris* (A. Adams) 12. *Nassarius fretorum* (Melvill & Standen) 13. *Nassarius splendidulus* (Dunker) 14. *Nassarius venustus* (Dunker) 15. *Nassarius crenoliratus* (A. Adams) 16. *Nassarius excellens* (Kuroda & Habe) 17. *Cyllene pulchella* Adams & Reeve 18. *Nassarius subspinosus* (Lamarck)

PLATE 42

4. Nassarius (Niotha) distortus (A. Adams, 1852)
22.0mm; Palawan.
Shell medium sized, solid and glossy, with a tall, acute spire and a large, globose body whorl; sculptured with slender, oblique axial ribs, sometimes partially obsolete on the body whorl; ribs ledged at the sutures and separated by a deep cleft from a row of subsutural granules; aperture ovate, outer lip thick and lirate within, columella smooth except for a posterior parietal denticle and 2 or 3 basal lirae; shell colour greenish brown with darker banding, aperture marginally white, blue and white banded within; reaches 35mm.
DISTRIBUTION: Throughout the Philippines.

5. Nassarius (Niotha) albescens albescens (Dunker, 1846)
18.6mm; Punta Engaño, Cebu.
Shell small, ovate, with a tall spire and weakly channeled, slightly ledged sutures; body whorl large, inflated, and finely granulose from overriding oblique axial cords and spiral lirae; aperture subquadrate, outer lip thick and lirate within, columella with an expansive callus, a parietal denticle, and numerous folds; cream coloured with a purple apex and a porcelaneous white aperture; reaches 20mm.
DISTRIBUTION: Sporadically found throughout the Philippines.

6. Nassarius (Zeuxis) siquijorensis (A. Adams, 1852)
27.3mm, 23.9mm; Palawan.
Shell medium sized, with a tall, acute spire and deeply channeled sutures; sculptured with axial cords which may be sparse, closely packed or almost obsolete on the body whorl dorsum; the axial cords are indented at the sutures forming a row of nodules; interstitial spaces with minute spiral striae and the base with 6 to 8 spiral cords; aperture elongate-ovate, outer lip lirate within, columella calloused and plicate; greyish brown to pale yellow with darker banding and a white aperture which is brown internally; reaches 35mm.
DISTRIBUTION: Throughout the Philippines.

7. Nassarius (Zeuxis) velatus (Gould, 1850)
17.9mm; Samar.
Shell small, conoidal, with a tall, acute spire, a large body whorl and weakly channeled sutures; sculptured with weak spiral threads over the entire surface, obsolete axial scratches and a row of weak, subsutural nodules; aperture ovate, outer lip thick and lirate within, columella heavily calloused, plicate along its entire length; black with white subsutural spots; reaches 20mm. Considered a form of *N. gaudiosus* (Pl. 41, fig. 16).
DISTRIBUTION: Sporadically found throughout the Philippines.

8. Nassarius (Zeuxis) vitiensis (Hombron & Jacquinot, 1853)
32.3mm; Manila Bay.
Shell medium sized, solid, with a tall, acute spire and a large, rounded body whorl; sculptured with numerous raised, closely spaced axial ribs which tend to become weaker on the body whorl dorsum; a narrow groove crosses the axial ribs below the sutures, forming a row of subsutural nodules; the interspaces are sculptured with weak spiral threads which become stronger anteriorly, and over-ride the axial ribs to form a nodulose surface; aperture elongate-ovate, outer lip thickened, dentate on the outer margin, lirate within, columella thickened, and plicate along its entire length; dark brown in colour with narrow white banding and a purplish aperture; reaches 35mm.
DISTRIBUTION: Sporadically found throughout the Philippines.

9. Nassarius (Niotha) gruneri (Dunker, 1846)
18.9mm; Cebu.
Shell small, squat, with a moderately tall spire, a rounded body whorl and deeply channeled sutures; entire surface bearing echinate nodules from intersecting axial and spiral cords, the subsutural row being separated from the rest by a deep trough; outer lip denticulo-plicate on the inner margin, columella calloused and irregularly plicate; cream coloured, occasionally brown-banded; reaches 20mm.
DISTRIBUTION: Sporadically found throughout the Philippines in limited quantity.

10. Nassarius (Hima) pauperus (Gould, 1850)
11.5mm; Punta Engaño, Cebu.
Shell small, conoidal with somewhat ventricose whorls; sculptured with numerous crowded, somewhat angulate axial ribs and overriding spiral striae; aperture small and ovate, outer lip denticulate, columella calloused, weakly crenulate along its entire length; cream coloured with brown bands, one on the spire whorls, 2 on the body whorl, aperture white; reaches 15mm.
DISTRIBUTION: Sporadically found throughout the Philippines.

11. Nassarius (Niotha) fissilabris (A. Adams, 1852)
9.1mm; Punta Engaño, Cebu.
Shell very small (seldom exceeding 10mm), ovate in outline, with a short acute spire, a large body whorl and the sutures obscured by a cord; sculptured with wide, rounded axial ribs, with no spiral sculpturing except for a few basal grooves; aperture ovate, outer lip dentate, columella heavily calloused with a single posterior parietal denticle and three anterior plicae; yellowish brown with a dark protoconch, brown spiral bands and a black blotch at the base of an all-white aperture.
DISTRIBUTION: Sporadically found throughout the Philippines.

12. Nassarius (Niotha) fretorum (Melvill & Standen, 1899)
8.5mm; Punta Engaño, Cebu.
Shell small, conoidal, with a tall spire and a ventricose body whorl; sculptured with numerous closely set, arcuate axial folds and weak spiral threads in the interspaces; base with 6 to 7 deep grooves

forming nodules with the axial folds; aperture ovate, lirate within, columella weakly calloused with a prominent parietal denticle and 3 to 4 basal plicae; white with brown maculations and subsutural brown spotting; reaches 12mm.
DISTRIBUTION: Sporadically found throughout the Philippines.

13. Nassarius (Niotha) splendidulus (Dunker, 1846)
14.2mm; Palawan.
Shell small, whorls slightly swollen, spire moderately tall and the sutures moderately incised; sculptured with numerous axial folds and overriding spiral grooves forming a heavily granulose surface; aperture ovate, lirate within, columella smooth except for a posterior parietal denticle; off-white with irregular brown spotting, aperture white; reaches 15mm.
DISTRIBUTION: Sporadically found throughout the Philippines.

14. Nassarius (Niotha) venustus (Dunker, 1847)
18.0mm, 18.7mm; Batangas.
Shell medium sized, squat, sutures weakly impressed with a row of granules and a shallow presutural groove; sculptured with thin axial cords and numerous overriding spiral threads; aperture ovate, outer lip thick, variced, marginally dentate and lirate within, columella heavily and expansively glazed with a few folds; greenish yellow with darker banding, aperture marginally white, purple within; reaches 20mm.
DISTRIBUTION: Throughout the Philippines.

15. Nassarius (Niotha) crenoliratus (A. Adams, 1852)
12.5mm; Punta Engaño, Cebu.
Shell small, rather squat, sculptured with a row of sutural nodules and numerous angulate axial ribs which are nodulose from overriding spiral threads; aperture elongate-ovate, outer lip thick and lirate within, columella heavily calloused with a posterior parietal denticle and irregular basal plicae; off-white to pale brown with darker banding, aperture white with a dark parietal blotch; reaches 15mm.
DISTRIBUTION: Sporadically found throughout the Philippines.

16. Nassarius (Zeuxis) excellens (Kuroda & Habe, 1961)
15.8mm; Samar.
Shell medium sized, elongate-ovate with a tall, acute spire and moderately convex spire whorls, the first few of which are weakly cancellate; sutures deeply channeled, body whorl large, rounded and smooth except for some basal spiral grooves; aperture ovate and lirate within, columella weakly calloused with a posterior parietal denticle and a few anterior plicae; off-white with nebulous brown blotching, subsutural brown spotting, a purple apex and a white aperture; reaches 17mm. Said to be a form of *N. comptus* (A. Adams).
DISTRIBUTION: Sporadically found throughout the Philippines.

17. Cyllene pulchella Adams & Reeve, 1850
13.9mm; Punta Engaño, Cebu.
Shell small, elongate-ovate with a short, acute spire and a large, elongate-ovate body whorl; sutures adpressed, spire whorls with weakly granulose spiral ribs, body whorl spirally grooved below the suture and at the base, smooth centrally; outer lip thick and lirate within, columella thinly glazed and weakly denticulate basally; white with pinkish brown maculations, a deep pink spire and a pink, marginally white aperture; reaches 15mm.
DISTRIBUTION: Sporadically found throughout the Philippines.

18. Nassarius (Hebra) subspinosus (Lamarck, 1822)
13.9mm; Punta Engaño, Cebu.
Shell small, ovate, sculptured with coarse axial folds which are crossed by spiral cords forming rows of spinose nodules, one on the spire whorls and 2 to 3 on the body whorl, the first being on the sharply angulate body whorl shoulder; aperture ovate, outer lip thick and lirate within, columella weakly calloused with 3 basal denticles; greyish white with darker banding, aperture purple with a white band; reaches 20mm.
DISTRIBUTION: Sporadically found throughout the Philippines.

Plate 43 Coralliophilidae

Class Gastropoda
Subclass Prosobranchia
Order Neogastropoda
Suborder Rachiglossa
Superfamily Muricacea
Family CORALLIOPHILIDAE

An extraordinarily complex group of carnivorous gastropods found in the shallow to deep waters of most tropical and sub-tropical seas. The family is fairly large and at present over 300 species have been described, although it is quite unlikely that more than 50% are actually valid taxa. Members of this family are mainly small to medium sized and characterized by their highly scabrous surface or by the numerous delicate spinose processes adorning the whorls. They have no radula and are generally found living in or on corals or sea fans. Because of their specialized habitat, members exhibit an unusually high degree of polymorphism which has undoubtedly given rise to the multitude of names currently in evidence. As a result, there has been a proliferation of genera and subgenera, the boundaries of which are often ill-defined. Within this section we have, in the main, restricted ourselves to the use of 3 main genera, although additional genera (including *Quoyula* and *Magilus*) have been used where applicable. The family is well represented in the Philippines, with most deep water species coming from Cebu, Bohol and Southern Mindanao.

Coralliophila H. & A. Adams, 1853: Small to medium sized, heavily sculptured shells with a large body whorl, a thick outer lip and a purple aperture.

Rapa Bruguiére, 1792: Thin, medium to large-sized shells with an inflated body whorl and a large aperture.

Latiaxis Swainson, 1840: The largest generic group within the family. The shells are normally small to medium sized and sculptured with elaborate scaly or spinose ridges.

1. Latiaxis (Latiaxis) pagodus (A. Adams, 1853)
40.8mm, 33.9mm; Punta Engaño, Cebu.
Shell small to medium sized, with a tall, acute spire and a strongly ventricose body whorl; spire whorls with a single row of 8 to 9 long upturned spines on a rounded, mainly smooth shoulder, body whorl with a second row below the shoulder row forming shallow pits; below this, 3 to 4 additional rows of short scabrous spines; aperture ovate, outer lip thin, lirate within, columella smooth, fasciolar cord highly spinose; white with squarish brown spots on the shoulder and within the pits; reaches 45mm. The left figure lacks the second spinal row, and, although nearly always figured as *pagodus,* may in fact be something different.
DISTRIBUTION: Cebu, Bohol and Southern Mindanao.

2. Latiaxis (Latiaxis) pilsbryi Hirase, 1908
17.0mm; Punta Engaño, Cebu.
Shell medium sized, triangular in outline, with an almost flat spire and a strongly carinate body whorl which gives rise to 8 or 9 flat triangular spines; surface crossed by fine axial and spiral striae, last whorl loosely coiled, umbilicus deep and wide, fasciolar cord with a spinose margin; aperture quadrangular, siphonal canal long and tubular; pale brown with orange blotching; reaches 45mm.
DISTRIBUTION: Bohol and Cebu.

3. Latiaxis (Babelomurex) dunkeri Kuroda & Habe, 1961
34.8mm; Bulan, Sorsogon.
Shell medium sized, somewhat ovate-biconic with a short spire; sculptured with a row of 8 to 9 broad, laterally open triangular spines on the sharply angled shoulder, rest of surface with numerous closely set, finely scabrous spiral cords, umbilicus deep, fasciolar cord heavily scaled; colour pure white; reaches 45mm.
DISTRIBUTION: Cebu, Bohol and Bulan, Sorsogon.

4. Latiaxis (Babelomurex) kawamurai Kira, 1954
28.5mm; Bohol.
Shell medium sized, fusiform-elongate, sutures deeply impressed, spire tall, whorls strongly carinate and bearing 9 to 10 narrow, slightly upturned triangular spines, surface with fine scabrous cords; aperture triangularly ovate, umbilicus narrow, fasciolar cord with weak, fluted folds; dirty white in colour, reaches 30mm.
DISTRIBUTION: Cebu, Bohol and Southern Mindanao.

5. Latiaxis (Echinolatiaxis) takahashii Kosuge, 1979
36.1mm, 29.1mm; Bohol.
Shell medium sized, tall spired with 8 rounded whorls each bearing 10 long, sharp shoulder spines, the size decreasing above and below the shoulder, interspaces with fine, scaly projections forming a densely echinate surface; aperture ovate, siphonal canal long, recurved and anteriorly constricted, umbilicus open, fasciolar cord scaled; pinkish purple in colour (occasionally white), with an internally white aperture; reaches 40mm.
DISTRIBUTION: Cebu and Bohol.

6. c.f. Latiaxis (Babelomurex) nakayasui Shikama, 1970
28.6mm; Bohol.
Shell medium sized, fusiform-ovate, with impressed, somewhat undulate sutures and a tall, acute spire; sculptured with 9 rounded axial folds and overriding spiral cords on the body whorl below the shoulder, forming pit-like interspaces; aperture ovate, outer lip thin, lirate within; pure white with some darker banding; reaches 30mm.
DISTRIBUTION: Cebu and Bohol.

7. Latiaxis sp.
30.9mm; Bohol.
Shell medium sized, fusiform-ovate, with a tall spire and impressed sutures; sculptured with 10 to 12 rounded axial folds and crossed by scabrous spiral cords, strongest on the presutural slope; a row of short, triangular, posteriorly curved spines is present on the roundly angulate shoulder; aperture wide, ovate, with a marginally serrate, internally lirate outer lip; the canal is short, open and recurved, and the umbilicus tubular with a heavily fluted fasciolar cord; pale brown with a pure white aperture; reaches 35mm. Appears undescribed.
DISTRIBUTION: Cebu and Bohol.

8. Latiaxis (Laevilatiaxis) fruticosus Kosuge, 1979
21.2mm, 15.5mm; Punta Engaño, Cebu.
Shell small to medium sized, distinctly elongate,

PLATE NO. 43 1. *Latiaxis pagodus* (A. Adams) 2. *Latiaxis pilsbryi* Hirase 3. *Latiaxis dunkeri* Kuroda & Habe 4. *Latiaxis kawamurai* Kira 5. *Latiaxis takahashii* Kosuge 6. *c.f. Latiaxis nakayasui* Shikama 7. *Latiaxis sp.* 8. *Latiaxis fruticosus* Kosuge 9. *Latiaxis pisori* D'Attilio & Emerson 10. *Latiaxis gemmatus* Shikama 11a. to 11b. *Latiaxis cristatus* Kosuge 12. *Latiaxis japonicus* (Dunker) 13. *Latiaxis purpuraterminus* Kosuge 14. *Latiaxis shingomarumai* Kosuge 15. *Latiaxis celinamarumai* Kosuge 16. *Latiaxis kiranus* Kuroda 17. *Coralliophila elvirae* D'Attilio & Emerson 18. *Latiaxis mucatanica* Kosuge 19. *Latiaxis deburghiae* (Reeve) 20. *Latiaxis purpuratus* (Chenu)

PLATE 43

surface smooth, whorls sculptured with 7 to 8 long, erect, bifurcate spines bearing semi-tubular scales, a second smaller row on the body whorl below the shoulder; aperture ovate, umbilicus narrow, siphonal canal long, tubular and recurved, fasciolar cord weakly scaled; pinkish white with darker blotches; reaches 30mm.
DISTRIBUTION: Cebu, Bohol and Southern Mindanao.

9. Latiaxis (Latiaxis) pisori D'Attilio & Emerson, 1980
28.9mm; Bulan, Sorsogon.
Shell medium sized, heavy, ovate with a depressed spire and impressed sutures; sculptured with 9 to 10 wide, roundly triangular curved spines on the sharply angled shoulder with an additional 2 slightly angled nodular cords on the lower body whorl, surface crossed with fine, scabrous spiral striae; aperture large, lirate within, irregularly ovate in shape, with a short, tubular, recurved siphonal canal, umbilicus open with a scaled fasciolar cord; white with a lavender hue, aperture deep purple; reaches 35mm.
DISTRIBUTION: Cebu, Bohol and Bulan, Sorsogon.

10. Latiaxis (Babelomurex) gemmatus Shikama, 1966
17.8mm, 15.5mm; Punta Engaño, Cebu.
Shell small to medium sized, broadly fusiform, with a tall spire and impressed sutures which are occluded by a scabrous cord; sculptured with rounded axial folds bearing a row of tubular, occasionally bifurcate, posteriorly curved spines, strongest on the body whorl shoulder, 1 to 2 additional scabrous cords below the body whorl shoulder, surface corrugated with fine scabrous lirae; aperture ovate, lirate within; umbilicus weak, fasciolar cord scaly; pale cream to reddish brown with a lavender aperture; reaches 30mm.
DISTRIBUTION: Cebu, Bohol and Southern Mindanao.

11. Latiaxis (Laevilatiaxis) cristatus Kosuge, 1979
24.5mm, 25.3mm (11a.), 19.8mm, 17.5mm (11b.); Bohol.
Shell small to medium sized, elongate-ovate in outline with a fairly tall spire; the body whorl has 7 posteriorly recurved shoulder spines which are flattened, somewhat irregular in shape and serrated on their leading edges; there are additional 2 or 3 spiral cords below the shoulder spines on the body whorl; aperture ovate, outer lip margin crenulate and lirate within, umbilicus weakly perforate with a heavily fluted fasciolar cord; younger specimens (11b) have longer, more curved spines, and lack the additional spiral cords on the lower body whorl; white to pinkish brown in colouration; reaches 25mm.
DISTRIBUTION: Cebu, Bohol and Southern Mindanao.

12. Latiaxis (Tolema) japonicus (Dunker, 1882)
31.0mm; Punta Engaño, Cebu.
Shell medium sized, widely biconic, with a large body whorl and a moderately tall spire; sculptured with highly scabrous spiral cords and a double row of broad, triangular, posteriorly recurved spines on the sharply angled shoulder, spines finely serrate on the leading edge, the second spiral row often hidden in the deeply channeled presutural area; aperture elongate-ovate, with a serrate outer lip margin, a partially open umbilicus, and a scabrous fasciolar cord; pure white; reaches 40mm.
DISTRIBUTION: Cebu, Bohol and Southern Mindanao.

13. Latiaxis (Babelomurex) purpuraterminus Kosuge, 1979
37.5mm; Bohol.
Shell medium sized, lightweight, fusiform-ovate in outline, and the sutures impressed with a scaly cord; sculptured with axial and spiral cords intersecting to produce numerous slender spines, strongest on the shoulder of each whorl, and pit-like interspaces, presutural slope very weakly spinose; aperture ovate, outer lip crenulate, umbilicus partially open, fasciolar cord scaly; off-white with a pinkish purple suffusion on the spines, aperture pure white; reaches 45mm.
DISTRIBUTION: Cebu, Bohol and Southern Mindanao.

14. Latiaxis (Babelomurex) shingomarumai Kosuge, 1981
25.0mm; Bohol.
Shell small to medium sized, fusiform-ovate, with impressed sutures and a moderately tall spire; sculptured with 8 to 10 rounded axial folds crossed by numerous finely scabrous spiral lirae, a single row of short, occasionally obsolete triangular spines on the roundly angulate shoulder, spines finely serrate on the leading edge; aperture ovate, lirate within, umbilicus weak, fasciolar cord scaly; shell pure white with a porcelaneous white aperture; reaches 30mm.
DISTRIBUTION: Cebu and Bohol.

15. Latiaxis (Echinolatiaxis) celinamarumai Kosuge, 1980*
35.7mm; Bohol.
Shell large, solid, pinkish purple with a deep pink aperture; elongate-ovate in shape with 7 to 8 whorls and sculptured with a row of large, open, upturned shoulder spines, and scabrous axial and spiral folds in the interstitial spaces; body whorl with 4 to 5 additional rows of spines; aperture elongate-ovate, weakly lirate within, umbilicus wide and perforate, fasciolar cord heavy and fluted; reaches 38mm. Named after Ms. Celina Galindes of Cebu and Mr. M. Maruma Suzuki of Japan.
DISTRIBUTION: Bohol.

16. Latiaxis (Babelomurex) kiranus Kuroda, 1959
26.8mm; Punta Engaño, Cebu.
Shell medium sized, lightweight, somewhat biconic in outline with impressed sutures and a tall spire; sculptured with 10 to 11 axial cords which are crossed by numerous scabrous spiral lirae forming long, thin, posteriorly curved spines, strongest on the roundly angulate shoulder, decreasing in strength anteriorly, and being absent on the presutural slope; aperture

ovate, canal long, open and recurved, umbilicus weak, fasciolar cord heavily fluted; colour dirty white; reaches 35mm.
DISTRIBUTION: Cebu and Bohol.

17. Coralliophila (Lepadomurex) elvirae D'Attilio & Emerson, 1980
23.9mm; Punta Engaño, Cebu.
Shell medium sized, fusiform with a short spire and a long triangular body whorl; sculptured with thick, rounded axial costae crossed by fine scabrous spiral lirae; aperture quadrangular, outer lip parallel to the columella, aperture interior smooth and violet coloured, umbilicus slightly open; surface colouration orange-brown, darker between the costae; reaches 30mm.
DISTRIBUTION: Cebu and Bohol.

18. Latiaxis (Laevilatiaxis) mucatanica Kosuge, 1979
20.5mm; Cebu.
Shell medium sized, tall spired, with sharply angled whorls and deeply excavated sutures; sculptured with 7 long, posteriorly curved spines on the shoulder, spines occasionally bifurcate, serrate and increasing in size anteriorly, with 1 or 2 weaker spines below the body whorl shoulder; entire surface with fine scabrous projections; aperture ovate, lirate within, canal short, open and recurved; colour cream to purple-brown with a pale lavender aperture; reaches 32mm.
DISTRIBUTION: Cebu and Bohol.

19. Latiaxis (Babelomurex) deburghiae (Reeve, 1857)
27.0mm; Bulan, Sorsogon.
Shell medium sized with a tall, acute spire and a large inflated body whorl; sculptured with a shoulder row of thin, triangular, posteriorly curved spines, and the whole surface encircled with fine, scabrous spiral cords; aperture ovate, anteriorly constricted, outer lip thin, marginally serrate and internally lirate, umbilicus open, fasciolar cord scaly; off-white with a lavender hue; reaches 35mm.
DISTRIBUTION: Cebu, Bohol and Bulan, Sorsogon.

20. Latiaxis (Babelomurex) purpuratus (Chenu, 1859)
25.5mm; Punta Engaño, Cebu.
A deep mauve-coloured species which has been synonymized with *L. deburghiae*, but differs in its more numerous, more elongated shoulder spines and its more finely scabrous surface; the body whorl is less angulate on the lower half and the umbilicus appears less deeply channeled; reaches 35mm.
DISTRIBUTION: Cebu, Bohol and Southern Mindanao.

* Valid emendation from *cerinamarumai* (I.C.Z.N. Code, Article 32a).

Plate 44 Coralliophilidae

1. Latiaxis (Babelomurex) fearnleyi Emerson & D'Attilio, 1965
54.0mm; Palawan.
Shell large, solid, fusiform-elongate, with a tall, acute spire and shallow sutures; sculptured with strong axial cords and numerous overriding scabrous spiral lirae, somewhat echinate at the angulate shoulder; aperture ovate, anteriorly compressed, outer lip foliate, lirate within, fasciolar ridge strongly corded; white with a violet hue, aperture lavenderish; reaches 60mm.
DISTRIBUTION: Sporadically found throughout the Philippines in limited quantity.

2. Coralliophila (Coralliophila) pyriformis Kira, 1954
46.1mm; Samar.
Shell large, short spired, with strongly convex whorls, body whorl large and bulbous, tapering to an almost obsolete anterior canal; entire surface covered with fine, scabrous spiral lirae; aperture irregularly elongate-ovate, outer lip thick, marginally serrate, flaring anteriorly; inner margin smooth, lirate within the aperture, umbilicus weakly channeled, siphonal fasciole scabrous; off-white with a purple hue, aperture deep purple; reaches 50mm.
DISTRIBUTION: Sporadically found throughout the Philippines.

3. Coralliophila (Coralliophila) erosa (Röding, 1798)
31.9mm; Palawan.
Shell medium sized, with a tall spire and a large, somewhat bulbous body whorl, which is roundly angulate centrally; sculptured with coarse axial folds, becoming obsolete at the sutures and crossed by numerous scabrous cords, 2 to 4 of which become prominent on the lower half of the body whorl; aperture ovate, marginally foliate, internally lirate, umbilicus open or closed; off-white with a white, occasionally lavender aperture; reaches around 35mm.
DISTRIBUTION: Sporadically found throughout the Philippines in limited quantity.

4. Coralliophila (Coralliophila) neritoidea (Lamarck, 1816)
Syn: *C. violacea* (Kiener, 1836)
23.7mm; Cebu.
Shell small, solid, globose to elongate-ovate with a short spire, entire surface encircled with scabrous spiral striae; aperture elongate-ovate, outer lip thin, foliate, heavily lirate, columella with a prominent, slightly expansive callus; umbilicus slightly open, fasciolar ridge strong; off-white with a purple aperture; reaches 40mm.
DISTRIBUTION: Throughout the Philippines.

5. Coralliophila (Latimurex) costularis (Lamarck, 1816)
30.5mm; Samar.
Shell medium sized, with a tall spire and a large, strongly convex body whorl; sculptured with rounded axial cords and overriding, closely-set, scabrous spiral lirae; aperture ovate, outer lip marginally serrate, internally lirate, columella smooth, adherent, umbilicus channeled, fasciolar ridge heavily corded; off-white with a mauve hue, aperture mauve; reaches 60mm.
DISTRIBUTION: Sporadically found throughout the Philippines.

6. Latiaxis (Pseudomurex) gyratus (Hinds, 1844)
Syn: *Latiaxis idoleum* (Jonas, 1847)
44.1mm; Punta Engaño, Cebu.
Shell large, heavy, tall-spired, with deeply impressed sutures and well shouldered, distinctly carinate whorls; sculptured with weak axial folds and crossed by finer, weakly scabrous spiral lirae, strongest and most sharply defined at the shoulder; aperture ovate, tapering to a short, open, slightly recurved canal, outer lip marginally serrate, lirate internally, umbilicus large with a heavily scabrous fasciolar cord; pure white; reaches 50mm.
DISTRIBUTION: Cebu, Bohol, Southern Mindanao and Bulan, Sorsogon.

7. Rapa rapa (Linné, 1758)
94.7mm; Cebu.
Shell medium sized to large, solid and bulbous, spire sunken, sutures deeply channeled and obscured by scabrous axial folds; surface sculptured with spiral cords becoming more widely spaced and scabrous towards the base, and fine interstitial axial lamellae; siphonal canal short or long with scabrous folds; aperture roundly quadrangular with an internally smooth, marginally serrate outer lip and a wide, posteriorly adherent columellar callus; off-white with a yellowish tinge; reaches 100mm.
DISTRIBUTION: Sporadically found throughout the Philippines.

8. Rapa incurva (Dunker, 1852)
52.8mm; Palawan.
Like a small *Rapa rapa* (Linné), but much smoother, and the sutures are only weakly channeled; shell light orange-brown, with weak, white axial stripes; reaches around 60mm.
DISTRIBUTION: Sporadically found throughout the Philippines.

9. Magilus antiquus Montfort, 1810
62.8mm (shell and tubular extension); Bohol.
A unique species where the mollusc lives in a non-descript globular shell, and successively adds calcareous matter to the last whorl, forming a long, tube-like extension; lives entwined in coral formations and may reach over 300mm; the larger the shell, the shorter and thicker the tube-like extension; colour controlled by environment, usually off-white.
DISTRIBUTION: Sporadically found throughout the Philippines.

10. Latiaxis (Latiaxis) mawae Griffith & Pidgeon, 1834
32.2mm; Panglao, Bohol.
Shell medium sized to large and uniquely sculptured; posterior whorl segments flat and bear a sharply carinate shoulder from which arises a series of 10 flat, inward curving spines, last whorl non-adherent to the shell body; aperture ovate, detached from the body whorl, terminating with a long, open, reflexed canal; shell surface encircled with fine striae, umbilicus flaring, with a heavily fluted fasciolar cord; colour off-white to brown, aperture pinkish white; reaches 70mm.
DISTRIBUTION: Cebu, Bohol and Southern Mindanao.

11. Latiaxis (Pseudomurex) eugeniae (Bernardi, 1853)
43.4mm; Panglao, Bohol.
Shell medium sized, fusiform-ovate, with a tall spire and well impressed sutures; whorls convex with a roundly angulate shoulder, sculptured with rounded axial folds (sometimes obsolete), and crossed by fine, weakly scabrous spiral cords; aperture ovate, tapering to a short, open, recurved canal, outer lip serrate, lirate within, umbilicus bordered by scabrous fasciolar cord; colour off-white; reaches 50mm. Incorrectly synonymized with *L. idoleum* Jonas, 1847 (=*gyratus* Hinds, 1844).
DISTRIBUTION: Cebu, Bohol and Southern Mindanao.

Class Gastropoda
Subclass Prosobranchia
Order Neogastropoda
Suborder Rachiglossa
Superfamily Buccinacea
Family BUCCINIDAE

A large group of carnivorous, rock and coral-dwelling gastropods inhabiting shallow to deep water in all world seas. The shells may be small or large, vary in shape from fusiform-ovate to elongate, have a sharp apex and a large body whorl, a large, smooth columella, a short anterior canal and may be smooth or axially and spirally sculptured. The operculum is corneous with either a lateral or terminal nucleus. Fairly well represented in the Philippines.

PLATE 44. 1. *Latiaxis fearnleyi* Emerson & D'Attilio 2. *Coralliophila pyriformis* Kira 3. *Coralliophila erosa* (Röding) 4. *Coralliophila neritoidae* (Lamarck) 5. *Coralliophila costularis* (Lamarck) 6. *Latiaxis gyratus* (Hinds) 7. *Rapa rapa* (Linné) 8. *Rapa incurva* (Dunker) 9. *Magilus antiquus* Montfort 10. *Latiaxis mawae* Griffith & Pidgeon 11. *Latiaxis eugeniae* (Bernardi) 12. *Cantharus leucotaeniatus* Kosuge 13. *Babylonia ambulacrum* (Sowerby I) 14. *Nassaria pusilla* (Röding) 15. *Nassaria sp.* 16. *Cantharus melanostomus* (Sowerby I) 17. *Nassaria fusiformis* Sowerby 18. *Nassaria recurva* Sowerby 19. *Nassaria amboynensis* Watson 20. *Nassaria sp.* 21. *Nassaria spinigera* (Hayasi & Habe)

PLATE 44

12. Cantharus leucotaeniatus Kosuge, 1985
56.4mm; Cebu.
Shell medium sized, elongate-ovate, with a tall spire, well impressed sutures, and a large body whorl; sculptured with axial folds and overriding spiral threads forming 2 rows of axially compressed nodules on the spire whorls, 4 on the body whorl; aperture elongate-ovate, outer lip serrate, internally lirate, columella with a slightly expansive callus, irregularly spaced plicae, and a basal fold; shell pale orange-brown, nodules deep brown, aperture and columella white with a yellow margin; reaches 60mm.
DISTRIBUTION: Cebu and Bohol.

13. Babylonia ambulacrum (Sowerby I, 1825)
47.8mm; Samar.
Shell medium sized, heavy, ovate, with a somewhat sunken spire and white coloured, deeply channeled sutures; body whorl large, smooth, aperture elongate-ovate and internally smooth; columella posteriorly calloused, umbilicus wide and deep, fasciolar cord heavy; off-white with greyish brown spotting forming 2 bands on the body whorl, early whorls reddish brown, aperture blue-white; reaches 55mm.
DISTRIBUTION: Sporadically found throughout the Philippines.

14. Nassaria (Nassaria) pusilla (Röding, 1798)
25.0mm; Palawan.
Shell small with a distinctly rotund appearance, 6 to 7 inflated whorls, deeply incised sutures and obsolete varices; sculptured with axial cords and overriding spiral striae forming a finely nodulose surface; aperture elongate-ovate, outer lip variced, lirate within and marginally denticulate; columella calloused and lirate, shell basally constricted with a short canal; off-white with brown banding; reaches 25mm.
DISTRIBUTION: Sporadically found throughout the Philippines.

15. Nassaria sp.
23.0mm; Leyte.
A fairly common species usually found attached to *Xenophora spp.*, but at present remains unidentifiable; shell small, probably a juvenile, elongate-ovate in outline, with a tall, partially eroded spire, well impressed sutures, and strongly convex whorls; sculptured with narrow, rounded axial cords (10 to 11 on the body whorl) which bear axially compressed nodules from overriding spiral striae; aperture elongate-ovate, outer lip variced and lirate within, columella irregularly plicate, canal short, open and recurved; off-white in colour. Most specimens examined seldom exceed 25mm.
DISTRIBUTION: Camotes Sea and Leyte.

16. Cantharus (Pollia) melanostomus (Sowerby I, 1825)
78.5mm; Sulu Archipelago.
Shell large, heavy, ovate, with a moderately tall spire and impressed sutures; sculptured with large axial folds and thick, grooved spiral cords; aperture ovate, outer lip thick, marginally serrate and internally lirate; columella calloused, burnt brown in colour; shell yellowish brown with dark brown on the ribs, aperture white internally, orange-brown marginally; reaches 80mm.
DISTRIBUTION: Camotes Sea and Sulu Sea.

17. Nassaria (Nassaria) fusiformis Sowerby, 1859
44.5mm; Bohol.
Shell medium sized with a tall, acute spire, impressed sutures, and a large, slightly ovate body whorl; sculptured with irregular varices, 5 to 6 thin axial ribs between the varices and overriding major and minor spiral threads, nodules forming where the major threads intersect the varices and axial ribs; aperture ovate, outer lip variced and lirate within, columella with irregular folds; reaches 45mm; synonymized with *N. pusilla* (fig. 14) but differs in shape and sculpture.
DISTRIBUTION: Sporadically found throughout the Philippines.

18. Nassaria (Nassaria) recurva Sowerby, 1859
35.9mm; Cebu.
Shell medium sized, with well-channeled sutures and a moderately tall spire, whorls convex but distinctly angulate on the lower portions; sculptured with irregular varices, oblique, rounded, well-spaced axial cords, and numerous overriding, weakly scabrous spiral threads; aperture elongate-ovate, outer lip variced and lirate within, columella irregularly plicate with 1 posterior and 2 anterior denticles; white with brown banding, lirae darker, aperture white; reaches 40mm.
DISTRIBUTION: Sporadically found throughout the Philippines.

19. Nassaria (Nassaria) amboynensis Watson, 1881
41.2mm; Palawan.
A medium-sized, triangularly ovate shell with a tall spire and sculptured with numerous rounded, arcuate axial cords which end in a shoulder below the suture, giving the shell a falsely channeled appearance; whorls encircled with smooth spiral lirae, fewer in number than seen in *N. recurva*, and producing weak nodules where they intersect the axial cords; white with darker banding; reaches around 45mm.
DISTRIBUTION: Sporadically found throughout the Philippines.

20. Nassaria sp.
39.5mm; Palawan.
Appears to be a somewhat more squat form of *N. amboynensis* (fig. 19), with the ribs forming more heavily sculptured shoulders; spiral sculpturing, apertural characteristics and colouring are the same.
DISTRIBUTION: Sporadically found throughout the Philippines.

21. Nassaria (Microfusus) spinigera (Hayashi & Habe, 1965)
43.5mm; Cebu.
Shell medium sized with a tall spire, moderately

channeled sutures, and a prominently constricted basal area; sculptured with 12 to 14 oblique axial cords, irregular varices, and encircled with fine spiral threads, whorls additionally sculptured with 2 major nodulose spiral cords on the spire whorls and 3 on the body whorl, giving the shell a distinctly bicingulate appearance; aperture ovate, outer lip with numerous marginal denticles, columella smooth (occasionally with 1 or 2 parietal denticles); pale yellowish orange, spiral cords darker; reaches 45mm.
DISTRIBUTION: Sporadically found throughout the Philippines.

Plate 45 Buccinidae

1. Phos (Phos) roseatus (Hinds, 1844)
Syn: *Phos varicosus* Gould, 1849
31.2mm; Samar.
Shell medium sized, fusiform-elongate, with rounded whorls, impressed sutures, and sculptured with varices, strong axial ribs and overriding spiral threads; aperture ovate, lavender coloured, outer lip thick and lirate within, columella white with 2 anterior denticles; yellow-white with 2 dark brown and white bands on the varices; reaches 35mm.
DISTRIBUTION: Sporadically found throughout the Philippines.

2. Phos (Phos) senticosus (Linné, 1758)
36.0mm; Manila Bay.
Shell medium sized, solid, with a tall, acute spire, impressed sutures and convex whorls; sculptured with varices, axial cords and overriding spiral threads forming echinate nodules at the intersection points; aperture ovate, outer lip thick, lirate, columella with one posterior and 3 anterior plicae; colour variable, usually cream with brown banding, aperture lavender; reaches 40mm.
DISTRIBUTION: Sporadically found throughout the Philippines.

3. Phos (Phos) hirasei Sowerby, 1913
31.5mm; Manila Bay.
Shell medium sized, elongate-ovate, with a tall, acute spire, strongly convex whorls, axial cords, overriding spiral striae, and irregularly spaced varices (when present); aperture ovate, outer lip thin, lirate within, columella with 2 basal folds; cream coloured with brown blotches and dark brown spiral lirae; reaches 40mm.
DISTRIBUTION: Sporadically found throughout the Philippines.

4. Pisania (Pisania) fasciculata (Reeve, 1846)
Syn: *P. crenilabrum* A. Adams, 1855
26.2mm; Palawan.
Shell medium sized, lightweight, fusiform-elongate; sculptured with strong spiral cords and small axial ribs on the early whorls, obsolete towards the base; aperture ovate, finely lirate within, columella with anterior and posterior denticles; pale orange-brown with dark-brown and white spiral bands, aperture white; reaches 35mm.
DISTRIBUTION: Sporadically found throughout the Philippines.

5. Pisania (Ecmanis) ignea (Gmelin, 1791)
35.0mm; Cebu.
Shell medium sized, lightweight, fusiform-ovate, earlier whorls with axial ribs, later whorls smooth, base of the body whorl with spiral grooves; aperture ovate, outer lip thin, weakly crenulate, columella smooth with 2 denticles, one basally, the other on the parietal area; orange-brown with darker brown and white spiral cords and flammules, aperture pale lavender; reaches 40mm.
DISTRIBUTION: Sporadically found throughout the Philippines.

6. Cantharus (Pollia) fumosus (Dillwyn, 1817)
30.5mm, 28.0mm; Samar.
Shell medium sized, solid, close to *C. undosus* but is more elongate, has pronounced axial folds, and more numerous, close-set spiral ribs which are crossed by fine axial scratches, sutures obscured by a thick cord; aperture the same as in *undosus* but more narrowly ovate; colour pale yellow-orange with dark brown cords, a white aperture, and an orange outer lip margin; reaches 40mm.
DISTRIBUTION: Sporadically found throughout the Philippines.

7. Cantharus (Clivipollia) wagneri (Anton, 1839)
26.4mm; Punta Engaño, Cebu.
Shell small to medium sized, solid, whorls convex and sculptured with broad, rounded axial folds and overriding spiral lirae; aperture ovate, outer lip thickened and lirate within, columella plicate, siphonal canal moderately long, open, dorsally recurved; yellowish brown with darker banding and purplish brown spots between the axial folds; reaches 30mm.
DISTRIBUTION: Sporadically found throughout the Philippines in limited quantity.

8. Cantharus (Pollia) undosus (Linné, 1758)
28.7mm; Cebu.
Shell medium sized, solid, fusiform-ovate, sutures ledged and weakly channeled; sculptured with raised, brown-coloured spiral cords and wide, white-coloured interspaces; aperture ovate, outer lip thick, variced and lirate within, columella with some basal plicae; aperture and columella white, outer lip margin orange; reaches 40mm.
DISTRIBUTION: Throughout the Philippines.

9. Cantharus (Clivipollia) pulcher (Reeve, 1846)
23.4mm; Samar.
Shell small to medium sized, fusiform-ovate and tall-spired; siphonal canal short, open and recurved; sculptured with axial folds and spiral grooves, intersecting to form elongate nodules; aperture triangulate, outer lip thick, internally denticulate, columella weakly plicate; colour pinkish brown with black grooves; aperture, denticles and plicae white, apertural margin pink; reaches 30mm.
DISTRIBUTION: Throughout the Philippines in limited quantity.

10. Engina alveolata (Kiener, 1836)
16.7mm; Punta Engaño, Cebu.
Shell small, fusiform-ovate and tall-spired; sculptured with nodulose axial cords, spiral grooves and a row of sutural nodules; aperture narrowly ovate, outer lip thick and denticulate, columella with lirae and denticles over its entire length; white with a double row of black and a single row of red nodules; reaches 20mm.
DISTRIBUTION: Sporadically found throughout the Philippines in limited quantity.

11. Cantharus (Pollia) wrightae Cernohorsky, 1974
25.5mm; Cebu.
Shell small, elongate-ovate, sutures moderately impressed, early whorls with axial ribs becoming obsolete on the last 2 whorls; spirally sculptured with elevated cords and wide interspaces containing fine spiral lirae which are sectioned by axial threads; aperture marginally denticulate, lirate within, columella basally plicate; reddish brown with a single white band on the body whorl; reaches 30mm.
DISTRIBUTION: Sporadically found throughout the Philippines in limited quantity.

12. Pisania (Caducifer) truncata (Hinds, 1844)
15.3mm; Punta Engaño, Cebu.
Shell small, apex bluntly truncated, sculptured with fine axial and spiral threads producing a finely cancellate surface; aperture ovate, outer lip and columella smooth; colour brown with thin white bands and axial streaks; reaches 20mm.
DISTRIBUTION: Sporadically found throughout the Philippines.

13. Engina lanceolata Kuroda & Habe, 1971
9.2mm, 9.8mm; Punta Engaño, Cebu.
Shell small, fusiform-ovate, with a tall spire and adpressed sutures; sculptured with 9 to 10 heavy axial cords which are cut by numerous spiral grooves, producing axially compressed nodules, strongest on the body whorl shoulder; canal long, open, recurved, aperture narrowly ovate, outer lip marginally denticulate, columella smooth with a few obsolete plicae; pale brown with pale yellow nodules bordered in dark brown; reaches 10mm.
DISTRIBUTION: Sporadically found throughout the Philippines.

14. Phos (Strongylocera) textilis A. Adams, 1851
13.6mm; Punta Engaño, Cebu.
Shell small, solid, with a tall spire, sharply angulate spire whorls and a large body whorl; sculptured with axial ribs which are nodulose at the shoulder and encircled with fine spiral threads, and heavier spiral cords on the lower body whorl; aperture ovate, outer lip thick, marginally crenulate and lirate within, columella weakly wrinkled and with 2 basal folds; colour pale brown; reaches 15mm.
DISTRIBUTION: Sporadically found throughout the Philippines in limited quantity.

15. Engina egregia (Reeve, 1844)
14.8mm; Punta Engaño, Cebu.
Shell small, fusiform-ovate, with impressed sutures, a tall spire and strongly convex whorls; sculptured with 14 to 16 axial cords bearing elongate nodules from overriding spiral lirae, interspaces finely cancellate from intersecting axial and spiral threads; aperture ovate, outer lip thin, variced and with 10 to 12 denticles, columella irregularly plicate; yellowish brown, dark brown on the nodules;
reaches 20mm.
DISTRIBUTION: Sporadically found throughout the Philippines.

16. Phos (Strongylocera) textum (Gmelin, 1791)
20.0mm; Samar.
Shell medium sized, with a tall, acute spire, a large body whorl and adpressed sutures; sculptured with axial cords and irregularly spaced spiral lirae intersecting to produce echinate tubercles, strongest at the sharply-angled shoulder; aperture ovate, outer lip lirate within and notched anteriorly, columella with a few folds, fasciolar cord strongly plicate; external colouration cream with a pinkish apex, aperture intensely orange; reaches 30mm.
DISTRIBUTION: Sporadically found throughout the Philippines.

17. Cantharus (Prodotia) iostomus (Gray, in Griffith and Pidgeon, 1834)
30.1mm, 23.5mm, 19.3mm; Punta Engaño, Cebu.
Shell small to medium sized, extremely variable, ranging from fusiform-ovate to fusiform-elongate and with impressed sutures; sculptured with axial cords and overriding spiral lirae forming rows of nodules,

PLATE NO. 45 1. *Phos roseatus* (Hinds) 2. *Phos senticosus* (Linné) 3. *Phos hirasei* Sowerby 4. *Pisania fasciculata* (Reeve) 5. *Pisania ignea* (Gmelin) 6. *Cantharus fumosus* (Dillwyn) 7. *Cantharus wagneri* (Anton) 8. *Cantharus undosus* (Linné) 9. *Cantharus pulcher* (Reeve) 10. *Engina alveolata* (Kiener) 11. *Cantharus wrightae* Cernohorsky 12. *Pisania truncata* (Hinds) 13. *Engina lanceolata* Kuroda & Habe 14. *Phos textilis* A. Adams 15. *Engina egregia* (Reeve) 16. *Phos textum* (Gmelin) 17. *Cantharus iostomus* (Gray, in Griffith & Pidgeon) 18. *Engina concinna* (Reeve) 19. *Engina mendicaria* (Linné) 20. *Engina obliquicostata* (Reeve) 21. *Engina alveolata* (Kiener)

PLATE 45

those on the lower portion of the spire whorls being larger than the rest, causing a slightly angulate whorl contour; aperture ovate, constricted anteriorly, outer lip plicate, columella smooth with some basal denticles; white with brown marbling and an occasional lavender tint; reaches 30mm.
DISTRIBUTION: Sporadically found throughout the Philippines in limited quantity.

18. Engina concinna (Reeve, 1846)
16.9mm; Punta Engaño, Cebu.
Shell small, biconic, with a tall spire and large body whorl; sculptured with rounded axial folds and overriding spiral cords forming laterally elongate nodules; aperture narrowly ovate, outer lip thick, variced and denticulate, columella with 4 denticles; colour pinkish white with black bands containing 2 rows of white nodules; reaches 17mm.
DISTRIBUTION: Sporadically found throughout the Philippines in limited quantity.

19. Engina mendicaria (Linné, 1758)
15.9mm; Palawan.
Shell small, ovate, with obsolete axial folds and weak spiral striae, the spire frequently eroded; aperture narrow, outer lip swollen and dentate, columella smooth with a posterior depression; black with yellowish banding, aperture white with a purple-brown margin; reaches 22mm.
DISTRIBUTION: Throughout the Philippines.

20. Engina obliquicostata (Reeve, 1846)
11.3mm, 11.0mm; Punta Engaño, Cebu.
Shell small, fusiform-elongate, with a tall spire, impressed sutures and a moderately long, open siphonal canal; sculptured with broad, oblique, closely-set axial ribs and encircled with numerous nodulose spiral cords; aperture narrowly ovate, outer lip thick with a large anal denticle and 7 weaker denticles, columella denticulate, parietal wall radially plicate; reddish brown with a white body whorl band and white spotting on the cords and at the sutures; reaches 15mm.
DISTRIBUTION: Sporadically found throughout the Philippines in limited quantity.

21. Engina alveolata (Kiener, 1836) (ecomorph)
14.5mm; Punta Engaño, Cebu.
A curious form of *alveolata* (fig. 10) where the shell is slightly more elongate, the whorls less angulate, the red nodules are replaced by white ones, and the black banding is discontinuous.
DISTRIBUTION: Sporadically found throughout the Philippines in limited quantity.

Columbellidae — Plate 46

Class Gastropoda
Subclass Prosobranchia
Order Neogastropoda
Suborder Rachiglossa
Superfamily Buccinacea
Family COLUMBELLIDAE

A large group of small, extremely variable shells with a worldwide distribution and inhabiting coral or coral sand. The shells may be ovate to elongate-ovate, smooth or sculptured with ribs or nodules, have a dentate outer lip and columella, a brown, opaque periostracum and a corneous operculum. The family is very poorly understood, and although well represented in the Philippines, most species are unavailable to collectors unless collected in the field. Accordingly only some of the more commonly encountered species are presented here.

1. Pyrene ocellata (Link, 1807)
16.2mm, 17.3mm; Samar.
Shell small with a large, ovate body whorl and a tall spire which is sometimes eroded or truncated; surface smooth except for some cords at the base of the body whorl; aperture narrow, outer lip with a central callosity bearing denticles, columella calloused and plicate; black in colour with white spots or axial zigzag bands; reaches 22mm.
DISTRIBUTION: Throughout the Philippines.

2. Pyrene punctata (Bruguiére, 1789)
21.2mm; Palawan.
Shell small, elongate-ovate with a short exserted spire and a large, elongate, somewhat ventricose body whorl; surface smooth with some cords at the body whorl base; aperture narrow, outer lip thick, incurved and marginally denticulate, columella calloused, concave, and weakly plicate on the anterior portion; colour variable, usually orange-brown to dark brown, with white subsutural blotching and axial white spotting on the lower body whorl, aperture violet; reaches 25mm.
DISTRIBUTION: Sporadically found throughout the Philippines in limited quantity.

PLATE NO. 46 1. *Pyrene ocellata* (Link) 2. *Pyrene punctata* (Bruguiére) 3. *Pyrene flava* (Bruguiére) 4a. *Pyrene testudinaria* (Link) 4b. streaked form 4c. form *nigropardalis* Habe & Kosuge 5. *Mitrella marquesa* (Gaskoin) 6. *Mitrella eximia* (Reeve) 7. *Mitrella ligula* (Duclos) 8. *Aesopus spiculum* (Duclos in Chenu) 9. *Columbella scripta* (Lamarck) 10. *Mitrella venulata* (Sowerby) 11. *Mitrella alabastrum* (Reeve) 12. *Pyrene epamelle* (Duclos) 13. *c.f. Nitidella sp.* 14. *Mitrella bella* (Reeve) 15. *Mitrella margarita* (Reeve) 16. *Mitrella puella* (Sowerby) 17. *Inella pyramidalis* (Adams & Reeve) 18. *Tetraphora princeps* (Sowerby) 19. *Viriola pagoda (Hinds)* 20. *Viriola corrugata* (Hinds) 21. *Inella maxillaris* (Hinds) 22. *Mastonia rubra* (Hinds)

PLATE 46

3. Pyrene flava (Bruguiére, 1789)
21.5mm; Cebu.
Shell small, elongate-ovate, with a tall, telescoped spire which is almost straight-sided, and a long ventricose body whorl; aperture narrow, extending approximately halfway up the body whorl, outer lip thick, internally dentate, columella with a central concavity and a few posterior plicae; colour usually orange-brown with white axial streaks and smaller spotting; reaches 25mm.
DISTRIBUTION: Sporadically found throughout the Philippines in limited quantity.

4. Pyrene testudinaria (Link, 1807)
Shell small, ovately biconic with a tall, acute spire; shell surface smooth except for some spiral cords on the body whorl base; aperture narrow, outer lip denticulate, columella with a few anteriorly placed plicae; colour variable, may be black, white, white with brown-and-black streaks, or brown with white dots and streaks; reaches 25mm.
4a. typical form; 16.8mm; Punta Engaño, Cebu.
4b. an axially-streaked form; 14.9mm; Punta Engaño, Cebu.
4c. form *nigropardalis* Habe & Kosuge, 1966: a white form with black dots; 14.4mm; Samar.
DISTRIBUTION: Sporadically found throughout the Philippines.

5. Mitrella (Mitrella) marquesa (Gaskoin, 1852)
10.8mm (smooth), 10.5mm (ribbed); Punta Engaño, Cebu.
Shell small, elongate-ovate, sculptured with thin, angulate axial ribs which may become obsolete on the last 3 or 4 whorls, spiral structure limited to a few basal cords; aperture elongate-ovate, outer lip anteriorly constricted and denticulate within, columella concave with a few anterior plicae; colour brown or brown with a white reticulate network, a white aperture and a pink spire; reaches 15mm.
DISTRIBUTION: Sporadically found throughout the Philippines.

6. Mitrella (Mitrella) eximia (Reeve, 1846)
10.0mm, 9.8mm; Punta Engaño, Cebu.
Shell small, elongate-ovate, with a tall spire, a large body whorl which is slightly angulate basally, and several oblique basal cords on an otherwise smooth surface; aperture ovate, outer lip constricted anteriorly and lirate within, columella plicate anteriorly; orange-brown in colour with weaker, darker axial stripes and a brown-spotted white band at the sutures and mid-body whorl; reaches 12mm.
DISTRIBUTION: Sporadically found throughout the Philippines.

7. Mitrella (Graphicomassa) ligula (Duclos, 1840)
14.7mm, 13.5mm; Punta Engaño, Cebu.
Shell small, solid, fusiform-elongate with a tall spire and strongly convex spire whorls; body whorl large, smooth, corded at the base; aperture narrowly quadrangular, outer lip thickened and denticulate, columella thickened and marginally plicate; colour variable, usually white with dark brown banding or reticulations, but occasionally all-white or brown; reaches 25mm.
DISTRIBUTION: Sporadically found throughout the Philippines.

8. Aesopus spiculum (Duclos in Chenu, 1846)
11.8mm, 11.4mm; Punta Engaño, Cebu.
Shell small, elongate, spire whorls weakly convex, initially clathrate but becoming spirally striated towards the truncated body whorl; aperture narrow, outer lip expanded and internally lirate, columella calloused with a single parietal denticle; purple with pale sutural banding; reaches 15mm.
DISTRIBUTION: Sporadically found throughout the Philippines.

9. Columbella scripta (Lamarck, 1822)
14.9mm, 14.9mm; Samar.
Shell small, solid, whorls rounded and angulate at the shoulder, spire moderately tall and acute, body whorl large; aperture narrow, outer lip thickened, marginally lirate and sharply angulate posteriorly, columella with marginal plicae; cream coloured with darker spots, streaks or mottling, aperture white or pinkish, columella often with brown spotting; reaches 20mm. We consider this species to be a *Columbella* and also consider this genus separable from *Pyrene* or *Mitrella,* whereby negating any homonymy with the Mediterranean species *Mitrella scripta* (Linné, 1758).
DISTRIBUTION: Throughout the Philippines.

10. Mitrella (Mitrella) venulata (Sowerby, 1894)
10.5mm, 9.7mm; Punta Engaño, Cebu.
Shell small, fusiform-ovate with a tall, sharply acute spire, a large body whorl somewhat angulate on the lower half, and weakly impressed sutures which have a presutural groove; whorls smooth except for some strong oblique cords on the base; aperture elongate-ovate, outer lip thin, lirate within, sharply angulated anteriorly, columella plicate; light to dark orange-brown in colour, with white and brown sutural spots and irregular white mottling over the rest of the shell; reaches 12mm.
DISTRIBUTION: Sporadically found throughout the Philippines.

11. Mitrella (Columbellopsis) alabastrum (Reeve, 1859)
13.5mm; Punta Engaño, Cebu.
Shell small, fusiform-elongate with a tall spire, slightly convex spire whorls, deeply channeled sutures and a large body whorl which is angulate towards the base, surface smooth with a series of strong ridges on the lower half of the body whorl; aperture ovate, outer lip thin, lirate and constricted basally, columella weakly plicate; pale brown in colour with brown subsutural spots and irregular spotting and mottling; reaches 15mm.
DISTRIBUTION: Sporadically found throughout the Philippines.

12. Pyrene epamella (Duclos, 1846)
Syn: *Pyrene philippinarum* Reeve, 1858
24.0mm; Sulu Archipelago
Shell large for the family, somewhat biconic in outline, with a moderately tall spire, impressed sutures, and a large body whorl which is sharply angulated at the shoulder; surface smooth, some spiral cords basally; aperture narrowly elongate, outer lip thick, heavily dentate, parallel to the columella then sharply angulate basally, columella weakly calloused and marginally plicate; white in colour with black axial streaks, some specimens orange or brown; reaches 27mm.
DISTRIBUTION: Sporadically found throughout the Philippines.

13. c.f. Nitidella sp.
9.2mm; Punta Engaño, Cebu.
Shell small, fusiform-ovate with a moderately tall spire, weakly convex whorls, weakly channeled sutures, and a large body whorl with some spiral cords at the base; aperture elongate-ovate, outer lip thin, lirate within, columella with some obsolete folds; pale brown in colour with a fine brown reticulate network over the entire surface; reaches 10mm. There appears to be no name currently available for this distinctive little species.
DISTRIBUTION: Sporadically found throughout the Philippines in limited quantity.

14. Mitrella (Mitrella) bella (Reeve, 1859)
11.2mm, 11.8mm; Punta Engaño, Cebu.
Shell small, fusiform-ovate with a tall spire and a large, somewhat angulate body whorl, whorls weakly convex, surface smooth except for some basal grooves; aperture narrowly elongate, outer lip straight then sharply angulate basally, internally denticulate, columella with 3 or 4 plicae; colour variable, usually ivory white with golden-brown splashes and reticulations, aperture white with a lavender margin; reaches 12mm.
DISTRIBUTION: Sporadically found throughout the Philippines.

15. Mitrella (Graphicomassa) margarita (Reeve, 1859)
9.5mm, 10.0mm; Punta Engaño, Cebu.
Shell small, fusiform-elongate, tall-spired, sutures weakly channeled, spire whorls almost straight-sided; aperture narrowly elongate, outer lip broadly arcuate, slightly angulate basally and denticulate within, columella calloused and weakly plicate, shell surface smooth with 8 to 9 basal grooves; shell greyish brown in colour with a lavender-coloured spire and a mauve aperture; reaches 10mm.
DISTRIBUTION: Sporadically found throughout the Philippines.

16. Mitrella (Mitrella) puella (Sowerby, 1844)
14.8mm, 14.5mm; Punta Engaño, Cebu.
Shell small, fusiform-elongate, with a tall spire and a large, slightly angulate body whorl, surface smooth with some basal grooving; aperture elongate-ovate with a thick, arcuate, basally constricted outer lip margin which is denticulate within, columella with some weak marginal plicae, siphonal canal short, open and recurved; golden brown in colour with brown-streaked white bands at the sutures and mid-body whorl; reaches 15mm.
DISTRIBUTION: Sporadically found throughout the Philippines.

Class Gastropoda
Subclass Prosobranchia
Order Heterogastropoda
Suborder Ptenoglossa
Superfamily Triphoracea
Family TRIPHORIDAE

A large group of small to medium-sized gastropods, mostly tropical in their distribution and living in the intertidal zone under coral and feeding on sponges. The shells are tall, slender, and immediately recognized by their sinistral coiling and the sculpturing of granulose spiral cords inter-connected by axial threads. The siphonal canal is straight but may be recurved in some species. The family is well represented in the Philippines but most species are hard to come by and, because of their small size, identification is often quite difficult. Only a few of the larger, well known species are illustrated here.

17. Inella (Cautotriphora) pyramidalis (Adams & Reeve, 1850)
17.4mm; Punta Engaño, Cebu.
Shell medium sized, elongate, with adpressed sutures; there are 4 carinae per whorl, the first normal sized, the second weak, the third heavy and over-hanging, the fourth weak and white-coloured with brown spots; some axial folds and grooves forming echinate nodules on some of the carinae; aperture ovate, marginally crenulate, canal closed, narrow and recurved; pale brown in colour with brown streaks and mottling; reaches 40mm.
DISTRIBUTION: Sporadically found throughout the Philippines.

18. Tetraphora princeps (Sowerby, 1904)
35.0mm; Palawan.
Shell extremely elongate, rather large for the family, whorls number from 20 to 25 and are straight-sided, sutures well incised with a strong cord; whorls sculptured with weak axial folds which are crossed by 4 spiral carinae and weaker interstitial spiral threads; aperture quadrangular, outer lip expanded and marginally crenulate, canal short and open; off-white in colour with dark reddish brown mottling; reaches 40mm.
DISTRIBUTION: Sporadically found throughout the Philippines.

19. Viriola pagoda (Hinds, 1843)
20.5mm; Punta Engaño, Cebu.
Shell small, elongate with a long, recurved, partially

open siphonal canal, weakly channeled sutures, and an ovate, marginally crenulate aperture; a single strong cord above each suture, becoming bicingulate on the body whorl; a second and third spiral cord are also seen below the suture and centrally, but much weaker in strength; off-white in colour with brown spots and axial streaking; reaches 30mm.
DISTRIBUTION: Sporadically found throughout the Philippines.

20. Viriola corrugata (Hinds, 1843)
17.5mm; Punta Engaño, Cebu.
Shell small, elongate with tricarinate whorls and a weak sutural cord; carinae smooth, the central one weakest, and the interspaces are sculptured with axial striae; aperture ovate, outer lip crenulate, canal long, narrow and recurved; pale brown in colour with darker cords; reaches 20mm.
DISTRIBUTION: Sporadically found throughout the Philippines.

21. Inella (Inella) maxillaris (Hinds, 1844)
18.5mm; Punta Engaño, Cebu.
Shell small, extremely elongate, sculptured with 3 flattened spiral cords, deep interspaces, and fine, semi-circular gemmules on either side of the cords and within the interspaces, subsutural cord strongest, central one weakest; aperture ovate, marginally crenulate; colour pale lavender to brown reaches 20mm.
DISTRIBUTION: Sporadically found throughout the Philippines.

22. Mastonia rubra (Hinds, 1843)
8.0mm; Punta Engaño, Cebu.
Shell small, purple-coloured, fusiform, somewhat ovate towards the base, sculptured with 2 rows of gemmules, 3 on the body whorl, sutures reasonably distinct; aperture quadrangular, marginally crenulate, canal short and open; reaches 10mm.
DISTRIBUTION: Sporadically found throughout the Philippines.

Class Gastropoda
Subclass Prosobranchia
Order Neogastropoda
Suborder Rachiglossa
Superfamily Buccinacea
Family FASCIOLARIIDAE

A large group of medium to large sized carnivorous gastropods inhabiting sand or coral rubble in the intertidal regions of most world seas. The shells are heavy and fusiform, with tall spires and long siphonal canals. Surface sculpturing consists of heavy axial cords, frequently nodulose at the shoulder, and spiral lirae. The aperture is ovate, frequently lirate within, the outer lip thin, and the columella may or may not bear folds. Most species are non-umbilicate. The periostracum is thick and fibrous, the operculum large, thick,ovate, tapering posteriorly, and with a spiral nucleus. Well represented in the Philippines with several smaller species apparently unnamed at this time.

1. Fusinus (Fusinus) undatus (Gmelin, 1791)
153.5mm; Cebu.
Shell large, fusiform-elongate, with a tall spire, a long siphonal canal, and impressed sutures; sculptured with large axial ribs, terminating at the sharply angled shoulder where they form axially compressed nodules, ribs obsolete on the presutural slope, surface spirally sculptured with fine cords; aperture ovate, outer lip marginally denticulate, internally lirate, columella finely plicate along its entire length; colour ivory-white; reaches 180mm.
DISTRIBUTION: Sporadically found throughout the Philippines.

2. Fusinus (Fusinus) forceps (Perry, 1811)
146.8mm; Samar.
Shell large, fusiform-elongate, with a tall spire, a long siphonal canal and deeply channeled sutures; whorls strongly convex, sculptured with large, rounded axial folds that are crossed by numerous elevated spiral cords; aperture ovate, lirate within, outer lip marginally serrate, columella non-adherent anteriorly with a few weak folds; white in colour with brown on and between the ribs; reaches 220mm.
DISTRIBUTION: Sporadically found throughout the Philippines.

3. Fusinus (Fusinus) nicobaricus (Röding, 1798)
140mm; Bohol.
Shell large, fusiform-elongate, with a tall, sharply acute spire, a long canal and adpressed sutures; sculptured with strong axial ribs on the earlier whorls, reduced to thick, rounded shoulder nodes on later whorls, additionally sculptured with thick spiral cords and weaker axial striae; aperture ovate, lirate within, outer lip marginally dentate, columella plicate; white in colour with purplish brown-coloured axial streaks; reaches 180mm.
DISTRIBUTION: Sporadically found throughout the Philippines.

PLATE NO. 47 1. *Fusinus undatus* (Gmelin) 2. *Fusinus forceps* (Perry) 3. *Fusinus nicobaricus* (Röding) 4. *Fusinus salisburyi* H. C. Fulton 5. *Fusinus longissimus* (Gmelin) 6. *Fasciolaria filamentosa* (Röding) 7. *Fasciolaria trapezium* (Linné) 8. *Latirus nodatus* (Gmelin) 9. *Fusinus colus* (Linné) 10. *Latirus lanceolatus* (Reeve) 11. *Latirus recurvirostratum* (Schubert & Wagner) 12. *Latirus gibbulus* (Gmelin) 13. *Latirus polygonus* (Gmelin) 14. *Latirus polygonus* (Gmelin) form *barclayi* Reeve 15. *Latirus belcheri* (Reeve) 16. *Latirus constrictus* (Koch, in Philippi) 17. *"Fasciolaria" walleri* Ladd

PLATE 47

4. Fusinus (Fusinus) salisburyi H. C. Fulton, 1930
239mm; Sulu Archipelago.
Shell very large, fusiform-elongate, with a tall spire, a long canal and deeply channeled sutures; early whorls sculptured with axial folds, becoming obsolete on the last 2 whorls, spiral sculpture consisting of numerous deeply incised grooves; aperture ovate, outer lip marginally crenulate, lirate within, columella weakly plicate, anteriorly non-adherent; shell pure white in colour; reaches 260mm.
DISTRIBUTION: Sporadically found throughout the Philippines in limited quantity.

5. Fusinus (Fusinus) longissimus (Gmelin, 1791)
289mm; Palawan.
Shell unusually large for the genus, fusiform-elongate, with adpressed sutures, a tall spire and a very long siphonal canal; sculptured with axial folds on the earlier whorls, reduced to a row of tuberculate nodules on the sharply-angled shoulder of the later whorls, entire surface encircled with numerous well defined spiral cords; aperture ovate, outer lip marginally serrate, internally lirate, columella weakly plicate and non-adherent anteriorly; shell creamy white in colour; reaches 300mm.
DISTRIBUTION: Sporadically found throughout the Philippines in limited quantity.

6. Fasciolaria (Pleuroploca) filamentosa (Röding, 1798)
154mm; Bohol
Shell large, broadly fusiform, with a tall spire, adpressed sutures and a short, open siphonal canal; whorls ventricose, slightly shouldered and bearing a row of nearly obsolete nodules, entire shell encircled with deeply incised spiral grooves; aperture ovate, outer lip marginally serrate, internally lirate, columella with 3 basal folds; shell a burnt orange-brown in colour with paler axial streaks, brown-coloured grooves, a pale pink aperture with orange lirae, and a dark brown anterior canal extremity; reaches 165mm.
DISTRIBUTION: Sporadically found throughout the Philippines.

7. Fasciolaria (Pleuroploca) trapezium (Linné, 1758)
185mm; Batangas.
Shell large, heavy, broadly fusiform with a tall conical spire, adpressed sutures and a short siphonal canal; whorls sculptured with a row of heavy raised tubercles, most pronounced on the last 2 whorls, spiral sculpturing obsolete; aperture large, ovate, outer lip thick, marginally dentate, internally lirate, columella heavily glazed with a few anterior folds, fasciolar ridge strong; off-white in colour, orange at the sutures and encircled with darker spiral lines, aperture purple, apertural lirae deep red, fasciolar cord brown; reaches 200mm.
DISTRIBUTION: Throughout the Philippines.

8. Latirus (Latirus) nodatus (Gmelin, 1791)
88.0mm; Samar.
Shell medium sized, narrowly fusiform with adpressed sutures, a tall spire and a moderately long canal; sculptured with large, rounded, smooth axial nodes and weak interstitial spiral striae; aperture ovate, weakly lirate within, outer lip finely crenulate, columella with 3 basal denticles; pale orange-brown in colour, nodes paler and encircled with rows of brown spots, aperture pink; reaches 100mm.
DISTRIBUTION: Sporadically found throughout the Philippines.

9. Fusinus (Fusinus) colus (Linné, 1758)
157mm; Manila Bay.
Shell medium sized, fusiform-elongate, with a tall spire, a long siphonal canal and impressed sutures; whorls rounded, becoming somewhat angulate on the last 2 or 3; sculptured with prominent axial cords on the earlier whorls, becoming short elongate nodules on later whorls and frequently disappearing on the body whorl, prominent spiral cords encircle the entire shell; aperture ovate, outer lip marginally dentate and internally lirate, columella smooth with a few basal plicae; white in colour with brown between the ribs of the earlier whorls and on the outer lip margin; reaches 200mm.
DISTRIBUTION: Sporadically found throughout the Philippines mostly from Manila Bay.

10. Latirus (Latirus) lanceolatus (Reeve, 1847)
44.0mm; Palawan.
Like a small version of *L. nodatus*, but is more fusiform and the nodes are less inflated; the colouration is also slightly different, mainly lacking the spiral rows of small brown dots; seldom exceeds 50mm.
DISTRIBUTION: Sporadically found throughout Philippines.

11. Latirus (Latirus) recurvirostratum (Schubert and Wagner, 1829)
81.2mm; Samar.
Shell medium sized, solid, fusiform, somewhat resembling *L. polygonus*, but more acuminate, with a taller spire, a wider siphonal fasciole and a more pronounced umbilical ridge; whorls with tuberculate axial cords bearing sharp ridges where they are crossed by strong spiral lirae; aperture widely ovate, lirate within, columella with 2 or 3 hidden plaits, canal long, tubular and recurved; pale brown in colour, spiral lirae white, dark brown between the sharp ridges on the axial cords, aperture orange, apertural lirae pale red; reaches 85mm.
DISTRIBUTION: Sporadically found throughout the Philippines.

12. Latirus (Latirus) gibbulus (Gmelin, 1791)
94.0mm; Mindoro.
Shell medium sized, solid, with rounded whorls bearing large, heavy axial nodes and obsolete spiral lirae; aperture triangularly ovate, constricted anteriorly, outer lip marginally dentate, lirate within and sharply angulate at the shoulder, columella smooth, canal short and recurved with a weak umbilical channel; brown in colour with dark brown

spiral lines and an orange aperture; reaches 100mm.
DISTRIBUTION: Sporadically found throughout the Philippines.

13. Latirus (Latirus) polygonus (Gmelin, 1791)
76.3mm; Samar.
Shell medium sized with a tall spire, adpressed sutures, and a short, open canal; whorls sharply angulate, sculptured with 2 rows of nodules on the body whorl which are interconnected by axial ribs, and the whole surface encircled with prominent spiral cords; aperture ovate, lirate within, columella with 3 to 4 basal cords; off-white to yellow in colour, spiral cords brown, interspaces dark brown, aperture and columella bluish white; reaches 80mm.
DISTRIBUTION: Sporadically found throughout the Philippines.

14. Latirus (Latirus) polygonus (Gmelin, 1791) form **barclayi** Reeve, 1847
76.7mm; Cebu.
In this form of *polygonus*, the shell is somewhat heavier, the nodes are much larger, the shell slightly more ovate, and the colour is predominantly orange-brown with white nodules.
DISTRIBUTION: Sporadically found throughout the Philippines.

15. Latirus (Latirus) belcheri (Reeve, 1847)
74.0mm; Samar.
Shell medium sized, fusiform-ovate in outline, with a tall spire, adpressed sutures and a short, open siphonal canal; sculptured with a row of nodules on the spire whorls and a double row on the body whorl, the second row weaker and sloping downwards; aperture ovate, outer lip bi-angulate, marginally dentate, lirate within, columella with obsolete basal plicae; off-white in colour, with brown axial bands, bright orange lirae, a bluish white aperture, brown-stained teeth, and a purple stained canal base; reaches 80mm.
DISTRIBUTION: Sporadically found throughout the Philippines.

16. Latirus (Latirus) constrictus (Koch in Philippi, 1845)
51.6mm; Sulu Archipelago.
Shell medium sized, solid, fusiform, with a tall spire, a relatively short, flattened, obliquely truncated siphonal canal, and impressed sutures which are demarcated by a prominent presutural slope; sculptured with large, rounded axial nodes terminating at the shoulder, and encircled with heavy spiral lirae; aperture ovate, lirate within, outer lip marginally serrate, columella smooth, fasciolar cord strong with a small umbilical channel; orange-brown in colour with a pale brown interior; reaches 60mm.
DISTRIBUTION: Sporadically found throughout the Philippines in limited quantity.

17. "Fasciolaria walleri" Ladd, 1976
54.5mm; Bohol.
Shell medium sized, fusiform, with a tall spire and a long, open, distally recurved canal; whorls rounded, slightly shouldered, bearing 8 to 9 strong axial folds, obsolete above the shoulder and crossed by fine spiral lirae; aperture ovate, lirate within, columella smooth with a single anterior and posterior denticle, greyish white in colour with discontinuous brown lines; reaches 65mm; described from a New Hebrides fossil, but living in the Philippines. [More recently available information provisionally assigns this species to the genus *Siphonofusus* Kuroda and Habe (Family: Buccinidae). Until living material is studied, the exact placement of this species will remain uncertain at both a generic and family level.]
DISTRIBUTION: Sporadically found throughout the Philippines in limited quantity.

Plate 48 Fasciolariidae

1. Pseudolathyrus c.f. kurodai Okutani and Sakurai, 1964
63.4mm; Bohol.
Shell medium sized, fusiform-elongate, with a tall, sharply acute spire, a long, slender anterior canal, and deeply channeled sutures; whorls sculptured with 7 rounded, widely spaced axial ribs which are crossed by numerous primary and secondary spiral lirae; aperture narrowly elongate, outer lip thin, variced and lirate within, columella moderately glazed, smooth except for a few basal plicae; colour pale creamy orange, spiral lirae off-white, aperture porcelaneous white; reaches 80mm. Typical *kurodai* has the main spiral ridges brown and the canal bent.
DISTRIBUTION: Cebu and Bohol.

2. Fusinus (Simplicifusus) hyphalus (M. Smith, 1940)
49.4mm; Punta Engaño, Cebu.
Shell medium sized, thin, lightweight, fusiform-elongate in outline, with a tall spire and a moderately long anterior canal; spire whorls sculptured with strong axial folds and spiral lirae, the axial folds becoming obsolete on the body whorl; aperture narrowly elongate, outer lip thin, lirate within, columella smooth and weakly calloused; shell colour pale fleshy white with a porcelaneous white aperture; reaches 60mm.
DISTRIBUTION: Cebu and Bohol.

3. Granulifusus nipponicus (Smith, 1879)
46.7mm; Lubang Island, Mindoro.
Shell medium sized, fusiform-elongate in outline, with moderately channeled sutures and strongly convex whorls; sculptured with arcuate axial folds (approximately 10 on the body whorl) which are crossed by major and minor spiral cords, the major cords forming large, elongate nodules on the folds; aperture elongate-ovate, lirate within, columella smooth; colour light brown, the axial cords

accentuated with dark brown, nodules and aperture white; reaches around 70mm.
DISTRIBUTION: Cebu, Bohol and Lubang Island, Mindoro.

4. Granulifusus kiranus Shuto, 1958
45.7mm; Punta Engaño, Cebu.
Very close to the previous species, but somewhat lighter in weight; the whorls are slightly more ventricose, there are more axial folds per whorl (11 to 12), and the nodules on those folds are much weaker in strength; the shell is pale brownish white with orange-yellow banding, the nodules lacking the white colouration of *G. nipponicus;* reaches 50mm.
DISTRIBUTION: Cebu and Bohol.

5. Latirus (Latirulus) craticulatus (Linné, 1758)
47.8mm; Bohol.
Shell medium sized, elongate, rather solid, with a tall spire and weakly channeled sutures which are obscured by a small lamellate cord; sculptured with weak axial folds that are crossed by strong spiral ridges and weaker intermediate threads; aperture ovate, outer lip thin, marginally denticulate and lirate within, columella anteriorly plicate, anterior canal short, open and bordered by a fasciolar ridge; colour pale brown, axial folds reddish orange, an off-white band on mid-body whorl, aperture and columella pale orange reaches 50mm.
DISTRIBUTION: Sporadically found throughout the Philippines.

6. Fusolatirus nana (Reeve, 1847)
31.5mm; Samar.
Shell small, fusiform-ovate, with a tall spire and a long, open, slightly recurved anterior canal; whorls strongly convex, with a somewhat angulate shoulder; sculptured with 8 to 9 rounded, well-elevated axial folds which are crossed by numerous spiral cords; sutures obscured by a finely scabrous cord; aperture ovate, lirate within, outer lip thin, columella smooth; surface colouration orange-brown with paler banding; reaches 40mm, and is extremely variable throughout its range.
DISTRIBUTION: Sporadically found throughout the Philippines.

7. Latirus (Latirus) pictus (Reeve, 1847)
45.0mm; Palawan.
Very close to *L. belcheri* (Reeve) but is smaller, distinctly more ovate and the canal is short and oblique; the colour pattern is different, being bluish-white in ground colour and encircled with brown lines; irregular orange blotching often occurs, and the aperture is bluish white; reaches 50mm.
DISTRIBUTION: Sporadically found throughout the Philippines.

8. Latirus (Latirulus) turritus (Gmelin, 1791)
38.5mm; Bohol.
Shell medium sized, solid, elongate, with adpressed sutures and weakly convex whorls; sculptured with weak axial folds that are crossed by thick undulate spiral cords, 4 on the penultimate whorl and 10 on the body whorl; aperture ovate, outer lip thin, lirate within and constricted anteriorly; columella smooth and calloused with 3 basal plicae, anterior canal extremely short; reddish orange in colour with deep reddish brown lirae, aperture deep orange; reaches 50mm.
DISTRIBUTION: Sporadically found throughout the Philippines.

9. Latirus (Latirus) formosior Melvill, 1891
43.3mm; Palawan.
Shell medium sized, solid, fusiform-elongate with a tall spire and undulate, well impressed sutures; sculptured with rounded axial folds which terminate just before the sutures and encircled with irregular spiral cords; aperture ovate, outer lip thin and lirate within, columella calloused with a few basal plicae; anterior canal moderately long, open, displaced to the left and distally recurved; colour pale cream with a pale brown band on the body whorl; very close to *Latirus pilsbryi* (Kuroda & Habe, in Habe, 1964), but that species has narrower axial folds and a straight siphonal canal; illustrated specimen the only one seen.
DISTRIBUTION: Sporadically found throughout the Philippines in limited quantity.

10. Latirus (Latirus) kandai Kuroda, 1950
39.7mm; Samar.
Shell medium sized, solid, fusiform in outline, with a tall, acute spire, and the sutures obscured by a band of small white nodules; whorls strongly convex, sculptured with 7 to 8 rounded axial folds and crossed by white spiral lirae, strongest on the folds, almost obsolete in the interspaces; aperture ovate, bluish white within, outer lip moderately thick, lirate within, columella calloused, anteriorly detached, and with some obsolete basal folds; anterior canal long, open, slightly curved and bordered by a strong fasciolar ridge; colour orange-brown, darkest on the folds; reaches 40mm.
DISTRIBUTION: Sporadically found throughout the Philippines in limited quantity.

11. Pseudolathyrus sp.
39.3mm; Punta Engaño, Cebu.
Shell small to medium sized, fusiform-ovate in outline,

PLATE NO. 48. 1. *Pseudolathyrus c.f. kurodai* Okutani and Sakurai 2. *Fusinus hyphalus* (M. Smith) 3. *Granulifusus nipponicus* (Smith) 4. *Granulifusus kiranus* (Shuto) 5. *Latirus craticulatus* (Linné) 6. *Fusolatirus nana* (Reeve) 7. *Latirus pictus* (Reeve) 8. *Latirus turritus* (Gmelin) 9. *Latirus formosior* (Melvill) 10. *Latirus kandai* Kuroda 11. *Pseudolathyrus sp.* 12. *Dolicholatirus lancea* (Gmelin) 13. *Latirolagena smaragdula* (Linné) 14. *Peristernia sp.* 15. *Peristernia nassatula* (Lamarck) 16. *Peristernia incarnata* (Kiener) 17. *Peristernia ustulata* (Reeve) 18. *Peristernia lirata* (Pease) 19. *Peristernia hesterae* Melvill 20. *Latirus noumeensis* (Crosse) 21. *Peristernia philberti* (Récluz) 22. *Latirus maculatus* (Reeve)

PLATE 48

with a tall spire, well impressed sutures, and strongly convex whorls; sculptured with 6 to 7 rounded, well-elevated, widely spaced axial folds, and crossed by strong spiral cords forming elongate nodules; aperture ovate, lirate within, columella calloused and smooth; shell pure white; largest size seen 45.7mm.
DISTRIBUTION: Cebu and Bohol.

12. Dolicholatirus lancea (Gmelin, 1791)
45.4mm; Samar.
Shell medium sized, fusiform-elongate, with a tall spire which is nearly always badly eroded, and a long, narrow siphonal canal; sutures adpressed, and the whorls which are strongly convex are sculptured with slender, widely spaced axial ribs and encircled by spiral cords of varying thickness; aperture small, ovate, lirate within, columella with 3 anterior plicae; colour brownish grey with darker interspaces and a brownish purple aperture; reaches 55mm.
DISTRIBUTION: Sporadically found throughout the Philippines in limited quantity.

13. Latirolagena smaragdula (Linné, 1758)
45.4mm; Samar.
Shell medium sized, fusiform-ovate, with a large, somewhat ventricose body whorl, sutures weakly channeled, whorls sculptured with numerous, nearly obsolete spiral cords and interstitial "pit-like" areas; aperture elongate-ovate, weakly denticulate and lirate within, columella heavily calloused and anteriorly plicate; pale bluish white with dark brown lirae, a black fasciolar cord, a bluish white aperture, and brown outer lip denticles; reaches 60mm.
DISTRIBUTION: Sporadically found throughout the Philippines.

14. Peristernia sp.
36.5mm, 35.4mm; Palawan.
Shell medium sized, fusiform-ovate, with a tall, conoidal spire and an unusually long, narrowly open, distally recurved siphonal canal which is displaced to the left; sutures obscured by a weakly nodulose cord, whorls convex, sculptured with large, rounded axial folds and overriding spiral lirae which become coarser on the siphonal canal; aperture ovate, outer lip thick, marginally serrate, internally lirate, columella calloused and irregularly plicate; white with brown or orange banding; reaches 40mm.
DISTRIBUTION: Sporadically found throughout the Philippines in limited quantity.

15. Peristernia nassatula (Lamarck, 1822)
34.8mm; Palawan.
Shell medium sized, solid, fusiform-ovate in outline, with adpressed sutures; sculptured with strong axial ribs, nodulose at the shoulders and crossed by strong spiral cords and weaker spiral striae; aperture elongate-ovate, outer lip angulate, marginally denticulate and internally lirate, columella calloused with 4 anterior plicae, anterior canal short; off-white with brown banding on the body whorl and between the ribs, aperture violet; reaches 45mm.
DISTRIBUTION: Sporadically found throughout the Philippines in limited quantity.

16. Peristernia incarnata (Kiener, 1840)
27.4mm; Punta Engaño, Cebu.
Shell medium sized, solid, fusiform-ovate, with a tall, acute spire, weakly convex whorls and a presutural ledge; sculptured with broad, rounded axial ribs which are crossed by numerous primary and secondary spiral lirae; aperture elongate-ovate, lirate within, columella calloused with a posterior parietal denticle and some anterior plicae; axial ribs deep orange-brown, interstitial spaces dark brown, aperture mauve; reaches 35mm. This name is pre-occupied by *Peristernia incarnata* (Deshayes, in Laborde & Linant, 1834) and no substitute name appears available.
DISTRIBUTION: Sporadically found throughout the Philippines.

17. Peristernia ustulata (Reeve, 1847)
19.7mm; Palawan.
Shell small, fusiform-ovate in outline, with adpressed sutures that are partially obscured by a weakly granulose cord; whorls convex, sculptured with 8 to 9 rounded, well spaced axial ribs, somewhat nodulose from overriding spiral cords; aperture elongate-ovate, outer lip thin, marginally dentate, internally lirate and constricted basally, columella calloused and weakly plicate, canal short, open and recurved; colour pale brown with a pinkish mauve hue and an occasional central band of spots, aperture pale pinkish orange, canal purple; reaches 25mm.
DISTRIBUTION: Sporadically found throughout the Philippines.

18. Peristernia lirata (Pease, 1868)
15.9mm; Punta Engaño, Cebu.
Shell small, fusiform-ovate, with a tall, acute spire, a large rounded body whorl, and adpressed sutures; sculptured with weak axial folds that are crossed by numerous scabrous spiral lirae, forming short spinose processes on the weakly angulate shoulder; aperture ovate, outer lip marginally serrate, internally lirate, columella with 2 to 3 basal plicae; colour brownish yellow with a porcelaneous white aperture and a brown stained anterior canal; reaches 20mm. Often figured as *P. squamosa* (Pease, 1863), but the type figure shows conspicuous shoulder spines, not seen in this species.
DISTRIBUTION: Camotes Sea.

19. Peristernia hesterae Melvill, 1911
22.5mm; Marinduque.
Shell small, fusiform-ovate, with weakly channeled sutures and sculptured with large rounded axial cords that are crossed by numerous fine spiral lirae; aperture ovate, outer lip variced, lirate within and anteriorly constricted, columella basally plicate, siphonal canal short, open and distally recurved; pale yellow in colour, with dark brown between the ribs and on the siphonal fasciole, aperture off-white, apex purple; reaches 25mm.

DISTRIBUTION: Sporadically found throughout the Philippines in limited quantity.

20. Latirus (Latirulus) noumeensis (Crosse, 1870)
11.3mm; Punta Engaño, Cebu.
Shell quite small, elongate, biconic, with weakly channeled sutures and moderately convex whorls which are sculptured with rounded axial nodes and encircled with numerous strong, undulate spiral lirae and microscopic axial threads in the interspaces; aperture ovate, anteriorly constricted and internally lirate, columella calloused, anteriorly plicate; deep purple in colour, with orange cords, white interspaces, and a brown columella; can reach 25mm, but Philippine specimens seldom exceed 15mm.
DISTRIBUTION: Camotes Sea.

21. Peristernia philberti (Récluz, 1844)
41.2mm; Jolo, Sulu Archipelago.
Shell medium sized, solid, fusiform-elongate in outline, with weakly channeled sutures, a tall spire, and sculptured with axial folds and spiral cords forming nodules at the sharply angled shoulder; aperture elongate-ovate, lirate within, columella smooth; colour reddish brown with black and white banding, aperture violet; reaches 50mm.
DISTRIBUTION: Sporadically found throughout the Philippines in limited quantity.

22. Latirus (Latirulus) maculatus (Reeve, 1847)
24.0mm; Punta Engaño, Cebu.
A small, solid, fusiform-elongate shell, with weakly channeled sutures bordered by a crenulate spiral cord, whorls bi-angulate and sculptured with 9 axial nodes per whorl which are crossed by strong spiral lirae, strongest on the nodes; aperture ovate, lirate within, columella smooth, anterior canal short, open, curved, and bordered by a strongly corded siphonal fasciole; colour pale yellow, streaked with brown; reaches 30mm.
DISTRIBUTION: Sporadically found throughout the Philippines in limited quantity.

Plate 49 Volutidae

Class Gastropoda
Subclass Prosobranchia
Order Neogastropoda
Suborder Rachiglossa
Superfamily Volutacea
Family VOLUTIDAE

A fairly large group of medium to large-sized carnivorous gastropods living in sandy areas and inhabiting shallow to deep water in most temperate and tropical seas. The shells are usually quite colourful and characterized by their cylindrical to fusiform-ovate shape, the wide elongate aperture, the heavily calloused columella, and a sculpturing of shoulder spines, nodulose spiral threads and/or axial folds. Relatively few Philippine species.

1. Melo (Melocorona) broderipii (Gray, in Griffith & Pidgeon 1834)
190mm; Tayabas Bay, Quezon.
A large, heavy, globose shell, with a low spire and a smooth, mammillate protoconch; body whorl large, with an elongate-ovate aperture, a thin, arcuate outer lip, a wide, shallow siphonal notch, and a heavy fasciolar ridge; sculptured with a shoulder row of numerous short, open, slightly curved spines, columella with 4 plaits; colour pale orange-yellow with lighter banding and a bright orange aperture; possibly a subspecies of *M. aethiopica* which has fewer spines, and a protoconch which is sunken below the level of the spines; reaches 350mm. Locally sold as "*M. aethiopica.*" Juvenile specimens have brown streaks.
DISTRIBUTION: Sporadically found throughout the Philippines.

2. Melo (Melo) melo (Lightfoot, 1786)
160mm; Palawan.
Shell large, ovate, the spire completely obliterated by the posterior outer lip margin; surface smooth with axial growth striae, columella with 3 to 4 long, well indented plaits, aperture large, widely ovate, outer lip thin and arcuate; colour pale orange with or without brown spotted banding, aperture light orange yellow with a pale orange margin; reaches 275mm.
DISTRIBUTION: Sporadically found throughout the Philippines.

3. c. f. Melo (Melocorona) georginae (Gray, in Griffith & Pidgeon, 1834)
170mm; Tayabas Bay, Quezon.
Shell similar in outline to *M. melo*, but the protoconch is sunken below a row of short, sharp, partially open spines on the body whorl shoulder which become more crowded towards the posterior edge of the outer lip margin; the aperture is somewhat wider and not as arcuate as in *melo*, columella with 3 to 4 plaits, and the colour is pale orange with or without brown blotches or streaks on the body whorl; the few seen seldom exceed 200mm.
DISTRIBUTION: Sporadically found throughout the Philippines in limited quantity.

4. Cymbiola (Aulica) nobilis (Lightfoot, 1786)
146mm; Southern Palawan.
Shell large, heavy, rectangularly ovate in outline, with a low, bluntly rounded protoconch; body whorl large, with a strong shoulder which is reflected into a wing-like structure on the posterior outer lip margin; aperture wide, somewhat rectangularly elongate, the outer lip is thin but solid, fairly straight then curved basally, columella with 4 plaits, and the parietal wall

is heavily calloused with a white glaze which frequently extends to the protoconch; colour cream or beige with brownish axial zigzag lines and blotches, frequently forming 2 indistinct spiral bands, aperture and columellar area bluish white, outer lip margin pinkish grey; reaches 190mm.
DISTRIBUTION: Sulu Sea in very limited quantity.

5. Cymbiola (Aulica) aulica (Sowerby I, 1825) form **cathcartiae** Reeve, 1856
A distinct form, both in terms of colour and morphology, which has often been considered a distinct species, but the lack of any anatomical differences from the normal *aulica*, and its sympatric occurrence with this species, show it to be a form only; shells are generally large with prominent, blunted tubercles on the shoulders, and a peculiar yellowish brown ground colour superimposed with brown to black blotches.
5a. typical form; 109.5mm; Sacol Is., Zamboanga.
5b. smooth form; 121mm; Basilan Is., Zamboanga.
5c. large, ponderous form with heavily developed tubercles; 131.4mm; Zamboanga.
DISTRIBUTION: Southern Zamboanga.

6. Lyria (Lyria) santoensis santoensis Ladd, 1975
Syn: *Lyria taiwanica* Lan, 1975 form *davaoensis* Kosuge, 1980
74.5mm; Davao.
Shell medium sized, broadly fusiform in outline, with a tall spire and weakly indented sutures; sculptured with 2 smooth protoconch whorls and 8 to 9 weakly convex whorls; the body whorl has between 8 and 12 narrow axial ribs, smooth interspaces, and some weak basal plicae; aperture narrowly elongate-ovate, outer lip marginally thickened, columella with around 10 folds, the 3 or 4 posterior ones stronger than the rest; external colour pale yellowish orange, with numerous close-set black spiral lines on and between the axial cords and frequently forming bands, aperture flesh-colored; reaches 90mm.
DISTRIBUTION: At present only known from Davao.

6a. Lyria (Lyria) santoensis taiwanica Lan, 1975
Syn: *Lyria taiwanica* Lan, 1975 form *boholensis* Kosuge, 1980; *Lyria kawamurai* Habe, 1975
70.6mm; Panglao, Bohol.
This subspecies extends from Bohol to Taiwan; it is somewhat lighter in weight than the nominate subspecies, has a paler ground colour, and the encircling black lines are sparser and restricted to the crests of the axial ribs; the columellar folds are weaker in construction and may number up to 13; size to 80mm.
DISTRIBUTION: So far only known from Bohol.

7. Lyria (Lyria) vicdani Kosuge, 1981 (October)
Syn: *Lyria (Lyria) reinai* Angioy & Biraghi, 1981 (November)
57.4mm, 59.0mm; Panglao, Bohol.
Shell medium sized, elongate, with a tall spire and a large body whorl, sutures impressed and undulate; whorls sculptured with numerous rounded axial folds and smooth interspaces, a few spiral grooves on the body whorl base; aperture narrowly elongate, somewhat truncated basally, outer lip thick, smooth within, columella moderately calloused with 3 prominent plaits and numerous weaker plicae above; colour yellowish orange, encircled with numerous fine brown lines, a discontinuous subsutural band of brown quadrate spots on the spire whorls and 3 on the body whorl aperture orange, plaits off-white; reaches around 60mm.
DISTRIBUTION: Bohol.

8. Lyria (Lyria) habei Okutani, 1979
Syn: *Lyria (Lyria) dondani* Angioy & Biraghi, 1982
66.0mm; Mactan, Cebu.
Shell medium sized, unusually light in weight, fusiform-ovate in outline, with 8 to 9 strongly convex spire whorls, the first 3 or 4 with numerous nodulose axial folds, becoming smooth towards the body whorl; aperture elongate-ovate, widening basally, outer lip smooth, thin and arcuate, columella weakly glazed with 3 strong plaits and several weaker plicae above; colour dark purplish brown with numerous brown spiral lines, nebulous axial streaks, a row of brownish subsutural spots on the spire whorls and 2 to 3 on the body whorl; reaches 70mm.
DISTRIBUTION: Cebu.

9. Cymbiola (Aulica) imperialis (Lightfoot, 1786)
Shell large, very heavy, cylindrically ovate in outline, with a tall spire, a smooth and mammillate apex, a weakly convex body whorl, and a row of strong, curved, open spines on the shoulder; aperture large, elongate, outer lip thick, smooth, almost straight, columella with 4 plaits, siphonal notch deep, fasciolar ridge strong; colour beige with brown axial zigzag lines and large triangular blotches, frequently forming bands, aperture and columella yellowish orange; reaches 250mm.
9a. form *robinsona* J. Q. Burch, 1954: a form with thin, non-coalescing, axial zigzag lines; 155mm; Sulu Archipelago.
9b. typical form; 148mm; Sulu Archipelago.
DISTRIBUTION: Sulu Sea.

10. Cymbiola (Aulicina) vespertilio (Linné, 1758)
Shell extremely polymorphic and polychromic; typical shell is medium sized, elongate-ovate in outline, with

PLATE NO. 49 1. *Melo broderipii* (Gray, in Griffith & Pidgeon) 2. *Melo melo* (Lightfoot) 3. *c.f. Melo georginae* (Gray, in Griffith & Pidgeon) 4. *Cymbiola nobilis* (Lightfoot) 5a. *Cymbiola aulica* (Sowerby I) form *cathcartiae* Reeve 5b. smooth form 5c. large, heavy tuberculate form 6. *Lyria santoensis santoensis* Ladd 6a. *s.s. taiwanica* Lan 7. *Lyria vicdani* Kosuge 8. *Lyria habei* Okutani 9. *Cymbiola imperialis* (Lightfoot) form *robinsona* J.Q. Burch 9b. typical form 10a. *Cymbiola vespertilio* (Linné) 10b. form *mitis* Lamarck 10c. orange form 10d. to 10e. pattern variations 10f. form *serpentina* Lamarck 10g. tuberculate form 10h. albinistic form 10i. melanistic form 10j. pattern variant 10k. sinistral form 11a. to 11g. *Cymbiola aulica* (Sowerby I) various morphological and pattern variations.

PLATE 49

1
2
3
4
5a
5b
5c
6
6a
7
8
9a
9b
10a
10b
10c
10d
10e
10f
10g
10h
10i
10j
10k
11a
11b
11c
11d
11e
11f
11g

adpressed sutures, a nodulose, mammillate protoconch and a row of spines just above the sutures of the earlier whorls and on the shoulder of the body whorl; aperture elongate, outer lip thick, simple and arcuate, columella with 4 oblique plaits, siphonal notch deep, fasciolar ridge moderately strong; base colour pale cream, with darker overlays of zigzag lines, blotches or streaks; very variable; reaches 115mm.
10a. typical form; 104.5mm; Sulu Archipelago.
10b. form *mitis* Lamarck, 1844: a cylindrical form with obsolete tubercles; 106.6mm; Samar.
10c. orange form; 97.8mm; Samar.
10d.·10e. pattern variations; 88.3mm, 93.5mm; Palawan.
10f. form *serpentina* Lamarck, 1844: an elongate form with weakly developed tubercles and a pale pink ground colour with dark-coloured, axial zigzag lines; 81.4mm; Sulu Archipelago.
10g. a heavily tuberculate form; 86.2mm; Palawan.
10h. albinistic form; 77.5mm; Albay Gulf.
10i. melanistic form; 87.1mm; Batangas, Southern Luzon.
10j. pattern variation; 68.5mm; Palawan.
10k. sinistral forms: brown coloured, 62mm, Sulu Sea; orange coloured, 53.5mm, Samar.
DISTRIBUTION: Throughout the Philippines.

11. Cymbiola (Aulica) aulica (Sowerby I, 1825)
A somewhat variable species, and apparently restricted in its range to a few discrete localities. The shell is medium sized, heavy, with a smooth rounded protoconch and 2½ to 3 teleoconch whorls which may be smooth or spinose; the spines when present, are above the adpressed sutures and on the shoulder of the body whorl; aperture elongate-ovate, outer lip thick, somewhat indented posteriorly, columella curved with 4 oblique plaits, a well defined fasciole and a deep fasciolar notch; colour usually pale orange with dark red blotches, frequently in bands; reaches 160mm.
Endemic to the Philippines.
11a. typical form; 145mm.
11b. pale pink form with darker banding; 107mm.
11c. a small, cylindrically ovate form, devoid of spines and cream in ground colour with spiral bands of red spots; 91mm, 57mm.
11d. colour form; 102mm.
11e. a heavy, dark orange form; 117mm.
11f. an almost pure red form, sharply spinose and with some pale spotting; 155mm.
11g. an extremely beautiful form, heavy, markedly inflated, with large tubercles on the body whorl; deep orange-brown in colour with white triangular blotches arranged in bands on the body whorl; 163mm.
DISTRIBUTION: All the above specimens are from the Sulu Archipelago and taken in intermediate depths, except the form 11c. which is from shallow water at Palawan, and represents a range extention for the species.

ADDITIONAL NOTES ON THE VOLUTIDAE

Lyria (Lyria) santoensis Ladd, 1975 (October) was originally described from fossiliferous marls of the Kere River on Santo Island, New Hebrides Island Group (Veliger 18:2). A short time later the species was described from Taiwanese waters under the names *Lyria (Lyria) taiwanica* Lan, 1975 (November) and *Lyria (Lyria) kawamurai* Habe, 1975 (December). The species was subsequently found living in Philippine waters, and in 1980, Kosuge described two forms to which he appended the names form *boholensis* (from Bohol) and form *davaoensis* (from Davao). Two morphologically different forms do indeed occur at these localities, but the use of 2 different names to describe them is ill-advised for 2 reasons: (1) Bohol and Davao are geographically very close in comparison with the total distribution range over which the species is known to occur. By appending 2 names to the forms found at these 2 areas, no consideration is given to genetic variation occurring within the taxon as a whole. (2) The use of 2 different form names fails to consider the morphology of the typical form.

Based on Ladd's original description and type figures, there is little doubt that the form from Davao (Kosuge's form *davaoensis*) is morphologically the same taxon described by Ladd. Further, the form from Bohol (Kosuge's form *boholensis*) is almost identical to the specimens described from Taiwan (i.e. *taiwanica* and *kawamurai*). Cernohorsky (1980) discussed the differences between these species and contended that the morphological variations separating them were well within the limtis expected in *Lyria spp.* There is no doubt that *santoensis, taiwanica* and *kawamurai* as well as Kosuge's *boholensis* and *davaoensis* are all conspecific, but certain evidence strongly suggests the existence of 2 races.

Kosuge's 2 forms, although morphologically very close, can be consistently separated on the basis of certain pattern and structural features which remain essentially constant in all the specimens thus far examined. Further, the 2 are not forms of a single taxon, because neither is known to co-exist in any sympatric community with the other. Presently available locality data confirms their geographic separation, though theoretically, their ranges may touch between Davao and Bohol, where possible interbreeding would occur. To date, we have seen no material from that area.

Based on this evidence, it is proposed that *santoensis* occurs in 2 races, *Lyria (Lyria) santoensis santoensis* Ladd, 1975, the nominate subspecies, with a range extending from the New Hebrides (the type locality) to Davao (Southern Philippines), and a second subspecies *Lyria (Lyria) santoensis taiwanica* Lan, 1975, with a range extending from Bohol (Southern Philippines) to Northern Borneo in the South and Taiwan in North West. The name *taiwanica* is here used as chronologically, it is the first available name applicable to this second subspecies. An earlier name for *santoensis* may be *planicostata* Sowerby, 1903

PLATE NO. 50 1a. *Oliva miniacea* Röding 1b. form *saturata* Dautzenberg 1c. form *marrati* Johnson 1d. form *johnsoni* Higgins 1e. form *sylvia* Duclos 2. *Oliva hirasei* Kira 3a. *Oliva tremulina Lamarck form oldi* Zeigler 3b. typical form 4a. *Oliva sericea* Röding-form *granitella* Lamarck 4b. typical form 4c. typical form, with pale ground colour 5a. *Oliva vidua* Röding 5b. form *aurata* Röding 5c. form *fulminans* Lamarck 5d. form *albofasciata* Dautzenberg 5e. form *sepulturalis* Lamarck

PLATE 50

Class Gastropoda
Subclass Prosobranchia
Order Neogastropoda
Suborder Rachiglossa
Superfamily Volutacea
Family OLIVIDAE

A large group of carnivorous, sand-dwelling molluscs inhabiting the shallow intertidal areas of most temperate and tropical seas. The shells are small to medium sized, cylindrically elongate, with a short spire, a large body whorl and deeply channeled sutures. The shells have a glossy, porcelaneous surface, are highly coloured, have a highly calloused, strongly lirate columella, and lack a periostracum. The family is well represented in the Philippines with around 75% of the species coming mainly from the Sulu Sea.

1. Oliva (Oliva) miniacea Röding, 1798
Shell large, cylindrically elongate, spire short with a slightly elevated callus and deeply canaliculate sutures; outer lip thick, columellar plaits calloused posteriorly; colour orange-brown with bluish reticulations and 2 or 3 dark transverse bands on the body whorl, aperture and columella deep orange; reaches 90mm.
1a. typical *miniacea*; 75.5mm, 72.7mm; Sulu Archipelago.
1b. form *saturata* Dautzenberg, 1927: similar to the basic form, but with denser lines and darker banding; 75.5mm; Sulu Archipelago.
1c. form *marrati* Johnson, 1871: an entirely brown or black form; 80.0mm, 60.2mm (grooved); Sulu Archipelago.
1d. form *johnsoni* Higgins, 1919: large white markings on a darker background; 76.5mm; Sulu Archipelago.
1e. form *sylvia* Duclos, 1844: deep orange colouration with some faint banding; 76.0mm, 71.5mm, 67.5mm; Sulu Archipelago.
DISTRIBUTION: Sporadically found throughout the Philippines, mostly from Sulu Sea.

2. Oliva (Oliva) hirasei Kira, 1959
47.7mm, 46.5mm, 40.0mm, 40.7mm, 37.9mm; Marinduque.
Shell medium sized, cylindrically elongate, spire short with a callosity, and the body whorl suture deeply canaliculate, outer lip thin, columellar plates fine and numerous; orange-brown in colour with darker axial zigzags forming flammules on top of the body whorl, and a second band centrally, area below the suture devoid of pattern; aperture marginally brown, blue-white within, spire whorls and callus with a violet tinge; reaches 45mm. The name is generally attributable to Kuroda & Habe (1952: *Check List and Bibliography of the Recent Marine Mollusca of Japan)*, based on a figure by Hirase (1909: *The Conchological Magazine*). Unfortunately, there was no type description appended to the figure citation, and the name was not validated until 1959, when Kira provided a brief, but nevertheless adequate description to differentiate this new taxon *(Coloured Illustrations of the Shells of Japan)*.
DISTRIBUTION: Mindoro, Marinduque, Batangas and Palawan.

3. Oliva (Oliva) tremulina Lamarck, 1811
Shell moderately large, cylindrically ovate, tall spired with a sharp apex, shallow canaliculate sutures, and a moderately high callosity; outer lip thick, columellar plaits weak and heavily calloused posteriorly but becoming stronger anteriorly; colour creamy yellow-white with darker blue mottling and dark brown, discontinuous spiral banding, aperture white, columella cream, red towards the base; reaches 90mm; almost identical to *O. miniacea* but with a white aperture (occasionally orange).
3a. form **oldi** Zeigler, 1969: a greyish form, with compact axial reticulation, and 2 darker, discontinuous spiral bands; the columella has a red basal blotch and a grey-brown fasciole; 64mm; Sulu Archipelago.
3b. typical form; 49.5mm; Sulu Archipelago.
DISTRIBUTION: Sporadically found throughout the Philippines.

4. Oliva (Oliva) sericea (Röding, 1789)
Shell large, heavy, cylindrically elongate, spire low with a prominently elevated callus, sutures canaliculate but close; outer lip thick, columellar plaits coarse and numerous; creamy salmon with a dense reticulate network and 2 darker bands, aperture and columella pale orange; reaches 100mm.
4a. form *granitella* Lamarck, 1811: form lacking dark banding; 55mm; Sulu Archipelago.
4b. typical *sericea* 85.5mm; Sulu Archipelago.
4c. typical *sericea* but with a paler ground colour; 101.4mm; Sulu Archipelago.
DISTRIBUTION: Sporadically found throughout the Philippines.

5. Oliva (Carmione) vidua Röding, 1798
Shell medium sized, solid, cylindrically elongate-ovate, spire depressed with a callosity; columellar plaits weak posteriorly, strong anteriorly and an oblique spiral groove crossing the base; shell pure black, aperture bluish white, columella cream with a reddish orange basal blotch; reaches 60mm.
5a. typical *vidua*; 54.2mm; Sulu Archipelago.
5b. form *aurata* Röding, 1798: a golden-orange colour form; 53.5mm; Sulu Archipelago.
5c. form *fulminans* Lamarck, 1811: a form with black axial zigzag markings; 51.9mm; Samar.
5d. form *albofasciata* Dautzenberg, 1927: the background is greenish grey, crossed posteriorly and centrally by 2 discontinuous black bands; the spiral groove of normal *vidua* is replaced by a wide, flat cord; 49.8mm; Batangas.
5e. form *sepulturalis* Lamarck, 1811: a greenish grey colour form with 3 spiral bands of black axial zigzags; 49.0mm; Sulu Archipelago.
DISTRIBUTION: Throughout the Philippines, but mainly from Sulu Sea.

1. Oliva (Oliva) caerulea Röding, 1798
Syn: *Oliva (Oliva) episcopalis* Lamarck, 1811
Shell medium sized, solid, elongate-ovate with deeply channeled sutures, a tall spire, strong columellar plaits and a thick outer lip; aperture deep purple, columella white, pale orange at the base, external colour pale cream, with black axial zigzag lines bordered with pale blue mottling; reaches 60mm.
1a. typical form; 41.9mm; Palawan.
1b. form *lugubris* Lamarck, 1811: a form with denser axial zigzags; 44.7mm; Palawan.
1c. form *emeliodina* Duclos, 1844: a form with a shorter spire; 39.5mm; Sulu Archipelago.
Although generally known as *O. episcopalis* there seems to be no reason not to utilize the earlier name. It is not a *nomen oblitum*, as evidenced by its use over the years and by its frequent application by several recent authors. Although Röding's figure citations refer to several species, Martini's figure 518 ("Das Kamelotchen") (Martini, F. H. W. 1773. *Neues systematisches Conchylien — Cabinet 2*. Nurnberg: G. N. Raspe. pp. 1-362, pls. 32-65), although badly drawn, clearly illustrates the species subsequently described as *O. episcopalis*, and has been designated as type figure by Kilburn (*Ann. Natal Mus.*, Vol. 24(1), pp. 221-231: 1980).
DISTRIBUTION: Throughout the Philippines.

2. Oliva (Galeola) bulbiformis Duclos, 1835
Shell small to medium sized, distinctly ovate, spire short and entirely calloused with a raised callosity, outer lip thick, columellar teeth fine, partially calloused and strengthening basally; yellow-grey in colour, with a darker reticulate network and 2 bands, aperture blue-white, columella with a red basal blotch; reaches 40mm.
2a.-2d. various colour forms; 38.3mm, 30.0mm, 33.0mm, 31.2mm; Marinduque.
DISTRIBUTION: Sporadically found throughout the Philippines.

3. Oliva (Carmione) reticulata Röding, 1798
Shell medium sized, cylindrically ovate, spire low with a callosity on the posterior end, outer lip thick, columellar plaits few in number, slightly calloused posteriorly and becoming coarser anteriorly; base colour orange-yellow, almost completely obscured by dense, closely packed axial lines, sometimes forming bands; aperture blue-white, columella blood red; reaches 50mm.
3a. typical form; 44.5mm, 43.8mm; Sulu Archipelago.
3b. form *azona* Dautzenberg, 1927: close to the typical form but lacks black banding; 45.5mm; Sulu Archipelago.
3c. form *evania* Duclos, 1835: a form with a more open network and darker banding; 45.5mm; Samar.
3d. form *zigzag* Perry, 1811: a pale form with axial zigzag lines; 46.5mm; Samar.
3e. form *pallida* Dautzenberg, 1927: a pale form with sparse pattern; 40.1mm; Sulu Archipelago.
DISTRIBUTION: Throughout the Philippines.

4. Oliva (Carmione) elegans Lamarck, 1811
42.7mm; Sulu Archipelago.
Shell medium sized, solid, cylindrically ovate, sutures canaliculate, spire short with a high callosity at the posterior end of the aperture, outer lip thick, columellar plaits fine posteriorly and coarse anteriorly; greyish yellow in colour with darker axial zigzags suffused with blue and orange, aperture blue-white, columella white with a red basal blotch; reaches 45mm.
DISTRIBUTION: Throughout the Philippines.

5. Oliva (Carmione) funebralis Lamarck, 1811
32.2mm, 34.0mm; Sulu Archipelago.
A problematical species; shell medium sized, elongate to elongate-ovate, spire short and calloused, sutures deeply canaliculate only on the last whorl, aperture with a slightly raised callosity, outer lip thick, columellar plaits coarse; greyish yellow in colour with darker axial zigzags and bands, aperture bluish, columella white with brown basal markings; reaches 45mm; close to *O. elegans* but differs in its bluish aperture, the lack of red columellar blotching, coarser columellar plaits, and a slightly different base colour.
DISTRIBUTION: Throughout the Philippines.

6. Oliva (Carmione) mustellina Lamarck, 1811
35.0mm, 33.5mm; Samar.
Shell medium sized, cylindrically elongate, spire low, apex acute, sutures deeply canaliculate, spire whorls calloused, forming a slightly elevated callus plate at the posterior end of the aperture, outer lip thick, columellar plaits fine posteriorly, becoming coarser anteriorly; yellow-grey in colour with a bluish hue, ornamented with close axial reticulations, aperture pale lavender, columella white; reaches 40mm.
DISTRIBUTION: Sporadically found throughout the Philippines.

7. Oliva (Neocylindrus) lignaria Marrat, 1868
Shell large, solid, cylindrically ovate, spire low and heavily calloused with a prominent callus extension, outer lip thick, columellar plaits weak posteriorly, becoming stronger anteriorly; creamy white in colour with fine axial zigzag lines, and 2 or 3 discontinuous dark brown spiral bands; aperture white, occasionally bluish, columella white with a faint orange tinge; reaches 70mm.
7a. typical form; 53.0mm; Sulu Archipelago.
7b. form *cryptospira* Ford, 1891: a golden orange form; 54.0mm; Sulu Archipelago.
7c. form *albescens* Johnson, 1914: a pure white form; 54.1mm; Sulu Archipelago.
7d. form *fordi* Johnson, 1910: A black or dark brown form; 55.2mm; Sulu Archipelago.
DISTRIBUTION: Mostly from Sulu Sea.

8. Oliva (Carmione) tricolor Lamarck, 1811
Shell medium sized, elongate-ovate, spire short with a basal callus, aperture moderately wide, blue-white within, columella with coarse basal folds and pinkish red in colour; shell colour blue, orange and green,

with a dense reticulate pattern and 2 green bands, spire with oblique, orange-tinged black lines; reaches 60mm.
8a. typical form; 51.5mm; Sulu Archipelago.
8b. form *philantha* Duclos, 1835: a light-coloured form with predominantly yellow spots and lacking spiral banding; 53.4mm; Sulu Archipelago.
DISTRIBUTION: Sporadically found throughout the Philippines.

9. Oliva (Carmione) rufula Duclos, 1835
32.3mm, 34.0mm; Sulu Archipelago.
Shell medium sized, cylindrically elongate, with a slightly convex body whorl, spire depressed with a callosity, columella with strong plaits; off-white in colour with an orange tinge anteriorly, aperture bluish white internally, marginally pale orange, body whorl creamy orange with oblique, brown-coloured spiral bands; reaches 35mm.
DISTRIBUTION: Sporadically found throughout the Philippines.

10. c. f. Oliva (Carmione) bulbosa Roding, 1798
31.0mm, 31.3mm; Balut Is., Southern Mindanao.
Shell medium sized, swollen, moderately ovate, with a heavily thickened outer lip and a wide, deeply excavated fasciolar furrow; spire short and heavily calloused, only the last whorl deeply canaliculate; columellar teeth weak posteriorly, becoming stronger anteriorly; pale cream in colour with broad, black triangular blotches in 2 or 3 spiral bands, columella white, aperture pale lavender; featured as *O. vidua* form *macleaya* Duclos in some books, but the deep fasciolar cord and the lack of the red basal blotch suggest something closer to *bulbosa,* even though not as bulbous as its South African representatives.
DISTRIBUTION: Southern Mindanao.

11. Oliva (Carmione) concavospira Sowerby, 1914
36.2mm, 37.1mm; Marinduque.
Shell medium sized, cylindrically elongate, with a short, calloused, well sunken spire, a high callosity, and a deeply canaliculate body whorl suture; outer lip thick, columellar plaits large and heavy; yellow-grey in colour with darker axial zigzags, aperture blue-white, columella white with a brown basal blotch; reaches 45mm.
DISTRIBUTION: Sporadically found throughout the Philippines.

12. Oliva (Oliva) annulata Gmelin, 1791
Shell medium sized, heavy, elongate with slightly convex sides; spire tall, conical, sutures deeply canaliculate, the inner margin of which is lined with a thick callus; upper columellar teeth fine, those towards the base longer and heavier; sutures lined with black dots, aperture marginally white, reddish orange within, and occasionally there is an oblique line of spots above the basal lirae; several colour and pattern forms; reaches 60mm.
12a. typical form: pinkish grey, lacking basal spots; 36.5mm, 33.7mm; Sulu Archipelago.
12b. form *carnicolor* Dautzenberg, 1927: pink to deep orange; 45.3mm; Sulu Archipelago.
12c. form *alba* Sowerby, 1825: an albinistic form; 33.9mm; Sulu Archipelago.
12d. form *intricata* Dautzenberg, 1927: yellow-pink base colour, with irregularly triangular purple and white spots, a dense reticulate network, an orange sutural callus, and a black-lined basal band; 54.3mm; Samar.
12e. form *amethystina* Röding, 1798: pale pink background with irregular purple spots and a pink sutural callus; 45.9mm; Cebu.
12f. form *nebulosa* Dautzenberg, 1927: small, with a yellowish pink background, and large, triangular, blue-coloured blotches; 31.8mm; Sulu Archipelago.
12g. form *mantichora* Duclos, 1835: pattern close to *intricata* but is smaller with a ridge on the upper body whorl; 34.5mm; Sulu Archipelago.
DISTRIBUTION: Sporadically found throughout the Philippines, form *carnicolor* mostly from Cebu and Sulu Sea.

Olividae Plate 52

1. Oliva (Omogymna) parkinsoni Prior, 1975
18.9mm, 19.3mm; Sulu Archipelago.
Shell small, solid, elongate-ovate and tall spired, sutures closely canaliculate, bordered with an orange callus, body whorl with a slight shoulder carina below the body whorl suture; outer lip thick, columellar plaits sharp and well defined; coloured with orange-brown axial zigzags over a yellowish base colour, aperture lemon, body whorl shoulder with axial stripes; reaches 25mm.
DISTRIBUTION: Southern Mindanao.

2. Oliva (Carmione) athenia Duclos, 1835
24.5mm, 24.6mm, 25.5mm; Cebu.
Shell medium sized, elongate-ovate, spire flat and centrally mucronate, sutures tightly canaliculate, outer lip thick and straight, columellar teeth fine and numerous, extending up the parietal wall; colour yellow-grey with axial zigzags and blotches of a darker colour, aperture marginally white, pale

PLATE NO. 51 1a. *Oliva caerulea* Röding 1b. form *lugubris* Lamarck 1c. form *emeliodina* Duclos 2a. to 2d. *Oliva bulbiformis* Duclos, various colour forms 3a. *Oliva reticulata* Röding 3b. form *azonata* Dautzenberg 3c. form *evania* Duclos 3d. form *zigzag* Perry 3e. form *pallida* Dautzenberg 4. *Oliva elegans* Lamarck 5. *Oliva funebralis* Lamarck 6. *Oliva mustellina* Lamarck 7a. *Oliva lignaria* Marrat 7b. form *cryptospira* Ford 7c. form *albescens* Johnson 7d. form *fordi* Johnson 8a. *Oliva tricolor* Lamarck 8b. form *philantha* Duclos 9. *Oliva rufula* Duclos 10. *c.f. Oliva bulbusa* Röding 11. *Oliva concavospira* Sowerby 12a. *Oliva annulata* Gmelin 12b. form *carnicolor* Dautzenberg 12c. form *alba* Sowerby 12d. form *intricata* Dautzenberg 12e. form *amethystina* Röding 12f. form *nebulosa* Dautzenberg 12g. form *mantichora* Duclos

PLATE 51

1a
1b
1c
2a
2b
2c
2d
3a
3b
3c
3d
3e
4
5
6
7a
7b
7c
7d
8a
8b
9
10
11
12a
12b
12c
12d
12e
12f
12g

lavender within; reaches 30mm; close to *O. sidelia* Duclos.
DISTRIBUTION: Sporadically found throughout the Philippines.

3. Oliva (Strephona) buloui Sowerby, 1887
19.8mm, 20.0mm; Southern Mindanao.
Shell small, cylindrically elongate, spire tall with closely canaliculate sutures, outer lip weakly thickened and straight, columellar plaits fine and numerous; base colour yellowish orange with purplish triangular blotches and thick transverse lines towards the base, aperture and columella bluish white; reaches 30mm.
DISTRIBUTION: Southern Mindanao.

4. Oliva (Carmione) rufofulgurata Schepman, 1911
20.9mm, 19.8mm, 18.2mm; Sulu Archipelago.
Shell small to medium sized, cylindrically elongate with a flattened spire and deeply canaliculate sutures, outer lip thick and straight, columellar teeth fine and numerous, and a thick callus band around the sutures of each whorl; shell pale yellow with fine brown axial zigzags, apical whorls violet, sutural callus band orange, aperture pale violet, columella white; reaches 25mm.
DISTRIBUTION: Sporadically found throughout the Philippines in limited quantity.

5. Oliva (Galeola) sidelia Duclos, 1835
Shell small, cylindrically ovate, spire short and centrally mucronate, suture deeply canaliculate at the body whorl, outer lip thick, columellar dentition the same as in *O. athenia*; colour variable, usually a pale background with axial zigzags and blotches, may be unicoloured.
5a. typical form; 18.5mm, 17.8mm; Samar.
5b. form *todosina* Duclos, 1835: a form with banding; 18.6mm, 21mm; Sulu Archipelago.
5c.-5e. form *volvaroides* Duclos, 1835: a unicoloured form; 19.5mm/19.5mm (white), 18.5mm/17.6mm (brown), 20mm/17.3mm (orange); Sulu Archipelago.
DISTRIBUTION: Sporadically found throughout the Philippines.

6. Oliva (Galeola) carneola Gmelin, 1791
Shell small, elongate-ovate, spire short and heavily calloused, sutures deeply canaliculate, outer lip thick, columellar plaits fine and numerous; shell colour variable, usually white with dense reticulations or red bands; reaches 25mm.
6a. typical form; 17.5mm; Samar.
6b. form *bizonalis* Dautzenberg, 1927: 2 white bands on a darker background; 16.8mm/16.0mm (orange), 17.5mm (lemon), 17.9mm (lavender); Sulu Archipelago.
6c. form *unizonalis* Dautzenberg, 1927: a single white band on a darker background; 17.5mm; Sulu Archipelago.
6d. albinistic form; 17.1mm; Sulu Archipelago.
6e. form *trichroma* Dautzenberg, 1927: pale orange with a central white band and a subsutural lavenderish band; 18.8mm; Sulu Archipelago.
DISTRIBUTION: Sporadically found throughout the Philippines.

7. Oliva (Oliva) oliva Linné, 1758
Shell small, elongate, sides slightly convex, spire low and conical, sutures channeled without thickened margins, columellar teeth fine, becoming coarse anteriorly; typical form dull white with a few irregularly arranged black spots, a white columella and a dark brown aperture; reaches 50mm.
7a. typical form; 32.8mm; Samar.
7b. form *algida* Vanatta, 1915: greyish white with pale longitudinal streaks; 30.2mm; Cebu.
7c. form *oriola* Lamarck, 1811: black or dark brown form; 28.1mm; Batangas.
7d. form *tigridella* Duclos, 1835: orange-grey with thick, black, axial zigzag lines; 22.9mm; Sulu Archipelago.
7e. form *stellata* Duclos, 1835: pale greyish white with dark dots and a short spire; 24.2mm; Marinduque.
7f. form *martini* Dautzenberg, 1927: pale form with a single dark band on the body whorl shoulder; 33.3mm; Sulu Archipelago.
7g. an albinistic form; 6.9mm; Sulu Archipelago. Frequently annotated as form *candida* Lamarck, 1811 but *Ancillaria candida* Lamarck, 1811 is a synonym of *Ancilla ampla ampla* (Gmelin, 1791) (an Indian Ocean endemic).
7h.-7k. various colour forms; 33mm, 25.8mm, 26.7mm, 28.6mm; Sulu Archipelago.
7l. form *flaviola* Duclos, 1835: a pure orange form with a pale aperture; 29.9mm; Sulu Archipelago.
DISTRIBUTION: Throughout the Philippines.

8. Oliva (Omogymna) duclosi Reeve, 1850
18.0mm, 18.8mm; Tayabas Bay, Quezon.
Shell small, elongate-ovate, spire tall with a pale smokey protoconch, sutures tightly canaliculate, outer lip thin, columellar plaits large and well defined; colour orange-yellow with thick olive-brown reticulations forming minute triangular patches and a lavender-and-yellow-spotted band below the sutures, aperture and columella pale blue-white; reaches 30mm.
DISTRIBUTION: Camotes Sea, Sulu Sea and Tayabas Bay, Quezon (deep water).

PLATE NO. 52 1. *Oliva parkinsoni* Prior 2. *Oliva athenia* Duclos 3. *Oliva buloui* Sowerby 4. *Oliva rufofulgurata* Schepman 5a. *Oliva sidelia* Duclos 5b. form *todosina* Duclos 5c. to 5e. *volvaroides* Duclos 6a. *Oliva carneola* Gmelin 6b. form *bizonalis* Dautzenberg 6c. form *unizonalis* Dautzenberg 6d. albinistic form 6e. form *trichroma* Dautzenberg 7a. *Oliva oliva* Linné 7b. form *algida* Vanatta 7c. form *oriola* Lamarck 7d. form *tigridella* Duclos 7e. form *stellata* Duclos 7f. form *martini* Dautzemberg 7g. form *candida* Lamarck 7h. to 7k. various colour forms 7l. form *flaviola* Duclos 8. *Oliva duclosi* Reeve 9. *Oliva dubia* Schepman 10. *Oliva panniculata* Duclos 11. *Oliva tessellata* Lamarck

PLATE 52

1
2
3
4
5a
5b
5c
5d
5e
6a
6b
6c
6d
6e
7a
7b
7c
7d
7e
7f
7g
7h
7i
7j
7k
7l
8
9
10
11

9. Oliva (Parvoliva) dubia Schepman, 1911
13.2mm, 14.2mm; Tayabas Bay, Quezon.
Shell small, smooth, glossy, cylindrically elongate with a short spire, a glassy, rose-coloured protoconch, tightly canaliculate sutures and a smooth columella with plaits only at the basal area; pale yellow in colour with fine brown axial zigzag lines, a row of obsolete spots at the sutures, and a pale lavender aperture; reaches 20mm.
DISTRIBUTION: Sulu Sea, Camotes Sea, Sorsogon and Tayabas Bay, Quezon.

10. Oliva (Omogymna) panniculata Duclos, 1835
20.0mm, 17.2mm; Sulu Archipelago.
Shell small, elongate-ovate, spire tall and acute, sutures weakly canaliculate, outer lip slightly thickened, entire columella with large, well defined plaits and a strong fasciolar groove at the base; off-white in colour with a row of subsutural dots, and covered with dark, closely-set axial zigzags, frequently coalescing to form bands, aperture and columella white; reaches 20mm.
DISTRIBUTION: Sorsogon, Camotes Sea, Sulu Sea and Tayabas Bay, Quezon (deep water).

11. Oliva (Neocylindrus) tessellata Lamarck, 1811
23.2mm, 24.5mm, 17.0mm; Sulu Archipelago.
Shell small to medium sized, cylindrically ovate, spire short and heavily calloused with a deeply canaliculate body whorl suture, outer lip thickened, somewhat thinner and angulate posteriorly, columella calloused, plaits large and well defined; shell pale yellow in colour with a lavender protoconch, a pale yellow spire callus, and purple spots which may form bands on the body whorl, aperture and columella deep purple outer lip margin and columellar base white; reaches 35mm.
DISTRIBUTION: Sporadically found throughout the Philippines.

Mitridae

Plate 53

Class Gastropoda
Subclass Prosobranchia
Order Neogastropoda
Suborder Rachiglossa
Superfamily Volutacea
Family MITRIDAE

A large group of carnivorous gastropods inhabiting silty sand, coral debris or under rocks in most geographical provinces. Most species prefer intertidal water, but a few are found in abyssal depths. The shells are cylindrical to fusiform-ovate, vary in size from small to large and are rarely smooth, the surface usually sculptured with grooves, pits, nodules or granulations. The aperture is narrow, the outer lip may be smooth or crenulate, and the columella bears from 3 to 11 folds, the first being the largest. The Mitridae also possess a distinct siphonal notch and the shell surface is normally quite colorful. The family is extremely well represented in the Philippines, with members exhibiting a diversity of morphology and colour unequaled by any other geographical locality.

1. Mitra (Mitra) papalis (Linné, 1758)
128.5mm; Bohol.
Shell large, solid, cylindrically elongate, with adpressed sutures, moderately strong, non-erect coronations, and sculptured with 5 to 6 columellar folds; white with rows of small red spots; reaches 165mm.
DISTRIBUTION: Sporadically found throughout the Philippines.

2. Mitra (Mitra) mitra (Linné, 1758)
144mm; Bohol
A large fusiform-elongate shell with weakly impressed sutures, smooth except for a few punctate spiral grooves on the earlier whorls; aperture large, equal to the spire height, outer lip thick, anteriorly denticulate, columella with 3 to 4 folds; white in colour with 3 rows of orange dots on the spire whorls, 7 to 8 on the body whorl; reaches 145mm.
DISTRIBUTION: Throughout the Philippines.

3. Mitra (Mitra) stictica (Link, 1807)
69.5mm; Samar.
Shell medium sized, spire whorls slightly stepped, sculptured with close-set sutural coronations and numerous punctate spiral grooves on all the whorls; aperture equal to the spire length, outer lip thin, marginally denticulate, columella with 5 folds; white in colour with rows of rectangular orange spots, aperture pale orange-yellow; reaches 75mm.
DISTRIBUTION: Throughout the Philippines.

4. Mitra (Mitra) cardinalis (Gmelin, 1791)
55.2mm; Samar.
Shell medium sized, elongate-ovate with moderately impressed sutures; sculptured with deep, punctate spiral grooves and fine axial threads; aperture narrow, longer than the spire, and widening rapidly toward

PLATE NO. 53. 1. *Mitra papalis* (Linné) 2. *Mitra mitra* (Linné) 3. *Mitra stictica* (Link) 4. *Mitra cardinalis* (Gmelin) 5. *Mitra incompta* (Lightfoot) 6. *Mitra triplicata* (von Martens) 7. *Mitra puncticulata* Lamarck 8. *Mitra acuminata* Swainson 9. *Mitra pyramis* (Wood) 10. *Mitra coarctata* Reeve 11. *Mitra eremitarum* Röding 12. *Mitra imperialis* Röding 13. *Mitra sacerdotalis* A. Adams 14. *Mitra ferruginea* Lamarck 15. *Mitra rubritincta* Reeve 16. *Mitra contracta* Swainson 17. *Mitra ustulata* Reeve 18. *Mitra fulgurita* Reeve form yaekoae (Habe & Kosuge) 19. *Cancilla isabella* Swainson 20. *Mitra inquinata* Reeve

PLATE 53

the base; outer lip thick with brown-marked denticles, columella with 4 to 5 oblique folds; pale greyish white in colour with rows of dark brown quadrate spots forming blotches on the spire whorls; aperture pale yellow; reaches 70mm.
DISTRIBUTION: Sporadically found throughout the Philippines.

5. Mitra (Mitra) incompta (Lightfoot,1786)
101.5mm; Samar.
Shell large, solid, with a tall spire, and sculptured with close-set axial folds, nodulose from overriding pitted spiral grooves, sutures moderately impressed and somewhat crenulate; aperture widening anteriorly, outer lip thin and crenulate, columella with up to 5 oblique folds; pale orange-yellow in colour with orange-brown axial stripes, aperture pale yellow, outer lip denticles brown; reaches 12mm.
DISTRIBUTION: Sporadically found throughout the Philippines in limited quantity.

6. Mitra (Mitra) triplicata von Martens, 1904
60.0mm; Siasi, Sulu Archipelago.
Shell medium sized, fusiform-elongate, with a tall spire, and moderately channeled sutures; sculptured with 3 or 4 strong subsutural grooves, weakening on the lower part of the whorl, and encircled by fine axial striae, producing a smoothly cancellate surface; body whorl with additional basal grooving; aperture narrow, outer lip thick and arcuate, columella with 4 to 5 folds; pure white in colour; reaches 60mm.
DISTRIBUTION: Sulu Archipelago.

7. Mitra (Nebularia) puncticulata Lamarck, 1811
45.5mm; Samar.
Shell medium sized, solid, elongate-ovate, with a moderately tall spire and well indented sutures bordered by irregular coronations; sculptured with axially striated spiral grooves; aperture longer than the spire, outer lip thick and crenulate, columella with 4 to 5 folds; colour burnt orange with white-spotted yellow banding, white coronations and a yellowish aperture; reaches 55mm.
DISTRIBUTION: Sporadically found throughout the Philippines.

8. Mitra (Strigatella) acuminata Swainson, 1824
21.6mm; Sulu Archipelago.
Shell small to medium sized, elongate-ovate, with a tall spire, an inflated body whorl, and incised sutures; whorls sculptured with punctate spiral grooves, obsolete on the middle of the body whorl; aperture narrow, widening anteriorly, outer lip thick and calloused, columella with 4 to 5 folds; orange-brown in colour with white banding; reaches 35mm.
DISTRIBUTION: Sporadically found throughout the Philippines in limited quantity.

9. Mitra (Nebularia) pyramis (Wood, 1828)
55.0mm; Samar.
Shell medium sized, fusiform-elongate with an inflated body whorl; sculptured with rounded axial riblets bisected by deep spiral grooves, 5 to 6 oblique spiral cords at the base; aperture narrowly elongate, half the shell height, with a thick, basally constricted outer lip; columella with 5 to 6 oblique folds; white in colour with orange blotches forming 2 bands on the spire whorls, 3 on the body whorl; reaches 55mm.
DISTRIBUTION: Sporadically found throughout the Philippines in limited quantity.

10. Mitra (Nebularia) coarctata Reeve, 1844
22.1mm; Samar.
Shell small, solid, elongate-ovate, with a tall spire and a large body whorl; sculptured with fine spiral striae and some axial growth lines, becoming obsolete on the body whorl; aperture narrow, longer than the spire, outer lip centrally thickened and glazed, columella with 4 to 6 folds, canal slightly twisted; yellowish orange in colour with brown blotches forming bands, aperture orange; reaches 25mm.
DISTRIBUTION: Sporadically found throughout the Philippines.

11. Mitra (Mitra) eremitarum Röding, 1798
95.2mm; Bohol.
Shell large, solid, elongate-ovate, with well defined, marginally crenulate sutures; sculptured with axial striae and overriding, finely punctate spiral grooves forming nodules; axial striae obsolete on the body whorl; aperture wide, longer than the spire, outer lip thick and marginally crenulate, columella with 5 oblique folds; orange-yellow in colour, with broad, brown, axial streaks, brown outer lip denticles and a pale orange aperture; reaches 100mm.
DISTRIBUTION: Throughout the Philippines.

12. Mitra (Mitra) imperialis Röding, 1798
56.5mm; Samar.
Shell medium sized, solid, elongate-ovate with a somewhat squat body whorl; sculptured with deeply punctate spiral grooves which are bisected by axial threads that become obsolete on the body whorl, sutures distinct with small white nodules; aperture wide, longer than the spire, with a thick, marginally dentate outer lip and 4 to 5 folds on the columella; tan coloured with white blotches and a brown body whorl band; reaches 55mm.
DISTRIBUTION: Sporadically found throughout the Philippines

13. Mitra (Mitra) sacerdotalis A. Adams, 1851
81.0mm; Punta Engaño, Cebu.
Shell large, broadly fusiform, with well channeled sutures; whorls smooth, glossy, with a subsutural groove, a few obsolete spiral grooves on earlier whorls, and 10 to 12 at the base; aperture narrowly elongate with a thickly calloused outer lip, and a columella bearing 4 to 5 folds; tan brown with subsutural white spots and a white band in the center on the body whorl; reaches 130mm.
DISTRIBUTION: Cebu and Bohol.

14. Mitra (Nebularia) ferruginea Lamarck, 1811
36.5mm; Samar.
Shell medium sized, solid, elongate-ovate, sculptured with prominent spiral cords and fine interstitial axial striae; aperture narrow, smooth, longer than the spire; outer lip thickened, centrally reflected and crenulate along its entire length, columella glazed with 5 folds; colour white with large brown axial blotches, aperture yellowish orange; reaches 55mm.
DISTRIBUTION: Sporadically found throughout the Philippines.

15. Mitra (Nebularia) rubritincta Reeve, 1844
34.2mm; Samar.
Shell medium sized, elongate-ovate with weakly incised, slightly ledged sutures; sculptured with raised spiral cords and axially striated interspaces; aperture equal to spire length, outer lip thick and crenulate, columella with 5 oblique folds; white in colour with 2 rows of red axial streaks on the body whorl; aperture and columella pale orange-yellow; reaches 40mm.
DISTRIBUTION: Sporadically found throughout the Philippines.

16. Mitra (Nebularia) contracta Swainson, 1820
Syn: *M. chrysostoma* Broderip, 1836
34.8mm, 22.5mm; Bohol.
Shell medium sized, solid, elongate-ovate with a tall spire and smooth, narrowly ledged, well indented sutures; sculptured with finely incised, punctate spiral grooves which become obsolete on the last 2 or 3 whorls; aperture narrow and half the shell length, outer lip smooth, thickened and slightly reflected, columella with 4 to 6 folds; yellowish white in colour with reddish brown blotches, fusing to form bands on the body whorl, aperture orange-yellow.
DISTRIBUTION: Sporadically found throughout the Philippines.

17. Mitra (Mitra) ustulata Reeve, 1844
39.0mm; Palawan.
Shell medium sized, solid, elongate-ovate with indistinct sutures; sculptured with deeply incised, punctate spiral grooves and occasional axial striae; aperture narrow, half the shell length, outer lip thick and smooth, columella with 6 to 8 oblique folds; cream-coloured with brown spotting in 2 to 3 discontinuous bands on the body whorl, aperture pale orange; reaches 60mm.
DISTRIBUTION: Sporadically found throughout the Philippines in limited quantity.

18. Mitra (Mitra) fulgurita Reeve, 1844 form **yaekoae** (Habe & Kosuge, 1966)
45.0mm; Palawan.
Shell medium sized, elongate-ovate, sculptured with punctate spiral grooves and fine axial folds; aperture less than the spire height, outer lip thickened and marginally crenulate, columella with 5 folds; colour yellowish with brown banding, white axial streaks and 4 to 5 rows of red-brown spots; reaches 55mm.
DISTRIBUTION: Palawan and Sulu Archipelago.

19. Cancilla (Cancilla) isabella Swainson, 1831
83.5mm; Tayabas Bay, Quezon.
Shell large, fusiform-elongate, with a tall spire and channeled sutures; sculptured with raised spiral cords, a single weak thread in the deep interspaces, and crossed by numerous close-set, overriding axial threads; aperture narrow, elongate, outer lip thin and marginally crenulate, columella with 6 oblique folds; pure white with some brown staining; reaches 50mm.
DISTRIBUTION: Camotes Sea, Sulu Sea and Tayabas Bay, Quezon.

20. Mitra (Mitra) inquinata Reeve, 1844
36.3mm; Samar.
Shell medium sized, elongate with a tall, acute spire, weakly incised, weakly ledged sutures, and sculptured with weak spiral grooves and obsolete interstitial axial striae; aperture narrowly elongate, half the shell length, outer lip thick, columella with 4 folds; off-white in colour with darker banding on the spiral cords and some brown axial streaks on earlier whorls, aperture white; reaches 45mm.
DISTRIBUTION: Sporadically found throughout the Philippines in limited quantity.

Plate 54 Mitridae

1. Mitra (Strigatella) paupercula (Linné, 1758)
35.3mm; Samar.
Shell medium sized, solid, elongate-ovate, smooth except for some basal cords; aperture longer than the spire, outer lip thick, occasionally crenulate, columella with 4 to 5 folds; shell black in colour with white axial stripes, aperture and columella brown, deep purple within; reaches 37mm.
DISTRIBUTION: Throughout the Philippines.

2. Mitra (Strigatella) scutulata (Gmelin, 1791)
32.8mm; Samar.
Shell medium sized, solid, elongate-ovate, somewhat inflated with well incised sutures; sculptured with spiral grooves that are absent on the center of the body whorl, and fine interstitial axial striae; aperture longer than the spire, outer lip thick, obsoletely crenulate, columella with 4 to 5 folds; black in colour with white axial streaks and a yellow subsutural band; reaches 45mm.
DISTRIBUTION: Throughout the Philippines.

3. Mitra (Strigatella) decurtata Reeve, 1844
32.9mm; Cebu.
Shell medium sized, solid, cylindrically ovate, with heavily ledged, deeply incised sutures; spire short, early whorls eroded, surface smooth except for some basal spiral cords; aperture narrow, longer than the spire, outer lip angulate, gibbous and thickened, columella with 5 folds; colour black with 1 or 2 discontinuous white bands, aperture blue-white; reaches 40mm.

DISTRIBUTION: Sporadically found throughout the Philippines in limited quantity.

4. Mitra (Strigatella) pica (Dillwyn, 1817)
26.0mm; Batangas.
Similar to *M. paupercula*, but more cylindrical in form, with a larger body whorl and sculptured with spiral grooves from base to apex; dark brown in colour with wavy, close-set, white-colored axial streaks, aperture marginally brown; reaches 45mm.
DISTRIBUTION: Sporadically found throughout the Philippines.

5. Mitra (Strigatella) retusa Lamarck, 1811
21.6mm; Samar.
Another species close to *M. paupercula* but distinctly ovate, with a large body whorl, a somewhat compressed spire, and sculptured all over with spiral grooves, strongest at the base; the shoulders are wider and the outer lip margin has characteristic crenulations along its entire length; dark brown with orange-stained white axial streaks, aperture bluish white, marginally brown; reaches 30mm.
DISTRIBUTION: Throughout the Philippines.

6. Mitra (Strigatella) litterata Lamarck, 1811
20.5mm; Samar.
Shell medium sized, solid, elongate-ovate, with a short spire and a large cylindrically-ovate body whorl; sculptured with punctate spiral grooves, becoming cords towards the base; aperture longer than spire, outer lip thick with a central triangular thickening, columella with 4 to 5 folds; blackish brown in colour with white streaks and blotches, aperture and columella bluish white; reaches 35mm.
DISTRIBUTION: Throughout the Philippines.

7. Mitra (Strigatella) colombelliformis Kiener, 1838
38.2mm; Bohol.
Shell medium sized, elongate-ovate, sutures narrowly incised, spire tall and sharply acute; sculptured with fine axial striae and punctate spiral grooves, somewhat obsolete mid-body whorl; aperture narrow, longer than the spire, outer lip thickened, gibbous and posteriorly constricted with a callosity on the inner wall margin, columella with 5 to 6 folds; light brown in colour with white subsutural blotching, aperture bluish white; reaches 40mm.
DISTRIBUTION: Sporadically found throughout the Philippines in limited quantity.

8. Mitra (Mitra) coffea Schubert & Wagner, 1829
39.0mm; Marinduque.
Shell medium sized, elongate-ovate with a large, somewhat ventricose body whorl and distinctly crenulate sutures; sculptured with punctate spiral grooves and weak axial threads; aperture wide, outer lip incurved posteriorly with obliquely aligned teeth on its outer margin, columella with 5 to 6 folds; deep brown in colour with obsolete white spots, aperture brown, columellar teeth white; reaches 60mm.
DISTRIBUTION: Sporadically found throughout the Philippines.

9. Mitra (Nebularia) aurora floridula Sowerby, 1874
38.5mm; Samar.
Shell medium sized, solid, with narrowly channeled, marginally crenulate sutures, and sculptured with deep, punctate spiral grooves which give rise to either closely-set or widely-spaced flat spiral cords; aperture longer than spire height, outer lip thick, marginally crenulate, incurved posteriorly, columella with 5 to 6 folds; orange in colour with white subsutural blotches and white speckling at the base, aperture bluish white; reaches 45mm.
DISTRIBUTION: Sporadically found throughout the Philippines in limited quantity.

10. Mitra (Mitra) aurantia aurantia (Gmelin, 1791)
35.8mm, 30.0mm; Palawan.
Shell medium sized, solid, elongate-ovate, with impressed, occasionally ledged sutures and sculptured with numerous shallow to deep spiral grooves and axially striate interspaces; aperture longer than the spire, outer lip thick and crenulate, columella with 5 to 6 folds; dark brown to black in colour, with a yellow subsutural band, aperture bluish white; reaches 45mm.
DISTRIBUTION: Sporadically found throughout the Philippines in limited quantity.

11. Mitra (Mitra) aurantia aurantia (Gmelin, 1791) form **nanus** Reeve, 1844
27.0mm; Samar.
A distinct form of *aurantia*, distinguished by its smaller size, a much narrower subsutural band, and the numerous nebulous white spots that occur within and around the band.
DISTRIBUTION: Sporadically found throughout the Philippines in limited quantity.

12. Mitra (Nebularia) coronata Lamarck, 1811
27.7mm, 26.3mm, 23.8mm; Batangas.
Shell medium sized, elongate-ovate, with heavy crenulate sutures; sculptured with spiral cords and pitted axial grooves, forming a slightly granulose surface; aperture narrow, longer than the spire, outer lip thin and crenulate, columella with 5 folds; colour light to dark brown with a white subsutural band and white sutural crenulations; aperture bluish white; reaches 30mm.

PLATE NO. 54 1. *Mitra paupercula* (Linné) 2. *Mitra scutulata* (Gmelin) 3. *Mitra decurtata* Reeve 4. *Mitra pica* (Dillwyn) 5. *Mitra retusa* Lamarck 6. *Mitra litterata* Lamarck 7. *Mitra colombelliformis* Kiener 8. *Mitra coffea* Schubert & Wagner 9. *Mitra aurora floridula* Sowerby 10. *Mitra aurantia aurantia* (Gmelin) 11. *Mitra aurantia aurantia* (Gmelin) form *nanus* Reeve 12. *Mitra coronata* Lamarck 13. *Ptergia scabricula* (Linné) 14. *Mitra vexillum* (Reeve) 15. *Mitra doliolum* Kuster 16. *Mitra fulvescens* Broderip 17. *Scabricola desetangsii* (Kiener) 18. *Mitra fastigium* Reeve 19. *Mitra ticaonica* Reeve 20. *Mitra luctuosa* A. Adams 21. *Mitra lugubris* Swainson 22. *Mitra rubiginosa* Reeve

PLATE 54

DISTRIBUTION: Sporadically found throughout the Philippines.

13. Pterygia scabricula (Linné, 1758)
28.2mm; Mindoro.
Shell medium sized, elongate-ovate, surface granulose from intersecting axial grooves and spiral cords, interspaces deeply punctate; aperture longer than the spire, outer lip thick, crenulate, slightly angulate posteriorly, columella with 4 to 5 folds; shell brown in colour with white blotches and spots, aperture pale brown; reaches 55mm.
DISTRIBUTION: Sporadically found throughout the Philippines.

14. Mitra (Strigatella) vexillum (Reeve, 1844)
27.4mm; Palawan.
Shell medium sized, elongate-ovate, sculptured with broad spiral cords and crossed by fine axial striae, obsolete on the lower part of the body whorl; aperture narrow, outer lip thick and crenulate, columella with 6 irregularly sized folds; reddish brown with darker interspaces, aperture pale brown; reaches 35mm.
DISTRIBUTION: Sporadically found throughout the Philippines in limited quantity.

15. Mitra (Nebularia) doliolum Küster, 1839
26.9mm; Punta Engaño, Cebu.
Shell medium sized, elongate-ovate, with slightly ledged sutures and sculptured with closely set, cord-like spiral ridges, axially striate in the interspaces; aperture equal to the spire height, outer lip thick and scalloped, columella with 4 to 5 folds; orange-brown in colour with darker grooves, columellar folds white to brown; reaches 30mm.
DISTRIBUTION: Sporadically found throughout the Philippines in limited quantity.

16. Mitra (Nebularia) fulvescens Broderip, 1836
32.2mm; Samar.
Shell medium sized, solid, elongate-ovate, body whorl varying from slender to slightly ventricose, the spire tall and acute, and the sutures weakly crenulate; sculptured with punctate spiral grooves which may be obsolete on the center of the body whorl, and some basal spiral cords; outer lip thickened, finely crenulate, posteriorly constricted, columella with 5 to 6 folds; pale brown in colour with irregular white spotting, aperture and columella bluish white; reaches 45mm.
DISTRIBUTION: Sporadically found throughout the Philippines in limited quantity.

17. Scabricola (Scabricola) desetangsii (Kiener, 1838)
27.5mm; Samar.
Shell medium sized, elongate-ovate, sculptured with spiral grooves and smooth interspaces, earlier whorls nodulose from bisecting axial cords which disappear on the lower section of the body whorl; aperture wide, greater than the spire height, outer lip thin and weakly crenulate, columella with 5 folds; pale grey in colour with white subsutural spotting and 2 irregular dark brown bands on the body whorl, spiral ribs with brown dashes, aperture bluish white; reaches 30mm.
DISTRIBUTION: Sporadically found throughout the Philippines.

18. Mitra (Strigatella) fastigium Reeve, 1845
21.9mm; Bohol.
Shell small to medium sized, elongate-ovate, with well incised, weakly crenulate sutures; sculptured with fine spiral grooves which are obsolete on the mid-body whorl; aperture longer than the spire, outer lip thick and crenulate, columella with 5 folds; colour dark tan with darker spiral threads, most obvious on the outer lip thick margin, aperture and columella white; reaches 30mm.
DISTRIBUTION: Sporadically found throughout the Philippines in limited quantity.

19. Mitra (Nebularia) ticaonica Reeve, 1844
23.5mm; Samar.
Shell medium sized, cylindrically ovate, with irregularly channeled sutures and sculptured with weak spiral grooves, occasionally absent mid-body whorl, and fine interstitial axial striae; aperture longer than the spire, outer lip thickened posteriorly and marginally crenulate, columella with 4 to 5 folds; pale yellow-tan in colour, darker towards the base, aperture pale brown, columellar teeth off-white to light brown; reaches 35mm.
DISTRIBUTION: Sporadically found throughout the Philippines in limited quantity.

20. Mitra (Nebularia) luctuosa A. Adams, 1853
24.8mm; Sulu Archipelago.
Shell medium sized, elongate, with distinctly crenulate sutures; sculptured with weakly pitted spiral grooves on the early whorls, giving rise to wide spiral cords on the body whorl and several distinct basal cords; aperture narrow, equal to spire height, outer lip thin and weakly crenulate, columella with 5 folds; dark brown in colour with subsutural white streaks and an orange-yellow spiral band; reaches 35mm.
DISTRIBUTION: Sporadically found throughout the Philippines.

21. Mitra (Nebularia) lugubris Swainson, 1821
28.0mm; Samar.
Shell medium sized, solid, cylindrically ovate with well incised, marginally crenulate sutures; sculptured with bisecting axial and spiral cords and deeply pitted interspaces; aperture longer than the spire, outer lip thick, straight and marginally crenulate, columella with 6 folds; dark brown in colour with a wide, white subsutural band and an additional band at the body whorl base, aperture bluish white; reaches 30mm.
DISTRIBUTION· Sporadically found throughout the Philippines.

22. Mitra (Nebularia) rubiginosa Reeve, 1844
42.9mm; Palawan.
Shell medium sized, solid, fusiform-elongate, with well

incised, weakly crenulate sutural margins; sculptured with wide, flat spiral cords and axially striate interspaces; aperture narrow, longer than the spire, outer lip thick and crenulate, columella with 4 to 6 folds; creamy white in colour with large brown axial streaks, aperture white; reaches 50mm.
DISTRIBUTION: Sporadically found throughout the Philippines in limited quantity.

Plate 55 — Mitridae

1. Cancilla (Domiporta) granatina (Lamarck, 1811)
49.2mm; Samar.
Shell medium sized, elongate, with a tall spire, convex whorls, and sculptured with spiral cords and axially striate interspaces; aperture equal to the spire length, outer lip thickened and crenulate, columella with 5 folds; colour off-white with 2 weak brown bands on the body whorl and interrupted brown spiral lines; reaches 55mm.
DISTRIBUTION: Sporadically found throughout the Philippines.

2. Cancilla (Domiporta) gloriola Cernohorsky, 1970
41.2mm; Palawan.
Shell medium sized, fusiform, elongate-ovate, with a tall spire and weakly channeled sutures; sculptured with strong spiral cords, weaker interstitial riblets, all crossed by fine axial grooves producing a highly granulose surface; aperture elongate, outer lip thin, weakly crenulate, columella with 4 folds; white with brown spots on the cords, forming 2 body whorl bands; reaches 60mm.
DISTRIBUTION: Sporadically found throughout the Philippines.

3. Neocancilla papilio (Link, 1807)
39.0mm; Samar.
Shell medium sized, elongate-ovate, with a narrow, orange coloured aperture, 5 columellar folds, a marginally crenulate outer lip, and convex whorls which are sculptured with strong spiral cords and bisected by axial grooves; white in colour with 2 brown bands on the body whorl and black spots on the nodulose spiral ribs; reaches 55mm.
DISTRIBUTION: Throughout the Philippines.

4. Mitra sp.
39.5mm; Bohol.
Shell medium sized, fusiform-elongate, with weakly channeled sutures, a tall spire, and a body whorl equal to the spire height; whorls smooth except for nearly obsolete spiral grooves, stronger towards the body whorl base; aperture narrowly elongate, outer lip smooth, columella with 5 folds; pale yellowish white in colour with irregular axial and spiral red lines forming a reticulate network where they cross, and some additional brown blotching; superficially resembling *M. cloveri* Cernohorsky in its pattern, but much larger, more elongate, and lacking the punctate spiral grooving; possibly an ecomorph of *M. ustulata,* Reeve but the shape is different, the surface lacks the deeply incised grooves of that species, and the outer lip is thin, not heavily thickened. The only specimen seen. Possibly *Mitra deynzeri* Cernohorsky, 1980.
DISTRIBUTION: Camotes Sea.

5. Subcancilla abyssicola (Schepman, 1913)
43.4mm; Tayabas Bay, Quezon.
Shell medium sized, fusiform-elongate, with an extremely tall spire and a small body whorl which is about 30% of the shell length; sutures deeply channeled and distinctly ledged; sculptured with strong, narrow spiral cords and crossed by axial threads, producing nodules on the cords and deeply pitted interspaces; aperture narrow, outer lip crenulate, columella with 5 folds; off-white in colour with brown streaks on the cords; reaches 50mm.
DISTRIBUTION: Sporadically found throughout the Philippines in limited quantity (deep water).

6. Scabricola (Scabricola) variegata (Gmelin, 1791)
35.2mm; Samar.
Shell medium sized, elongate-ovate, with convex whorls, slightly angulate at the sutures; sculptured with strong spiral cords and fine, axially striate interspaces; aperture smooth, outer lip thick and crenulate, columella with 4 to 5 oblique folds; creamy white in colour with orange-brown bands and streaks, and some orange-flushed white areas, aperture cream to orange; reaches 50mm.
DISTRIBUTION: Sporadically found throughout the Philippines in limited quantity.

7. Pterygia nucea (Gmelin, 1791)
43.0mm; Palawan.
Shell medium sized, solid, cylindrically elongate-ovate, with a short spire, a large, somewhat inflated body whorl, crenulate sutures, and fine encircling spiral grooves; aperture wide, outer lip thin but solid, columella with 6 folds; white in colour with 1 or 2 brown bands and spiral rows of brown spots, aperture white; reaches 55mm.
DISTRIBUTION: Sporadically found throughout the Philippines.

8. Pterygia dactylus (Linné, 1767)
46.7mm; Samar.
Shell medium sized, rather heavy, with a low spire and a large ventricose body whorl; sculptured with numerous spiral grooves, bisected on the spire whorls and on the shoulder of the body whorl by axial striae; aperture narrow, slightly angulate posteriorly, outer lip thick and smooth, columella with 6 to 7 folds; shell white in colour with interrupted brown bands, grooves deep brown, aperture white; reaches 55mm.

DISTRIBUTION: Sporadically found throughout the Philippines.

9. Pterygia conus (Gmelin, 1791)
33.0mm; Samar.
Shell medium sized, coniform, with a large body whorl, a short spire and nodulose, somewhat angulate whorls; sculptured with well incised spiral grooves on the body whorl; aperture narrow, smooth, outer lip thickened, slightly expansive anteriorly, columella with 6 to 7 folds; colour pale brown with dark brown streaks and lines, aperture white; reaches 55mm.
DISTRIBUTION: Sporadically found throughout the Philippines.

10. Pterygia fenestrata (Lamarck, 1811)
Shell medium sized, cylindrically elongate, with a short spire and a large body whorl; sculptured with punctate spiral grooves and axial folds, forming a finely nodulose surface; aperture long, outer lip marginally crenulate, columella with 8 to 9 folds; colour orange-brown with white nodules and aperture; reaches 35mm.
10a. typical form; 24.0mm; Palawan.
10b. smooth form *undulosa* Reeve, 1844; 23mm; Samar.
DISTRIBUTION: Sporadically found throughout the Philippines in limited quantity.

11. Imbricaria punctata (Swainson, 1821)
14.7mm; Punta Engaño, Cebu.
Shell small, yellow-coloured, coniform, with a short mammillate apex, a large cylindrical body whorl, and sculptured with widely spaced, punctate spiral grooves; aperture long and narrow, outer lip smooth, columella with 5 to 6 folds; reaches 25mm.
DISTRIBUTION: Sporadically found throughout the Philippines in limited quantity.

12. Imbricaria vanikorensis (Quoy and Gaimard, 1833)
12.0mm; Punta Engaño, Cebu.
A small cone-shaped shell, with a short, stepped spire, and a large body whorl; sculptured with flat spiral ridges and axial striae in the interspaces; aperture narrow, outer lip marginally crenulate, columella with 6 folds; colour brown to bluish grey with white sutural bands and spots, aperture blue-grey; reaches 25mm.
DISTRIBUTION: Sporadically found throughout the Philippines in limited quantity.

13. Pterygia arctata (Sowerby, 1874)
16.8mm; Palawan.
Shell small to medium sized, cylindrically elongate with a short spire; whorls angulate at the sutures and sculptured with numerous well incised, punctate spiral grooves and very fine axial striae; aperture narrowly elongate, smooth within, outer lip thick and weakly crenulate, columella with 6 folds; colour white with red axial and spiral lines forming a reticulate network, and a thick band on the body whorl; reaches 25mm.
DISTRIBUTION: Sporadically found throughout the Philippines in limited quantity.

14. Scabricola (Swainsonia) casta (Gmelin, 1791)
29.3mm; Albay Gulf, Southern Luzon.
Shell medium sized, rather elongate, with straight sided whorls; sculptured with punctate spiral grooves and axial ribs on the spire whorls, smooth on the body whorl; aperture narrow, longer than the spire, outer lip smooth, columella with 6 folds; colour white with a thin brown sutural band and a wider band in the centre of the body whorl; reaches 50mm.
DISTRIBUTION: Sporadically found throughout the Philippines in limited quantity.

15. Pterygia sinensis (Reeve, 1844)
29.8mm; Palawan.
Shell medium sized, cylindrically elongate, with a short spire and a large body whorl; sculptured with numerous axial and spiral cords intersecting to produce a pitted and somewhat granulose surface; aperture narrow, long (75% of the shell length), outer lip thick and crenulate, columella with 10 folds; colour white to brown, aperture light brown; reaches 50mm.
DISTRIBUTION: Sporadically found throughout the Philippines in limited quantity.

16. Mitra (Nebularia) rosacea Reeve, 1845
Shell medium sized, elongate to elongate-ovate with prominent sutures, convex spire whorls which increase in size towards the base, and a long inflated body whorl which exceeds the spire height; sculptured with thin, elevated spiral cords and axial grooves, giving a filleted appearance to the surface; aperture narrow and smooth within, outer lip thick and weakly scalloped, columella with 4 to 6 folds; cream coloured with a rose-violet tint, occasionally banded, aperture pale pinkish lavender; reaches 40mm.
16a. typical form; 26mm; Corregidor, Western Luzon.
16b. tall-spired form; 39mm; Sulu Archipelago.
DISTRIBUTION: Sporadically found throughout the Philippines in limited quantity.

17. Pterygia crenulata (Gmelin, 1791)
25.8mm; Punta Engaño, Cebu.
Shell medium sized, cylindrically elongate, with a short spire and a long body whorl; sculptured with punctate spiral grooves which are bisected by fine

PLATE NO. 55 1. *Cancilla granatina* (Lamarck) 2. *Cancilla gloriola* Cernohorsky 3. *Neocancilla papilio* (Link) 4. *Mitra sp.* 5. *Subcancilla abyssicola* (Schepman) 6. *Scabricola variegata* (Gmelin) 7. *Pterygia nucea* (Gmelin) 8. *Pterygia dactylus* (Linné) 9. *Pterygia conus* (Gmelin) 10a. *Pterygia fenestrata* (Lamarck) 10b. form *undulosa* Reeve 11. *Imbricaria punctata* (Swainson) 12. *Imbricaria vanikorensis* (Quoy & Gaimard) 13. *Pterygia arctata* (Sowerby) 14. *Scabricola casta* (Gmelin) 15. *Pterygia sinensis* (Reeve) 16a. *Mitra rosacea* Reeve 16b. tall-spired form 17. *Pterygia crenulata* (Gmelin) 18. *Imbricaria conularis* (Lamarck) 19. *Imbricaria olivaeformis* (Swainson) 20. *c.f. Scabricola newcombii* (Pease) 21. *Cancilla incarnata* (Reeve)

PLATE 55

axial striae; aperture narrowly elongate, outer lip thick and marginally crenulate, columella with 8 to 9 folds; colour white with large, orange-coloured axial blotches and a white aperture; reaches 35mm.
DISTRIBUTION: Sporadically found throughout the Philippines.

18. Imbricaria conularis (Lamarck, 1811)
21.3mm; Palawan.
Shell medium sized, coniform, with a tall spire, a large, cylindrically elongate body whorl, and sculptured with shallow, widely spaced spiral grooves; aperture narrowly elongate, outer lip smooth, columella with 6 to 7 folds; greyish white in colour with faint brown bands, white spots and brown spiral lines; aperture deep purple, with a purple blotch on the columellar base; reaches 25mm.
DISTRIBUTION: Sporadically found throughout the Philippines.

19. Imbricaria olivaeformis (Swainson, 1821)
16.2mm; Samar.
Shell small, cylindrically ovate, with a short spire and a large body whorl; surface mostly smooth but with a few punctate spiral grooves; aperture narrow, longer than the spire, outer lip smooth, columella with 5 folds; colour yellow, spire and base purple, aperture pale yellow; reaches 25mm.
DISTRIBUTION: Sporadically found throughout the Philippines.

20. c. f. Scabricola (Swainsonia) newcombii (Pease, 1869)
24.1mm; Sulu Archipelago.
Shell medium sized, cylindrically elongate, the spire whorls finely clathrate from intersecting axial cords and punctate spiral grooves; the axial cords become obsolete and the spiral grooves become smooth on the lower part of the body whorl; aperture narrowly elongate, outer lip thin and arcuate, columella with 4 to 5 folds; colour white with obsolete brown banding; extremely close to the Hawaiian endemic *newcombii*, but lacks the brown-banded, centrally smooth body whorl of that species; reaches 30mm.
DISTRIBUTION: Sporadically found throughout the Philippines in limited quantity.

21. Cancilla (Domiporta) incarnata (Reeve, 1845)
26.2mm; Punta Engaño, Cebu.
Shell medium sized, elongate-ovate, with convex whorls; sculptured with small, crowded spiral lirae and bisecting axial grooves, producing a finely cancellate surface; aperture narrowly elongate, equal to the spire length, outer lip crenulate, columella with 4 to 5 folds; colour off-white with occasional brown banding, aperture orange to pale pink; reaches 30mm.
DISTRIBUTION: Sporadically found throughout the Philippines in limited quantity.

Mitridae Plate 56

1. Mitra (Mitra) hilli Cernohorsky, 1975
81.8mm; Bohol.
Shell moderately large, with a tall spire and a large ventricose body whorl; aperture half the shell length; whorls smooth except for some weak spiral grooving centrally and basally on the body whorl; aperture narrowly elongate, outer lip thick, slightly angulate anteriorly, columella with 5 to 7 folds; white in ground colour with brown axial flammules and 2 brown bands on the body whorl, aperture creamy white; reaches 90mm.
DISTRIBUTION: Cebu, Bohol and Southern Mindanao (deep water).

2. Scabricola (Scabricola) vicdani Cernohorsky, 1981
37.6mm; Punta Engaño, Cebu. (Paratype)
Shell medium sized, elongate-ovate, with a moderately tall spire of 7 to 8 weakly convex whorls, and a large body whorl; sculptured with elevated spiral cords, weakly granulose from crowded, overriding axial striae, interspaces deeply concave and wider than the cords; aperture narrowly elongate, taller than the spire, outer lip thickened, crenulate, columella with 5 to 6 folds, bordered by a distinct siphonal fasciole and a narrow umbilicus; creamy orange in colour with a brownish band on the body whorl, cords with white and brown spots, aperture pale orange; reaches 40mm. Named after V. Dan of Manila.
DISTRIBUTION: Cebu (deep water)

3. Cancilla (Ziba) rehderi (Webb, 1958)
35.4mm; Punta Engaño, Cebu.
Shell medium sized, fusiform-elongate, with a tall spire, spire whorls convex, body whorl large and elongate; sculptured with numerous wide, flattened spiral cords, and fine, axially striated interspaces, the axial striae obsolete on the spire cords; aperture narrowly elongate, half the shell length, outer lip thin,

PLATE NO. 56 1. *Mitra hilli* Cernohorsky 2. *Scabricola vicdani* Cernohorsky 3. *Cancilla rehderi* (Webb) 4. *Neocancilla circula* (Kiener) 5. *Cancilla filaris* (Linné) 6. *Cancilla praestantissima* (Röding) 7. *Subcancilla insculpta* (A. Adams) 8. *Cancilla aegra* (Reeve) 9. *Subcancilla flammea* (Quoy & Gaimard) 10. *Neocancilla clathrus* (Gmelin) 11. *Mitra multiplicata* (Pease) 12. *Subcancilla annulata* (Reeve) 13. *Cancilla bacillum* (Lamarck) 14. *Subcancilla verrucosa* (Reeve) 15. *Cancilla astyagis* (Dohrn) 16. *Cancilla fulgetrum* (Reeve) 17. *Mitra edentula* Swainson 18a. *Subcancilla interlirata* (Reeve) elongate form 18b. ovate form 18c. typical form 18d. form *pia* Dohrn 19a. *Mitra fraga* Quoy & Gaimard form *rubiginea* A. Adams 19b. broad form 19c. elongate form 20. *Mitra lienardi* Sowerby

PLATE 56

columella with 5 to 6 folds; orange brown in colour with white axial streaks, aperture white; reaches 37mm.
DISTRIBUTION: Camotes Sea.

4. Neocancilla circula (Kiener, 1838)
34.0mm; Samar.
Shell medium sized, with a tall spire and weakly convex whorls; sculptured with spiral cords, the interspaces finely reticulate from overriding axial and spiral threads; aperture narrow, outer lip thin, marginally crenulate, columella with 4 folds; colour white or brown with white banding, aperture brown; reaches 50mm.
DISTRIBUTION: Sporadically found throughout the Philippines.

5. Cancilla (Domiporta) filaris (Linné, 1771)
27.4mm; Palawan.
Shell medium sized, elongate-ovate, encircled with spiral ridges and bisecting axial striae, the first ridge on each whorl being well below the sutures; aperture moderately wide, outer lip thin, marginally crenulate, columella with 5 folds; shell and aperture off-white with brown ridges; reaches 30mm.
DISTRIBUTION: Sporadically found throughout the Philippines.

6. Cancilla (Domiporta) praestantissima (Röding, 1798)
40.8mm; Samar.
Shell medium sized, fusiform-elongate, sculptured with spiral ridges and intermediate spiral threads which are all bisected by fine axial grooves; aperture narrow, less than the spire length, outer lip crenulate, columella with 5 folds; colour white with red ribs; similar to *C. filaris,* but larger, more elongate with a narrower aperture, and the gap between the suture and the first subsutural spiral cord is less pronounced; reaches 45mm.
DISTRIBUTION: Sporadically found throughout the Philippines.

7. Subcancilla insculpta (A. Adams, 1853)
21.5mm; Punta Engaño, Cebu.
Shell small, broadly fusiform, with a moderately tall, acute spire, a large, elongate body whorl, and a narrow aperture which is greater than half the shell length; sculptured with thin, raised spiral cords and wide, deep interspaces which are finely cancellate from intersecting axial and spiral grooves; outer lip marginally crenulate, columella with 4 to 5 folds; creamy white in colour with weakly spotted brown ribs, aperture light brown; reaches 25mm.
DISTRIBUTION: Sporadically found throughout the Philippines.

8. Cancilla aegra (Reeve, 1845)
27.1mm; Punta Engaño, Cebu.
Shell medium sized, fusiform-elongate, with a tall, acute spire, a large, elongate-ovate body whorl, and a narrow aperture equal to half the shell length; sculptured with moderately heavy spiral cords, and fine, closely-spaced overriding axial striae, producing a weakly cancellate, finely nodulose surface; outer lip thin, crenulate, columella with 4 folds; colour white with brown axial streaks or bands; reaches 45mm.
DISTRIBUTION: Sporadically found throughout the Philippines in limited quantity.

9. Subcancilla flammea (Quoy & Gaimard, 1833)
26.4mm; Palawan.
Shell medium sized, elongate-ovate, with a tall, acute spire, and a large, slightly inflated body whorl; sculptured with slightly flattened spiral cords and fine axial striae in the interspaces; aperture narrow, half the shell length, outer lip finely crenulate, columella with 5 to 6 folds; off-white in colour with brownish stripes and a brown aperture; reaches 35mm.
DISTRIBUTION: Sporadically found throughout the Philippines in limited quantity.

10. Neocancilla clathrus (Gmelin, 1791)
33.2mm; Samar.
Shell medium sized, elongate-ovate, with a tall, acute spire, convex whorls and a narrowly elongate aperture, equal to half the shell length; surface finely granulose from overriding axial and spiral cords; outer lip thick and crenulate, columella with 5 folds; white with brown subsutural blotches and brown bands on the body whorl, aperture white; reaches 55mm.
DISTRIBUTION: Sporadically found throughout the Philippines.

11. Mitra (Dibaphus) multiplicata (Pease, 1865)
28.3mm; Marinduque.
Similar to *M. (D) edentula,* (fig. 17) but is more fusiform, has a longer spire, and is sculptured with much finer spiral striae than *edentula* (up to 50 on the body whorl); aperture long, narrow and smooth, outer lip thick and centrally inflected, columella with 10 to 11 fine folds; white with orange-brown spiral lines; reaches 50mm.
DISTRIBUTION: Sporadically found throughout the Philippines.

12. Subcancilla annulata (Reeve, 1844)
23.8mm; Samar.
Shell small to medium sized, fusiform-elongate, with a tall, acute spire; sculptured with elevated spiral cords and axially striate interspaces; aperture narrow, longer than the spire, outer lip thick, marginally crenulate, columella with 5 to 6 folds; creamy white in colour with a central brown band, the sutures and cords streaked or spotted with brown, aperture white; reaches 35mm.
DISTRIBUTION: Sporadically found throughout the Philippines.

13. Cancilla (Ziba) bacillum (Lamarck, 1811)
27.5mm; Samar.
Shell medium sized, solid, elongate-ovate, with a large body whorl and moderately channeled, slightly ledged sutures; sculptured with numerous fine, punctate spiral grooves and axially striate interspaces; aperture narrow, longer than the spire, outer lip crenulate,

columella with 6 folds; greenish brown in colour with white streaks and spots, aperture brown within, outer margin and columella white; reaches 30mm.
DISTRIBUTION: Sporadically found throughout the Philippines.

14. Subcancilla verrucosa (Reeve, 1845)
23.4mm; Palawan.
Shell medium sized, fusiform-ovate, sculptured with spiral cords and threads which are bisected by axial ridges, forming echinate nodules at the points of intersection; aperture shorter than the spire, outer lip crenulate, columella with 6 to 7 folds; colour white with brown ridges and a brown band on the body whorl; reaches 30mm.
DISTRIBUTION: Sporadically found throughout the Philippines.

15. Cancilla (Ziba) astyagis (Dohrn, 1860)
18.0mm; Palawan.
Shell small, fusiform, with a short spire and a large body whorl; whorls smooth except for a few lines on the penultimate whorl, and 5 to 6 grooves on the body whorl base; aperture narrow, longer than the spire, outer lip thick, columella with 4 to 5 folds; colour variable, usually brownish with subsutural brown dots and whitish axial streaks, aperture brown, reaches 30mm.
DISTRIBUTION: Sporadically found throughout the Philippines in limited quantity.

16. Cancilla (Ziba) fulgetrum (Reeve, 1844)
19.0mm, 17.1mm, 17.5mm; Palawan.
Shell small, fusiform-elongate, with a tall acute spire equal to half the shell length, and weakly incised sutures; sculptured with wide, flat spiral ribs and shallow, axially striate spiral grooves; aperture white, narrow, outer lip smooth, columella with 4 to 5 folds; off-white in colour with discontinuous brown-coloured axial streaks; reaches 45mm.
DISTRIBUTION: Sporadically found throughout the Philippines in limited quantity.

17. Mitra (Dibaphus) edentula Swainson, 1823
19.1mm; Samar.
Shell medium sized, cylindrically elongate, with a short spire and long body whorl; sculptured with distinct, narrowly ledged sutures, narrow spiral cords (up to 40 on the body whorl), and fine, axially striate interspaces; aperture narrow, longer than the spire, outer lip thickened, incurved and denticulate, columella with 10 to 11 folds; white with brown axial streaks forming 2 body whorl bands, aperture orange; reaches 45mm.
DISTRIBUTION: Sporadically found throughout the Philippines.

18. Subcancilla interlirata (Reeve, 1844)
Appears to be a somewhat variable species; shell medium sized, solid, elongate to elongate-ovate, with a tall spire and a large body whorl equal to or greater than the spire height; whorls weakly or strongly convex and sculptured with narrow, raised spiral cords and interspaces which are axially striate, and may show an additional spiral thread; aperture narrowly elongate, outer lip weakly crenulate, columella with 5 to 6 folds; creamy white with brown axial flammules and bands; reaches 55mm.
18a. elongate form; 40.0mm; Samar.
18b. ovate form with slightly heavier cords; possibly a juvenile form; 35.2mm; Palawan.
18c. typical form closely resembling Reeve's holotype; 29.4mm; Samar.
18d. form *pia* Dohrn, 1860, with white ground colour and golden brown cords; 25mm; Punta Engaño, Cebu.
DISTRIBUTION: Sporadically found throughout the Philippines in limited quantity.

19. Mitra (Nebularia) fraga Quoy & Gaimard, 1833
Shell small to medium sized, elongate-ovate with weakly impressed sutures; sculptured with elevated, sometimes undulate spiral cords and axially striate interspaces; aperture equal to spire height, outer lip thick and crenulate, columella with 4 folds; colour orange-brown with paler spots on the cords; reaches 45mm.
19a. form *rubiginea* A. Adams, 1855: a large slender form with few pale spots on the ribs; 36.2mm; Palawan.
19b. broad form; 21.3mm; Samar.
19c. a more elongate form; 21.5mm; Samar.
DISTRIBUTION: Sporadically found throughout the Philippines.

20. Mitra (Nebularia) lienardi Sowerby, 1874
22.9mm; Punta Engaño, Cebu.
Shell small, fusiform-ovate, with a tall spire, weakly incised sutures, and a large body whorl; sculptured with a weak subsutural cord, followed by a series of stronger cords and axially striate interspaces; aperture narrowly elongate, outer lip arcuate, thin and marginally crenulate, columella with 4 folds; shell colour reddish brown, a white band at the sutures, and one on the body whorl, cords red and white-spotted, aperture pale orange to white; reaches 25mm; often synonymised with *M. rubritineta*, but differs in shape and apertural characteristics.
DISTRIBUTION: Sporadically found throughout the Philippines in limited quantity.

Plate 57 Mitridae

1. Mitra (Strigatella) telescopium Reeve, 1844
20.0mm; Cebu.
Shell small to medium sized, cylindrically elongate with shouldered, well channeled sutures; sculptured with fine punctate spiral grooves and cords on the body whorl base; aperture narrow, longer than the spire, outer lip thin, angulate, posteriorly constricted and marginally crenulate, columella with 5 folds; white in colour with a broad reddish brown band on the lower body whorl, bordered by a yellow line,

aperture brown with white columellar folds; reaches 23mm.
DISTRIBUTION: Sporadically found throughout the Philippines in limited quantity.

2. Mitra (Nebularia) avenacea Reeve, 1845
16.8mm; Cebu.
Shell small to medium sized, fusiform-elongate, solid, with adpressed sutures; sculptured with spiral cords, 3 on the penultimate whorl and 15 to 20 on the body whorl; the interspaces of the cords are axially striate; aperture narrow, smooth within, outer lip thick, angulate and marginally crenulate, columella with 5 folds; colour orange-brown with paler interspaces, a white apex and a pale orange aperture; reaches 25mm.
DISTRIBUTION: Sporadically found throughout the Philippines in limited quantity.

3. Scabricola (Swainsonia) ocellata (Swainson, 1831)
18.2mm; Samar.
Shell medium sized, fusiform-elongate, the spire whorls and upper portion of the body whorl sculptured with intersecting axial and spiral threads, producing a clathrate appearance; lower body whorl encircled with closely set or widely spaced spiral grooves, axial structure being obsolete; aperture narrow, smooth within, outer lip thick, columella with 5 folds; colour white with brown flecks, the lower body whorl being greenish brown in colour and spotted with white; reaches 30mm.
DISTRIBUTION: Sporadically found throughout the Philippines in limited quantity.

4. Mitra (Nebularia) chrysalis Reeve, 1844
17.2mm; Palawan.
A medium sized, ovate shell with a short spire; sculptured with numerous shallow, punctate spiral grooves forming wide, closely-spaced cords; aperture narrow, longer than the spire, outer lip thickened and marginally crenulate, columella with 4 to 5 folds; orange-brown to dark brown in colour with an interrupted white band mid-body whorl, aperture orange; reaches 25mm.
DISTRIBUTION: Sporadically found throughout the Philippines.

5. Mitra (Nebularia) cucumerina Lamarck, 1811
19.2mm; Samar.
Shell medium sized, fusiform, markedly globose, with impressed sutures and sculptured with shallow, V-shaped, axially striated spiral grooves; aperture narrow, smooth within, outer lip thick, marginally crenulate, columella with 3 to 4 folds; colour a deep red to reddish orange with an interrupted white band mid-body whorl, aperture pale orange; reaches 30mm.
DISTRIBUTION: Sporadically found throughout the Philippines.

6. Mitra (Strigatella) pellisserpentis pellisserpentis Reeve, 1844
16.0mm; Samar.
Shell variable in form, from elongate to elongate-ovate, with moderately incised sutures and weakly convex whorls which may be smooth or covered with granulose spiral lirae; aperture narrow, smooth within, outer lip thick and marginally crenulate, columella with 5 to 6 oblique folds; colour pale yellow to orange-brown; reaches 40mm.
DISTRIBUTION: Sporadically found throughout the Philippines in limited quantity.

7. Mitra (Nebularia) tabanula Lamarck, 1811
Shell small, ovate to elongate-ovate, biconic, with weakly impressed sutures; sculptured with numerous raised, equi-spaced spiral cords and axially striate, V-shaped interspaces; aperture narrow, equal to spire height, smooth within, outer lip thick, marginally crenulate, columella with 3 to 4 folds; reaches 15mm.
7a. normal deep red form; 13.8mm; Samar.
7b. form *pediculus* Lamarck, 1811: reddish brown with white cords; 12.5mm; Palawan.
DISTRIBUTION: Sporadically found throughout the Philippines

Class Gastropoda
Subclass Prosobranchia
Order Neogastropoda
Suborder Rachiglossa
Superfamily Volutacea
Family COSTELLARIIDAE

A large group of "Mitrid-like" shells, which were originally classified with the Mitridae, but have now been relegated a family status. Separation is based mainly on radula characteristics, but morphological differences which characterize the members of the Costellariidae are the surface sculpturing of axial ribs and spiral grooves, and the lirate aperture interior which, in the Mitridae, is smooth; the members are mainly sand dwelling carnivores and have a worldwide distribution. Extremely well represented in the Philippines.

8. Vexillum (Pusia) cavea (Reeve, 1844)
19.3mm; Batangas.
A medium sized, ovate-biconic shell with rounded shoulders and sculptured with thick, rounded axial ribs and some obsolete spiral striae in the interspaces, and several heavy cords at the base; aperture narrow,

PLATE NO. 57 1. *Mitra telescopium* Reeve 2. *Mitra avenacea* Reeve 3. *Scabricola ocellata* (Swainson) 4. *Mitra chrysalis* Reeve 5. *Mitra cucumerina* Lamarck 6. *Mitra pellisserpentis pellisserpentis* Reeve 7a. *Mitra tabanula* Lamarck 7b. form *pediculus* Lamarck 8. *Vexillum cavea* Reeve 9. *Vexillum suavis* (Souverbie) 10. *Vexillum rubrum* (Broderip) 11. *Vexillum salisburyi* Cernohorsky 12. *Vexillum michaui* (Crosse & Fischer)-colour variation 13. *Vexillum lautum* (Reeve) 14. *Vexillum semicostatum* (Anton) 15. *Vexillum flexicostatum* (Garrett) 16. *Vexillum scitulum* (A. Adams) 17. *Vexillum militaris* (Reeve) 18. *Vexillum consanguinea* (Reeve) 19. *Vexillum microzonias* (Lamarck) 20. *Vexillum fortiplicatum* (Pease)

PLATE 57

lirate within, columella with 4 cream-coloured folds; brownish grey in colour with white banding and an additional 1 or 2 thin bands at the base of the body whorl, aperture purple; reaches 20mm.
DISTRIBUTION: Sporadically found throughout the Philippines in limited quantity.

9. Vexillum (Pusia) suavis (Souverbie, 1875)
6.3mm, 6.4mm; Punta Engaño, Cebu.
Shell extremely small, elongate to elongate-ovate, with weakly convex whorls and moderately incised sutures; sculptured with axial cords, weakly nodulose at the sutures and crossed by numerous spiral grooves; aperture narrow, lirate within, outer lip thin and angulate, columella with 4 to 5 folds, some oblique spiral cords towards the base; pinkish orange in colour with a white band enclosed by two brown spiral lines which are occasionally replaced by brown spotting, aperture pink; reaches 10mm.
DISTRIBUTION: Camotes Sea.

10. Vexillum (Pusia) rubrum (Broderip, 1836)
6.2mm, 6.3mm; Punta Engaño, Cebu.
Shell extremely small, elongate-ovate, with inflated, prominently convex whorls and a peculiar protoconch of 3 whorls which projects abruptly above the shell; sculptured with angulate axial ribs and numerous spiral grooves in the interspaces; aperture narrow, lirate within, outer lip thin, sharply angulate, columella with 4 folds and a prominent parietal denticle; reddish in colour with a presutural white band on the spire whorls and centrally on the body whorl, the interspaces within the white banding being brown-spotted, aperture pink, reaches 8mm.
DISTRIBUTION: Camotes Sea.

11. Vexillum (Pusia) salisburyi Cernohorsky, 1976
11.3mm, 11.1mm; Punta Engaño, Cebu.
Shell small, elongate-ovate, with weakly convex whorls, angulate at the sutures; sculptured with rounded axial ribs which may become obsolete on the body whorl dorsum, and fine spiral striae in the interspaces; outer lip thin, sharply angulate anteriorly and lirate within, columella with 4 folds, extending onto the body whorl base as strong cords; deep reddish brown in colour with paler banding and a rose-coloured aperture; reaches 12mm.
DISTRIBUTION: Camotes Sea.

12. Vexillum (Costellaria) michaui (Crosse & Fischer, 1854)
20.6mm; Samar.
A curious colour variation of the species figured and described on Pl. 58 fig. 4.

13. Vexillum (Pusia) lautum (Reeve, 1845)
14.1mm; Palawan.
Shell small, ovate to elongate-ovate with a short spire and a large, ovate body whorl; sculptured with broad axial ribs, almost obsolete on the body whorl, and encircled by fine spiral threads; aperture narrow, lirate within, outer lip thin and angulate, columella with 4 to 5 folds, a few cords on the body whorl base; ribs orange and black, with a white band at the center of the body whorl, aperture lavender; reaches 15mm.
DISTRIBUTION: Sporadically found throughout the Philippines.

14. Vexillum (Costellaria) semicostatum (Anton, 1838)
17.7mm; Punta Engaño, Cebu.
A medium sized, fusiform-elongate shell, sculptured with broad axial ribs, obsolete towards the base of the body whorl, and encircled by fine spiral grooves which become obsolete mid-body whorl; aperture narrow, lirate within, outer lip thin, slightly constricted basally, columella with 4 folds; colour reddish brown, with 1 or 2 white-spotted bands on the whorls, aperture orange; reaches 20mm.
DISTRIBUTION: Camotes Sea.

15. Vexillum (Costellaria) flexicostatum(Garrett, 1880)
21.5mm; Palawan.
Shell medium sized, fusiform-elongate, whorls weakly convex and angulate at the sutures; sculptured with rounded axial cords and spiral grooves in the interspaces; aperture narrow, outer lip thin, angulate and lirate within, columella with 4 folds; colour dark orange-brown with orange cords, a white sutural band, and another band mid-body whorl, aperture bluish white; reaches 25mm.
DISTRIBUTION: Sporadically found throughout the Philippines in limited quantity.

16. Vexillum (Costellaria) scitulum (A. Adams, 1853)
17.7mm; Cebu.
Shell medium sized, fusiform-elongate, with a moderately tall spire and weakly convex whorls, slightly angulate at the sutures; sculptured with numerous axial cords and punctate spiral grooves in the interspaces, the grooves obsolete just below the suture, but becoming stronger on progressing down each whorl; aperture narrow, lirate within, outer lip thin and angulate, columella with 4 folds continuing as grooves on the body whorl base; light brown in colour with irregular dark brown spotting; reaches 20mm.
DISTRIBUTION: Sporadically found throughout the Philippines in limited quantity.

17. Vexillum (Costellaria) militaris (Reeve, 1845)
20.0mm; Mindoro.
Shell medium sized, fusiform-elongate, with sharply shouldered straight-sided whorls, somewhat stepped in appearance; sculptured with wide, rounded axial cords and numerous fine lirae which either override the cords or are restricted to the interspaces; aperture narrow, lirate within, outer lip thin, angulate, indented anteriorly, columella with 4 to 5 folds, a few strong, granulose spiral cords on the base; white in colour with a broad reddish brown band on the body whorl base, aperture white and brown-banded; reaches 22mm.
DISTRIBUTION: Sporadically found throughout the Philippines in limited quantity.

18. Vexillum (Pusia) consanguinea (Reeve, 1845)
14.5mm; Bohol.
Shell small to medium sized, fusiform-ovate, with strongly convex whorls and a tall spire; sculptured with broad, rounded axial ribs, weaker axial riblets and spiral striae in the interspaces; aperture narrow, lirate within, outer lip thin, angulate and weakly crenulate on the anterior margin, columella with 4 folds, continuing as cords on the body whorl base; orange in colour with a central row of white spots on the ribs, and another row of smaller spots at the sutures and on the base; reaches 20mm.
DISTRIBUTION: Sporadically found throughout the Philippines in limited quantity.

19. Vexillum (Pusia) microzonias (Lamarck, 1811)
17.1mm; Sulu Archipelago.
Shell small to medium sized, fusiform-elongate, with slightly convex to almost straight-sided whorls; sculptured with numerous rounded axial ribs which become obsolete towards the shell base, and fine spiral grooves in the interspaces, base with a few nodulose cords; aperture narrow, lirate within, outer lip thin and arcuate, columella with 4 folds; dark brown in colour with white banding, aperture bluish white within, marginally brown; reaches 25mm.
DISTRIBUTION: Sporadically found throughout the Philippines in limited quantity.

20. Vexillum (Costellaria) fortiplicatum (Pease, 1868)
8.8mm; Punta Engaño, Cebu.
Shell very small, elongate, sculptured with numerous axial ribs (up to 26 on the body whorl), fine spiral grooves in the interspaces, and 3 to 4 granulose cords at the base; aperture narrow, lirate within, columella with 4 folds; dark brown in colour with a thin white line above the sutures on the spire whorls, and a second mid-body whorl; aperture bluish white within, marginally brown; reaches 10mm.
DISTRIBUTION: Sporadically found throughout the Philippines in limited quantity.

Plate 58 — Costellariidae

1. Vexillum (Costellaria) exasperatum (Gmelin, 1791)
23.1mm; Marinduque.
Shell small to medium sized, elongate, with a tall, acute, slightly turreted spire and impressed sutures; sculptured with strong, narrow axial ribs and weaker interstitial axial riblets, all crossed by spiral grooves; aperture white, narrowly elongate with an angulate outer lip and 4 columellar folds; white in colour, occasionally with 1 presutural brown band on the spire whorls and 2 on the body whorl, axial cords orange-brown; reaches 25mm.
DISTRIBUTION: Sporadically found thoughout the Philippines.

2. Vexillum (Costellaria) rubrocostatum (Habe & Kosuge, 1966)
24.2mm; Palawan.
Shell medium sized, fusiform-elongate, with a tall, acute spire and moderately impressed sutures; sculptured with solid, rounded axial ribs and encircled with numerous fine spiral grooves, forming a weakly nodulose surface; aperture narrowly elongate, lirate within, columella with 4 to 5 folds; off-white in colour with red axial ribs, a purple-coloured spire and a pale orange aperture; reaches 40mm.
DISTRIBUTION: Sporadically found throughout the Philippines.

3. Vexillum (Costellaria) virgo (Linné, 1767)
21.7mm; Samar.
Shell medium sized, elongate to elongate-ovate in outline, with convex whorls; sculptured with either large, widely spaced or narrow, closely-set axial ribs (10 to 14 on the body whorl), and crossed by deeply incised spiral grooves; aperture narrow, lirate within, columella with 4 folds; deep brown to black in colour, axial ribs orange, a single white presutural band on the spire whorls and centrally on the body whorl, aperture and columellar teeth bluish white; reaches 25mm.
DISTRIBUTION: Sporadically found throughout the Philippines.

4. Vexillum (Costellaria) michaui (Crosse & Fischer, 1864)
21.5mm; Cebu.
Shell medium sized, elongate, with a tall, acute spire and slightly shouldered convex whorls; sculptured with 10 to 12 smooth, rounded, widely-spaced axial ribs and numerous fine interstitial spiral grooves; aperture narrowly elongate, outer-lip thick, angulate, lirate within, columella with 4 folds; bluish grey in colour, with 2 white bands and 3 to 4 discontinuous red spiral bands; reaches 25mm. See also Pl. 57 fig. 12.
DISTRIBUTION: Sporadically found throughout the Philippines.

5. Vexillum (Costellaria) exasperatum (Gmelin, 1791)
22.2mm; Marinduque.
A curious ecomorph of fig. 1 which might be taken for an entirely different taxon; shell medium sized, elongate, with a tall spire which is longer than the aperture, and moderately convex whorls which are slightly shouldered below the sutures; sculptured with numerous very closely set axial cords and overriding spiral grooves, producing a weakly granulose surface; aperture narrowly elongate, outer lip thin, weakly lirate within, columella with 3 to 4 folds; white in colour, with a brown subsutural band on the spire whorls, 2 or 3 on the body whorl; reaches 25mm.
DISTRIBUTION: Sporadically found throughout the Philippines in limited quantity.

6. Vexillum (Costellaria) spicatum (Reeve, 1845)
24.0mm; Palawan.
Shell medium sized, fusiform-elongate, glossy, with a tall spire and large body whorl; sculptured with widely spaced, angulate axial ribs which are nodulose on the presutural slope, and minutely pitted interspaces; sculpture becoming almost obsolete on the body whorl except for a few basal spiral cords; aperture narrow, outer lip thin, angulate and lirate within, columella with 4 folds; shell entirely flesh-coloured, nodules occasionally paler; reaches 25mm.
DISTRIBUTION: Sporadically found throughout the Philippines in limited quantity.

7. Vexillum (Vexillum) gruneri (Reeve, 1844)
27.1mm; Mindoro.
Shell small to medium sized, with a large body whorl, and a tall, acute, slightly stepped spire; sculptured with 10 to 12 widely spaced axial ribs, nodulose at the sharply angled shoulder, and smooth interspaces; aperture narrowly elongate, outer lip thick, angulate, lirate within, columella with 4 folds; colour bluish grey with 2 to 3 white bands, and 2 to 3 brown spiral threads, aperture brown within, marginally white; illustrated specimen largest seen.
DISTRIBUTION: Sporadically found throughout the Philippines.

8. Vexillum (Costellaria) sculptile (Reeve, 1845)
32.3mm; Palawan.
Shell medium sized, fusiform-elongate, with a tall spire somewhat longer than the aperture, and well incised, slightly ledged sutures; sculptured with a subsutural groove on each whorl, numerous axial ribs, and thin spiral lirae, producing laterally elongate pits in the interspaces; aperture narrowly elongate, outer lip thin, arcuate and lirate within, columella with 4 folds; white in colour with a single subsutural orange band on the spire whorls, 2 to 3 on the body whorl; reaches around 32mm.
DISTRIBUTION: Sporadically found throughout the Philippines in limited quantity.

9. Vexillum (Vexillum) melongena auct. (non Lamarck, 1811)
32.1mm; Batangas.
Shell medium sized, fusiform-ovate to fusiform-elongate, with a short spire less than half the shell length; sculptured with close-set axial ribs and spiral grooves, the grooves crossing the ribs below the sutures, but then remaining restricted to the interspaces, base with several oblique, granulose spiral cords; aperture narrow, outer lip thin, straight-sided and lirate within, columella with 4 folds; off-white in colour, with a brown sutural band, a second subsutural band and brown spiral threads, body whorl with 3 or 4 additional brown bands, aperture bluish white, outer lip margin with brown spotting; reaches 35mm. Lamarck's holotype is untraceable, and his type figure cannot be identified with certainty. The taxon should be considered a *nomen dubium.*
DISTRIBUTION: Sporadically found throughout the Philippines.

10. Vexillum (Vexillum) buriasense (Tomlin, 1920)
27.0mm; Samar.
Shell medium sized, fusiform, with a fairly tall spire greater than half the shell length, whorls strongly convex with deeply channeled sutures; sculptured with numerous axial ribs and spiral grooves in the interspaces which cross the ribs as fine threads, base with 5 to 6 nodulose spiral cords; aperture narrow, lirate within, columella with 4 folds; off-white in colour with 1 or 2 brown spiral threads on the spire whorls and up to 4 on the body whorl, aperture white with brown spots on the outer lip margin; reaches 35mm.
DISTRIBUTION: Sporadically found throughout the Philippines in limited quantity.

11. Vexillum (Vexillum) lyratum (Lamarck, 1811)
31.5mm; Manila Bay.
Shell medium sized, fusiform, with moderately incised sutures and a tall, acute spire; sculptured with thin axial ribs and deep, widely spaced grooves in the interspaces; aperture narrow, lirate within, columella with 4 folds; shell colour grey, apical whorls dark brown, later whorls with a brown band, and up to 4 on the body whorl, bands frequently bordered with white, aperture violet, marginally brown; reaches 45mm.
DISTRIBUTION: Sporadically found throughout the Philippines.

12. Vexillum (Costellaria) macrospirum (A. Adams, 1853)
36.8mm; Albay Gulf, Southern Luzon.
Shell medium sized, with a tall, acute spire which is greater than half the shell length; sculptured with numerous close-set axial cords and encircled with numerous punctate spiral grooves; aperture moderately wide, outer lip thin, angulate and lirate within, columella with 4 folds; fleshy pink in colour with a pale brown band and a row of nebulous brown spots, aperture pale pinkish orange, columellar folds bright pink; reaches 40mm.
DISTRIBUTION: Sporadically found throughout the Philippines in limited quantity.

PLATE NO. 58 1. *Vexillum exasperatum* (Gmelin) 2. *Vexillum rubrocostatum* (Habe & Kosuge) 3. *Vexillum virgo* (Linné) 4. *Vexillum michaui* (Crosse & Fisher) 5. *Vexillum exasperatum* (Gmelin)-ecomorph 6. *Vexillum spicatum* (Reeve) 7. *Vexillum gruneri* (Reeve) 8. *Vexillum sculptile* (Reeve) 9. *Vexillum melongena* (Lamarck) *auct.* 10. *Vexillum buriasense* (Tomlin) 11. *Vexillum lyratum* (Lamarck) 12. *Vexillum macrospirum* (A. Adams) 13. *Vexillum acuminatum* (Gmelin) 14. *Vexillum semifasciatum* (Lamarck) 15. *Vexillum cadaverosum* (Reeve) 16. *Vexillum pacificum* (Reeve) 17a. *Vexillum curviliratum* (Sowerby) form *adornata* Tomlin 17b. form *formosensis* Sowerby 18. *Vexillum turben* (Reeve) 19. *Vexillum coronatum* (Helbling) 20. *Vexillum takakuwai* (Cernohorsky & Azuma) 21. *Vexillum obeliscus* (Reeve) 22. *Vexillum funereum* (Reeve)

PLATE 58

13. Vexillum (Costellaria) acuminatum (Gmelin, 1791)
23.0mm; Palawan.
Shell medium sized, somewhat fusiform in outline with well incised sutures, a tall spire, and weakly convex spire whorls; sculptured with numerous axial cords and rows of pitted grooves in the interspaces, both of which become obsolete on the body whorl, lower body whorl with several spiral cords; aperture narrowly elongate, lirate within, outer lip thick and arcuate, columella with 4 to 5 folds; colour variable, usually dark brown to black, with paler bands, streaks or blotches; reaches 30mm.
DISTRIBUTION: Sporadically found throughout the Philippines.

14. Vexillum (Costellaria) semifasciatum (Lamarck, 1811)
24.0mm; Samar.
Shell small to medium sized, elongate with a tall spire; sculptured with arcuate, slightly shouldered axial ribs, smooth interspaces, and a few rows of oblique spiral cords at the base; aperture narrow, outer lip thick, angulate and lirate within, columella with 4 folds; off-white in colour with orange and grey banding, some additional red spiral bands on the spire and body whorls; reaches 30mm.
DISTRIBUTION: Sporadically found throughout the Philippines.

15. Vexillum (Costellaria) cadaverosum (Reeve, 1844)
18.0mm; Bohol.
Shell small, elongate-ovate, with sharply shouldered whorls, bi-angulate on the body whorl and sculptured with 10 to 12 axial ribs, echinate at the shoulders, and with spiral grooves in the interspaces; aperture narrow, lirate within, outer lip thin and shouldered posteriorly, columella with 4 folds; off-white in colour with a brown band, aperture white; reaches 20mm.
DISTRIBUTION: Sporadically found throughout the Philippines in limited quantity.

16. Vexillum (Costellaria) pacificum (Reeve, 1845)
18.8mm; Samar.
Very close to the preceeding species, but differs in the following ways: the shoulder nodules are rounded and not sharply echinate, the encircling spiral grooves are more deeply incised, and the body whorl lacks the bi-angulate contours of *V. cadaverosum;* colour off-white with a row of brown spots; reaches 20mm.
DISTRIBUTION: Sporadically found throughout the Philippines in limited quantity.

17. Vexillum (Vexillum) curviliratum (Sowerby, 1874)
A variable species with several forms being recognized; shell medium sized, fusiform, elongate to elongate-ovate and sculptured with narrow axial ribs and numerous closely set spiral grooves in the interspaces, base with numerous oblique spiral cords; aperture narrow, outer lip thick, angulate and lirate within, columella with 4 folds, colour brown with white banding, aperture bluish white with a brown margin; reaches 70mm; typical form is illustrated on Pl. 59 (fig. 6).
17a. form *adornata* Tomlin, 1920: a fusiform-ovate shell where the axial ribs are closely set; 33.0mm; Palawan.
17b. form *formosensis* Sowerby, 1889: a more elongate form, with wider spaced axial ribs which become obsolete on the body whorl dorsum; 30.9mm; Samar.
DISTRIBUTION: Sporadically found throughout the Philippines in limited quantity.

18. Vexillum (Pusia) turben (Reeve, 1844)
20.4mm; Mindoro.
Shell small to medium sized, extremely ovate, sutures well incised and slightly ledged, spire short but sharply acute; sculptured with numerous close-set axial ribs, and fine spiral striae which remain in the interstitial spaces on the spire whorls, but then override on the body whorl base producing granulose cords; aperture narrowly elongate, outer lip thin, reflected and lirate within, columella with 5 folds; orange-brown in colour with a paler spire; reaches 23mm.
DISTRIBUTION: Sporadically found throughout the Philippines in limited quantity.

19. Vexillum (Costellaria) coronatum (Helbling, 1779)
22.3mm; Samar.
Shell small to medium sized, fusiform-elongate, with shouldered whorls; sculptured with 10 to 12 axial ribs noduled at the shoulder, and spiral grooves in the interspaces; aperture narrow, outer lip thin, angulate, lirate within, and slightly shouldered posteriorly, columella with 4 folds; colour variable, usually white with brown banding on the body whorl and spotting on the spire whorls; reaches 25mm.
DISTRIBUTION: Sporadically found throughout the Philippines.

20. Vexillum (Costellaria) takakuwai Cernohorsky & Azuma, 1974
36.3mm; Bohol.
Shell medium sized, fusiform-elongate with a very tall spire; sculptured with numerous fine, slightly arcuate axial ribs which are crossed by spiral grooves, producing small rounded nodules, sutures canaliculate and slightly ledged; aperture narrow, constricted basally, outer lip thin and lirate within, columella with 4 to 5 folds; white in colour with 2 interrupted reddish bands on the spire whorls, 3 on the body whorl, aperture white; reaches 40mm.
DISTRIBUTION: Camotes Sea.

21. Vexillum (Costellaria) obeliscus (Reeve, 1844)
26.3mm; Sulu Archipelago
Shell medium sized, fusifom, with deeply channeled, somewhat ledged sutures; sculptured with numerous close-set axial and spiral threads intersecting to produce a finely cancellate surface; aperture narrow, shorter than the spire, outer lip thin, arcuate and constricted basally, forming a short canal, columella

with 4 to 5 folds; off-white in colour with orange-brown spiral banding; reaches 30mm.
DISTRIBUTION: Sporadically found throughout the Philippines in limited quantity.

22. Vexillum (Costellaria) funereum (Reeve, 1844)
20.5mm; Samar
Shell medium sized with concave, slightly shouldered whorls; sculptured with fine axial ribs and overriding spiral grooves which become obsolete just above the aperture on the body whorl, and then a smooth area followed by a series of oblique spiral cords on the body-whorl base; aperture narrow, outer lip thick, lirate within, notched posteriorly, columella with 3 to 4 folds; external colour brown with a thin white band, aperture pale blue, marginally brown, columellar folds bluish white; reaches 30mm.
DISTRIBUTION: Sporadically found throughout the Philippines.

Plate 59 — Costellariidae

1. Vexillum (Costellaria) costatum (Gmelin, 1791)
48.1mm; Samar.
Shell medium sized, fusiform-elongate, with a tall spire and weakly convex to almost straight-sided whorls; sculptured with slender, closely-set axial ribs and crossed by numerous overriding spiral grooves, a prominently deep groove subsuturally; aperture narrowly elongate, lirate within, outer lip thin, arcuate, columella with 5 folds; orange-brown in colour with white blotches and streaks; reaches 45mm.
DISTRIBUTION: Sporadically found throughout the Philippines in limited quantity.

2. Vexillum citrinum (Gmelin, 1791)
An extremely beautiful and distinctly marked species; shell large, fusiform-elongate, with a tall spire and impressed sutures; sculptured with numerous sharp axial ribs (variable in number) and fine overriding spiral riblets (also variable); aperture narrow, lirate within, outer lip thickened, incurved and angulate on the anterior margin, columella with 5 folds; colour variable, usually white with a subsutural red band bordered by a thin black band and fine red threads on the white area, body whorl with an additional wide red band on the base and 3 additional black bands, the red areas occasionally grading into brown or black, aperture orange to yellowish white; reaches 70mm. Commonly known under the name *Vexillum regina* (Swainson in Sowerby, 1828) which is a *nomen nudum*.
2a. form *compressa* Sowerby, 1874: characterized by its more numerous and more heavily sculptured axial and spiral elements; 57.3mm; Sulu Archipelago. Actually, this specimen agrees with the lectotype figure of Gmelin as selected by Cernohorsky (1979), and should be considered as "typical" *citrinum*.
2b. 64.5mm; Balabac, Palawan.
2c. form *filiareginae* J. Cate, 1961: an orange and white-banded form where the axial ribs are fewer and less prominent and the spiral sculpturing tends to be incised grooves rather than raised riblets; 68.7mm; Palawan.
2d. 55.5mm; Palawan.
2e. 63.5mm; Palawan.
DISTRIBUTION: 2a. Cebu, Bohol and Sulu Sea; 2b.-2e. Palawan and Bantayan, Northern Cebu.

3. Mitra (Mitra) nivea (Broderip, 1836)
91.0mm; Bohol.
Clearly a mitrid species but inadvertently placed here within the Costellariidae. Shell large, fusiform-elongate, with moderately incised sutures and a tall spire, equal to half the shell length; whorls weakly convex, sculptured with numerous flat spiral ribs and punctate grooves in the interspaces; aperture narrowly elongate, smooth within, outer lip thick, weakly arcuate and slightly indented posteriorly, columella glazed with 4 folds; pale orange-brown in colour, with a weak, pale yellow subsutural band on the spire whorls and a second one mid-body whorl; reaches around 90mm.
DISTRIBUTION: Camotes Sea.

4. Vexillum (Costellaria) coccineum (Reeve, 1844)
74.7mm; Palawan.
A medium to large-sized shell with convex whorls and a tall, acute spire; sculptured with numerous axial ribs which become weak on the body whorl and fine spiral grooves in the interspaces which override the axial ribs on the lower portion of the body whorl; aperture narrow, lirate within, outer lip thick, angulate, with a slight posterior notch, columella with 4 to 5 folds; orange-brown in colour with a central blackish brown or yellowish band, aperture white, marginally orange; reaches 80mm.
DISTRIBUTION: Sulu Sea.

5. Vexillum (Costellaria) mirabile (A. Adams, 1853)
51.1mm; Mindoro.
Shell medium sized, fusiform, with sharply shouldered whorls which are crossed by regular spiral grooves and broad axial cords, producing flat, squarish nodules on the shell surface, most prominent on the angulate shoulder; aperture narrowly elongate, lirate within, outer lip thick and sharply angulate anteriorly, columella with 4 to 5 folds, siphonal fasciole thickened; colour white with brown sutural blotches, a brown band mid-body whorl, and brown-lined grooves; reaches 50mm.
DISTRIBUTION: Sporadically found throughout the Philippines in limited quantity.

6. Vexillum (Vexillum) curviliratum (Sowerby, 1874)
56.0mm, 47.6mm; Palawan.
Shell fusiform-ovate to fusiform-elongate, with

moderately incised sutures, a tall, acute spire, and moderately convex spire whorls; sculptured with numerous axial cords (variable in number) and fine spiral lirae in the interspaces which override the axial cords on the body whorl base; aperture narrowly elongate, lirate within, outer lip thick and angulate, straight sided, columella with 4 folds; off-white in colour with a brown subsutural band and a thin golden spiral thread on the white area, body whorl with additional light and dark areas; reaches 70mm. This is the typical form of the species (see also Pl. 58 fig. 17).
DISTRIBUTION: Sporadically found throughout the Philippines in limited quantity.

7. Vexillum (Vexillum) dennisoni (Reeve, 1844)
Shell medium sized, elongate, with a tall, acute spire and weakly convex whorls; sculptured with wide, closely set axial ribs and fine, well incised spiral grooves which cross the axial ribs; aperture narrowly elongate, lirate within, outer lip thick and angulate, columella with 4 to 5 folds; pale pinkish white in colour, with a subsutural orange band on the spire whorls and 3 on the body whorl, with black dots between the ribs at the band junction, aperture whitish, marginally orange, spire deep purple; reaches 62mm.
7a. typical form; 59.0mm; Palawan.
7b. colour form; 61.3mm; Albay Gulf.
DISTRIBUTION: Palawan, Sulu Sea, and Albay Gulf, Southern Luzon.

8. Vexillum (Costellaria) subdivisum (Gmelin, 1791)
37.0mm; Sulu Archipelago.
Shell medium sized, elongate, with a tall, acute spire of 8 to 9 shouldered whorls; sculptured with 18 to 20 axial ribs, slightly recurved basally on the body whorl, and crossed with numerous spiral grooves; aperture narrow, lirate within, outer lip slightly thickened and sharply shouldered, columella with 4 to 5 folds; dark brown in colour with a white band on the spire whorls, 2 to 3 on the body whorl, aperture bluish white within, marginally brown; reaches 50mm.
DISTRIBUTION: Sporadically found throughout the Philippines in limited quantity.

9. Vexillum (Costellaria) stainforthi (Reeve, 1841)
51.2mm; Palawan.
Shell medium sized, with a tall, acute spire equal to half the shell length; sculptured with 12 to 14 rounded, widely-spaced axial cords and numerous overriding spiral grooves which bisect the axial cords to produce a finely nodulose surface; aperture narrowly elongate, lirate within, outer lip thick and angulate, columella with 4 folds; bluish white in colour with deep red spots arranged in spiral rows, 2 on the spire whorls, 5 on the body whorl, aperture bluish white within, marginally brown, shell base dark brown; reaches around 55mm.
DISTRIBUTION: Sulu Sea and Palawan.

10. Vexillum (Vexillum) plicarium (Linné, 1758)
49.2mm, 51.4mm; Samar.
Shell medium sized, solid, somewhat variable ranging from elongate to elongate-ovate in outline; sutures weakly incised, whorls convex and sculptured with axial ribs which start on the sharply angled shoulder of each whorl, interspaces with spiral grooves which become overriding spiral cords towards the body whorl base; aperture narrow, lirate within, outer lip thick and angulate, columella with 4 to 5 folds; creamy white in colour with a black band (occasionally interrupted) on the spire whorls, and an additional two (occasionally orange-tinted) bands centrally and basally on the body whorl, aperture white, columellar fold interspaces sometimes brown; reaches 60mm.
DISTRIBUTION: Sporadically found throughout the Philippines.

11. Vexillum (Vexillum) vulpecula (Linné, 1758)
Extremely variable in terms of sculpture and colour; shell medium sized, elongate to elongate-ovate, sculptured with broad, rounded axial ribs which may be strong or nearly obsolete, and deep, overriding spiral grooves which may be partially absent in some cases; aperture narrow to moderately wide, lirate within, columella with 4 to 5 folds; colour orange to white with darker banding, aperture white to violet, marginally black; reaches 60mm.
11a. 60.5mm; **11b.** 57.8mm; **11c.** 65mm; and **11d.** 55.5mm; all from Sulu Archipelago.
DISTRIBUTION: Sporadically found throughout the Philippines, mostly from Sulu Sea.

12. Vexillum (Vexillum) rugosum (Gmelin, 1791)
48.5mm; Samar.
Shell medium sized, solid, broadly fusiform, with adpressed sutures and a short, acute spire; sculptured with heavy, arcuate axial ribs, heavily nodulose on the shoulder and obsolete on the presutural ramp; encircled with numerous, deeply incised spiral grooves which become spiral cords on the presutural ramp and on the bodywhorl base; aperture narrow, lirate within, outer lip thick, angulate and heavily shouldered; columella with 4 folds, the first horizontal, the rest obliquely aligned; white in colour with a black presutural band on the spire whorls and 3 on the body whorl, aperture and columellar teeth bluish white, columella brown; reaches 50mm.
DISTRIBUTION: Sporadically found throughout the Philippines.

PLATE NO. 59 1. *Vexillum costatum* (Gmelin) 2a. *Vexillum citrinum* (Gmelin) form *compressa* Sowerby 2b. colour variations 2c. form *filiareginae* J. Cate 2d. to 2e. colour variations 3. *Mitra nivea* (Broderip) 4. *Vexillum coccineum* (Reeve) 5. *Vexillum mirabile* (A. Adams) 6. *Vexillum curviliratum* (Sowerby) 7a. *Vexillum dennisoni* (Reeve) 7b. colour form 8. *Vexillum subdivisum* (Gmelin) 9. *Vexillum stainforthi* (Reeve) 10. *Vexillum plicarium* (Linné) 11a. to 11d. *Vexillum vulpecula* (Linné) 12. *Vexillum rugosum* (Gmelin)

PLATE 59

1. Vexillum (Costellaria) tankervillei (Melvill, 1888)
31.7mm; Palawan.
Shell medium sized, elongate, with a tall spire slightly greater than half the shell length, and impressed sutures; whorls sculptured with regular axial cords, sharply echinate on the whorl shoulder, weaker interstitial axial striae, and crossed by regular spiral grooves producing a granulose surface; aperture narrowly elongate, outer lip thin, columella with 4 folds; white in colour with an orange sutural band and a second band mid-body whorl; reaches 37mm.
DISTRIBUTION: Sporadically found throughout the Philippines in limited quantity.

2. Vexillum (Pusia) crocatum (Lamarck, 1811)
Shell small to medium sized, elongate to elongate-ovate, with moderately channeled sutures and a tall spire which is greater than half the shell length; sculptured with numerous shouldered axial ribs and overriding spiral lirae intersecting to form nodules, and large echinate projections on the shoulder (not always seen); aperture lirate within, with 4 to 5 folds on the columella; orange to pale yellow in colour with darker banding, aperture orange; reaches 25mm.
2a. typical form with sharp sutural nodules; 23.2mm; Palawan.
2b. typical form lacking sutural nodules; 23.2mm; Samar.
2c. form *cumingii* Reeve, 1844, with deeply incised spiral grooves, weak axial cords and the banding replaced by pinkish grey blotches all over the surface; 25.2mm; Samar.
DISTRIBUTION: Sporadically found throughout the Philippines in limited quantity.

3. Vexillum (Costellaria) sanguisugum (Linné, 1758)
25.0mm, 29.0mm, 33.2mm; Palawan.
Shell medium sized, fusiform elongate with convex whorls; sculptured with 16 to 18 narrow, evenly spaced axial ribs, and numerous well incised, overriding spiral grooves forming nodules which are strongest towards the base; aperture narrowly elongate, weakly lirate within, columella with 4 folds; colour variable, usually greyish brown with an interrupted red spiral band on the spire whorls and 2 on the body whorl; spiral grooves lavender-coloured as is the apex and outer lip margin; aperture, nodules and columellar teeth off-white; reaches 50mm.
DISTRIBUTION: Sporadically found throughout the Philippines. Mostly from Palawan and Sulu Sea and always in limited quantity.

4. Vexillum (Costellaria) polygonum (Gmelin, 1791)
37.2mm; Mindoro.
Shell medium sized, slightly fusiform, with a tall spire and a large body whorl; sculptured with numerous regularly spaced axial cords which are crossed by spiral grooves; earlier whorls straight-sided, but the later ones become more angulate with a row of rounded shoulder nodules; aperture narrow, lirate within, columella with 4 to 5 folds; white to orange-brown in colour with darker spots and streaks, aperture and outer lip margin white, brown within; reaches 40mm.
DISTRIBUTION: Sporadically found throughout the Philippines in limited quantity.

5. Vexillum (Costellaria) granosum (Gmelin, 1791)
34.5mm, 25.0mm; Samar.
Shell medium sized, elongate, with a tall, acute spire and weakly convex whorls; sculptured with numerous close-set axial ribs and fine, evenly spaced overriding spiral grooves, producing a coarsely nodulose surface; aperture lirate within, columella with 4 folds; grey in colour with paler banding and purple-coloured grooves, columella and outer lip margin; aperture dark purple within; reaches 35mm.
DISTRIBUTION: Sporadically found throughout the Philippines.

6. Vexillum (Pusia) unifascialis (Lamarck, 1811)
22.9mm, 18.5mm; Cebu.
Shell small, somewhat polymorphic, and may be squat to elongate-ovate, with a tall spire and convex whorls which are shouldered at the sutures; sculptured with numerous closely-set axial ribs that are crossed by flat spiral cords or weakly incised spiral grooves, producing either a reticulate or nodulose surface; aperture narrow, lirate within, columella with 4 folds; colour variable, reddish brown to yellow with white or black spiral bands; reaches 23mm.
DISTRIBUTION: Sporadically found throughout the Philippines in limited quantity.

7. Zierliana anthracina (Reeve, 1844)
21.5mm; Palawan.
Shell small, elongate-ovate, with a tall spire and a large body whorl which is greater than half the shell length; whorls smooth except for some oblique grooves towards the base; aperture ovate, outer lip thin, flattened, with about 20 fine lirae on the inner margin, columella with 4 to 5 folds; dark brown to black in colour with a brownish black aperture and white columellar teeth; reaches 25mm.
DISTRIBUTION: Sporadically found throughout the Philippines.

PLATE NO. 60 1. *Vexillum tankervillei* (Melvill) 2a. *Vexillum crocatum* (Lamarck) 2b. *Vexillum crocatum* (Lamarck) 2c. form *cumingii* Reeve 3. *Vexillum sanguisugum* (Linné) 4. *Vexillum polygonum (Gmelin)* 5. *Vexillum granosum* (Gmelin) 6. *Vexillum unifascialis* (Lamarck) 7. *Zierliana anthracina* (Reeve) 8. *Vexillum ficulinum* (Lamarck) 9. *Zierliana woldemarii* (Kiener) 10. *Zierliana oleacea* (Reeve) 11. *Zierliana ziervogelii* (Gmelin) 12. *Vexillum acupictum* (Reeve) 13. *Vexillum millecostatum* (Broderip) 14. *Vexillum cancellarioides* (Anton) 15. *Vexillum patriarchalis* (Gmelin) 16. *Vexillum bernhardina* (Röding) 17. *Vexillum amabilis* (Reeve)

PLATE 60

8. Vexillum (Pusia) ficulinum (Lamarck, 1811)
19.0mm; Samar.
Shell small, somewhat squat in outline, with strongly convex whorls, a moderately tall, somewhat telescoped spire, and moderately incised sutures; sculptured with numerous close-set axial ribs and crossed by numerous spiral cords, producing a finely granulose surface; aperture narrowly elongate, outer lip thin, angulate and lirate within, columella with 4 folds; dark brown in colour with a light purple aperture; reaches 20mm.
DISTRIBUTION: Sporadically found throughout the Philippines in limited quantity.

9. Zierliana woldemarii (Kiener, 1838)
23.0mm; Samar.
Shell small to medium sized, elongate-ovate in outline, with a tall, straight-sided spire and a large, somewhat inflated body whorl; surface covered with numerous thin, oblique axial ribs and pitted spiral grooves which are normally confined to the interstitial spaces, but occasionally override the axial ribs to produce a finely reticulate, sometimes plicate surface; aperture elongate, outer lip thick and finely lirate along the inner margin, columella heavily calloused with 4 large folds; surface colouration brown, aperture dark purple within, apertural lirae and columellar folds bluish white; reaches 25mm.
DISTRIBUTION: Sporadically found throughout the Philippines.

10. Zierliana oleacea (Reeve, 1844)
23.2mm; Palawan.
Shell medium sized, lightweight, with a short spire and a large, elongate-ovate body whorl; comparatively smooth save for some fine axial growth striae on the earlier whorls which are crossed on the body whorl by fine spiral grooves, most pronounced on the anterior portion; aperture wide, lirate within, columella with 3 to 4 folds; deep bluish black in colour, with a bluish aperture and white columellar folds; reaches 35mm.
DISTRIBUTION: Sporadically found throughout the Philippines.

11. Zierliana ziervogelii (Gmelin, 1791)
25.5mm; Palawan.
Shell medium sized, heavy, with a short spire and a large, globose body whorl; sutures weakly impressed with a deep presutural groove on each whorl; sculptured with strong axial ribs which become obsolete on the lower body whorl and are replaced by strong spiral cords; aperture narrow, outer lip thick and flattened, with numerous lirate denticles on the inner margin, columella with 4 strong folds; shell blackish brown with a light brown outer lip margin, a white aperture and bluish white columellar folds; reaches 35mm.
DISTRIBUTION: Sporadically found throughout the Philippines in limited quantity.

12. Vexillum (Costellaria) acupictum (Reeve, 1844)
35.5mm; Bohol.
Shell medium sized, elongate, with a tall spire and moderately incised sutures; whorls convex, sculptured with numerous very closely set axial ribs and crossed by spiral grooves, producing a finely granulose surface; aperture narrow, outer lip thin, angulate and lirate within, columella with 4 to 5 folds; pale cream in colour with brown banding and blotching, outer lip and columella white, aperture interior lemon-coloured; reaches 40mm.
DISTRIBUTION: Sporadically found throughout the Philippines.

13. Vexillum (Pusia) millecostatum (Broderip, 1836)
25.5mm; Mindoro.
Shell small to medium sized, elongate-ovate, with a tall, somewhat telescoped spire and weakly shouldered convex spire whorls; sculptured with numerous coarse axial ribs and overriding spiral grooves; aperture narrow, outer lip thin, angulate and lirate within, columella with 4 folds; pale orange-yellow in colour with dark brown blotching and banding; reaches 25mm.
DISTRIBUTION: Sporadically found throughout the Philippines in limited quantity.

14. Vexillum (Pusia) cancellarioides (Anton, 1839)
16.2mm; Cebu.
Shell small, elongate-ovate, with a tall stepped spire and a large body whorl; sculptured with irregular axial cords and crossed with deeply incised spiral grooves forming a nodulose surface; a deep pre-sutural furrow forms a row of large nodules below the sutures; aperture narrow, outer lip thin, lirate within, columella with 4 folds; creamy white in colour, aperture yellowish and brown within; reaches 20mm.
DISTRIBUTION: Sporadically found throughout the Philippines.

15. Vexillum (Pusia) patriarchalis (Gmelin, 1791)
22.2mm; Batangas.
Shell small to medium sized, slightly squat with a stepped spire; sculptured with numerous rounded, closely-set axial ribs which are nodulose on the shoulder, deeply incised, overriding spiral grooves, a row of finely echinate granules mid-body whorl and rows of coarser granules towards the base; aperture narrow, lirate within, columella with 4 folds; off-white in colour with a dark band on the body whorl, spiral grooves brown, columellar base and aperture orange; reaches 25mm.
DISTRIBUTION: Sporadically found throughout the Philippines.

16. Vexillum (Pusia) bernhardina (Röding, 1798)
21.4mm; Samar.
Shell medium sized, with a tall, turreted spire bearing a row of large, rounded nodules below the sutures; sculptured with rows of nodulose axial cords and bisecting spiral grooves; aperture narrowly elongate, outer lip thin, marginally crenulate and lirate within,

columella with 3 folds; pale yellowish brown in colour with white coronations; reaches 22mm.
DISTRIBUTION: Sporadically found throughout the Philippines in limited quantity.

17. Vexillum (Pusia) amabilis (Reeve, 1845)
17.6mm; Cebu.
Shell small, ovate in outline and sculptured with broad axial ribs that are bisected by deeply incised spiral grooves, producing a highly nodulose surface; sutures weakly channeled, below which is a row of large nodules and a deep furrow; aperture narrowly elongate, lirate within, columella with 4 folds; pale yellowish white in colour, with orange spiral grooves, a green sutural band and another at mid-body whorl, aperture greyish green; reaches 20mm.
DISTRIBUTION: Sporadically found throughout the Philippines.

Plate 61 — Conidae

Class Gastropoda
Subclass Prosobranchia
Order Neogastropoda
Suborder Toxoglossa
Superfamily Conacea
Family CONIDAE

The Conidae represents a large family (approximately 600 species) of marine gastropods with a worldwide distribution, inhabiting intertidal to deep water in sand, amongst coral or under rocks. The shells may be small to large sized, conical, with a spire that may be short, tall or turreted. The body whorl is large, with a rounded or sharply angled shoulder which may be smooth, coronate or carinate. The spire whorls are generally sculptured with spiral and/or axial striae and the body whorl is smooth (occasionally granular), with a few basal sulci. The body whorl surface frequently has a very colourful and intricately designed pattern which may be obscured by a thick or translucent "skin" or periostracum. The aperture is narrow, the outer lip thin but strong, the columella smooth, and the operculum small and "claw-like."

The physiology and anatomy of the Conidae is well documented in specialized literature devoted to this family and accordingly, only a very brief resumé is presented here. The Conidae are a group of highly successful predaceous gastropods with a long evolutionary history and belong to a specialized sub-order of Neogastropods called the Toxoglossa (literally, "poison tongue"). Along with the Turridae, the Terebridae, and the Cancellariidae, the Conidae have evolved a specialized radula system which is used for spearing intended prey with a poisoned barb. Essentially, this system is divided into 3 basic components:
(a) A bulb or sack which doesn't appear to be a storage organ, but rather, is concerned with propulsion of the venom along a so-called poison duct;
(b) The poison duct, which in reality is a secretory organ, liberates the venom, and transports it along to the gut area where, by an as yet poorly understood process, the radula tooth is primed and loaded;
(c) The radula sack, which is essentially a storage organ for radular teeth.

Radular teeth can be grouped into 3 types and this reflects the type of prey which certain species prefer.
1. Piscivorous (fish-hunting) Species: this group includes such species as *C. magus, C. striatus* and *C. geographus,* and are considered the most dangerous to man. The radular teeth of this group generally have a long, smooth shaft with pronounced terminal barbs.

2. Molluscivorous (mollusc-hunting) Species: included in this group are *C. textile, C. gloriamaris* and presumably other *textile* cones such as *C. bengalensis* and *C. canonicus.* The radular teeth are characterized by a serrated shaft and a prominent basal barb.

3. Vermivorous (worm-hunting) Species: Most *Conus* species fall into this group and representatives include *C. leopardus, C. betulinus* and *C. quercinus.* The radular teeth are usually short, with poorly developed terminal barbs, a heavily serrated shaft, and an extremely large basal barb. Members of this group are generally considered the least dangerous to man, with most reported incidents detailing a sting of no more intensity than that received from a bee.

Over the past 10 or 15 years, considerable biochemical research has been undertaken in the study of marine toxins. Perhaps no group of marine animals have received as much attention in this respect as the Conidae. With the Philippine archipelago containing the largest and richest source of *Conus* species in the world, it is not surprising that most of the more recent advances in Cone venom research have taken place here. Under the direction of Dr. Baldomero Olivera; Jr. (Filipino scientist working in the Department of Biology, University of Utah, U.S.A.), a molecular biology research laboratory has been set up in the Department of Biochemistry of the University of the Philippines, College of Medicine. At present, a Conus Research Program is actively underway, under the supervision of Dr. Lourdes Cruz.

Over the past 3 or 4 years, much has been learned about the pharmacology and biochemistry of certain *Conus* venoms, particularly that of *Conus geographus,* generally considered to be the most toxic to man.

Detailed studies of *C. geographus* venom has shown that it is a mixture of 5 individual components and not just a single entity. These components are essentially short-chained polypeptide units, steriochemically orientated with disulphide bonds. It is the unique arrangement of these units which convey on them their unique potency. The mode of action of these toxins is still poorly understood and is still being researched, but in man, their lethal effect is generally produced by inhibiting neuronal transmission at the skeleto-neuromuscular junction of the diaphragm, producing paralysis and eventually, respiratory failure. The toxins are also known to act directly on muscle tissue, have a depressant effect on the central nervous system, and produce blood haemolysis.

Cone collectors should be aware that contrary to popular belief, cone venom is, under normal circumstances, heat stable and boiling will not appreciably reduce the potency of the venom although the activity may be reduced as a result of dissipation through the boiling medium. It is therefore expedient to regard even boiled specimens as potentially dangerous, and to handle them with due care.

Although studies in cone venom toxicology may only evoke a fleeting interest to the avid *Conus* collector, the medical profession is viewing the results of such studies with ever increasing interest. The high toxicity of some *Conus* venoms reflect an extremely selective mode of action and, by understanding how they are composed chemically and where they actually act in the body, scientists hope to gain a better understanding of certain disease states. With such understanding will hopefully come the development of newer and more effective drugs to counteract the effects produced by such diseased states.

The taxonomy of the Conidae is today still in a state of flux and uncertainty, with considerable controversy surrounding the validity of many of the names currently in use. Over the years, a number of subgeneric groups have been employed by systematists, in an attempt to group members into discrete morphological units. Although such subdivisions may be fully justified in many cases, far too little is known about the family as a whole to be able to make use of such a system with any degree of confidence. Until this problem is satisfactorily resolved, we have described all species in this family under a single genus, *Conus* Linné, 1758. In the words of J.R. Le B. Tomlin, "the grouping of the cones is as yet so little understood that I have made no attempt to deal with the growing number of subgenera and genera into which the *Conus* is being dismembered." The authors completely endorse this opinion.

There are around 185 valid species of *Conus* to be found in the Philippines with around 5 or 6 currently under revision. This number represents about one third of the known *Conus* species worldwide, and certainly provides the specialist in this group with plenty of challenge. The identifications featured in this section are based on the latest research and information currently available and it is hoped that collectors of this extremely popular group of gastropods will now be able to identify Philippine species with more confidence.

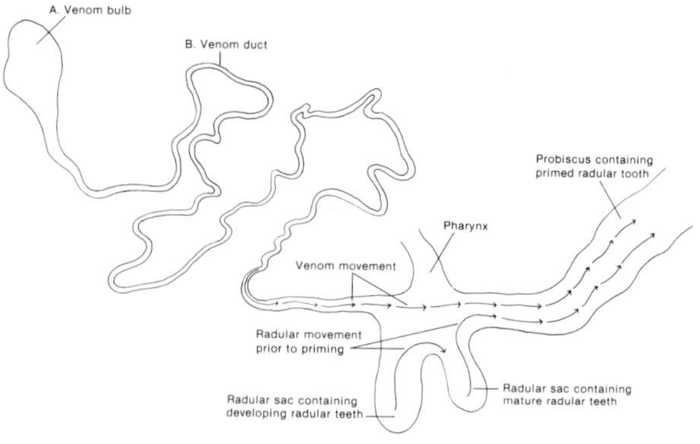

Diagramatic Representation of the Conus Radular System

1. Conus geographus Linné, 1758
104mm; Albay Gulf, Southern Luzon.
Shell moderately large, lightweight, elongate-ovate with a low, somewhat stepped, coronated spire; body whorl with a wide, sharply angulate, nodulose shoulder, a smooth surface, and a wide bluish white aperture; shell colour bluish white to pinkish with a brown reticulate network and 2 discontinuous brown bands; reaches 160mm.
DISTRIBUTION: Throughout the Philippines.

2. Conus tulipa Linné, 1758
63.0mm; Eastern Samar.
Shell medium sized, lightweight, glossy, elongate-ovate with a low, rounded spire and smooth spire whorls with undulate margins; body whorl with a rounded shoulder and a smooth surface with some basal spiral grooves; aperture narrow, flaring anteriorly, violet within; shell bluish white with black and white spiral bands and reddish brown blotches; reaches over 80mm.
DISTRIBUTION: Mainly Sulu Sea and along the Eastern coast of the archipelago, but occasionally found in other areas.

3. Conus obscurus Sowerby I, in Sowerby II, 1833
31.2mm; Eastern Samar.
Shell small, lightweight, elongate-ovate with a low, blunt spire and flat, spirally striated whorls; body whorl roundly shouldered with convex sides and some obsolete basal lirae; aperture wide, extremely so anteriorly, purplish within; external colour purple with axially aligned brown streaks and blotches; similar to

PLATE NO. 61 1. *Conus geographus* Linné 2. *Conus tulipa* Linné 3. *Conus obscurus* Sowerby I, in Sowerby II 4. *Conus emaciatus* Reeve 5. *Conus virgo* Linné 6a. *Conus coccineus* Gmelin 6b. to 6c. colour variations 7a. to 7b. *Conus caracteristicus* G. Fischer 8. *Conus mustelinus* Hwass, in Bruguiére 9. *Conus capitaneus* Linné 10a. *Conus glaucus* Linné 10b. unusual form 11a. *Conus quercinus* Lightfoot 11b. slightly sub-adult specimen 12. *Conus capitanellus* Fulton 13. *Conus tessulatus* Born 14. *Conus nussatella* Linné 15. *Conus betulinus* Linné 16. *Conus suratensis* Hwass, in Bruguiére 17a. *Conus figulinus* Linné 17b. bandless specimen

PLATE 61
1
2
3
4
5
6a
6b
6c
7a
7b
8
9
10a
10b
11a
11b
12
13
14
15
16
17a
17b

a young *C. geographus;* reaches 35mm.
DISTRIBUTION: Mostly from Eastern Samar and Sulu Sea.

4. Conus emaciatus Reeve, 1849
34.9mm; Bohol.
Shell small to medium sized, elongate, low conic, with a flat spire; body whorl sharply angled at the shoulder and encircled with granulose spiral ridges at the base; aperture narrow, blue-white within and a purple blotch on the base; shell orange to pale yellow with 1 or 2 lighter bands; similar to a juvenile *C. virgo;* reaches 35mm.
DISTRIBUTION: Sporadically found throughout the Philippines in limited quantity.

5. Conus virgo Linné, 1758
95.4mm; Samar.
Shell large, heavy, with a low gloss, an obtuse spire and weakly threaded spire whorls; body whorl with a sharply angled shoulder and sculptured with fine axial and spiral striations and spiral ridges at the base; aperture and shell colour yellowish white with a purple basal stain; reaches 150mm.
DISTRIBUTION: Throughout the Philippines.

6. Conus coccineus Gmelin, 1791
Shell medium sized, moderately glossy, with a short, acuminate spire; body whorl initially straight-sided, suddenly constricted towards the base and encircled with fine, occasionally granulated spiral ridges; colour deep red-brown, more rarely orange or yellow with a central white band which may or may not (rarely) contain black spots and blotches; reaches 85mm. Gem specimens scarce.
6a. typical form; 46.5mm; Northern Samar.
6b. unusual yellowish orange colour form; 41.6mm; Northern Samar.
6c. deep orange form with central band lacking black spots; 44mm; Northern Samar.
DISTRIBUTION: Catarman, Northern Samar and Palawan (rarely).

7. Conus caracteristicus G. Fischer, 1807
Shell medium sized, heavy, moderately glossy with a low, smooth, concave-sided spire and a wide, rounded, sharply angulate shoulder; body whorl smooth with a few basal grooves and frequently scarred with heavy growth flaws; white or cream in colour with 3 discontinuous brown bands, connected with axial streaks, aperture yellow within; reaches 70mm.
7a. 53.1mm; Manila Bay.
7b. smaller form with more distinct banding; 45.5mm; Manila Bay.
DISTRIBUTION: Manila Bay and Lamon Bay, Quezon.

8. Conus mustelinus Hwass, in Bruguière, 1792
70.3mm; Cebu.
Shell medium sized, weakly glossed, low conical with a short, acuminate spire and concave, spirally striated whorls; body whorl smooth with some basal sulci, shoulder rounded but sharply angled above; aperture uniformly narrow, light purple-and-white banded within; yellow to greenish brown in colour with 2 white bands bordered with black dots and occasionally secondary black-spotted white bands, spire with black axial streaks; reaches 75mm.
DISTRIBUTION: Sporadically found throughout the Philippines.

9. Conus capitaneus Linné, 1758
51.1mm; Cebu.
Shell medium sized, lightweight, glossy, with a low, smooth spire; body whorl with a wide, roundly-angled shoulder, the base with punctate spiral cords which may extend the length of the body whorl; spire white with brown flammules, body whorl tan to greenish yellow with a central white band bordered by a row of black spots and additional rows of black spots on rest of the whorl; reaches 70mm.
DISTRIBUTION: Throughout the Philippines.

10. Conus glaucus Linné, 1758
Shell medium sized, heavy, pyriform, with a low spire, a smooth, wide, roundly-shouldered body whorl, and a few basal spiral grooves; aperture wide, bluish white within; shell colour blue-grey with transverse rows of black flecks; reaches 70mm.
10a. typical form; 53mm; Lingayen Gulf, Western Luzon.
10b. unusual form lacking rows of black flecks; 46mm; Lingayen Gulf.
DISTRIBUTION: Lingayen Gulf, W. Luzon and Lamon Bay, South Eastern Luzon.

11. Conus quercinus Lightfoot, 1786
Shell medium sized, heavy, low gloss, biconic, with an exserted spire and flat sided, spirally striated whorls; body whorl smooth but with fine spiral threads on the lower half; shoulder wide and angulate, aperture uniformly wide, white within; shell yellowish brown with darker brown spiral lines, becoming weaker in larger specimens; reaches 70mm.
11a. typical adult; 63.7mm; Batangas.
11b. smaller, slightly sub-adult specimen showing spiral lines; 49.2mm; Batangas.
DISTRIBUTION: Sporadically found throughout the Philippines.

12. Conus capitanellus Fulton, 1938
21.7mm, 20.6mm; Punta Engaño, Cebu.
Shell small to medium sized, lightweight and glossy with a low spire and spirally striated whorls; body whorl rounded and slightly inflated at the shoulder, smooth, except for weak sulci at the base; white, with 1 or 2 brown bands and axial streaks; reaches 45mm.
DISTRIBUTION: Cebu and Bohol (deep water).

13. Conus tessulatus Born, 1778
41.1mm, 41.3mm; Batangas.
Shell medium sized, heavy, glossy, conical, with a low, sharply pointed, concave-sided spire; spire

whorls with strong ridges and grooves, body whorl smooth, with a sharply-angled shoulder and spiral grooves at the base; aperture narrow, white within; shell colour white with orange to reddish brown rectangles, spirally arranged and occasionally fusing to form 2 bands, violet-stained at the base; reaches 60mm.
DISTRIBUTION: Samar, Batangas, Negros and Sulu Sea.

14. Conus nussatella Linné, 1758
62.1mm; Cebu.
Shell medium sized, low gloss, cylindrically elongate; spire moderately tall, straight to slightly convex with a pointed apex, spire whorls flat, axially and spirally striated; body whorl straight-sided, with a rounded shoulder and sculptured with numerous close-set spiral grooves; aperture uniformly narrow, bluish white within; spire black-spotted, body whorl with brown blotching and rows of small, black dots; reaches 80mm.
DISTRIBUTION: Sporadically found throughout the Philippines in limited quantity.

15. Conus betulinus Linné, 1758
118.6mm; Manila Bay.
A large, broadly pyriform shell with a flat, sharply-pointed spire and a large, round-shouldered body whorl with a few sulci towards the base; yellow-orange in colour with black dots on white spiral bands; aperture uniformly wide, white within; reaches 150mm.
DISTRIBUTION: Sporadically found throughout the Philippines in limited quantity.

16. Conus suratensis Hwass, in Bruguiére, 1792
95.8mm; Lingayen Gulf, Pangasinan.
Shell large, heavy, moderately glossy, obconic with a flat spire and weakly sculptured spire whorls; body whorl large, inflated, with a rounded, sharply-angulate shoulder, straight sides, and a few weak basal sulci; aperture wide, white within; colour cream with brown longitudinal streaky maculations; very close to *C. betulinus*, except for the different pattern; reaches 150mm.
DISTRIBUTION: Lingayen Gulf (W. Luzon), Quezon and Manila Bay.

17. Conus figulinus Linné, 1758
Shell moderately large, heavy, bulbous, with a low to flat spire, which may be bluntly or sharply acute; body whorl roundly shouldered and inflated with weak basal sulci; aperture wide, bluish white within; colour tan to dark brown, encircled with continuous, dark spiral lines, frequently with yellowish bands at the shoulder and occasionally at mid-body whorl, spire whorls a deeper brown; reaches 100mm.
17a. large specimen with yellowish shoulder band; 91mm; Lingayen Gulf, W. Luzon.
17b. a smaller specimen lacking yellowish shoulder band; 59.9mm; Lingayen Gulf.
DISTRIBUTION: Lingayen Gulf (W. Luzon) and Quezon.

Plate 62 Conidae

1. Conus aurisiacus Linné, 1758
82.0mm; Balut Is., Southern Mindanao.
Shell medium sized, solid, glossy, broadly cylindrical with a low, rounded spire; spire whorls canaliculate, weakly ridged; body whorl attenuated, rounded at the shoulder and encircled with primary and secondary spiral ridges compacting towards the base; white to pinkish in colour with brown dots on the whorl margins, and 2 to 3 pink-brown bands on body whorl with large dots; reaches 72mm. Featured specimen is unusually large.
DISTRIBUTION: Samar, Sulu Sea, and Southern Mindanao.

2. Conus dusaveli (H. Adams, 1872)
Shell solid, glossy and cylindrically elongate with a short, exserted, straight spire; body whorl with a narrow subangulate shoulder, concave above, convex below, a few sulci at the base; aperture narrow, widening anteriorly, outer lip thick, incurved anteriorly; colour variable, usually white to orange with axial streaks on the spire, and 2 to 3 darker brown bands on the body whorl, with white and black flecks between; reaches 78mm.
2a. dark-banded specimen; 77.5mm; Cebu.
2b. orange-banded specimen; Cebu.
2c. typical form; 76.7mm; Cebu.
DISTRIBUTION: Cebu, Bohol, and Southern Mindanao.

3. Conus thomae Gmelin, 1791
Shell moderately large, glossy, elongate-conical, with a low spire, an acute apex and concave, slightly stepped spire whorls; body whorl with a carinate shoulder and a few basal sulci; aperture narrow, flaring basally; shell white with brown spots and streaks on the spire and rows of spiral dots merging to form several bands on the body whorl, aperture white; reaches 92mm. Specimens coming from Davao are usually darker in colour.
3a. typical pattern; 91.3mm; Davao Gulf, S. Mindanao.
3b. specimen with bands; 73.5mm; Davao Gulf.
DISTRIBUTION: Balut Island and Davao Gulf, S. Mindanao.

4. Conus armadillo Shikama, 1971
67.3mm; Balut Is., Southern Mindanao.
Shell thick, medium-sized and moderately glossy, spire concave and spirally striated; body whorl somewhat inflated, finely striated and encircled with deep

grooves over the entire surface; white with brown axial lines which are interrupted by the spiral grooves; reaches 70mm.
DISTRIBUTION: Southern Mindanao.

5. Conus kuroharai (Habe, 1965)
61.2mm; Balut Is., Southern Mindanao.
Some resemblance to *C. armadillo*, but the shell is smaller, the spire much lower, the body whorl somewhat more angulate and the spiral grooves at the shoulder are weaker and more numerous; white with irregular brown blotches forming two bands and numerous uninterrupted, densely packed, dark brown axial lines; reaches 70mm.
DISTRIBUTION: Southern Mindanao.

6. Conus lynceus Sowerby II, 1857
Shell medium sized, glossy, elongate-ovate, with a turbinated, concave spire the whorls of which are slightly stepped; body whorl with a narrow rounded shoulder and encircled with numerous equi-spaced, deep, punctate spiral grooves; aperture narrow, widening anteriorly, violet within; colour variable, usually creamy white with spiral rows of brown spots, frequently forming 1 or 2 bands; reaches 100mm.
6a. typical form; 71mm; Manila Bay.
6b. orange-brown specimen with white spotting; 63mm; Manila Bay.
6c. dark specimen; 62mm; Manila Bay.
DISTRIBUTION: Manila Bay and Tayabas Bay, Quezon.

7. Conus crocatus Lamarck, 1810
49.8mm; Samar.
Shell medium sized, glossy, with a conic spire and an attenuated body whorl; body whorl with a rounded shoulder and encircled with fine spiral striae, strongest towards the base; aperture parallel in width, white within; shell colour a rich golden brown with irregular white triangular spots; reaches 60mm.
DISTRIBUTION: Samar and S. Mindanao.

8. Conus teramachii (Kuroda, 1956)
74.5mm; Lubang Is., Mindoro.
Shell lightweight, spire weakly concave, spire whorls slightly stepped, flat and encircled with a few weak spiral threads; body whorl large and roundly angulate, rapidly tapering towards the base; surface with weak axial and spiral striae and a few sulci at the base; aperture narrowly extended, outer lip thin; colour chalky white with a dull greenish brown periostracum, aperture white; reaches 100mm.
DISTRIBUTION: Known only from Lubang Island, Mindoro.

9. Conus pertusus Hwass, in Bruguiére, 1792
Shell small to medium sized, glossy, elongate-ovate, with a short, acuminate spire the whorls of which are flat and spirally striated; body whorl with a roundly angulate shoulder and sulcated at the base; aperture narrow, violet to pink within; shell slightly variable in colour, normally pink to deep red with nebulous white blotchings; reaches 60mm.
9a. small typical pink form; 32.4mm; Sulu Archipelago.
9b. pink, spirally-corded form; 43.3mm; Punta Engaño, Cebu.
9c. somewhat large, pinkish orange, corded specimen; 52.4mm; P. Engaño, Cebu.
DISTRIBUTION: Sulu Sea, Cebu, Bohol, Samar and Balut Island, Southern Mindanao.

10. Conus sulcocastaneus Kosuge, 1981
48.8mm, 48.7mm; Punta Engaño, Cebu.
This is a *nomen novum* for *Conus vicdani* Kosuge, 1980 (*non* Lan, 1978). Shell medium sized, moderately heavy, glossy, conic with concave spire whorls which are threaded and coronated; body whorl with a coronate, moderately angled shoulder and sculptured with numerous widely-spaced grooves; aperture narrow, white within; shell white with brown spire blotches and irregular orange-brown blotches and streaks on the body whorl; reaches 50mm.
DISTRIBUTION: Only found in Cebu.

11. Conus circumcisus Born, 1778
Shell moderately large, heavy, cylindrical, with a low, somewhat domed spire, and beaded postnuclear whorls; body whorl attenuated, roundly shouldered and encircled with fine spiral striae which are frequently granulose; aperture pinkish blue, spire bluish white with brown dots, body whorl pinkish violet with brown banding and several rows of encircling black dots; reaches 85mm.
11a. large specimen; 85.6mm; Mactan Island, Cebu.
11b. specimen lacking black spotting; 72.6mm; Samar.
11c. a small beautifully patterned specimen; 40.9mm; P. Engaño, Cebu.
DISTRIBUTION: Batangas, Cebu, Bohol, Samar and Sulu Sea.

12. Conus neptunus Reeve, 1843
58.1mm; Punta Engaño, Cebu.
Shell medium sized, conic with a waxy gloss; spire concave with a sharp apex and the spire whorls undulate and spirally lirate; body whorl smooth with a wide, sharply angulate shoulder and a few weak spiral grooves; aperture narrow, pink within; surface colour white with dark brown filamentous lines and irregular brown blotching; reaches 60mm.
DISTRIBUTION: Cebu and Bohol.

PLATE NO. 62 1. *Conus aurisiacus* Linné 2a. *Conus dusaveli* (H. Adams) dark-banded specimen 2b. orange-banded specimen 2c. typical form 3a. *Conus thomae* Gmelin 3b. specimen with bands 4. *Conus armadillo* (Shikama) 5. *Conus kuroharai* (Habe) 6a. *Conus lynceus* Sowerby II 6b. to 6c. colour variations 7. *Conus crocatus* Lamarck 8. *Conus teramachii* (Kuroda) 9a. *Conus pertusus* Hwass, in Bruguiére 9b. to 9c. corded forms 10. *Conus sulcocastaneus* Kosuge 11a. to 11c. *Conus circumcisus* Born 12. *Conus neptunus* Reeve 13. *Conus cervus* Lamarck 14. *Conus bullatus* Linné 15a. to 15b. *Conus floccatus* Sowerby I, in Sowerby II 15c. unusual yellow colour, known as *Conus magdalenae* Kiener

PLATE 62

1
2a
2b
2c
3a
3b
4
5
6a
6b
7
8
9a
9b
9c
10
11a
11b
11c
12
13
14
15a
15b
15c

13. Conus cervus Lamarck, 1822
108.6mm, 91.9mm; Balut Is., Southern Mindanao.
Shell moderately large, bulbous and glossy with a low, weakly concave spire; body whorl long, widest on the upper third, aperture narrow, widening anteriorly, outer lip thin, slightly incurved centrally; colour orange-brown with darker brown blotches and white flecks, frequently in bands; reaches 120mm.
DISTRIBUTION: Bohol, Samal Is. (Davao) and Southern Mindanao.

14. Conus bullatus Linné, 1758
56.0mm; Samar.
Shell smooth, glossy, lightweight, elongate-ovate in outline with a short, acute spire, body whorl ventricose with a few basal sulci; aperture narrow, widening anteriorly and pink to orange in colour; shell pinkish white to orange with irregular white blotches; reaches 70mm.
DISTRIBUTION: Cebu, Bohol, Samar, Sulu Sea and Negros Oriental.

15. Conus floccatus Sowerby I in Sowerby II, 1839
Shell medium sized, elongate, with a moderately acuminate spire and weakly ledged sutures; spire whorls flat, becoming slightly channeled on later whorls; body whorl sculptured with numerous axial and spiral striae, aperture narrow, widening basally, white coloured within; external colour white, overlain with a pinkish brown suffusion, orange-brown blotching on the spire whorls, and dark brown axial streaks and blotches on the body whorl, frequently forming 3 interrupted bands.
15a. large specimen; 54mm; Sulu Archipelago.
15b. 51.5mm; Sulu Archipelago.
15c. unusual colour form, yellowish with brownish marks; often called *Conus magdalenae* Kiener, 1845; 40.1mm; Sulu Archipelago.
DISTRIBUTION: Sulu Sea.

Conidae — Plate 63

1. Conus excelsus Sowerby III, 1908
77.1mm, 77.2mm; Balut Is., Southern Mindanao.
Shell medium sized, heavy, elongate, conic, with an exceptionally long spire the whorls of which are concave, centrally shouldered and spirally grooved; body whorl with a moderately angulate shoulder encircled with deep, narrow spiral grooves; aperture narrow, white within; base colour white with 2 brown bands which may be discontinuous and have brown axial streaks radiating outwards; reaches 100mm.
DISTRIBUTION: Southern Mindanao and Bohol.

2. Conus ochroleucus Gmelin, 1791
63.1mm, 59.0mm; Batangas.
Shell medium sized, heavy, cylindrical, smooth, with a conic spire and weakly coronate, spirally-threaded whorls; body whorl with a wide, sharply-angled shoulder, aperture uniformly narrow, yellow within; yellowish brown with indistinct paler banding; reaches 65mm.
DISTRIBUTION: Batangas and Marinduque.

3. Conus pilkeyi Petuch, 1974
52.6mm, 52.5mm; Batangas.
Shell medium sized, glossy, conic, with a moderately acuminate spire the whorls of which are lirate and weakly carinate; body whorl roundly shouldered and sculptured with spiral grooves on the lower half of the whorl; spire white with brown dots, body whorl golden yellow with darker spiral bands or axial streaks, aperture deep orange; reaches 55mm. (For a time locally known as *"Conus daullei"*.)
DISTRIBUTION: Batangas and Marinduque.

4. Conus varius Linné, 1758
37.2mm, 38.9mm; Marinduque.
Shell small to medium sized, solid, glossy, elongate-ovate, with a moderately pointed, conic, straight spire and heavily coronated, spirally-threaded whorls; body whorl encircled with granulose spiral cords, shoulder rounded, sharply angulate and coronate, aperture narrow, marginally white and yellow within; spire white, body whorl white with black dotted cords, brown blotches and occasionally interrupted bands; reaches 40mm.
DISTRIBUTION: Sporadically found throughout the Philippines in limited quantity.

5. Conus colorovariegatus Kosuge, 1981
Shell medium sized, solid and glossy; spire tall, pointed, concave-sided with 11 to 12 concave, slightly stepped whorls which have deep spiral grooves and numerous fine axial threads; body whorl large, sharply angled at the shoulder and sculptured with fine spiral grooves, becoming stronger towards the base; aperture long, narrow, widening slightly towards the base, outer lip thin, slightly curved to the right basally; deep chestnut brown with white

PLATE NO. 63 1. *Conus excelsus* Sowerby III 2. *Conus ochroleucus* Gmelin 3. *Conus pilkeyi* Petuch 4. *Conus varius* Linné 5a. to 5b. *Conus colorovariegatus* Kosuge 6. *Conus mucronatus* Reeve 7. *Conus scabriusculus* Dillwyn 8. *Conus glans* Hwass, in Bruguiére 9. *Conus tenuistriatus* Sowerby II 10. *Conus viola* Cernohorsky 11. *Conus luteus* Sowerby I, in Sowerby II 12. *Conus mucronatus* Reeve 13. *Conus monachus* Linné 14. *Conus mitratus* Hwass, in Bruguiére 15. *Conus cylindraceus* Broderip & Sowerby I 16. *Conus achatinus* Gmelin 17. *Conus parius* Reeve 18. *Conus proximus* Sowerby II 19a. to 19c. *Conus papillosus* Kiener 20. *Conus moluccensis* Küster 21a. to 21e. *Conus generalis* Linné 22a. *Conus furvus* Reeve 22b. albinistic form 22c. yellow form 22d. brown with continuous lines 22e. unusual pink form 23. *Conus litoglyphus* Röding

PLATE 63

1
2
3
4
5a
5b
6
7
8
9
10
11
12
13
14
15
16
17
18
19a
19b
19c
20
21a
21b
21c
21d
21e
22a
22b
22c
22d
22e
23

triangular blotching, aperture and early spire whorls white; reaches 65mm.
It has recently been suggested (H.S.N., Vol. XXXII, No. 8, p. 9 : 1984) that this is merely a colour form of *C. neptunus* Reeve. Although a congener in structure, *C. colorovariegatus* appears restricted to the Philippines in distribution and is polytipic in colour and pattern. *C. neptunus* on the other hand also occurs in the S.W. Pacific, and the patterning is constantly the same. On this basis, separation is probably justified.
5a. chestnut-coloured specimen lacking white markings; 64mm; Punta Engaño, Cebu.
5b. specimen with white blotching; 60mm; Punta Engaño, Cebu.
DISTRIBUTION: So far, only found in Punta Engaño, Cebu.

6. Conus mucronatus Reeve, 1843
34.6mm, 35.9mm; Bulan, Sorsogon.
Shell medium sized, lightweight, with a low, sharply-pointed, concave spire; spire whorls spirally threaded, body whorl with a broad sharply-angled shoulder and sculptured with heavy spiral ridges and axial cords; aperture narrow with a blue-white tint; shell colour white with irregular brown streaks and blotches; reaches 50mm.
DISTRIBUTION: Cebu, Sorsogon, Bohol and Tayabas Bay, Quezon.

7. Conus scabriusculus Dillwyn, 1817
41.0mm; Eastern Samar.
Shell small to medium-sized, obconic, with a low, slightly convex-sided spire and pointed apex; body whorl with a well-rounded shoulder and sculptured with granulose spiral lirae over the entire body whorl; shell colour bluish white with purple-brown blotches, sometimes in bands which occasionally obliterate the pale background, shell base purple; aperture violet with white banding; reaches 45mm.
DISTRIBUTION: Eastern Samar and Sulu Sea.

8. Conus glans Hwass, in Bruguière, 1792
40.0mm; Eastern Samar.
Shell small to medium sized, solid, elongate-ovate, with a short, convex, weakly nodulose, sharply-pointed spire and a slightly inflated body whorl encircled with nodulose spiral cords; aperture narrow, violet within; white to violet with brown banding; reaches 40mm.
DISTRIBUTION: Eastern Samar and Sulu Sea.

9. Conus tenuistriatus Sowerby II, 1858
37.4mm; Eastern Samar.
Shell medium sized, heavy, glossy, elongate; spire moderately tall, domed, spire whorls weakly threaded occasionally beaded; body whorl with a wide, rounded shoulder and sculptured with fine axial striae and closely-set spiral grooves; aperture narrow, pale purple in colour; shell colour deep brown with white spire spots and pale white bands on the body whorl; reaches 45mm.
DISTRIBUTION: Eastern Samar and Sulu Sea.

10. Conus viola Cernohorsky, 1977
40.5mm; Punta Engaño, Cebu.
Shell medium sized, lightweight, low gloss, conic, cylindrically ovate with a tall, dome-shaped spire, a sharp apex, and spirally striated whorls; body whorl long, narrow and smooth, with a rounded shoulder and sculptured with sulci at the base; aperture narrow, widening anteriorly, lavender within; colour mauve with 3 interrupted brown bands, some lines and spots; reaches 60mm.
DISTRIBUTION: Cebu, Bohol and Sulu Sea.

11. Conus luteus Sowerby I, in Sowerby II, 1833
24.8mm, 21.0mm; Punta Engaño, Cebu.
Shell small, glossy, cylindrically ovate, with an obtuse apex on a slightly domed spire; body whorl with a narrow, rounded shoulder and a few basal sulci; aperture narrow, pink to violet within; shell pinkish to purple, with brown spire blotches and a brown band centrally on the body whorl; reaches 40mm.
DISTRIBUTION: Cebu, Bohol and Sulu Sea.

12. Conus mucronatus Reeve, 1843
33.5mm; Punta Engaño, Cebu.
An albinistic variety of *mucronatus* (see also fig. 6) which is often erroneously figured as *C. alabaster* (Pl. 68, fig. 4).
DISTRIBUTION: Camotes Sea.

13. Conus monachus Linné, 1758
43.5mm, 43.0mm; Marinduque.
Similar to *C. achatinus* in shape and colouration but lacks the rows of brown and white dashes; it also possesses a more convex spire, a narrower, more roundly angulate body whorl shoulder, and is more attenuated in general outline; reaches 65mm.
DISTRIBUTION: Marinduque, Batangas and Palawan.

14. Conus mitratus Hwass, in Bruguiére, 1792
43.9mm; Samar.
Shell small to medium sized, cylindrically elongate, conic, with a pointed spire; spire whorls rounded, encircled with fine spiral lirae, body whorl long, encircled with granulose spiral cords extending partially or totally up the whorl; aperture narrow, white within; shell off-white to yellow with brown blotched bands interconnected with brown axial streaks; reaches 45mm.
DISTRIBUTION: Samar and Sulu Sea.

15. Conus cylindraceus Broderip & Sowerby I, 1830
28.3mm, 30.0mm; Samar.
Shell small, solid and cylindrically fusiform; spire acuminate, spire whorls finely threaded, body whorl rounded below the shoulder with a few basal sulci; brown to orange in colour with white axial streaks; reaches 30mm.
DISTRIBUTION: Uncommon, usually washed ashore after a big storm. Found in Samar and Sulu Sea.

16. Conus achatinus Gmelin, 1791
44.0mm, 49.3mm; Marinduque.
Shell medium sized, solid, dull to glossy, ovate with a short, concave spire; spire whorls encircled with punctate grooves; body whorl large, convex at the shoulder and with heavy sulci at the base which may extend to the mid-body whorl line; blue-grey, with deeper blue and brown axial flammules, encircled with rows of brown and white dashes; reaches 65mm.
DISTRIBUTION: Marinduque, Batangas and Palawan.

17. Conus parius Reeve, 1844
35.9mm, 36.9mm; Batangas.
Shell small, glossy, elongate-ovate with a short, conic spire; spire whorls slightly stepped, concave and threaded, body whorl with a smooth, rounded shoulder and basal grooving; aperture uniformly narrow, blue-white in colour with a posterior purple blotch; shell uniformly grey to pale yellow; reaches 40mm.
DISTRIBUTION: Batangas and Palawan.

18. Conus proximus Sowerby II, 1859
32.9mm, 41.7mm; Punta Engaño, Cebu.
Shell small to medium sized, solid, glossy, elongate-ovate; spire low, concave and weakly coronated, apex acute, body whorl with numerous close-set, elevated spiral threads, smooth interstices and a wide, coronate, carinated shoulder; aperture narrow, blue-white within; shell surface cream coloured with dark brown lines on the spiral lirae and white spots and blotches in 2 or 3 rows; reaches 45mm.
DISTRIBUTION: Sorsogon, Cebu, Bohol and Southern Mindanao.

19. Conus papillosus Kiener, 1845
Shell small, solid, elongate-ovate; spire conical, concave with weakly coronated spire whorls which are sculptured with axial and spiral threads; body whorl with a roundly angulate, somewhat coronated shoulder and sculptured with finely pustulose spiral cords; aperture narrow, outer lip broadly arcuate; colour variable from white, yellowish, to deep pink, spire and coronations white, occasionally with brown spots, body whorl pink, white or grey, spiral cords occasionally with dark brown spots; reaches 22mm. Locally known as *"Conus puncturatus"*
19a. pink form lacking brown spots, 21mm; Palawan.
19b. white form; 18mm; Palawan.
19c. white with brown dots; 14.3mm; Palawan.
DISTRIBUTION: Palawan, Cebu and Balut Is., Southern Mindanao.

20. Conus moluccensis Küster, 1838
41.7mm, 42.8mm; Punta Engaño, Cebu.
Shell medium sized, solid and glossy, with a short, sharply-pointed, concave spire the whorls of which are threaded; body whorl with a wide, rounded, somewhat coronated shoulder and convex sides that are weakly "pinched-in" centrally and attenuated towards the base; surface smooth with some nodulose cords on the lower third of the whorl; aperture wide, flaring anteriorly, white within; shell colour variable, usually off-white with dark brown streaks and blotches, occasionally in bands; reaches 60mm.
DISTRIBUTION: Cebu, Bohol, Sorsogon and Balut Is., Southern Mindanao.

21. Conus generalis Linné, 1767
Shell medium sized, elongate, glossy, conic with an exserted apex and a smooth body whorl, widely angulate at the shoulder; aperture uniformly narrow, blue-white with a basal purple blotch; extremely variable in colour and pattern, basically white with narrow to wide orange or brown banding, connected with axial streaks; reaches 90mm.
21a. 64.4mm; Samar.
21b. 67.0mm; Samar.
21c. unusual specimen with dense, axial streaks; 68.1mm; Palawan.
21d. young, sharply patterned specimen; 62.3mm; Cebu.
21e. 65mm; Palawan.
DISTRIBUTION: Throughout the Philippines.

22. Conus furvus Reeve, 1843
Shell medium sized, lightweight, elongate-ovate in outline with a low gloss; spire low, slightly concave, spire whorls weakly stepped and encircled with two spiral grooves; body whorl straight-sided, with a rounded, sharply-angulate shoulder and a few narrow, occasionally granulose sulci at the base; colour and pattern extremely variable, usually greyish white, brown, or pink with numerous rows of brown dots; reaches 60mm. The brown form is known locally as *"Conus lignarius."*
22a. typical form; 59.2mm; Samar.
22b. albinistic form; 54.3mm; Cebu.
22c. yellow form; 50mm; Palawan.
22d. brown with continuous lines; 50.5mm; Samar.
22e. unusual pink form; 35.6mm; Palawan.
DISTRIBUTION: Sporadically found throughout the Philippines, but most of the unusual color varieties come from Palawan.

23. Conus litoglyphus Röding, 1798
40.0mm, 46.2mm; Samar.
Shell medium sized, glossy, with a low spire and a pointed apex; spire whorls threaded, body whorl with a wide, sharply angled shoulder, minute axial and spiral striae, and a few occasionally nodulose basal sulci; orange to dark brown with two bands of white spots and purple blotch towards the base; reaches 70mm.
DISTRIBUTION: Throughout the Philippines.

Plate 64 Conidae

1. Conus radiatus Gmelin, 1791
68.5mm, 66.2mm; Manila Bay.
Shell medium sized, glossy and cylindrical; spire short, concave, with an acute apex; spire whorls concave and spirally striated, body whorl with a

wide, rounded, sharply-angulate shoulder and widely spaced, deep spiral grooves at the base; aperture uniformly wide, white within; colour light to dark brown with a white shoulder band; reaches 80mm.
DISTRIBUTION: Manila Bay.

2. Conus spectrum Linné, 1758
Shell small to medium sized, lightweight, glossy, strongly ventricose, with a flat, concave spire; body whorl with a wide, rounded, sharply angulate shoulder and widely spaced spiral grooves towards the base; aperture wide, flaring anteriorly, white within; white with orange or black mottling and occasionally thin spiral lines; reaches 60mm.
2a. form *filamentosa* Reeve, 1849; a form with thin spiral lines; 51.2mm; Sulu Archipelago. Locally known as *"Conus conspersus."*
2b. form *pica* A. Adams & Reeve, 1848; dark blotching on a white shell; Palawan.
2c. typical form; 32.1mm, 30.9mm; Balabac, Palawan.

3. Conus boeticus Reeve, 1844
33.5mm, 30.0mm; Samar.
Shell small and conic, spire whorls coronated and finely cancellate from overriding axial and spiral striae; body whorl coronate, roundly shouldered, with nodulose spiral cords on the lower third; colour pattern variable, usually a white base colour, with dark brown banding which may be solid, discontinuous or composed of axially orientated streaks and spots, aperture bluish white; reaches 40mm. This figured specimen resembles *C. ruppellii* Reeve, 1848, and maybe a valid taxon.
DISTRIBUTION: Marinduque, Batangas and Samar.

4. Conus chaldaeus (Röding, 1798)
23.5mm, 24.8mm; Eastern Samar.
Shell small, squat, glossy, with a low convex spire which is weakly coronate and spirally striated; body whorl with a wide, rounded, slightly coronate shoulder and spiral (occasionally granular) ridges at the base; spire white with black streaks, body whorl white with black axial streaks in 2 discontinuous bands; reaches 25mm.
DISTRIBUTION: Throughout the Philippines.

5. Conus collisus Reeve, 1849
35.6mm, 45.4mm; Marinduque.
Shell medium sized, solid, elongate-ovate and moderately glossy; spire moderately tall, concave, with a sharply-pointed apex, spire whorls flat, axially and spirally striated; body whorl with a rounded, sharply angled shoulder and encircled with finely striated spiral grooves which become crowded towards the base; aperture narrow, outer lip thin, marginally white, internally pale orange; shell pale brown to light orange with brown spiral blotches and rows of brown dots, arranged in 2 or 3 bands; reaches 50mm.
DISTRIBUTION: Marinduque.

6. Conus broderipii Reeve, 1844
41.2mm, 39.0mm; Tara Tara Is., Sulu.
Shell medium sized, lightweight, glossy, ovate, with a short, weakly stepped, concave spire and flat, spirally striated whorls; body whorl convex, roundly shouldered and sculptured with wide spiral grooves, becoming deeper and more compact basally; aperture wide, marginally white, pink within, outer lip thin; colour off-white with irregular, interrupted brown bands and rows of dark spots; reaches 50mm; incorrectly figured as *"Conus conspersus"* in some books. Locally known as *"Conus daphne."*
DISTRIBUTION: Known only from Tara Tara Island, Sulu.

7. Conus stramineus Lamarck, 1810
47.6mm, 48.2mm; Marinduque.
Shell medium sized, lightweight and glossy; spire moderately acuminate, concave, spire whorls threaded; body whorl roundly shouldered, smooth, with some basal sulci; aperture narrow, outer lip sharp, marginally white, brown within; bluish white with rows of brown streaks and blotches; reaches 60mm.
DISTRIBUTION: Marinduque and Sorsogon.

8. Conus ebraeus Linné, 1758
Shell small to medium sized, solid, rhomboidal, with a low spire and an inflated body whorl; aperture narrow with a pinkish hue; base colour white with black spots on the spire and 3 rows of black squares on the body whorl, base stained with black; reaches 30mm.
8a. typical form; 36.6mm; Samar.
8b. specimen with a reduced central band; 36.7mm; Palawan.
8c. specimen lacking central band; 34mm; Cebu.
DISTRIBUTION: Throughout the Philippines.

9. Conus rattus Hwass, in Bruguiére, 1792
35.9mm, 37.7mm; Eastern Samar.
Shell small to medium sized, heavy, glossy, elongate-ovate with a moderately tall spire; body whorl smooth, with sulci at the base, shoulder roundly to sharply angulate; aperture uniformly narrow, purple within; shell deep brown with white mottling on the spire and body whorl, sometimes forming interrupted bands, base violet; reaches 60mm.
DISTRIBUTION: Throughout the Philippines.

PLATE NO. 64 1. *Conus radiatus* Gmelin 2a. *Conus spectrum* (Linné) form *filamentosa* (Reeve) 2b. form *pica* A. Adams & Reeve 2c. typical form 3. *Conus boeticus* Reeve 4. *Conus chaldaeus* (Röding) 5. *Conus collisus* Reeve 6. *Conus broderipii* Reeve 7. *Conus stramineus* Lamarck 8a. to 8c. *Conus ebraeus* Linné 9. *Conus rattus* Hwass, in Bruguiére 10a. *Conus cinereus* Hwass, in Bruguiére 10b. *Conus bernardi* Kiener 11. *Conus gubba* Kiener 12a. *Conus arenatus* Hwass, in Bruguiére form *granulosa* Lamarck 12b. typical form 13a. *Conus pulicarius* Hwass, in Bruguiére 13b. to 13c. pattern variations 14a. *Conus catus* Hwass, in Bruguiére 14b. to 14c. colour variations 15. *Conus stercusmuscarum* Linné 16a. *Conus eburneus* Hwass, in Bruguiére 16b. typical form 16c. albinistic form 16d. unusual yellow-banded specimen 17. *Conus polyglotta* Weinkauff

PLATE 64

1
2a
2b
2c
3
4
5
6
7
8a
8b
8c
9
10a
10b
11
12a
12b
13a
13b
13c
14a
14b
14c
15
16a
16b
16c
16d
17

10. Conus cinereus Hwass, in Bruguiére, 1792
Shell medium sized, glossy, with a low, sharply exserted, concave-sided spire with smooth whorls; body whorl long, with a rounded shoulder and weakly convex sides which taper gradually towards the base; the surface is smooth with a few weakly incised spiral grooves at the base which may on occasion, extend halfway up the body whorl; aperture narrow, widening anteriorly, marginally brown-coloured and pale blue within; the colour is fairly constant and is usually bluish grey with brown axial streaks or blotches, and rows of brown dots flecked with white; there is normally a thin, light coloured band on the shoulder which is completely devoid of markings and effectively separates the pattern of the spire whorls from that of the body whorl; reaches 70mm.
10a. 54.8mm; Marinduque.
DISTRIBUTION: Batangas and Marinduque.
10b. Conus bernardi Kiener, 1845: Shell medium sized, glossy, with a low, exserted spire the whorls of which are flattened and spirally threaded; body whorl moderately long, with a slightly subangulate shoulder, and convex sides; the surface is basically smooth, but with several widely spaced, well incised spiral grooves which may extend halfway up the body whorl; colour variable, usually dark brown with paler spots and blotches or off-white to orange, with darker streaks and blotches, occasionally in bands, aperture white; reaches 70mm. This species is very close to *cinereus* but can be separated by the spirally threaded spire whorls, the slightly more convex sides, the apertural colouration, and the surface colouration and pattern which, although variable, never approaches the bluish grey of *cinereus.* It is also allopatric with *cinereus* populations; 44.3mm, 45.1mm; Sulu Archipelago.
DISTRIBUTION: Sulu Sea.

11. Conus gubba Kiener, 1845
52.9mm; Marinduque.
Shell medium sized, fairly glossy, with a moderately tall, concave-sided spire the whorls of which are weakly ledged and spirally threaded; body whorl elongate, with a roundly angulate shoulder and weakly convex sides which gradually taper towards the base; deeply incised, equi-spaced spiral grooves encircle the lower part of the body whorl and may extend upwards along its entire length; ground colour reddish brown to dark brown with white spots along the sutures of the spire whorls, at the shoulder, and towards the base; aperture uniformly narrow, slightly dilated basally, white within; reaches 60mm. Erroneously synonymized with *C. cinereus* to which it bears no morphological resemblance whatsoever.
DISTRIBUTION: Marinduque.

12. Conus arenatus Hwass, in Bruguiére, 1792
Shell medium sized, heavy, turbinate, with a moderately flat apex; spire whorls coronate and spirally threaded; body whorl smooth with convex sides, a wide, smooth angulate shoulder, and some basal sulci; aperture narrow, widening anteriorly, outer lip thin; colour cream with black spots forming axial bands; reaches 55mm.
12a. form *granulosa* Lamarck, 1822: a highly granulose form; 37.8mm; Palawan.
12b. typical form; 50.5mm; Palawan.
DISTRIBUTION: Throughout the Philippines.

13. Conus pulicarius Hwass, in Bruguiére, 1792
Shell medium sized, heavy, glossy, elongate-ovate, with a flat spire and concave, spirally striated, heavily coronated spire whorls; body whorl with a rounded, sharply angulate, coronated shoulder, smooth, except for some fine basal sulci; aperture narrow, white within; shell white with small black dots, sometimes in diffused bands; reaches 55mm.
13a. typical form; 40.5mm; Eastern Samar.
13b-13c. freakish pattern variations; 42.8mm, 52mm; both from Eastern Samar.
DISTRIBUTION: Throughout the Philippines.

14. Conus catus Hwass, in Bruguiére, 1792
Shell small to medium sized, glossy, with a low spire; spire whorls threaded, body whorl somewhat rounded at the shoulder, smooth posteriorly with several punctate spiral ridges on the anterior half; aperture narrow, widening anteriorly, blue-white within; white to brown with darker blotches; reaches 50mm.
14a. typical form; 35.5mm; Samar.
14b. orange-brown form; 34.8mm; Palawan.
14c. bluish white form; 32mm; Cebu.
DISTRIBUTION: Throughout the Philippines.

15. Conus stercusmuscarum Linné, 1758
45.2mm, 54.3mm; Samar.
Shell medium sized, heavy, glossy, cylindrically ovate; spire low, somewhat domed, with an acuminate apex; whorls spirally striated, body whorl smooth, sharply angled at the shoulder with some basal sulci; aperture wide, flaring anteriorly, marginally white, orange within; blue-white with rows of crowded black dots and black axial blotches; reaches 65mm.
DISTRIBUTION: Throughout the Philippines.

16. Conus eburneus Hwass, in Bruguiére, 1792
Shell medium sized, lightweight, glossy, obconic with a low, almost flat spire; body whorl smooth with a few pronounced sulci at the base; variable in colour and pattern, normally white with rows of black dots and occasional yellow banding; normally reaches around 55mm, but may grow as large as 70mm (see 16a)
16a. large, heavy specimen; 70.9mm; Batangas.
16b. typical form; 49mm; Samar.
16c. albanistic form with a single row of dots; 43.5mm; Palawan.
16d. unusual yellow banded form lacking black dots; 38.5mm; Cebu.
DISTRIBUTION: Throughout the Philippines.

17. Conus polyglotta Weinkauff, 1874
55.0mm, 50.6mm, 35.5mm; Camarines Sur (Bicol region).
Morphologically identical to *C. eburneus* (fig. 16) but

immediately distinguished by its characteristic surface pattern which consists of a white ground colour suffused with yellow, over which is superimposed several black spiral bands which may either be continuous or broken into short axial dashes. Normally considered a form of *eburneus* but is never found in any sympatric populations of that species, and intergrades which might link the two have not, as yet, been observed; reaches 55mm.
DISTRIBUTION: Presently known only from Camarines Sur, Southern Luzon.

Plate 65 — Conidae

1. Conus miles Linné, 1758
79.6mm; Eastern Samar.
Shell medium sized to large, heavy, solid, glossy; spire low, bluntly rounded, body whorl with a wide, roundly angulate shoulder and widely spaced spiral sulci at the base; aperture narrow, white and black banded within; white with numerous fine, brown filamentous axial lines, a deep solid brown base and 2 wide orange-brown bands; reaches 110mm.
DISTRIBUTION: Throughout the Philippines.

2. Conus distans Hwass, in Bruguiére, 1792
91.2mm; Bohol.
Shell large, heavy, with a low, apically flattened spire; spire whorls heavily noduled, body whorl with a round to sharply-angled shoulder, initially convex, becoming concave, with a few deeply incised basal sulci; aperture uniformly wide, bluish white within; shell greyish brown with 1 or 2 pale bands, brown spots on a white spire and a purple stained base; reaches 130mm.
DISTRIBUTION: Sporadically found throughout the Philippines.

3. Conus imperialis Linné, 1758
81.2mm; Bohol.
Shell medium sized to large, glossy, with a flat, obtuse spire and an attenuated body whorl, coronate and sharply angled at the shoulder, with a few spiral, occasionally granulose sulci at the base; aperture narrow, blue-white within and purple stained at the base; colour and pattern variable, usually white with spiral bands of black spots and 2 wide brown bands; reaches 110mm.
DISTRIBUTION: Sporadically found throughout the Philippines.

4. Conus vexillum Gmelin, 1792
89.7mm; Samar.
Shell medium sized to large, heavy, glossy, broadly turbinate, with a low, straight, exserted spire, the whorls of which are flat and spirally threaded; body whorl with a wide, rounded shoulder and sculptured with axial and spiral striae and a few basal granular sulci; aperture wide, blue-white within; shell brown with white streaks on the spire and a median white band on the body whorl; reaches 85mm.
DISTRIBUTION: Throughout the Philippines.

5. Conus sulcatus Hwass, in Bruguiére, 1792
64.9mm; Cebu.
Shell medium sized, solid, low glossed; spire low, concave-sided with an acute apex, spire whorls with 4 to 5 spiral threads crossed by weaker, arcuate axial striae, sutures undulate and slightly coronate; body whorl with a rounded, weakly coronated shoulder and sculptured with coarse, granular spiral cords, becoming weak on the shoulder; aperture wide, white within, outer lip thin; yellow brown with brown spire patches and darker brown-stained cords; reaches 90mm.
DISTRIBUTION: A few years ago when trawling was still active in Tayabas Bay, Quezon, specimens were plentiful in that area. Specimens were also found in the Manila Bay and Batangas area. Lately, most specimens are found in the Visayan region particularly Cebu.

6. Conus bocki Sowerby III, 1881
72.0mm, 65.5mm; Cebu.
Once considered a form of *C. sulcatus* but is actually a distinct species; it differs from *C. sulcatus* in that it is generally larger in size, has a more ovate outline and a wider, more heavily coronated shoulder, with folds extending onto the upper part of the body whorl; the spire is less concave, the apex more sharply pointed, and the spiral cords become obsolete on the upper body whorl; the aperture is blue-white within, has a narrow brown margin, and the shell is a dark golden brown, frequently with darker staining; reaches 75mm.
DISTRIBUTION: Initially thought to be endemic to Thailand waters but started to appear in Philippine waters around the 70's, first in Tayabas Bay, Quezon, then lately in the Visayan islands.

7. Conus samiae da Motta, 1982
54.8mm, 51.9mm; Balut Is., Southern Mindanao.
Shell medium sized, heavy, spire turbinate, concave with a sharply pointed apex, spire whorls spirally threaded and nodulose; body whorl with a coronate, round to sharply angulate shoulder, encircled with equi-spaced, nodulose spiral cords and fine axial striae in the interstices; aperture narrow, white within; shell white with brown dots and dashes frequently forming bands; reaches 60mm.
DISTRIBUTION: Only from Balut Island, Southern Mindanao.

8. Conus striatus Linné, 1758
81.7mm; Samar.
Shell moderately large, oblong-ovate, with a low,

weakly concave spire; spire whorls initially flat, becoming deeply concave and sculptured with arcuate axial striae and overriding spiral threads; body whorl with convex sides, greatest below the sharply angled, carinate shoulder, and sculptured with fine axial and spiral striations; aperture narrowly elongate, widening basally; colour pale pinkish white with orange-brown axial streaks on the spire whorls and 3 to 4 interrupted, transverse bands of axially orientated brown maculations on the body whorl; variable; reaches 120mm.
DISTRIBUTION: Throughout the Philippines.

9. Conus laterculatus Sowerby III, 1870
58.3mm; Tayabas Bay, Quezon.
Shell medium sized, turbinated, with a low, concave spire and a somewhat carinated shoulder; body whorl slightly inflated below the shoulder, tapering gradually towards the base and encircled with numerous equi-spaced, deeply incised spiral grooves and obsoletely granulose ridges; surface pale brown, covered with rows of irregularly spaced brown spots frequently forming 2 or 3 discontinuous bands; the aperture is narrow, widening basally and pale lavender within; erroneously synonymized with *C. australis* Holten; reaches 60mm.
DISTRIBUTION: Tayabas Bay (Quezon), Cebu, Manila Bay and Sorsogon.

10. Conus australis Holten, 1802
67.8mm; Tayabas Bay, Quezon.
Shell medium sized, solid, subcylindrical, with a low gloss; spire tall, concave, with an acute apex, its whorls stepped and spirally lirate; body whorl carinate below the suture and sculptured with widely spaced, flattened, occasionally granular ribbings, which frequently appear paired, and interstitial axial threads; aperture narrow, widening anteriorly, white to violet within; shell cream coloured with fine, brown axial streaks joining larger brown blotches, frequently forming 2 or 3 bands; reaches 75mm.
DISTRIBUTION: Tayabas Bay (Quezon), Manila Bay, Cebu, Bohol, and Sorsogon.

11. Conus terebra Born, 1778
71.5mm; Eastern Samar.
Shell medium sized, heavy, with a low gloss; spire obtuse, spire whorls smooth, body whorl attenuated with a rounded shoulder and sculptured with numerous close-set spiral ridges and axially striated interspaces; aperture narrow, flaring basally, white within; shell pale yellow, occasionally lavender, with darker banding and a violet base; reaches 85mm.
DISTRIBUTION: Sporadically found throughout the Philippines.

12. Conus asiaticus da Motta, 1985
37.2mm; Tayabas Bay, Quezon.
Shell small to medium sized, lightweight and turbinate; spire moderately tall, concave and weakly stepped with a pointed apex; spire whorls threaded, body whorl with a wide, sharply-angled shoulder and sculptured with smooth to granular, widely-spaced spiral ribbings, weaker transverse striae and interstitial axial threads; aperture white, shell pale cream with brown banding; reaches 45mm.
DISTRIBUTION: Tayabas Bay (Quezon) and Punta Engaño, Cebu.
N.B. Erroneously figured by most authors as *Conus pseudosulcatus*, a Japanese Miocene fossil, but bears no resemblance to the type figure, and has until now, been unnamed.

13. Conus batheon Sturany, 1904
55.1mm; Punta Engaño, Cebu.
Shell medium sized, solid, glossy; spire moderately tall, convex with a sharp apex; spire whorls deeply concave, axially and spirally striated and with coronate margins; body whorl markedly inflated below the wide, coronate, roundly angulate shoulder and sculptured with flat, widely spaced, occasionally granular spiral cords, a single interstitial thread, and coarse axial striae; aperture uniformly wide, white within; shell pale cream with brown axial streaks on the spire and brown spots on the spiral cords; reaches 60mm. Sold locally as *"C. granifer"* and *"C. grangeri."*
DISTRIBUTION: Tayabas Bay (Quezon), Cebu and Bohol.

14. Conus pagodus Kiener, 1845
36.3mm, 36.8mm; Tayabas Bay, Quezon.
Shell small, biconic, with a tall, stepped, concave spire and sharply-pointed apex, its whorls weakly concave and spirally striated; body whorl carinate below the suture, with a wide, sharply-angled shoulder, almost parallel sides, and then "pinched-in" on the lower third; encircled with flat spiral cords and axially striated interspaces; off-white with brown bands of axial streaks; reaches 40mm.
DISTRIBUTION: Tayabas Bay (Quezon), Cebu, Bohol and Sorsogon.

15. Conus eximius Reeve, 1849
37.2mm; 41.1mm; Cebu.
Shell medium sized, turbinate and glossy; spire concave, acuminate and slightly stepped; body whorl with a sharply angled shoulder and encircled with fine sulci at the base; aperture narrow, widening anteriorly, white within, outer lip thin, posteriorly incurved; white with brown axial streaks and blotches, frequently forming discontinuous bands; reaches 45mm.

PLATE NO. 65 1. *Conus miles* Linné 2. *Conus distans* Hwass, in Bruguiére 3. *Conus imperialis* Linné 4. *Conus vexillum* Gmelin 5. *Conus sulcatus* Hwass, in Bruguiére 6. *Conus bocki* Sowerby III 7. *Conus samiae* da Motta 8. *Conus striatus* Linné 9. *Conus laterculatus* Sowerby III 10. *Conus australis* Holten 11. *Conus terebra* Born 12. *Conus asiaticus* da Motta 13. *Conus batheon* Sturany 14. *Conus pagodus* Kiener 15. *Conus eximius* Reeve 16a. to 16c. *Conus thalassiarchus* Sowerby I, in Sowerby II unusual black & white form 16d. to 16f. pattern & colour variations 16g. typical form

PLATE 65

DISTRIBUTION: Manila Bay, Sorsogon, Cebu and Bohol.

16. Conus thalassiarchus Sowerby I, in Sowerby II, 1834
Shell medium sized, elongate, solid and glossy; spire low and concave, its whorls concave, with undulate margins and spiral ridges; body whorl smooth with a wide, sharply-angled shoulder and some basal sulci; aperture narrow, blue-white within; colour pattern extremely variable; ground colour creamy white with brown to black dots and lines forming tents and arranged in 2 bands, spire white with dark brown spotting; reaches 100mm.
16a.-16g. A series showing extreme variation in colour and pattern. All from Palawan except for 16a to 16c; unusual black and white form from Bulan, Sorsogon; 16g — typical form. Sizes: 89.8mm, 48.3mm, 70.8mm, 57.8mm, 67.9mm, 66.9mm and 88.6mm, respectively.
DISTRIBUTION: Found in the Sulu Sea; Palawan has yielded the most unusual varieties. The unusual black and white variety comes from Bulan (Sorsogon) and Palawan.

Conidae — Plate 66

1. Conus magus Linné, 1758
Shell medium sized, glossy and subcylindrically elongate; spire low to moderately tall, concave, with an acute apex, its whorls initially flat, becoming concave and spirally threaded, occasionally noduled; body whorl moderately wide, sharply angulate at the shoulder, with long, straight sides and a few weak, occasionally granulose sulci at the base; aperture narrow, widening anteriorly and white within; extremely variable in colour and pattern but normally white to orange-tan and crossed with rows of dots, axial streaks or blotches, and some darker banding; reaches 80mm.
1a. typical *magus;* Albay Gulf, Southern Luzon.
1b. a large brown-coloured form with white bands and streaks; 63.5mm; Albay Gulf.
1c. an unusual black-and-white-coloured form; 49mm; Albay Gulf.
1d. Conus carinatus Swainson, 1822; Albay Gulf. One of several species erroneously lumped with *magus* and to which it bears no morphological resemblance whatsoever; the shell is subcylindrically elongate with slightly convex sides which are attenuated basally; the spire is quite flat and the colour, although variable, is usually orange-tan with axial streaks and 1 or 2 transverse bands; reaches around 75mm.
1e. 42.6mm; Albay Gulf.
1f. orange form; 32.4mm; Palawan.
1g. pink form; 32.2mm; Palawan.
1h. 80mm; Albay Gulf.
1i. an unusual form with wider shoulders and approaching *C. raphanus* (fig. 2); 58.8mm; Albay Gulf. This shell may possibly be identified as *Conus daullei* Cross, 1858.
DISTRIBUTION: Throughout the Philippines.

2. Conus raphanus Hwass, in Bruguière, 1792
58.5mm; Albay Gulf, Southern Luzon.
Nearly always found mixed with *C. magus* Linné, which it often resembles in colour and pattern. In shape, however, it has a wider, more sharply-angled shoulder, the sides of which are sort of "rolled up"; the spire whorls are more deeply concave and the spire contour is distinctly more concave than *magus;* the shell is consistently heavier and has a more ovate shape about it; reaches 80mm.
DISTRIBUTION: South Eastern Luzon and Palawan.

3. Conus cernohorskyi da Motta, 1983
54.7mm; Palawan.
Shell medium sized, heavy, ovately inflated, with a moderately tall, straight spire and a large, somewhat swollen body whorl which is peculiarly truncated anteriorly; spire whorls flat and spirally striated, body whorl may be smooth or encircled with heavily granular spiral cords extending its whole length; aperture wide, flaring anteriorly with a straight-sided outer lip; colour and pattern variable, usually off-white with brown to black longitudinal streaks arranged in two discontinuous bands; reaches 60mm. This is another species often lumped with *C. magus* but from which it can be consistently separated. Most specimens come from Samar and are sold locally as *"C. vinctus."*
DISTRIBUTION: Samar and Palawan.

4. Conus consors Sowerby I, in Sowerby II, 1833
Shell heavy, rather large, glossy with a low, concave, sharply pointed spire, its whorls narrowly concave, axially and spirally striated; body whorl smooth, straight sided, and with some basal sulci; shoulder unusually wide and roundly angulate; aperture uniformly wide, white within, outer lip thin, with a

PLATE NO. 66 1a. *Conus magus* Linné 1b. to 1c. pattern and colour variations 1d. *Conus carinatus* Swainson 1e. to 1i. pattern and colour variations 2. *Conus raphanus* Hwass, in Bruguiére 3. *Conus cernohorskyi* da Motta 4a. *Conus consors* Sowerby I, in Sowerby II 4b. form *anceps* A. Adams 5. *Conus moreleti* Crosse 6. *Conus sugillatus* Reeve 7. *Conus flavidus* Lamarck 8a. to 8b. *Conus lividus* Hwass, in Bruguiére 9a. *Conus muriculatus* Sowerby I, in Sowerby II 9b. smooth form 10. *Conus ranunculus* Hwass, in Bruguiére 11. *Conus frigidus* Reeve 12. *Conus striatellus* Link 13. *Conus balteatus* Sowerby I, in Sowerby II 14. *Conus sanguinolentus* Quoy & Gaimard 15. *Conus parvulus* Link 16. *Conus vulpinus* Hwass, in Bruguiére 17. *Conus planorbis* Born 18. *Conus vitulinus* Hwass, in Bruguiere 19. *Conus circumactus* Iredale

PLATE 66

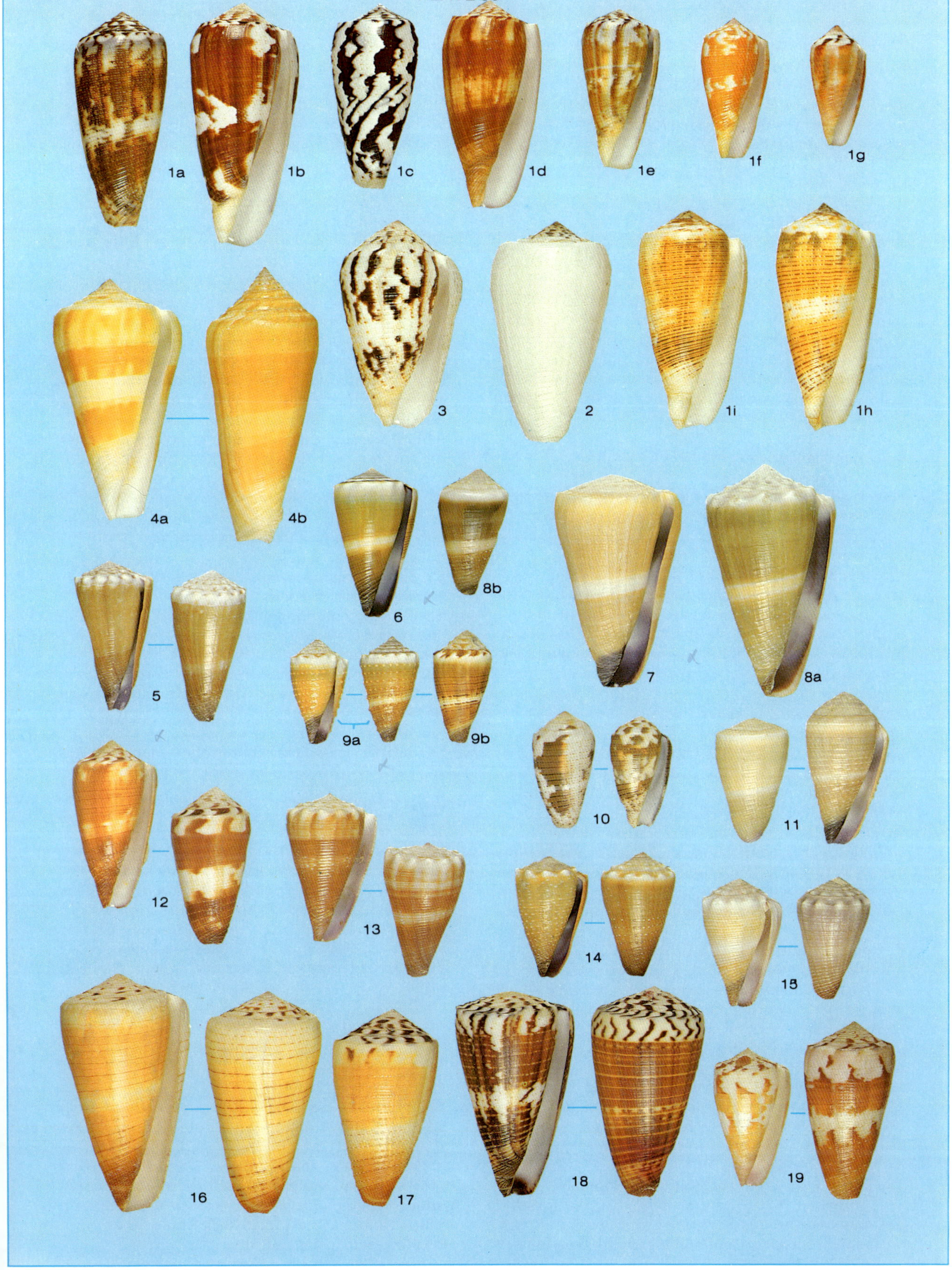
1a
1b
1c
1d
1e
1f
1g
4a
4b
3
2
1i
1h
6
8b
7
8a
5
9a
9b
10
11
12
13
14
15
16
17
18
19

"pinched-in" look centrally; colour off-white to yellow with 2 or 3 brown transverse bands; reaches 80mm.
4a. typical form; 63.8mm; Samar.
4b. form *anceps* A. Adams, 1854: a distinctly narrower form which is more elongate and has a more intense colour pattern; 72.6mm; Palawan.
DISTRIBUTION: Sporadically found throughout the Philippines.

5. Conus moreleti Crosse, 1858
40.2mm, 40.9mm; Marinduque.
Shell small to medium sized, subcylindrical and lightly glossy; spire low, apex obtuse and coronate with straight to slightly concave sides; spire whorls with spiral cords, body whorl attenuated, smooth, with granular sulci at the base, shoulder wide, sharply angulate and coronate; aperture narrow, violet-coloured with a white band, shell violet-brown to yellow-grey with a weak central band; reaches 45mm.
DISTRIBUTION: Marinduque and Bohol.

6. Conus sugillatus Reeve, 1844
40.0mm; Marinduque.
Shell medium sized, solid, moderately glossy; spire low, concave, its whorls obsoletely coronate with undulate margins and fine, arcuate axial striae; body whorl large with fine axial and spiral striations, shoulder wide, rounded and sharply angulate; aperture narrow, internally purple with a white band, colour bluish white with 2 pale orange bands and orange spiral lines, purple at the base; reaches 45mm.
DISTRIBUTION: Sporadically found throughout the Philippines.

7. Conus flavidus Lamarck, 1810
57.0mm; Eastern Samar.
Shell medium sized, solid, with a weakly threaded surface; spire low, slightly convex, its whorls spirally striated; body whorl large with an inflated, sharply-angled shoulder and straight sides that are slightly "pinched-in" centrally; aperture narrow, violet-coloured with a white band, shell surface orange to yellow with 1 or 2 bands and a purple-stained base; reaches 60mm.
DISTRIBUTION: Throughout the Philippines.

8. Conus lividus Hwass, in Bruguiére, 1792
Shell small to medium sized, solid, turbinate and slightly glossy; spire low and coronated; body whorl with straight sides, a wide, sharply angulate, coronated shoulder and encircled with widely spaced granular cords which may extend to the center of the body whorl; spire bluish white, body olive-brown with white banding, aperture narrow, purple and white banded within; reaches 75mm.
8a. large specimen; 62mm; Eastern Samar.
8b. unusually small, weakly coronate specimen; 33.5mm; Palawan.
DISTRIBUTION: Throughout the Philippines.

9. Conus muriculatus Sowerby I, in Sowerby II, 1833
Shell small, turbinate and moderately glossy; spire of medium height, concave and coronate, its whorls slightly stepped, with weak spiral threads and numerous fine arcuate axial striae; body whorl subcylindrically attenuated, with a wide, nodulose, sharply angled shoulder and encircled with heavily pustulose spiral ridges; aperture narrow, purplish and white banded within, spire white with brown dots, body whorl with two brown bands, red spiral lines and purple stains at the base; reaches 30mm.
9a. typical form; 28.2mm, 28.2mm; Cebu.
9b. smooth form with some pustules on the ventral side (not visible); 30.5mm; Cebu.
DISTRIBUTION: Sorsogon, Samar and Cebu.

10. Conus ranunculus Hwass, in Bruguiére, 1792
31.5mm, 30.1mm; Palawan.
Incorrectly figured as *C. nigropunctatus* in most books, being confused with a similar looking Middle East shell; shell medium sized, turbinate, elongate-ovate in outline, with a low, acuminate, convex spire, its whorls spirally striated with a large carinate ridge at the sutures; body whorl smooth, roundly shouldered, spirally sulcated at the base; aperture wide, blue-white within, shell creamy-white with bands of black dashes and darker blue and brown blotching; reaches 45mm.
DISTRIBUTION: Palawan and Samar.

11. Conus frigidus Reeve, 1848
33.2mm, 40.9mm; Palawan.
Shell medium sized, solid, heavy and glossy; spire low and straight, its whorls spirally striated; body whorl large, straight-sided, roundly shouldered, slightly bulbous posteriorly and frequently granulated on the lower half; aperture narrow, purple within, purple-stained at the base, spire pale tan, body whorl olive tan; reaches 50mm.
DISTRIBUTION: Throughout the Philippines, usually sympatric with *C. lividus* and *C. flavidus.*

12. Conus striatellus Link, 1807
44.9mm, 41.9mm; Samar.
Shell medium sized, solid, heavy, glossy and elongate; spire low and straight, its whorls flat, axially and spirally striated; body whorl smooth with convex sides, a sharply angled shoulder and a few basal sulci; aperture narrow, flaring anteriorly, white within; shell white with axial streaks on the spire and 1 or 2 interrupted transverse bands and brown axial streaks on the body whorl; reaches 60mm.
DISTRIBUTION: Sporadically found throughout the Philippines.

13. Conus balteatus Sowerby I, in Sowerby II, 1833
39.5mm, 35.9mm; Samar.
Shell small to medium sized, heavy with a low gloss; spire low and straight with a pointed apex, its whorls flat, coronate and spirally striated; body whorl with a wide, coronate, sharply angled shoulder, tapering anteriorly and sculptured with heavy spiral cords which become obsolete posteriorly; aperture narrow, purple and white banded, shell blue-white with brown banding and white flecks; reaches 45mm.

DISTRIBUTION: Marinduque, Batangas and Samar.

14. Conus sanguinolentus Quoy & Gaimard, 1834
33.4mm; Eastern Samar.
Shell small to medium sized, heavy, moderately glossy, turbinate, with a low, coronate, sharply acuminate spire and a body whorl with a rounded, sharply angulate, coronated shoulder, and granulations extending to the center of the body whorl; aperture narrow, violet within, spire white with brown dots, body yellow to dark brown; reaches 45mm.
DISTRIBUTION: Sporadically found throughout the Philippines.

15. Conus parvulus Link, 1807
35.0mm, 32.9mm; Marinduque.
Shell medium sized, heavy, glossy and turbinated; spire short, straight and sharply pointed, its whorls coronated and threaded; body whorl with a rounded, coronate, sharply angled shoulder and sculptured with spiral cords and axial threads; aperture wide, light purple and white banded within, shell grey to pale yellow, with bluish banding and stained with purple at the base; reaches 40mm.
DISTRIBUTION: Marinduque and Batangas.

16. Conus vulpinus Hwass, in Bruguiére, 1792
63.5mm, 59.1mm; Palawan.
Shell medium sized, heavy, glossy, elongate-ovate; spire short and straight, with a pointed apex, its whorls weakly concave and spirally striated; body whorl wide, with a rounded, sharply angulate shoulder and sulcated at the base; aperture narrow, white within, spire white with brown spotting, body whorl pale tan with lighter banding, a dark basal area, and brown spiral lines; reaches 65mm.
DISTRIBUTION: Sporadically found throughout the Philippines.

17. Conus planorbis Born, 1778
52.0mm; Palawan.
Extremely close to the preceding species but differs in the following ways: the spire is taller and is more heavily marked with brown axial streaks (spotting in *C. vulpinus)*; the body whorl is widest at the shoulder but in *vulpinus*, it is widest below the shoulder; the brown and white bandings are far sharper than those of *vulpinus* and it lacks the strong dark spiral lines; the granular spiral cords, restricted to the base in *vulpinus*, extend well onto the body whorl in *C. planorbis;* reaches 65mm.
DISTRIBUTION: Throughout the Philippines.

18. Conus vitulinus Hwass, in Bruguiére, 1792
55.5mm, 55.8mm; Samar.
Shell medium sized, solid, glossy, with a low, weakly concave spire and a flattened apex, its whorls axially and spirally striated; body whorl long, tapering, with a rounded, sharply angulate shoulder and granulose sulci towards the base; spire white with brown spots, body whorl white with 2 broad brown bands, white spiral dashes and purple-stained at the base; reaches 80mm.
DISTRIBUTION: Throughout the Philippines.

19. Conus circumactus Iredale, 1929
Syn: *C. connectens* A. Adams, 1855 (*nomen dubium*)
37.9mm, 42.2mm; Sulu Archipelago.
Shell medium sized, lightweight, glossy and turbinate with a low, concave spire, its whorls spirally striated; body whorl with a moderately wide, rounded, sharply angulate shoulder, axially and spirally striated, with a few sulci at the base; aperture narrow and purplish tinted, spire white with brown axial maculations; body whorl with a purplish tinge, and encircled with 2 wide, irregular brown bands, brown spots and purple stained at the base; reaches 50mm.
DISTRIBUTION: Sulu Sea and Samar.

Plate 67 Conidae

1. Conus gloriamaris Chemnitz, 1777
Shell medium to large sized, subcylindrically elongate and glossy; spire acuminate and conic, body whorl long, smooth, subangulate at the shoulder; aperture wide, blue-white within, shell colour bluish white, with golden brown axial zigzags forming white tents, and 2 to 3 orange-brown, somewhat discontinuous bands of flammules; reaches 168mm.
1a. 145mm; Punta Engaño, Cebu.
1b. pale form; 91mm; Punta Engaño, Cebu.
1c. beige coloured with ultra-fine tenting; 93.5mm; Punta Engaño, Cebu.
DISTRIBUTION: Bohol, Cebu, Corregidor and Negros.

2. Conus aureus Hwass, in Bruguiére, 1792
Shell medium sized, elongate, subcylindrical with rounded shoulders, a tall, acute, slightly concave spire, and a long body whorl, slightly swollen below the shoulder and encircled with strong spiral ridges; base colour white with irregularly placed, medium sized tents and golden to dark brown blotches, frequently in 2 or 3 discontinuous bands; reaches 65mm.
2a. typical, live-taken specimen; 54.5mm; Eastern Samar.
2b. orange-coloured beach specimen; 53.8mm; Eastern Samar.
DISTRIBUTION: Eastern Samar and Sulu Sea.

3. Conus auricomus Hwass, in Bruguiére, 1972
44.2mm, 43.7mm; Eastern Samar.
Shell small to medium sized, glossy, cylindrical, with parallel sides; spire short, straight, its whorls weakly striated; body whorl attenuated, roundly shouldered, with strong, closely-spaced spiral striae; spire light brown with irregular tents, body whorl white with numerous small crowded tents and 2 solid brown bands; reaches 50mm.
DISTRIBUTION: Eastern Samar and Sulu Sea.

4. Conus retifer Menke, 1829
Shell small to medium sized, heavy, squat, very glossy and rhomboidal; spire conic, its whorls spirally striated, body whorl with a rounded shoulder and spiral grooves; aperture uniformly wide, marginally white, pink within; shell colour off-white, pink or orange with black axial bands, white triangular tents and occasionally, some light banding; reaches 50mm.
4a. light pink form lacking banding; 33.9mm; Eastern Samar.
4b. pink form with typical pattern; 28.9mm; Eastern Samar.
4c. unusually large specimen; 48.9mm; Eastern Samar.
DISTRIBUTION: Eastern Samar and Sulu Sea.

5. Conus telatus Reeve, 1848
Shell medium sized, lightweight, elongately turbinated; spire concave and sharply pointed, its whorls nodulose and concave, with weak spiral striae, body whorl convex with a rounded, sharply angled coronate shoulder, sculptured with numerous fine spiral grooves and axial threads; aperture wide, white within, outer lip thin; shell bluish white with brown bordered white tents and large orange-brown axial blotches divided into bands by tenting marks; reaches 85mm.
5a. smooth form; 54.1mm; Palawan.
5b. light brown form; 49.2mm; Palawan.
5c. uncommon golden form; 53.5mm; Palawan.
5d. typical specimen but unusually large; 83.3mm; Palawan.
DISRIBUTION: Palawan.

6. Conus ammiralis Linné, 1758
Shell medium sized, heavy, rather solid and highly glossed; spire moderately raised, straight or slightly concave, apex sharply acute; body whorl smooth, roundly angled at the shoulder; white to yellow in ground color with 2 broad brown bands marked with white tenting and lighter brown tenting between; reaches 80mm.
6a. pustulose form; 50.5mm; Sulu Archipelago.
6b. pale form with large tents; 55.5mm; Negros Oriental.
6c. dark form with large tents; 61.2 mm; Sulu Archipelago.
6d. very dark form; 59mm; Sulu Archipelago.
6e. unusually large, dark form; 73.9mm; Albay Gulf, Southern Luzon.
DISTRIBUTION: Mostly from Southern Philippines, but a few have been found in other areas.

7. Conus legatus Lamarck, 1810
27.8mm, 37.5mm; Eastern Samar.
Shell small, glossy, subcylindrically elongate, spire acuminate, straight, its whorls flat-sided, body whorl roundly shouldered with some basal sulci; aperture narrow, white within; shell colour pinkish with numerous irregularly sized tents, orange-brown blotches and axial streaks; reaches 35mm.
DISTRIBUTION: Eastern Samar and Sulu Sea.

8. Conus textile Linné, 1758
Shell medium sized to large, moderately heavy, glossy, oblong-ovate; spire conic, slightly concave, its whorls axially and spirally striated; body whorl smooth except for some basal sulci, shoulder may be narrow or wide and is roundly angulate; aperture wide, blue-white within; colour and pattern extremely variable, basic colour blue-white with axial lines forming irregular-sized tents and irregular yellowish brown blotches in discontinuous spiral bands; reaches 130mm.
8a. typical *textile;* 91mm; Palawan.
8b. a somewhat turbinate, roundly inflated form called *ponderosa* Dautzenberg, 1932 (*non* Brocchi, 1814); 89mm; Palawan.
8c. a form with unusually large tenting; 59mm; Samar.
DISTRIBUTION: Throughout the Philippines.

9. Conus canonicus Hwass, in Bruguiére, 1792
47.5mm, 56.0mm; Samar.
Shell medium sized, solid, glossy and turbinate; spire concave and conic, its whorls axially and spirally striated; body whorl smooth except for a few basal sulci, shoulder narrow, roundly angulate; aperture narrow, widening anteriorly, marginally white and deep pink within; shell white to pale pink with small and large brown-bordered tents, 2 indistinct bands normally present; reaches 50mm. An uncommon species and known locally under the name *"Conus tigrinus".*
DISTRIBUTION: Eastern Samar and Sulu Sea.

Conidae Plate 68

1. Conus marmoreus Linné, 1758
Shell medium sized, moderately glossy and heavy; spire low, weakly concave and nodulose; body whorl with a wide, sharply angled shoulder and fine basal sulci; aperture narrow posteriorly, widening slightly towards the base, pink within; shell colour black, with large, irregular white tents; reaches 150mm.
1a. typical form; 94.3mm Albay Gulf, Southern Luzon.
1b. a pale, salmon-coloured (albinistic) specimen with reduced tenting; 102.2mm; Albay Gulf, Southern Luzon.
1c. Conus vidua Reeve, 1843: Originally considered to be merely a form of *marmoreus*, this species is now provisionally considered to be valid although morphologically, there is very little to separate C.

PLATE NO. 67 1a. to 1c. *Conus gloriamaris* Chemnitz 2a. to 2b. *Conus aureus* Hwass, in Bruguiére 3. *Conus auricomus* Hwass, in Bruguiére 4a. to 4c. *Conus retifer* Menke 5a. *Conus telatus* Reeve smooth form 5b. to 5c. colour variations 5d. typical form 6a. *Conus ammiralis* Linné pustulose form 6b. to 6e. colour variations 7. *Conus legatus* Lamarck 8a. and 8c. *Conus textile* Linné 8b. form *ponderosa* Dautzenberg (non Brocchi) 9. *Conus canonicus* Hwass, in Bruguiére

PLATE 67

marmoreus and *vidua*. The distinctive dark bands, one centrally on the body whorl and the other, basally, are consistently present. Reeve noted distinctive differences when describing *vidua* which is known to occur in isolated communities as well as sympatrically with normal *marmoreus* where no intergrades have been observed. Taxonomically, separation appears fully justifiable; 70.5mm; Puerto Princesa, Palawan. Erroneously sold locally as *"Conus nocturnus"*, a species endemic to Indonesia.
1d. Conus bandanus Hwass, in Bruguiére, 1792: Another species which has been considered a form of *C. marmoreus*, but whose taxonomic separation can be fully justified on the same basis as *C. vidua*. Morphologically almost identical to *marmoreus*, it can be separated on the basis of its distinctive surface pattern which consists of 2 dark, rather diffused black bands, somewhat intermediate between typical *marmoreus* (1a) and *C. vidua* (1c); 77mm; Cebu.
1e. Conus vidua Reeve, 1843: A curious form with a surface pattern close to that of *C. nicobaricus* (fig. 5); 61.1mm; Palawan.
1f. Conus vidua Reeve, 1843: An unusual orange colour form; 51.2mm; Puerto Princesa, Palawan.
1g. Conus vidua Reeve, 1843: Another unusual form where the tents are replaced by thin axial lines; 63.9mm; Puerto Princesa, Palawan.
1h. Conus marmoreus Linné, 1758: An unusual form with much reduced tenting; 67.3mm; Albay Gulf, Southern Luzon.
DISTRIBUTION: Sporadically found throughout the Philippines, except for *Conus vidua* which occurs only in Palawan.

2. Conus nobilis Linné, 1758
Shell medium sized, glossy and subcylindrically turbinated; the spire is low, almost flat, its whorls concave and spirally striated; body whorl smooth with some basal sulci; the shoulder is wide and sharply angulate, the aperture narrow posteriorly, widening basally; shell colour orange-brown, with numerous small white tents which occasionally tend to fuse, forming 3 indistinct white bands, one below the shoulder, another at mid-body, the third at the lower end, aperture white; reaches 60mm.
2a. 51.8mm, 47.2mm; Balabac Is., Palawan.
2b. a smaller, heavier, paler specimen with a finer pattern and larger tent marks; 30.5mm; Sulu Archipelago.
DISTRIBUTION: Palawan and Sulu Sea.

3. Conus spiculum Reeve, 1849
30.1mm; Punta Engaño, Cebu.
Shell small, elongate, with a sharply acuminate spire, a long tapering body whorl, straight sides, a sharply angled shoulder and sculptured with oblique spiral sulci at the base; white with rows of brown spots and larger brown blotches; reaches 30mm. Like young *C. generalis*, but has a biconic spire (*C. generalis* has concave lines to some extent), and is fully mature at 20mm. (*C. generalis* shows evidence of juvenility at this size.)
DISTRIBUTION: Cebu.

4. Conus alabaster Reeve, 1849
35.5mm, 37.1mm; Tayabas Bay, Quezon.
Shell medium sized, non-glossy, spire moderately tall, concave and sharply pointed, its whorls spirally striated; body whorl attenuated, wide at the sharply-angulate shoulder and sculptured with regularly spaced spiral cords, weaker interstitial threads and numerous closely set axial striae; aperture narrow, widening anteriorly, white within, outer lip thin; creamy-white with brown axial streaks on the spire and longer brown streaks and blotches on the body whorl; reaches 40mm.
DISTRIBUTION: Tayabas Bay, Quezon and Punta Engaño, Cebu.

5. Conus nicobaricus Hwass, in Bruguiére, 1792
72.3mm, 61.1mm; Coron Is., Palawan.
Shell medium sized, solid, glossy, spire low, concave, its whorls coronate and distinctly concave; body whorl large, smooth, except for some basal sulci, shoulder moderately wide, coronate, rounded and sharply angulate; aperture uniformly narrow, marginally white, lemon coloured within; shell pinkish white with 3 interrupted black bands between which are numerous crowded, medium sized white tents; reaches 100mm. A distinct species with little similarity to the Indian Ocean endemic species *C. araneosus* Lightfoot, 1786 with which it is sometimes erroneously synonymized.
DISTRIBUTION: Sulu Sea.

6. Conus kintoki Coomans & Moolenbeek, 1982
94.5mm, 70.0mm, 52.5mm; Punta Engaño, Cebu.
Shell large, solid, rather elongated with a matt gloss; spire low with almost flat sides and a sharp mamillate apex, its whorls flat to weakly concave, spirally threaded and with numerous fine, arcuate axial striae; body whorl smooth, shoulder wide and sharply angulate, sculptured with weak axial and spiral threads and stronger spiral cords towards the base; aperture uniformly wide, bluish white to lavender within; colour variable, pink, lavender or orange with a paler central band; reaches 100mm.
DISTRIBUTION: Cebu.

7. Conus ione Fulton, 1938
A beautiful glossy-white cone with rows of small brown dots and larger blotches forming one or two bands on the body whorl; shell medium sized, turbinate with a low, sharply pointed spire and

PLATE NO. 68 1a. *Conus marmoreus* Linné 1b. a pale, salmon-coloured (albinistic) specimen with reduced tenting 1c. *Conus vidua* Reeve 1d. *Conus bandanus* Hwass, in Bruguiére 1e. to 1g. *Conus vidua* Reeve 1h. *Conus marmoreus* Linné 2a. to 2b. *Conus nobilis* Linné 3. *Conus spiculum* Reeve 4. *Conus alabaster* Reeve 5. *Conus nicobaricus* Hwass, in Bruguiére 6. *Conus kintoki* Coomans & Moolenbeek 7a. to 7b. *Conus ione* Fulton

PLATE 68

concave spire whorls which are axially and spirally striated; body whorl is sharply angled with a somewhat carinate shoulder, smooth, except for some sulci at the base; reaches 75mm.

7a. typical *C. ione*; 66.8mm, 62.7mm; Cebu and Southern Mindanao.
7b. specimen with very reduced pattern; 60.5mm; Cebu.
DISTRIBUTION: Cebu and Southern Mindanao.

Conidae — Plate 69

1. Conus litteratus Linné, 1758
104.3mm; Bohol.
A large heavy shell, very similar to *C. leopardus* Röding in general structure, but can be separated by its solid columella, the violet-tinged base, and the presence of yellow banding (not a constant feature); the spotting is larger, squarish and regularly spaced; reaches 150mm.
DISTRIBUTION: Throughout the Philippines.

2. Conus leopardus (Röding, 1798)
108.6mm; Cebu.
Shell large, heavy, moderately glossy, with a low spire; body whorl large and smooth with a wide, sharply angled shoulder and a few obsolete sulci near the base; aperture wide, white within; shell off-white with numerous closely-spaced rows of short longitudinal spots; reaches 220mm.
DISTRIBUTION: Throughout the Philippines.

3. Conus tribblei Walls, 1977
Shell moderately large, lightweight, glossy, obconic, with a sharply pointed spire, its whorls concave and spirally striated; body whorl attenuated, with straight sides, a sharply angled shoulder, and heavy, occasionally nodular spiral striae on the lower third; columella straight but curiously twisted on the anterior portion; shell colour white with brown spire blotches, 2 brown bands on the body whorl, and interrupted brown spiral lines on the banding, aperture white; reaches 100mm.
3a. typical form; 75mm; Tayabas Bay, Quezon.
3b. an unusually granulated form; 68.5mm; Tayabas Bay, Quezon.
3c. specimen with a somewhat broken banded pattern; 59mm; Tayabas Bay, Quezon.
DISTRIBUTION: Tayabas Bay (Quezon), Cebu and Sorsogon.

4. Conus lenavati da Motta & Röckel, 1982
41.4mm, 52.7mm; Punta Engaño, Cebu.
Shell medium sized, solid, glossy, with a low, concave and sharply pointed spire, its whorls weakly concave and spirally striated; body whorl with a wide, rather swollen, sharply angled shoulder tapering rapidly anteriorly; surface smooth except for a few basal sulci; aperture uniformly narrow, white within; shell white with a few spiral maculations; body whorl with axial streaks and blotches forming 1 or 2 discontinuous circular bands; reaches 65mm.
DISTRIBUTION: Cebu.

5. Conus sugimotonis Kuroda, 1928
62.5mm, 54.9mm; Bohol.
Shell medium sized, solid, moderately glossy, obconic, spire almost flat, with a low, sharply pointed apex, its whorls flat and spirally striated with numerous arcuate axial striae; body whorl smooth with a few sulci at the base; shoulder wide, slightly swollen, sharply angled and tapering rapidly anteriorly; shell white with a characteristically lemon tinted apex and an occasional pale yellow band on the body whorl; reaches 110mm.
DISTRIBUTION: Bohol.

6. Conus kinoshitai (Kuroda, 1956)
67.8mm, 48.2mm, 68.1mm; Panglao, Bohol.
Shell medium sized, lightweight, glossy, subcylindrically elongate with a conic spire; body whorl attenuated, with slightly convex sides, a sharply angulate shoulder and some basal sulci; aperture uniformly wide and white within; shell usually violet with dark brown blotches and bands, but can be yellowish or bluish; reaches 90mm.
DISTRIBUTION: Cebu, Bohol and Tayabas Bay, Quezon.

7. Conus fulgetrum Sowerby I, in Sowerby II, 1834
30.8mm, 23.5mm, 25.2mm; Samar.
A small, squat shell with a low, coronated, straight-sided spire; body whorl widely shouldered, coronated, encircled with spiral grooves over the entire surface and granulated spiral cords on the lower half; aperture violet, shell orange-brown with white zigzag flammules; reaches 35mm.
DISTRIBUTION: Samar.

8. Conus miliaris Hwass, in Bruguière, 1792
Shell small, glossy, solid, rather squat in outline; spire low, coronate, straight, with a pointed apex; body whorl large, ovate, sharply angled and coronate at the shoulder and encircled with granulated spiral ridges on the lower half; aperture narrow, brown and white banded within; colour bluish grey to orange-tan with

PLATE NO. 69 1. *Conus literatus* Linné 2. *Conus leopardus* (Röding) 3a. *Conus tribblei* Walls 3b. unusually granulated form 3c. unusual pattern 4. *Conus lenavati* da Motta & Röckel 5. *Conus sugimotonis* Kuroda 6. *Conus kinoshitai* (Kuroda) 7. *Conus fulgetrum* Sowerby I, in Sowerby II 8. *Conus miliaris* Hwass, in Bruguiére 8b. unusual form 9. *Conus sp.* 10. *Conus recluzianus* Bernardi 11. *Conus macarae* Bernardi 12. *Conus voluminalis* Reeve 13. *Conus sp.*

PLATE 69

2 darker bands and rows of brown and white spots; reaches 40mm.
8a. 30.9mm, 31.4mm; Batangas, Southern Luzon.
8b. an unusual form with strong rows of brown spots and lacking any banding; 30.9mm; Punta Engaño. Cebu.
DISTRIBUTION: Throughout the Philippines but of infrequent occurrence.

9. Conus sp.
47.0mm, 48.2mm; Cebu.
Shell medium sized, lightweight, obconic with a distinctive waxy gloss; spire very low to flat with a sharply pointed apex, its whorls flat and spirally striated with wavy margins; body whorl smooth, with a distinctly coronate, sharply angulate shoulder and a few obsolete sulci at the base; aperture narrow, lavender within, outer lip thin; shell colour deep pink with white shoulder coronations, some brown spiral blotches and 2 or 3 darker bands on the body whorl; reaches 60mm. Locally sold as *"Conus gloriakiiensis,"* a species not yet seen in the Philippines.
DISTRIBUTION: Cebu, Bohol and Balut Island, Southern Mindanao.

10. Conus recluzianus Bernardi, 1853
63.7mm, 64.2mm; Tayabas Bay, Quezon.
Shell medium sized, heavy with a low gloss; spire low, concave with a pointed apex, its whorls slightly concave and threaded; body whorl sharply angulate at the shoulder with a few sulci at the base; aperture narrow, white within; white with brown spire spots and 2 bands of brown blotches interconnected by axial streaks; reaches 85mm.
DISTRIBUTION: Tayabas Bay (Quezon) and Cebu.

11. Conus macarae Bernardi, 1857
40.9mm, 40.3mm; Turtle Islands.
This species resembles a pale, small-sized *C. voluminalis* but differs in its proportionately taller spire, its more angulate shoulder and its single central white band which lacks the brown spotting of typical *C. voluminalis;* usually pale yellow to orange and seldom exceeds 45mm.
DISTRIBUTION: Turtle Islands, Sulu Sea.

12. Conus voluminalis Reeve, 1843
48.3mm, 57.1mm; Tayabas Bay, Quezon.
Shell medium sized, lightweight with a waxy gloss; spire low with a sharply pointed apex, its whorls flat with arcuate axial striae; body whorl attenuated, tapering anteriorly, shoulder wide, sharply angulate, the base with a few obsolete sulci; aperture narrow, white within, outer lip thin, very slightly "pinched-in" centrally; shell white with brown transverse streaks, 2 broad bands on the body whorl, fine brown spiral lines, and a central white zone bordered with brown spots; reaches 80mm.
DISTRIBUTION: Tayabas Bay (Quezon), Bulan (Sorsogon) and Manila Bay.

13. Conus sp.
43.5mm, 64.0mm; Cebu.
Shell medium sized, heavy with a waxy gloss; spire flat with a sharply pointed apex, wavy margins and threaded on the first few whorls; body whorl attenuated and tapering, with a sharply angulate, somewhat undulate shoulder and a few sulci at the base; aperture uniformly narrow, light pink within; colour pale purple with brown streaks on the spire, 2 brown bands on the body whorl and fine brown spiral lines; reaches 70mm; similar to *C. recluzianus* in shape and *C. voluminalis* in colour pattern but appears unnamed.
DISTRIBUTION: Bohol, Cebu and Tayabas Bay, Quezon.

Conidae Plate 70

1. Conus aulicus Linne, 1758
Shell moderately large, well glossed, elongate, with an elevated, pointed spire and flat spire whorls which are axially and spirally lirate; body whorl ventricose, encircled with fine spiral striae and weaker axial growth lines; aperture wide, yellowish orange within, body whorl with irregular white tents and large brown blotches, irregularly placed, but occasionally in 2 or 3 bands; reaches 170mm.
1a. large, reddish brown form; 136mm; Albay Gulf, Southern Luzon.
1b. typical form; 115.2mm; Albay Gulf.
1c. pinkish form *rosea* Sowerby, 1834; 83.2mm; Samar.
1d. specimen with large, white areas; form *propenudus* Melvill, 1900; 104.7mm; Albay Gulf.
DISTRIBUTION: Sporadically found throughout the Philippines.

2. Conus magnificus Reeve, 1843
81.3mm; 60.0mm; Samar.
Shell medium sized, glossy, ovately elongate, with a short, straight spire, obtuse apex and weakly threaded spire whorls; body whorl with a rounded, occasionally keeled shoulder and encircled with fine spiral cords, numerous small white tents and large maroon-brown blotches in an irregular pattern; reaches 90mm.
DISTRIBUTION: Uncommon, specimens mostly come from Sulu Sea and Eastern Samar.

3. Conus omaria Hwass, in Bruguiére, 1792
Shell medium sized, glossy, subcylindrically elongate with a short, straight, pink-tipped acuminate spire; body whorl long with either a roundly or sharply

PLATE NO. 70 1a. *Conus aulicus* Linné reddish brown form 1b. typical form 1c. form *rosea* Sowerby 1d. form *propenudus* Melvill 2. *Conus magnificus* Reeve 3a. to 3f. *Conus omaria* Hwass, in Bruguiére 4. *Conus auratinus* da Motta 5a. *Conus episcopatus* da Motta 5b. unusual form

PLATE 70
1a
1b
1c
1d
2
3a
3b
3c
3d
3e
3f
4
5a
5b

angulate shoulder, almost straight sides, and an open, basally flaring, white-colored aperture; ground colour may be yellow, orange or brown with moderately sized white tents, axially arranged in spiral bands; reaches 80mm.

3a. unusually large specimen with typical pattern and colour; 81.2mm; Mactan Is., Cebu.

3b. smaller, slightly more bulbous specimen; 47.2mm; Mactan Is., Cebu. This resembles *Conus episcopatus* da Motta, 1982.

3c. similar to 3b. in shape but with dense, brown axial bands; 47.4mm; Mactan Is., Cebu.

3d. golden form, with a pinkish aperture; 46.9mm; Sulu Archipelago.

3e. a beautifully patterned form where the tents are enlarged; 57.5mm; Mactan Is., Cebu.

3f. a beautiful form where the tenting is reduced to small white dots and the brown blotching forms 2 dense bands. Sold locally as *"C. elisae,"* a species not endemic to the Philippines; 42.7mm; Mactan Is., Cebu.

DISTRIBUTION: Sporadically found throughout the Philippines.

4. Conus auratinus da Motta, 1982
103.3mm; E. Samar.
A replacement name for *C. auratus* Hwass, in Bruguiére, 1792 which has been shown to apply to a golden form of *C. aulicus.* Very close to *aulicus,* but more cylindrically elongate, with a taller, obtuse spire and a smoothly rounded shoulder and flat sides, the shell is sculptured with fine spiral striae; colour similar to a chestnut brown *aulicus,* but the maculations are larger and the tenting marks smaller and more isolated; normally reaches around 85mm, the above specimen being the largest we have seen. Elsewhere, *C. auratinus* is invariably found in a reddish colour.
DISTRIBUTION: Eastern Samar and Sulu.

5. Conus episcopatus da Motta, 1982
Shell medium sized, solid, subcylindrical with a low gloss; spire low, somewhat turbinate with an obtuse, pink coloured apex, shoulder wide and subangulate; body whorl elongate with moderately convex sides and sculptured with fine spiral cords and weak axial striae; aperture wide, white within, flaring noticeably at the anterior end; shell pattern consists of large white areas with numerous irregularly sized chestnut brown blotches which may be thin and form white triangular tents, or thick and coalescing in an irregular pattern; reaches 80mm. Popularly known as *"C. episcopus auct."* and until recently was nameless.

5a. typical form showing pattern variations; 67.5mm, 71.6mm, 81.3mm; Bohol.

5b. a curious form somewhat more cylindrical than normal, 90.2mm; Samar.

DISTRIBUTION: Sporadically found throughout the Philippines.

Conidae — Plate 71

1. Conus sowerbii Reeve, 1849
53.5mm, 58.4mm; Tayabas Bay, Quezon.
Shell medium sized, moderately glossy, biconic, spire acuminate, its whorls carinate and spirally striated; body whorl with broad, flat, spiral cords and faint axial striae, shoulder wide, carinate and sharply angulate, aperture narrow and white within; white with brown axial spiral blotches and rectangular spotted bands on the body whorl; incorrectly figured as *C. praecellens* in many books; reaches 60mm.
DISTRIBUTION: Tayabas Bay (Quezon), Sorsogon and Cebu.

2. Conus subaequalis Sowerby III, 1870
34.8mm; Punta Engaño, Cebu.
Shell medium sized, biconic, matt surface, with an unusually tall, carinate, straight and sharply pointed spire, its whorls weakly concave and spirally striated; body whorl straight-sided, sharply angled at the shoulder, encircled with shallow, punctated transverse grooves; pale cream with orange maculations on the spire whorls and banding on the body whorl; close to *C. sowerbii* but has a straight conic spire and lacks the interstitial spiral grooving of that species; reaches 60mm.
DISTRIBUTION: Cebu and Southern Mindanao.

3. Conus acutangulus Lamarck, 1810
20.9mm, 19.3mm; Cebu.
Shell small, lightweight, conical, low gloss; spire tall, weakly concave and sharply pointed, its whorls coronate and spirally striated; body whorl weakly convex, tapering anteriorly and sculptured with deep, narrow, punctate spiral grooves which form broad, rounded, elevated spiral cords; shoulder carinate, occasionally nodulose and sharply angled; aperture narrow, blue-white within, shell light brown with white coronations and white body whorl spotting; reaches 35mm.
DISTRIBUTION: Tayabas Bay (Quezon), Marinduque, Sorsogon and Cebu.

PLATE NO. 71 1. *Conus sowerbii* Reeve 2. *Conus subaequalis* Sowerby III 3. *Conus acutangulus* Lamarck 4. *Conus memiae* (Habe & Kosuge) 5. *Conus kimioi* (Habe) 6. *Conus comatosa* Pilsbry 7. *Conus delicatus* Schepman 8. *Conus sp.* 9. *Conus hypochlorus* Tomlin 10. *Conus sp.* 11. *Conus ichinoseana* (Kuroda) 12. *Conus pseudorbignyi* Röchel and Lan 13. *Conus orbignyi* Audouin 14. *Conus vimineus* Reeve 15a. *Conus floridulus* Adams & Reeve 15b. highly pustulose form 16. *Conus montillai* Röckel 17. *Conus scalptus* Reeve 18. *Conus musicus* Hwass, in Bruguiére 19. *Conus sponsalis* Hwass, in Bruguiére 20. *Conus aristophanes* Sowerby II 21a. to 21b. *Conus coronatus* Gmelin

PLATE 71
1
2
3
4
5
6
7
8
9
10
11
12
13
14
15a
15b
16
17
18
19
20
21a
21b

4. Conus memiae (Habe & Kosuge, 1970)
24.7mm, 22.5mm; Punta Engaño, Cebu.
A small, glossy, turbinate shell with a tall, acute, convex spire; body whorl with a wide, sharply carinate shoulder and encircled with wide, flat ridges; aperture narrow, pink within; a pinkish shell with brown spots on the spire and 2 or 3 interrupted brown spotted white bands on the body whorl; reaches 40mm.
DISTRIBUTION: Cebu and Bohol.

5. Conus kimioi (Habe, 1965)
16.3mm; Punta Engaño, Cebu.
A small glossy cone with a short, almost flat spire and a wide, sharply-angled shoulder; body whorl attenuated, tapering slightly, pinched-in centrally and sulcated at the base; white with an overlay of dense brown blotches and encircled with brown spiral lines; reaches 30mm.
DISTRIBUTION: Cebu and Bohol.

6. Conus comatosa Pilsbry, 1904
38.1mm; Cebu.
A medium sized, moderately glossy shell, subcylindrical, with a tall, sharply acute spire; body whorl with a sharply carinate shoulder and encircled with deep, punctate spiral grooves; aperture narrow, pinkish white within; white with brown spire streaks and interrupted brown bands and interstitial rows of brown spots on the body whorl; reaches 60mm.
DISTRIBUTION: Bohol, Cebu and Tayabas Bay, Quezon.

7. Conus delicatus Schepman, 1913
37.1mm; Punta Engaño, Cebu.
Shell medium sized, lightweight, low gloss, fusiform and fairly elongate; spire tall and sharply pointed; body whorl with a narrow, sharply-angulate shoulder, tapering rapidly towards the base; sculptured with deep, narrow, axially striated grooves, becoming closer towards the shoulder; aperture narrow, white within; colour white with brown spiral blotches and 2 or 3 interrupted brown bands on the body whorl; reaches 40mm.
DISTRIBUTION: Cebu, Bohol and Tayabas Bay, Quezon.

8. Conus sp.
32.6mm; Punta Engaño, Cebu.
Shell small, fusiform, with an acuminate spire and an attenuated body whorl which is straight-sided; spire whorls flat and spirally striated, body whorl sculptured with widely spaced, equidistant, shallow spiral grooves forming flat transverse ridges; aperture uniformly narrow, outer lip thin with a sinuous contour; shell colour white with brown spire streaks, brown spotting on the body whorl and 2 broken brown bands; reaches 40mm; may be a subspecies of *C. comatosa* (fig. 6).
DISTRIBUTION: Known only from Punta Engaño, Cebu.

9. Conus hypochlorus Tomlin, 1937
30.5mm; Punta Engaño, Cebu.
A small, fusiform-elongate shell with a tall, sharply pointed spire; body whorl straight-sided with a carinate, sharply angled shoulder and encircled with broad, flat spiral ridges and axial riblets in the interstices; aperture narrow, tan within; shell uniformly yellow to orange-brown; reaches around 30mm.
DISTRIBUTION: Presently known only from Punta Engaño, Cebu and Marinduque.

10. Conus sp.
34.5mm; Cebu.
Shell thin, moderately glossy, fusiform with a tall, sharply acute, weakly-concave spire; body whorl weakly convex, subangulate at the shoulder and encircled with transverse grooves, forming strong spiral cords; aperture narrow, violet within, outer lip thin; colour off-white with brown spots and streaks, sometimes in bands; reaches 50mm. Often incorrectly figured as *C. aculeiformis* Reeve, 1944 (endemic to India) and *C. longurionis* Kiener, 1845 (endemic to East Africa). See inserts.
DISTRIBUTION: Cebu, Bohol and Tayabas Bay, Quezon.

11. Conus ichinoseana (Kuroda, 1956)
48.2mm, 55.5mm; Punta Engaño, Cebu.
Shell medium sized, elongate, subcylindrical; spire weakly concave, sharply acuminate; body whorl long and tapering, shoulder wide and sharply angulate; encircled with deep, narrow, punctate spiral grooves, becoming obsolete on the posterior third of the body whorl; aperture narrow, white within; shell white with brown spots and streaks, occasionally forming bands; reaches 70mm.
DISTRIBUTION: Tayabas Bay (Quezon), Cebu and Bohol.

12. Conus pseudorbignyi Röckel and Lan, 1982
36.5mm; Punta Engaño, Cebu.
Shell small to medium sized, subcylindrical, with a tall, weakly-concave, sharply acute spire and a long, slightly convex body whorl tapering towards the base; spire whorls carinate, nodulose and spirally threaded, body whorl with a wide, carinate, sharply angled shoulder and encircled with wide, deep, axially striated grooves; aperture narrow, white to brown within; shell pale brown with darker spotting, spire whorls with brown spots between the nodules; reaches around 40mm. Often incorrectly figured as *C. insculptus* Kiener, 1845 (see insert), which has a wider shoulder and a recurved anterior end. Not recorded from the Philippines as yet.
DISTRIBUTION: Tayabas Bay (Quezon), Bohol, Manila Bay and Turtle Islands, Sulu Sea.

13. Conus orbignyi Audouin, 1831
47.0mm; Tayabas Bay, Quezon.
Shell medium sized, subcylindrical and lightweight; spire sharply acuminate, weakly concave, and spirally

striated; body whorl with a wide, nodulose, sharply-angled shoulder and encircled with numerous flat spiral ribs and deep, transverse grooves; aperture uniformly narrow, white within; body whorl white with brown dots and blotches, axially arranged, occasionally in bands; reaches 70mm.
DISTRIBUTION: Tayabas Bay (Quezon), Cebu and Bohol.

14. Conus vimineus Reeve, 1849
49.5mm, 46.1mm; Punta Engaño, Cebu.
Shell medium sized, lightweight, cylindrical, with a sharply acuminate, straight spire, its whorls sculptured with 2 or 3 deep marginal grooves; body whorl attenuated with a subangulate shoulder, grooved above, and encircled below with numerous transverse grooves; aperture narrow, marginally white, lavender within; shell entirely pale brown; reaches 70mm.
DISTRIBUTION: Tayabas Bay (Quezon), Turtle Islands (Sulu Sea), Cebu, Bohol and Manila Bay.

15. Conus floridulus Adams & Reeve, 1848
Shell small to medium sized, solid, broadly turbinate; spire concave and sharply pointed, its whorls coronate and weakly threaded; body whorl straight-sided, with a wide, nodulose, sharply angled shoulder; surface smooth except for some pustulose sulci at the base which may occasionally extend to the shoulder; shell colour white with 2 broad orange-brown zones, spots and a lavender-coloured base; reaches 50mm.
15a. typical smooth form; 38.9mm; Mactan Is., Cebu.
15b. highly pustulose form; 34.3mm; Mactan Is., Cebu.
DISTRIBUTION: Bulan (Sorsogon), Cebu, Bohol and Southern Mindanao.

16. Conus montillai Röckel, 1985
20.3mm, 20.5mm; Palawan.
Shell small, lightweight, glossy, ovately turbinate; spire concave and sharply pointed, its whorls weakly coronate and spirally striated; body whorl with a wide, subangulate shoulder, convex sides, tapering rapidly towards the base; surface smooth with a few sulci at the base, aperture uniformly narrow, white within; shell colour white with brown spots between the spire coronations and large brown blotches and streaks on the body whorl, often in 2 bands; reaches 25mm. Known locally as *"Conus sphacelatus,"*
DISTRIBUTION: Palawan.

17. Conus scalptus Reeve, 1843
34.5mm; Punta Engaño, Cebu.
Shell medium sized, lightweight, glossy and turbinate; spire conic, its whorls flat and threaded; body whorl with a narrow, subangulate shoulder and sculptured with flat spiral incisions on the lower half; aperture white, narrow, widening anteriorly; shell white with red transverse lines on the body whorl; reaches 40mm. Frequently confused with *C. mucronatus* which is a broader shell with a more angulate shoulder and has a different surface sculpturing.
DISTRIBUTION: Cebu, Bohol, Sorsogon, and Quezon.

18. Conus musicus Hwass, in Bruguiére, 1792
11.4mm, 17.9mm, 19.5mm; Batangas.
Shell small and weakly glossed with a low spire; body whorl with a wide, sharply angled, undulate shoulder and sculptured with fine spiral threads; aperture narrow, white with 2 purplish bands; shell greyish white with black dots on the spire and shoulder, and encircled with rows of fine, black spots and darker banding, base purple-stained; reaches 20mm.
DISTRIBUTION: Throughout the Philippines.

19. Conus sponsalis Hwass, in Bruguiére, 1792
17.9mm, 17.8mm, 18.8mm; Batangas.
Shell small, glossy, with a flat spire, a large ventricose body whorl, a sharply angulate shoulder and sculptured with granulose spiral cords on the lower half; aperture narrow, purple within; white to pale yellow, with a purple-stained base and brown figurations often in interrupted bands; reaches 30mm.
DISTRIBUTION: Sporadically found throughout the Philippines.

20. Conus aristophanes Sowerby II, 1857
26.9mm; Samar.
Shell medium sized, solid, with a low gloss; spire short, straight and sharply pointed, its whorls coronate with 1 or 2 spiral threads; body whorl with a wide, sharply angled, coronate shoulder, straight sides, oblique, granulose spiral cords at the base and a uniformly wide, violet-coloured aperture; colour greyish with white banding at the shoulder and mid-section, and white spiral rows of black dots; reaches 50mm.
DISTRIBUTION: Throughout the Philippines.

21. Conus coronatus Gmelin, 1791
This and the previous species are often confused but can be separated quite easily: *C. coronatus* is much more bulbous at the shoulder and has a taller, more heavily coronate spire; the spire whorls have more spiral threads than *C. aristophanes* and the aperture widens rapidly towards the base; the shell colour is usually bluish with a distinct pink tinge, and the brown blotching is more pronounced; reaches 50mm.
21a. 27.2mm; Samar.
21b. 29.6mm, 27.0mm; Samar.
DISTRIBUTION: Throughout the Philippines.

Plate 72 Conidae

1. Conus aphrodite Petuch, 1979
13.5mm; Panglao, Bohol.
Shell small, glossy, turbinate with a moderately tall, concave spire, the early spire whorls beaded and sculptured with numerous arcuate axial striae; body whorl with a wide, sharply-angled, carinate shoulder and straight sides which taper rapidly towards the base; surface smooth with some spiral incisions on the anterior third; colour pale lavender with 3 bands of brown axial streaks and numerous spiral rows of

brown and white dots, the central one strongest, aperture violet; reaches 30mm.
DISTRIBUTION: Panglao (Bohol) and Balut Island, Southern Mindanao.

2. Conus hirasei (Kira, 1956)
52.5mm; Balut Is., Southern Mindanao.
Shell medium sized, with a low spire and a sharply acute apex; spire whorls flat and encircled with 5 to 6 spiral threads; body whorl with a stout shoulder tapering rapidly towards the base in a "goblet-like" fashion; aperture narrowly elongate, outer lip parallel to the columellar surface; pure white, the spire whorls with rows of brown spots along the sutures and at the body whorl shoulder, body whorl completely encircled with continuous brown lines; reaches around 85mm.
DISTRIBUTION: Southern Mindanao.

3. Conus corallinus Kiener, 1845
18.4mm; Sulu Archipelago.
Shell small, subcylindrical with a conic spire and a sharp mamillate apex; spire whorls slightly concave, encircled with fine spiral striae and weak, arcuate threads; body whorl with a sub-angulate shoulder, somewhat convex sides tapering gradually towards the base and a few basal sulci; aperture narrow, outer lip straight; colour deep pink, the spire whorls with pale brown and white maculations, body whorl with a central double band of pale pink and white tesselations; reaches 20mm.
DISTRIBUTION: Sulu Sea.

4. Conus fulvobullatus da Motta, 1982
62.2mm; Palawan.
Shell medium sized, oblong-ovate, with a low, concave spire and a pink-tinted apex, its whorls weakly convex, encircled with 5 deep spiral grooves; body whorl large with inflated sides, widest below the sharply angled shoulder; aperture wide, flaring anteriorly, columella twisted with weak basal cords; colour white with orange-brown maculations, arranged in 2 indistinct transverse bands and numerous spiral rows of small brown dots, aperture pure white; no larger specimens seen.
DISTRIBUTION: Palawan, Sulu Sea.

5. Conus leobrerai da Motta & Martin, 1982
26.0mm; between Malapascua and Bantayan Is., Visayan Sea.
Shell small, subcylindrical with a tall, sharply acute spire, its whorls slightly concave, carinate and sculptured with deeply incised spiral grooves; body whorl with a carinate shoulder and convex sides, widest at the shoulder and tapering to a slightly recurved base; sculptured with numerous equidistant spiral ridges and deeply pitted interstices; aperture narrow, widening basally; colour white, the spire with several brown axial maculations, the body whorl with brown axial streaks, frequently arranged in 2 bands, aperture white; reaches 31mm. Named after Carlos and Fely Leobrera of Manila.
DISTRIBUTION: Visayan Sea.

6. Conus eugrammatus Bartsch & Rehder, 1943
25.5mm; Punta Engaño, Cebu.
Shell small, turbinate, spire whorls turreted, concave, sculptured with fine arcuate axial ridges, weaker spiral threads and some granules on the earlier whorls; body whorl with a broad, carinate shoulder, straight sides which are attenuated towards the base and encircled with deep, axially striated spiral grooves; shell colour white with brown spots and streaks on the spire whorls, and on the body whorl, small brown spots, frequently fusing into larger blotches or axial streaks, occasionally forming indistinct, interrupted spiral bands, aperture uniformly narrow, white within; reaches 25mm.
DISTRIBUTION: Cebu.

7. Conus articulatus Sowerby III, 1873
20.5mm; Punta Engaño, Cebu.
Shell small, turbinate with a waxy gloss, a concave spire and a weakly convex body whorl which tapers slightly towards the base; spire whorls sometimes granular, sculptured with weak axial striae, body whorl with a sharply angled, carinate shoulder; surface smooth but some deep punctate spiral grooves on the lower half; colour variable, usually light brown to yellow with a central and basal brown band containing white spots, spire off-white with brown axial streaks, aperture uniformly narrow, white within and purple stained at the base; reaches 30mm.
DISTRIBUTION: Cebu.

8. Conus dayriti Röckel & da Motta, 1983
18.3mm; Palawan.
Shell small, smooth, glossy, obconic, with a sharp, projecting apex; spire whorls flat, weakly stepped, encircled with a few deeply incised spiral grooves on the upper part of each whorl; body whorl with a wide, sharply angled shoulder, tapering rapidly towards the base; surface smooth with a few oblique sulci basally; colour pale purplish-white, with light and dark brown suffusion, spire whorls with indistinct white spotting along the carinae, the body whorl with two paler bands, one below the shoulder the other centrally, and 3 groups of brown and white spotted spiral threads, aperture pale lavender; reaches 23mm (illustrated specimen sub-adult).
DISTRIBUTION: Palawan.

PLATE NO. 72 1. *Conus aphrodite* Petuch 2. *Conus hirasei* (Kira) 3. *Conus corallinus* Kiener 4. *Conus fulvobullatus* da Motta 5. *Conus leobrerai* da Motta & Martin 6. *Conus eugrammatus* Bartsch & Rehder 7. *Conus articulatus* Sowerby III 8. *Conus dayriti* Röckel & da Motta 9. *Conus chiangi* (Azuma) 10. *Conus baileyi* Röckel & da Motta 11. *Conus boholensis* Petuch 12. *Conus lapulapui* da Motta & Martin 13. *Conus dondani* Kosuge 14. *Conus wakayamaensis* (Kuroda) 15. *Conus polongimarumai* Kosuge 16. *Conus albicans* Sowerby II 17. *Conus cebuganus* da Motta & Martin 18. *Conus saecularis* Melvill 19. *Conus sazanka* Shikama 20. *Conus fischoederi* Röckel & da Motta

PLATE 72

9. Conus chiangi (Azuma, 1972)
19.5mm; Panglao, Bohol.
Shell small, solid, glossy, conic, with sharply pointed spire; spire whorls stepped, the earlier ones with weakly nodular coronations, becoming sharply spinose on the shoulder; body whorl straight-sided, tapering towards the base and encircled with wavy, punctate spiral grooves; aperture narrow, flaring basally; spire whorls mainly creamy white, body whorl pale tan, with a brown-blotched, white-coloured central band, and the entire surface with numerous rows of brown spotted ridges; reaches 25mm.
DISTRIBUTION: Bohol.

10. Conus baileyi Röckel & da Motta, 1979 (October)
Syn.: *C. nereis* Petuch, 1979 (December)
20.1mm; Panglao, Bohol.
Shell small, glossy, turbinate with a tall, weakly stepped, concave spire the whorls of which are flat and sculptured with numerous arcuate axial striae and weak spiral threads; body whorl with a sharply angled, carinate shoulder, tapering gradually towards the base and sculptured with numerous punctate spiral grooves and flattened spiral ridges; colour white, the spire with large brown blotches extending onto the shoulder, body whorl with brown spots on the ridges, frequently fusing to form three interrupted bands, aperture narrow and white; reaches 30mm.
DISTRIBUTION: Bohol.

11. Conus boholensis Petuch, 1979
22.9mm; Panglao, Bohol.
Shell small, moderately glossy, turbinate with a tall, weakly scalariform spire, bearing deeply excavated spire whorls, finely coronated and sculptured with arcuate striae; body whorl sharply angled and kneeled at the shoulder, somewhat constricted below and encircled with up to 30 axially striated grooves, densely packed below the shoulder carina, more widely spaced basally; white with a few brown flammules on the upper body whorl area; reaches 40mm.
DISTRIBUTION: Bohol.

12. Conus lapulapui da Motta & Martin, 1982
19.5mm; between Malapascua and Bantayan Is., Visayan Sea.
Shell small, turbinated, with a spire consisting of 2 globose protoconch whorls, and 8 distinctly turreted spire whorls, minutely tuberculate at the sutures and sculptured with deeply incised arcuate striae; body whorl with a sharply angulate, carinated shoulder, and weakly convex sides which taper rapidly towards the base; sculptured with numerous weakly incised spiral grooves; colour off-white, spire whorls with brown dots and streaks, body whorl with irregular brown flecks and blotches, frequently fusing to form 3 indistinct bands; aperture with a thin, undulate outer lip, white within; reaches 26.5mm. Named after the first Filipino hero, Lapu-Lapu of Cebu.
DISTRIBUTION: Visayan Sea.

13. Conus dondani Kosuge, 1981
22.9mm; Panglao, Bohol.
Shell small, glossy, turbinate with a moderately raised spire which is weakly stepped; early spire whorls beaded and spirally striated but becoming smooth towards the penultimate whorl; body whorl with a narrow subangulate shoulder, long tapering sides and a smooth surface with a few spiral grooves on the lower third of the whorl; ground colour pale brown, spire whorls with large whitish blotches, the body whorl with 2 paler bands and encircled over its entire length with brown and white spotted spiral threads; aperture narrow, bluish white within; reaches 25mm. Named after Donald Dan, a prominent local collector now residing in the U.S.
DISTRIBUTION: Bohol.

14. Conus wakayamaensis (Kuroda, 1956)
22.9mm; Visayan Sea.
Close to *C. eugrammatus* Bartsch & Rehder (fig. 6) but differs in the following ways: it generally reaches a much larger size than does *eugrammatus* and in relation to its size, is a much broader shell and is far more attenuate towards the base; the colour pattern is one of much denser brownish maculations, the brown spotting of *eugrammatus* being replaced by solid brown bands; may reach 32mm.
DISTRIBUTION: Visayan Sea.

15. Conus polongimarumai Kosuge, 1980
19.9mm; Punta Engaño, Cebu.
Shell small, glossy, turbinate with a tall, straight spire; spire whorls with a nodulose surface and encircled with fine spiral grooves and obsolete, arcuate axial striae; body whorl with a nodulose, moderately angled shoulder and straight sides which taper gradually towards the base; sculptured with weakly striated grooves, becoming stronger basally, aperture uniformly narrow; spire whorls white with brown maculations, body whorl chocolate brown posteriorly, fading to light yellow anteriorly with a pale, white-spotted median band; aperture pale lavender; reaches 23mm.
DISTRIBUTION: Cebu.

16. Conus albicans Sowerby II, 1857
33.0mm; Samar.
Shell medium sized, solid, turbinate with a vitreous lustre; spire concave, its whorls undulate, encircled with a few spiral grooves; body whorl with a wide, sharply angled, distinctly coronate shoulder, and convex sides which taper gradually towards the base; spire colour pale creamy white with brown spots between the coronations; body whorl with numerous thin, axial streaks and stained with brown at the base; reaches 45mm; often synonymized with *C. furvus* but is easily distinguished by its wavy sutures, coronate whorls and the characteristic colour pattern.
DISTRIBUTION: Samar and Palawan.

17. Conus cebuganus da Motta & Martin, 1982
26.9mm; between Malapascua and Bantayan Is., Visayan Sea.
Shell cylindrical, acuminately turbinated, with a pointed apex and around eleven spire whorls, strongly trellised by intersecting axial and spiral striae; shoulders angular, carinated and the sides are almost flat, tapering down its attenuated length; body whorl completely incised with around 26 equidistant grooves of axial riblets separating the shallow flat ridges which are sculptured with evenly spaced, continuous rows of granules over the entire body surface; colour off-white with rows of pale brown quadrate spots, encircling the surface completely; these spots coalesce to form two solid brown bands, one medially, the other basally, spire with brown streaks and the sutural edges with regularly spaced brown and white spots; aperture uniformly narrow, dilating slightly at the base, white within; reaches around 35mm.
DISTRIBUTION: Visayan Sea.

18. Conus saecularis Melvill, 1898
28.2mm; Punta Engaño, Cebu.
Shell small to medium sized, subcylindrical, with a tall spire; early spire whorls finely beaded, becoming carinate and encircled with deeply incised spiral grooves that are crossed by weaker axial threads producing a finely latticed appearance to the deeply concave surface; body whorl with a wide, carinate, sharply angled shoulder and weakly convex sides which taper rapidly towards a recurved base; surface with deeply incised, punctate spiral grooves; aperture uniformly narrow, flaring basally; off-white in ground colour, the spire with large brown blotches, the body whorl with irregular brown blotches, forming a central and basal spiral band; aperture white within; reaches around 28mm at maturity.
DISTRIBUTION: Most specimens from Cebu.

19. Conus sazanka Shikama, 1970
33.8mm; Davao.
Shell medium sized, solid, glossy, obconic; spire straight to weakly concave, its whorls slightly convex and encircled with weak spiral ridges and axial striae; body whorl with a sub angulate, weakly coronate shoulder and straight sides tapering gradually towards the base; colour a deep brownish orange with a pale central band and white base; the spire is somewhat pinkish with brown maculations, the aperture deep pink; reaches 50mm.
DISTRIBUTION: Davao.

20. Conus fischoederi Röckel & da Motta, 1983
37.5mm; Samar.
Shell medium sized, with a large body whorl and a moderately tall, straight spire, its whorls weakly convex and encircled with a few spiral grooves; body whorl weakly carinate, slightly bulging and widest below the shoulder, tapering gradually towards the base; aperture moderately wide, flaring basally; colour variable, usually white in ground colour with brown or orange blotches on larger areas of the body whorl, forming two indistinct bands; entire body whorl surface encircled with fine black threads, aperture bluish white; reaches around 40mm.
DISTRIBUTION: Samar and Coron, Palawan.

ADDITIONAL NOTES ON THE CONIDAE

1. Conus kintoki Coomans & Moolenbeek, 1982
(Pl. 68, fig. 6)
This species has a curious taxonomic history. Named and figured as *Conus kintoki* by Habe & Kosuge (1970: *Pacific Shell News* 2:9), the taxon was subsequently rejected as a *nomen nudum*, as no formal description was ever published. In 1972, a closely allied taxon was described by Veillard (*Of Sea and Shore*, 3(4) : 176) to which the name *Conus berdulinus* was applied. For some time, it was held that the Philippine shells (i.e. "*C. kintoki*") were conspecific with Veillard's taxon, and Walls (1979, p. 232 lower figure) further illustrated the Philippine shells under the name of *Conus coelinae berdulinus*. Veillard treated the taxon as being quite different structurally, colour and pattern-wise from *Conus coelinae* Crosse, 1858, a fact which was further endorsed by Kosuge when he synonymized Philippine shells with *berdulinus* (1979 : *Bull. Inst. Malac. Tokyo* 1(2) : 21-22, pl. 4, figs. 8 - 9). Recent studies of the holotype of *C. berdulinus* have shown it to be distinct from Philippine shells, and the name *kintoki* was subsequently validated by Coomans & Moolenbeek (1982, *Bulletin Zoologisch Museum, Universiteit Van Amsterdam*, Vol. 8, no. 15).

N. B. In a recent article by W. K. Emerson (1984: Veliger, Vol. 98(2):58-60), it has been stated that this taxon was validly proposed by Habe & Kosuge, and is not a *nomen nudum* as commonly accepted by most authors.

2. Conus sulcatus Hwass, in Bruguiére, 1792
(Pl. 65, fig. 5)
Philippine shells have recently been named *Conus sulcatus brettinghami* Coomans, Moolenbeek & Wils, 1982 [see *Basteria* 46 (1-4)]. This name was introduced as a replacement for the homonymous *Conus undulatus* Sowerby II, 1857-1858 (non Solander, 1786) and specimens are said to differ from the nominate subspecies (which occurs in the Formosa Straits) in their lighter weight, more slender outline, their yellowish brown colouration, and the smooth surface of the upper half of the last whorl (although granulated specimens are frequent). The designated type locality is Marinduque. The authors have seen far too little material from intermediate areas to comment on the validity of *brettinghami*, and provisionally retain the name *sulcatus* for Philippine shells.

3. Conus pagodus Kiener, 1845
(Pl. 65, fig. 14)
Figured under the name of *Conus cancellatus* Hwass, 1792, by Walls (1978: p. 185). Examination of the holotype of *cancellatus* in the Musee d' Histoire Naturelle, Geneva, has shown it to be identical to *Conus austini* Rehder & Abbott, 1951, a West Atlantic endemic, and which must be considered a junior synonym of the former species. The West Pacific *C. pagodus* is well figured by Kiener (1845: pl. 70, fig. 4).

4. Conus circumactus Iredale, 1929
(Pl. 66, fig. 19)
Frequently figured under the name *C. connectens* A. Adams

(Walls 1978: p. 240), which should be considered a *nomen dubium* as the holotype in the B.M. (Nat. Hist.) is a poor, beachworn specimen, too indistinct to be unequivocally identified. *C. circumactus* is the next available name and replaces *C. pulchellus* Swainson, 1822 (non Röding, 1798) (see *Mem. Queensland Mus.* ix, p. 281).

5. Conus sowerbii Reeve, 1849
(Pl. 71, fig. 1)
Figured by Walls (1978: p. 552) under the name *praecellens*, a non-Philippine endemic shell which bears no morphological resemblance to *sowerbii* whatsoever. The name *sowerbii* is not pre-occupied nor can it be considered a homonym because of the various other spellings used: *sowerbei* Nyst, 1836, and *sowerbyi* Bronn, 1848. Reeve was apparently well aware of this when he renamed *C. sinensis* Sowerby, 1833 (non Gmelin, 1791), and took pains to spell it differently as "*sowerbii*" to distinguish it from the other versions. There appear to be at least two good taxa lumped under the name *sowerbii* and which are currently undergoing investigation. (see *Carfel Philippine Shell News* 1981, 3(5) : 7-8).

6. Conus subaequalis Sowerby, 1870 (Pl. 71, fig. 2)
Considered a "variation" of *C. sowerbyi* Sowerby III, 1870 and as a juvenile *C. praecellens* A. Adams, 1854 by Walls (1978: pp. 807, 808). The species bears some morphological resemblance to *sowerbii* (Wall's *praecellens*), but is separable by numerous structural differences clearly annotated within the text. The sympatric occurrence of *subaequalis* and *sowerbii* further justifies its taxonomic separation.

7. Conus armadillo Shikama, 1971/**Conus kuroharai** (Habe, 1965)
(Pl. 62, figs. 4, 5)
Both these distinctly different species are figured by Walls (1978: p. 269) under the name *C. duplicatus* Sowerby I, 1823. Walls cites an earlier publication (1823), but Sowerby in *Thes, Conch,* Vol. III, in the Index to the Genus *Conus*, p. 51, notes simply in italics (meaning a synonym, not a species) "duplicatus So. - *australis*". Kiener and Reeve made no mention of the name. Tomline was more specific: "*duplicatus* Sowerby I, *Gen. Rec. & Foss. Shells*, pl. XVI pl. 267, figs. 5, 31/7/1823=*australis* Lam."

Terebridae

Plate 73

Class Gastropoda
Subclass Prosobranchia
Order Neogastropoda
Suborder Toxoglossa
Superfamily Conacea
Family TEREBRIDAE

A large group of tropical, sand dwelling, toxoglossate gastropods, closely allied to the Conidae and the Turridae in that they have a poison apparatus to kill their prey which is mainly marine worms. The shell is characterized by its long, slender shape, acute spire, and the whorls which may be smooth, axially or spirally corded and which may or may not possess a presutural groove. All possess a short aperture, a simple outer lip, and a columella which is generally folded. The taxonomy of this group is still under revision, but 4 genera are normally recognized: *Terebra, Hastula, Duplicaria* and *Impages.* The family is well represented in the Philippines.

1. Terebra maculata (Linné, 1758)
196mm; Samar.
Shell large, thick, heavy, with a tall, acute spire, and a long rounded body whorl; sutures wavy and slightly impressed; colour cream with pale brown axial streaks and 2 rows of purple-black bands; reaches 250mm.
DISTRIBUTION: Sporadically found throughout the Philippines.

2. Terebra guttata (Röding, 1798)
158mm; Palawan.
Shell large, heavy, solid, with a tall, acute spire and wavy, impressed sutures; whorls swollen on their upper half and sculptured with axial scratches; colour orange-brown with a subsutural row of white spots on the spire whorls, 2 on the body whorl; reaches 160mm.
DISTRIBUTION: Sporadically found throughout the Philippines.

3. Terebra areolata (Link, 1807)
102mm; Samar.
Shell large, solid, with a tall, acute spire and a weak subsutural groove, axially ribbed on the earlier whorls, but becoming obsolete towards the base; colour creamy orange with 3 bands of brown dots on the spire whorls, 4 on the body whorl; reaches 180mm.
DISTRIBUTION: Sporadically found throughout the Philippines.

4. Terebra dimidiata (Linné, 1758)
108.2mm; Cebu.
Shell moderately large, heavy, with a tall, acute spire; sculptured with strongly incised sutures and a strong presutural groove, the area between the suture and the presutural groove being smooth and constricted, causing a convexity in the rest of the whorl; colour orange with white axial bands starting below the presutural groove and continuing to the next suture; reaches 135mm.

PLATE NO. 73 1. *Terebra maculata* (Linné) 2. *Terebra guttata* (Röding) 3. *Terebra areolata* (Link) 4. *Terebra dimidiata* (Linné) 5. *Terebra crenulata* (Linné) 6. *Terebra areolata* (Link) 7. *Terebra cerithina* Lamarck 8. *Terebra chlorata* Lamarck 9. *Terebra consors* Hinds 10. *c.f. Terebra maculata* (Linné) 11. *Terebra pertusa* (Born) 12. *Terebra triseriata* Gray 13. *Hastula anomala* (Gray) 14. *Duplicaria raphanula* (Lamarck) 15. *Terebra argus* Hinds 16. *Terebra felina* (Dillwyn) 17. *Hastula lanceata* (Linné) 18. *Terebra pretiosa* Reeve 19. *Duplicaria baileyi* Bratcher & Cernohorsky

PLATE 73

DISTRIBUTION: Sporadically found throughout the Philippines.

5. Terebra crenulata (Linné, 1758)
106.6mm; Palawan.
Shell medium sized to large, solid, heavy, with a tall, acute spire and weakly channeled sutures; sculptured with a row of subsutural tubercles and weak axial folds which become more pronounced on the body whorl; beige in color with 2 rows of fine dots on the spire whorls and 3 on the body whorl, tubercles white with brown streaks between; reaches 130mm.
DISTRIBUTION: Sporadically found throughout the Philippines.

6. Terebra areolata (Link, 1807)
143mm; Samar. (see also fig. 3)
Extremely close to *T. subulata* (Linné, 1767) (not illustrated), which differs in having 2 rows of brown quadrate spots on the spire whorls and 3 on the body whorl.

7. Terebra cerithina Lamarck, 1822
38.9mm; Cebu.
Shell small to medium sized, solid, with a tall, acute spire and inflated lower whorls; sutures impressed with a presutural groove which becomes weaker on the body whorl but visible as a ledge on the posterior outer lip margin; sculptured with weak sutural plicae on the earlier whorls but becoming smooth on later whorls; external colour greyish white, stained and streaked with orange-coloured axial and spiral lines; reaches 80mm.
DISTRIBUTION: Sporadically found throughout the Philippines.

8. Terebra chlorata Lamarck, 1822
83.4mm; Batangas.
Like a small *maculata*, but with a narrower spire angle and a weak presutural groove which is nearly obsolete on the body whorl; creamy white in colour with irregular purplish blotches and brown axial streaks and spots; reaches 80mm.
DISTRIBUTION: Sporadically found throughout the Philippines.

9. Terebra consors Hinds, 1844
74.7mm; Sulu Archipelago.
Shell medium sized, lightweight, with a tall, acute spire; whorls sculptured with axial cords which become obsolete towards the base, a deep presutural groove, and a heavily noduled presutural area; columella thick with 3 folds; light orange in colour with darker axial bands between the nodules and on the whorls, a white band on the body whorl; reaches 80mm.
DISTRIBUTION: Sporadically found throughout the Philippines.

10. c.f. Terebra maculata (Linné, 1758)
62.8mm; Marinduque.
A curious specimen having the shape and colour pattern of *T. maculata* and possessing some characteristics of *T. chlorata,* namely, the axial cords and presutural grooving, both of which become obsolete towards the body whorl; this latter characteristic is seen to some extent in young specimens of *maculata,* but not as intense as is the case here.
DISTRIBUTION: Marinduque.

11. Terebra pertusa (Born, 1778)
85.8mm; Palawan.
Shell medium sized, glossy, with a tall, acute spire and moderately incised sutures; sculptured with a distinct row of axially elongated sutural nodules, arcuate axial cords, and 3 or 4 punctate spiral grooves on the whorls; orange-yellow in colour with white nodules demarcated by brown stripes, and a white band on the lower body whorl; reaches 90mm.
DISTRIBUTION: Sporadically found throughout the Philippines.

12. Terebra triseriata Gray, 1834
92.3mm; Tayabas Bay, Quezon.
Shell medium sized, very slender, with a tall spire; sutures well incised with a double row of rounded, equi-sized nodules; whorls sculptured with fine axial and spiral cords forming a densely puncticulate surface; colour creamy white to orange-brown; reaches 100mm.
DISTRIBUTION: Sporadically found throughout the Philippines in limited quantity.

13. Hastula anomala (Gray, 1834)
53.0mm; Bohol.
Shell medium sized, glossy, with a tall, acute spire and a slightly rounded body whorl; sculptured with fine axial cords which become obsolete on the lower section of each whorl, and a deep, well incised presutural groove; pale grey in colour with a purple apex, white sutural nodules, brown spots between the nodules, and 2 brown bands on the base;reaches 60mm.
DISTRIBUTION: Sporadically found throughout the Philippines.

14. Duplicaria raphanula (Lamarck, 1822)
64.0mm; Batangas.
Shell medium sized, glossy, with a tall spire, moderately channeled sutures, and a deep presutural groove; sculptured with fine axial riblets which become obsolete on the lower body whorl; off-white in colour with 2 rows of orange dots on the spire whorls, 3 on the body whorl; reaches 65mm.
DISTRIBUTION: Sporadically found throughout the Philippines.

15. Terebra argus Hinds, 1844
62.5mm; Samar.
Shell medium sized, with a tall, acute spire and slightly inflated lower whorls; sutures weakly channeled and the presutural groove is weakly incised, earlier presutural areas with weak axial ribs becoming obsolete towards the base; shell surface smooth with

weak axial scratches; greyish white in colour with 3 pale bands of orange quadrate spots on the spire whorls, 4 on the body whorl; reaches 70mm.
DISTRIBUTION: Sporadically found throughout the Philippines.

16. Terebra felina (Dillwyn, 1817)
49.7mm; Palawan.
Shell medium sized, with a tall, acute, straight-sided spire; sutures moderately impressed with a pronounced presutural groove; shell surface smooth, white in colour, with a row of brown spots on the spire whorls and 2 on the body whorl; reaches 90mm.
DISTRIBUTION: Sporadically found throughout the Philippines.

17. Hastula lanceata (Linné, 1767)
63.0mm; Samar.
Shell medium sized, lightweight, with a tall, narrowly acute spire; sutures moderately impressed, shell surface smooth and white with brown axial lines; reaches 65mm.
DISTRIBUTION: Sporadically found throughout the Philippines.

18. Terebra pretiosa Reeve, 1842
95.2mm; Mindoro.
Shell large, solid, with a tall spire, weakly convex whorls, deeply channeled sutures and a weak presutural groove; sculptured with numerous arcuate axial cords and fine spiral striae in the interstitial spaces below the groove; orange-yellow in colour with a row of brown quadrate blotches on the spire whorls, 2 on the body whorl, smaller spots on the subsutural zone; reaches 140mm.
DISTRIBUTION: Sporadically found throughout the Philippines.

19. Duplicaria baileyi Bratcher & Cernohorsky, 1982
33.8mm; Punta Engaño, Cebu.
Shell medium sized, glossy, with a tall, acute, purple-tipped spire, and a deep presutural groove; sculptured with numerous oblique axial ribs and smooth interspaces; pale grey in colour with brown interspaces, white sutural nodules and a brown basal band; reaches 65mm.
DISTRIBUTION: Camotes Sea (deep water).

Plate 74 Terebridae

1. Terebra cumingi Deshayes, 1857
82.0mm; Samar.
Shell moderately large, lightweight, rather slender, with a tall, acute spire and impressed sutures, a shallow presutural groove with a row of axially elongate nodules above, and a smaller row below; whorls with fine axial striae and punctate spiral grooves; variable, usually greyish white in colour with orange-brown nodules; reaches 85mm.
DISTRIBUTION: Sporadically found throughout the Philippines in limited quantity.

2. Terebra mariesi E. A. Smith, 1880
47.2mm; Cebu.
Shell medium sized, with a tall, acute spire; sculptured with a double row of sutural nodules, those above axially elongate, and those below, smaller and rounder; rest of the whorl with fine axial and spiral cords, slightly nodulose where they cross; pale brown in colour with darker brown axial bands; reaches 70mm.
DISTRIBUTION: Sporadically found throughout the Philippines in limited quantity.

3. Terebra boucheti Bratcher, 1981
48.5mm; Samar.
Shell medium sized, with a tall, acute spire and a strong presutural groove which has a row of axially elongate nodules above and a row of smaller nodules below; sculptured with numerous arcuate axial cords, finely nodulose from intersecting spiral grooves (2 to 3); creamy white in colour with brown axial streaks; reaches 60mm.
DISTRIBUTION: Sporadically found throughout the Philippines in limited quantity.

4. Terebra amanda Hinds, 1844
53.5mm; Marinduque.
Shell medium sized, lightweight, with a tall spire and impressed sutures; there is a double row of sutural nodules, those above axially elongate, those below smaller and rounder, whorls sculptured with punctate spiral grooves and fine axial striae; orange in colour with white nodules; reaches 55mm.
DISTRIBUTION: Sporadically found throughout the Philippines.

5. Terebra undulata Gray, 1834
42.4mm; Bohol.
Shell medium sized, with a tall acute spire, weakly impressed sutures and a weak presutural groove; whorls sculptured with sinuous axial cords and fine interstitial spiral striae; presutural area slightly nodulose where the groove intersects the axial cords; orange-brown in colour, nodules white, interstitial spaces darker brown; reaches 45mm.
DISTRIBUTION: Sporadically found throughout the Philippines.

6. Terebra conspersa Hinds, 1844
36.0mm; Palawan.
Shell small to medium sized, non-glossy, with a tall acute spire, incised sutures, a weakly punctate presutural groove, and sculptured with arcuate axial riblets and fine spiral striae, intersecting to produce a punctate surface below the level of the groove; pale

lavender in colour with brown axial streaks and a brown band at the base of the body whorl; reaches 50mm.
DISTRIBUTION: Sporadically found throughout the Philippines.

7. Terebra sp.
64.8mm; Cebu.
Shell moderately large, solid, with a tall acute spire and deeply incised sutures; sculptured with a strong presutural groove forming axially orientated nodules above and smaller, rounder nodules below; lower whorl area with arcuate axial ribs and fine overriding spiral grooves; pale orange-brown in colour, presutural nodules white, a white band on the body whorl, and dark reddish brown in the interspaces; largest seen 73.4mm.
DISTRIBUTION: Camotes Sea.

8. Terebra pertusa (Born, 1778) (ecomorph)
63.0mm; Sulu Archipelago.
Shell moderately large, slender, glossy, with a tall acute spire and a double row of sutural nodules separated by a deeply incised presutural groove; sculptured with arcuate axial cords and fine interstitial spiral grooves; orange-yellow in colour, sutural nodules white with brown interspaces and a discontinuous red-brown band on the body whorl; reaches 65mm. Appears to be an unusual form of this species where the sutural band is double and there is an additional band on the body whorl periphery. See also Pl. 73 fig. 11 (typical form).
DISTRIBUTION: Sporadically found throughout the Philippines in limited quantity.

9. Terebra affinis Gray, 1834
32.2mm; Samar.
Shell small, with a tall acute spire, moderately incised sutures and a row of axially elongate sutural nodules bordered by an almost obsolete presutural groove; sculptured with fine, arcuate axial riblets and fine interstitial spiral lirae; creamy white in colour with brown axial streaks and a brown band at the base; reaches 35mm.
DISTRIBUTION: Sporadically found throughout the Philippines.

10. Terebra punctostriata Gray, 1834
Syn: *Terebra pallida* Deshayes, 1857
57.2mm; Palawan.
Shell moderately large, glossy, with a tall acute spire and a slightly inflated body whorl; sculptured with a double sutural row of nodules, both very weak and separated by a presutural groove; whorls with spiral grooving and weak axial growth striae; colour orange-brown; reaches 100mm.
DISTRIBUTION: Sporadically found throughout the Philippines in limited quantity.

11. Hastula caliginosa (Deshayes, 1859)
38.2mm; Cebu.
Shell small to medium sized, glossy, with a tall spire and moderately impressed sutures; sculptured with rounded axial ribs, smooth between, and a punctate presutural cord; aperture elliptically ovate, columella folded; totally grey in colour, aperture orange-brown; reaches 40mm.
DISTRIBUTION: Sporadically found throughout the Philippines.

12. Terebra babylonia Lamarck, 1822
51.5mm; Batangas.
Shell medium sized, moderately glossy, with a tall acute spire and 2 rows of sutural nodules which fuse into one on later whorls; sculptured with deeply incised, arcuate axial growth striae, two closely set spiral grooves mid-whorl, and additional spiral threads on the base of the body whorl; off-white in colour with reddish grooves and a reddish brown basal band; reaches 80mm.
DISTRIBUTION: Sporadically found throughout the Philippines.

13. Terebra mactanensis Bratcher & Cernohorsky, 1982
48.8mm; Punta Engaño, Cebu.
Shell medium sized, with a tall, acute, straight spire and moderately impressed sutures; sculptured with narrow axial ribs, a weak presutural groove which is deeply punctate in the interspaces, and 4 or 5 weaker spiral grooves on the lower half of the whorls; aperture narrowly elongate, columella thick and twisted; colour deep orange-brown, subsutural nodules white, black between, and a white band on the body whorl base; reaches 62mm.
DISTRIBUTION: Cebu.

14. c.f. Terebra punctostriata Gray, 1834
33.8mm; Punta Engaño, Cebu.
A small pure white shell with a tall acute spire, moderately impressed sutures and a double row of sutural nodules separated by a deep presutural groove; whorls with fine axial threads and numerous finely punctate spiral grooves. Possibly a form of *T. punctostriata*, but possesses punctate spiral grooves; largest seen 53.6mm.
DISTRIBUTION: Sporadically found throughout the Philippines in limited quantity.

PLATE NO. 74 1. *Terebra cumingi* Deshayes 2. *Terebra mariesi* E. A. Smith 3. *Terebra boucheti* Bratcher 4. *Terebra amanda* Hinds 5. *Terebra undulata* Gray 6. *Terebra conspersa* Hinds 7. *Terebra sp.* 8. *Terebra pertusa* (Born) — ecomorph 9. *Terebra affinis* Gray 10. *Terebra punctostriata* Gray 11. *Hastula caliginosa* (Deshayes) 12. *Terebra babylonia* Lamarck 13. *Terebra mactanensis* Bratcher & Cernohorsky 14. *c.f. Terebra punctostriata* Gray 15. *Terebra quoygaimardi* Cernohorsky & Bratcher 16. *Terebra columellaris* Hinds 17. *Terebra montgomeryi* R. D. Burch 18. *Hastula solida* (Deshayes) 19. *Terebra funiculata* Hinds 20. *c.f. Terebra jenningsi* R. D. Burch 21. *Terebra fortunei* Deshayes 22. *Terebra mactanensis* Bratcher & Cernohorsky — ecomorph

PLATE 74

15. Terebra quoygaimardi Cernohorsky & Bratcher, 1976
Syn: *T. monile* Quoy & Gaimard, 1833 (non Linné 1771)
39.9mm, 29.2mm; Palawan.
Shell small to medium sized, glossy, with a tall acute spire and a row of heavy sutural nodules; whorls sculptured with 2 or 3 deep spiral grooves and weak overriding axial scratches; colour orange-brown with white sutural nodules and brown staining between; rarely pure white; reaches 55mm.
DISTRIBUTION: Sulu Sea.

16. Terebra columellaris Hinds, 1844
37.9mm; Samar.
Shell small to medium sized, with a matt surface and a tall spire; sculptured with a row of white rounded sutural nodules, a moderately deep presutural groove, numerous rounded, slightly arcuate axial cords, and deep interstitial spiral grooves; pale orange in colour with white axial flecks, white-coloured nodules, and a white band on the body whorl; reaches 40mm.
DISTRIBUTION: Sporadically found throughout the Philippines in limited quantity.

17. Terebra montgomeryi R. D. Burch, 1965
30.5mm; Cebu.
Shell small, glossy, with a tall spire and moderately incised sutures; sculptured with a sutural cord which is finely granulose, a deep presutural groove, and 3 to 4 punctate spiral grooves on the rest of the whorls; creamy white in colour with orange maculations; reaches 35mm.
DISTRIBUTION: Sporadically found throughout the Philippines in limited quantity.

18. Hastula solida (Deshayes, 1857)
34.3mm; Mindoro.
Shell small, glossy, with impressed sutures and weakly stepped whorls; sculptured with numerous weak axial ribs which become obsolete towards the base; pale grey in colour, brown between the ribs, and crossed with a brown spiral band on the spire whorls and 3 on the body whorl; reaches 35mm.
DISTRIBUTION: Sporadically found throughout the Philippines.

19. Terebra funiculata Hinds, 1844
37.0mm; Marinduque.
Shell small, with a tall acute spire; sculptured with 2 sutural cords, the upper one small and lightly grooved, the lower one larger and smoother; whorls with 2 or 3 deeply incised grooves and crossed by weak axial growth striae; creamy white in colour with orange stained grooves; reaches 45mm.
DISTRIBUTION: Sporadically found throughout the Philippines.

20. c.f. **Terebra jenningsi** R. D. Burch, 1965
31.0mm; Punta Engaño, Cebu.
Shell small, glossy, with a tall acute spire and weakly incised sutures; sculptured with a double row of sutural nodules fusing to become a single, weakly nodulose cord towards the body whorl; sutural cord well raised above the shell surface; spire whorls weakly concave, sculptured with weak axial striae, body whorl with a corded base; shell yellowish brown in colour with white nodules and base; reaches 40mm; not as slender as *T. jenningsi,* and bears a close similarity to a young *T. guttata,* but lacks the white presutural spots, and appears fully mature at this size.
DISTRIBUTION: Camotes Sea.

21. Terebra fortunei Deshayes, 1857
37.9mm; Cebu.
Shell medium sized, with a tall, acuminate spire and finely incised sutures; sculptured with a sutural row of nodules, a deep presutural groove, and numerous closely-set axial cords which are crossed by spiral striae; colour pale brown with a dark brown band below the nodules of the spire whorls, 2 on the body whorl; reaches 100mm.
DISTRIBUTION: Camotes Sea.

22. Terebra mactanensis Bratcher & Cernohorsky, 1982 (ecomorph)
80.0mm; Cebu.
Shell moderately large, with a matt surface, a tall acute spire and weakly convex whorls; sutures impressed, with a weak, punctate presutural groove; sculptured with rounded axial cords and fine interstitial striae; aperture elliptically elongate, columella almost straight; pure white in colour with occasional brown banding; reaches 100mm.
Structurally identical to the typical *T. mactanensis* (fig. 13), but the sutures appear more deeply incised, the whorls more strongly convex, and the colour is white.
DISTRIBUTION: Camotes Sea.

Terebridae Plate 75

1. Hastula albula (Menke, 1843)
31.0mm; 31.5mm; Samar.
Shell small to medium sized, shiny, variable, with a tall acute spire and weakly ledged sutures; sculptured with fine, oblique axial ribs, prominent at the sutures and obsolete on the rest of the whorl; off-white in colour with or without darker axial streaks and blotches; reaches 35mm.
DISTRIBUTION: Sporadically found throughout the Philippines.

PLATE NO. 75 1. *Hastula albula* (Menke) 2. *Hastula strigilata* (Linné) 3. *Hastula matheroniana* (Deshayes) 4. *Hastula plumbea* (Quoy & Gaimard) 5. *Hastula strigilata* (Linné) form *verreauxi* (Deshayes) 6. *Terebra laevigata* Gray 7. *Terebra anilis* (Röding) 8. *Impages hectica* (Linné) 9. *Terebra sp.* 10. *Hastula albula* (Menke) 11. *Terebra parkinsoni* Cernohorsky & Bratcher 12. *Hastula swainsoni* (Deshayes) 13. *Hastula mera* (Hinds) 14. *Terebra succincta* (Gmelin)

PLATE 75

2. Hastula strigilata (Linné, 1758)
30.0mm, 26.5mm, 22.4mm; Marinduque.
A small, shiny, narrow shell, with a tall acute spire, a long, tapering body whorl, and slightly impressed sutures; sculptured with fine axial cords which extend the length of the whorls; pale grey in colour with a brown-spotted yellow band at the sutures and a white band on the body whorl, visible in the pale brown aperture; reaches 30mm.
DISTRIBUTION: Sporadically found throughout the Philippines.

3. Hastula matheroniana (Deshayes, 1859)
32.4mm, 28.5mm, 28.8mm; Palawan.
Very similar to *H. strigilata,* but normally a paler shell (not always) with an off-white aperture; there are fewer ribs on each whorl which become obsolete on the lower half of the body whorl; reaches 33mm.
DISTRIBUTION: Sporadically found throughout the Philippines.

4. Hastula plumbea (Quoy & Gaimard, 1832)
21.2mm, 20.4mm; Palawan.
Shell small, glossy, with a tall acute spire, impressed sutures, and weakly convex, almost straight-sided whorls; sculptured with numerous narrow axial ribs and weak spiral striae, both of which become obsolete on the lower half of the body whorl; pale greyish brown in colour, with a pale presutural band, and deep brown on the body whorl and within the aperture; reaches 25mm.
DISTRIBUTION: Sporadically found throughout the Philippines.

5. Hastula strigilata (Linné, 1758) form **verreauxi** (Deshayes, 1857)
40.0mm; Samar.
A distinct form of *H. strigilata* which reaches a larger size (over 45mm.), has a distinct and fairly consistent colouration (greyish orange with a pale green overtone), and the ribs are wider and much flatter.
DISTRIBUTION: Sporadically found throughout the Philippines.

6. Terebra laevigata Gray, 1834
36.7mm, 34.0mm; Cebu.
Shell small to medium sized, with a tall spire of around 25 whorls; sculptured with 2 sutural cords separated by a deep spiral groove, the upper cord strongest, and below, 6 to 8 fine spiral lirae crossed by weak axial growth folds; colour pale orange-brown; reaches 50mm.
DISTRIBUTION: Sporadically found throughout the Philippines.

7. Terebra anilis (Röding, 1798)
35.5mm, 28.8mm; Samar.
Shell medium sized with a white apex and a non-glossy, pale brown surface; whorls sculptured with fine axial and spiral cords and a double row of sutural nodules separated by a deep presutural groove; the nodules of the upper row are large and axially elongate, those of the lower row smaller and rounder; reaches 45mm.
DISTRIBUTION: Sporadically found throughout the Philippines.

8. Impages hectica (Linné, 1758)
41.2mm; Batangas.
Shell medium sized, glossy, pure white with or without bands of black axial streaks; slightly broad in outline with straight-sided whorls, the body whorl slightly convex; a weak depression is seen just before each suture, as well as weak axial growth striae; reaches 75mm.
DISTRIBUTION: Sporadically found throughout the Philippines.

9. Terebra sp.
32.5mm, 24.8mm; Palawan.
A small, glossy, pale grey shell with a tall acute spire, 12 to 14 whorls, and 3 purple-coloured protoconch whorls; sculptured with 15 to 20 axial ribs, a deeply punctate presutural groove and 5 to 6 rows of non-overriding spiral grooves in the interspaces; axial ribs terminating midway on the body whorl and demarcated by a white spiral band; reaches 35mm.
DISTRIBUTION: Sporadically found throughout the Philippines.

10. Hastula albula (Menke, 1843)
21.5mm, 24.5mm; Samar.
Shell small, shiny, with a tall, slightly obtuse spire, a dark brown protoconch and moderately impressed sutures; sculptured with fine axial riblets which become obsolete on the body whorl, and smooth interspaces; greyish white with a white subsutural band; reaches 35mm; Extremely variable.
DISTRIBUTION: Sporadically found throughout the Philippines.

11. Terebra parkinsoni Cernohorsky & Bratcher, 1976
32.2mm, 28.9mm; Sulu Archipelago.
A small, slender, shiny shell, with a tall acute spire and weakly channeled sutures; sculptured with thick axial ribs, straight on the earlier whorls, becoming slightly arcuate on later whorls; interspaces with fine, non-overriding spiral grooves; pale orange-yellow in colour with dark brown interspaces; reaches 35mm; close to *T. undulata* but lacks the subsutural nodulation of that species, and the spiral grooves do not ascend the ribs as in *T. undulata.*
DISTRIBUTION: Sulu Sea.

12. Hastula swainsoni (Deshayes, 1859)
21.5mm; Mindoro.
Shell small, glossy, with a tall acute spire and straight-sided spire whorls; sculptured with rounded axial cords, deep punctate grooves in the interspaces under the slightly undulate sutures, and obsolete spiral striae between the ribs, the axial structure becoming obsolete on body whorl dorsum; deep greyish brown in colour; reaches 30mm.

DISTRIBUTION: Sporadically found throughout the Philippines.

13. Hastula mera (Hinds, 1844)
19.3mm, 19.4mm; Batangas.
Shell small, glossy, with a tall spire and sculptured with numerous weak, closely-set axial riblets, becoming obsolete on the body whorl; colour dark grey, protoconch purple, followed by 1 or 2 white coloured whorls, remaining whorls with a brown-and-white spotted subsutural band; reaches 20mm; said to be conspecific with the extremely variable *H. albula* (fig. 10).
DISTRIBUTION: Sporadically found throughout the Philippines.

14. Terebra succincta (Gmelin, 1791)
44.0mm; Bohol.
Shell medium sized with a matt gloss, a tall, acute spire, and sculptured with numerous arcuate axial cords (20 to 30) on the body whorl, fine overriding spiral striae, and a deep presutural groove; dark brown to black in colour, sutures paler; reaches 50mm.
DISTRIBUTION: Sporadically found throughout the Philippines.

Plate 76 — Turridae

Class Gastropoda
Subclass Prosobranchia
Order Neogastropoda
Suborder Toxoglossa
Superfamily Conacea
Family TURRIDAE

An extremely versatile group of carnivorous gastropods occurring in all geographical localities and occupying all types of marine environment. They are truly toxoglossate, possessing a poison radular system allied to that seen in the Conidae and the Terebridae. They range in size from a few millimeters to over 150mm and although extremely polymorphic, are generally fusiform in outline, have a tall spire, are sculptured with nodules, cords, ribs or spines and all possess a labial sinus or "turrid notch" on the posterior margin of the outer lip, which may be deep or barely visible. The columella may be smooth or plicate, the outer lip simple or dentate, and the siphonal canal long or short. Members of this family live in sand or under rocks and may inhabit shallow or deep water. This is by far the largest gastropod family with numerous subfamily groups having been proposed to encompass over 500 genera and subgenera. At present, identification of different taxa is based on differences in morphology and radular characteristics, but current research clearly indicates the inadequacy of such diagnostic features. The family is clearly in need of revision. Extremely well represented in the Philippines, with most of the smaller species coming from deep water.

1. Turris (Turris) babylonia (Linné, 1758)
96.6mm, 82.5mm; Balabac, Palawan.
Shell large, heavy, fusiform-elongate, with a tall acute spire; sculptured with broad, rounded spiral ribs and smaller intermediate cords, sutures deep and weakly lirate, labial sinus deep; surface smooth, white in colour with large dark brown spots on the major cords and smaller dots on the minor cords, occasionally some diffuse banding, aperture white; reaches 100mm.
DISTRIBUTION: Sporadically found throughout the Philippines.

2. Turris (Turris) crispa crispa (Lamarck, 1816)
171mm, 96.2mm; Albay Gulf, Southern Luzon.
Shell large with a tall spire and a long siphonal canal; sutures marked with scabrous spiral cords below which is a spirally bisected cord, followed by a large rounded cord, then numerous cords of intermediate strength, labial sinus deep; creamy white in colour with thin brown axial bands or spots on the cords, aperture white; reaches 175mm.
DISTRIBUTION: Sporadically found throughout the Philippines.

3. Gemmula (Unedogemmula) unedo (Kiener, 1839)
84.2mm; Tayabas Bay, Quezon.
Shell large, heavy, fusiform-elongate, sutures deep with a spirally striated, concave upper shoulder, lower half of spire whorls with 2 or 3 cords, largest on the shoulder, lower body whorl cords roughly noduled, sinus deep and "U-shaped"; creamy white in colour with brown spots and dashes on the cords; reaches 110mm.
DISTRIBUTION: Sporadically found throughout the Philippines.

4. Turris (Turris) undosa (Lamarck, 1816)
85.5mm; Mindoro.
Shell large, tall spired with a moderately short body whorl; whorls are flat to weakly concave with a massive subsutural fold which bears 4 to 5 sharp spiral threads. Below this is a fairly prominent, slightly rounded sinus rib which may be spirally threaded. On the body whorl there are 4 strong spiral cords below the sinus rib, somewhat compacted towards the periphery; canal moderately long with 8 to 10 weaker cords and intermediaries; colour yellowish brown with dark brown axial maculations, aperture violet; reaches 85mm.
DISTRIBUTION: Sporadically found throughout the Philippines.

4a. Turris (Turris) nadaensis Azuma, 1973
70.2mm; Palawan.
Nearly always confused with *T. undosa* to which it bears a close resemblance. It can be separated from the latter species by the following criteria: the spire angle is greater, the whorls distinctly convex and sharply angulate, the subsutural fold is weak, the sinus rib more prominent, the spiral cords below the sinus rib on the body whorl are stronger, more elevated, equally spaced, threaded in brown, and number between 8 and 10; overall colouration more yellowish than in *undosa*, and the brown axial maculations are replaced with thinner axial streaks; reaches 85mm.
DISTRIBUTION: Sporadically found throughout the Philippines in limited quantity.

5. Gemmula (Gemmula) kieneri (Doumet, 1840)
50.7mm; Samar.
Frequently confused with *G. speciosa* (fig. 17), but is slightly less elongate, less sharply contracted on the body whorl, has a larger, more heavily sculptured fold; the colour pattern is one of red dashes between the granules of the sinus keel and broken dots and dashes on the primary spiral lirae; in *speciosa*, the colour patterns are continuous; reaches 75mm.
DISTRIBUTION: Sporadically found throughout the Philippines.

6. Gemmula (Gemmula) congener congener (E. A. Smith, 1894)
63.5mm; Sulu Archipelago.
A curious form of this species (see fig. 21) in which the outer lip has become fluted, possibly due to injury and/or disease; it bears a superficial resemblance to members of the genus *Pinguigemmula*, but members of that group have a much broader spire angle. This outer lip fluting is also seen in the subspecies *congener diomedea* Powell, 1954.
DISTRIBUTION: Sporadically found throughout the Philippines in limited quantity.

7. Lophiotoma (Lophiotoma) albina (Lamarck, 1822)
Shell medium sized, elongate, biconic, with indistinct sutures and a steep, spirally striated shoulder area; sinus rib triple corded, crossed by thick axial folds; there are 2 weaker ribs below the sinus rib on the spire whorls, 4 below on the body whorl, canal with finer riblets; pale brown in colour with regularly spaced brown spots on the sinus rib; reaches 75mm.
7a. typical form; 53.6mm; Cebu.
7b. similar in structure and colour pattern, but more angulate than normal. May be an abberant form of *L. acuta* (Perry, 1811); 49.0mm; Cebu.
DISTRIBUTION: Sporadically found throughout the Philippines.

8. Lophiotoma (Lophioturris) polytropa (Helbling, 1779)
51.5mm, 51.2mm; Davao.
Shell medium sized, heavy, with a tall spire and a short canal; sculptured with a primary subsutural cord, followed by a deeply concave, spirally lirate shoulder, a bicingulate sinus rib, and a further 2 cords with an interstitial thread, base with several primary cords and numerous intermediary threads, sinus deep and "U-shaped"; dark blackish brown in colour with white bands; reaches 55mm.
DISTRIBUTION: Sporadically found throughout the Philippines in limited quantity.

9. Turris (Turris) garnonsii (Reeve, 1843)
95.0mm; Zamboanga.
Shell moderately large, fusiform-elongate, very similar in structure to *T. babylonia* but more elongate and less angulate in outline; the cords are narrower, more strongly carinate, and the surface is crossed by fine axial threads; white in colour with brown oblong maculations on the subsutural fold, brown spots on the cords, and a brown band on the base; reaches 100mm.
DISTRIBUTION: Cebu, Bohol and Zamboanga.

10. Lophiotoma (Lophioturris) indica (Röding, 1798)
77.3mm, 72.4mm, 62.9mm; Palawan.
Shell moderately large, fusiform-elongate, with a tall acute spire, a long, slightly recurved siphonal canal and indistinct sutures; presutural ramp concave, spirally corded, followed by a raised, sharply angled peripheral carina, then 2 or 3 further cords; siphonal canal with weaker cords, sinus deep and "U-shaped"; colour off-white with brown axial streaks on the shoulder and dots on the cords, both of varying intensities; reaches 90mm.
DISTRIBUTION: Sporadically found throughout the Philippines.

11. c.f. Comitas kamakurana (Pilsbry, 1895)
80.5mm; Panglao, Bohol.
Shell moderately large, fusiform-elongate, with a tall acute spire, wavy adpressed sutures, and a relatively long, open, slightly recurved siphonal canal; sculptured with large arcuate axial folds, becoming obsolete prior to the sutures and on the lower part of the body whorl, and the entire surface crossed by strong spiral threads, outer lip smooth and recurved, sinus deep; off-white in colour with orange-brown banding; reaches 90mm.

PLATE NO. 76 1. *Turris babylonia* (Linné) 2. *Turris crispa crispa* (Lamarck) 3. *Gemmula unedo* (Kiener) 4. *Turris undosa* (Lamarck) 4a. *Turris nadaensis* Azuma 5. *Gemmula kieneri* (Doumet) 6. *Gemmula congener congener* (E. A. Smith) 7a. *Lophiotoma albina* (Lamarck) 7b. *a more angulate specimen* 8. *Lophiotoma polytropa* (Helbling) 9. *Turris garnonsii* (Reeve) 10. *Lophiotoma indica* (Röding) 11. *c.f. Comitas kamakurana* (Pilsbry) 11a. *Comitas sp.* 12. *Turris cryptorrhaphe* (Sowerby) 13. *Ptychobela flavidula* (Lamarck) 14. *c.f. Brachytoma carmen* (Sowerby) 15. *Drillia enna* (Dall) 16. *Thatcheria mirabilis* Angas 17. *Germula speciosa* (Reeve) 18. *Ptychobela flavidula* (Lamarck) — ecomorph 19. *c.f. Inquisitor sterrha* (Watson) 20. *c.f. Inquisitor kurodai* (Habe & Kosuge) 21. *Gemmula congener congener* (E. A. Smith) 22. *Xenoturris cingulifera cingulifera* (Lamarck)

PLATE 76

11a. Comitas sp.: A pure white specimen which is less elongate, has more inflated whorls and a rounder aperture; structurally somewhere between *C. kaderlyi* (Lischke, 1872) and *C. kamakurana* (Pilsbry, 1895); 70mm; Palawan.
DISTRIBUTION: Camotes Sea and Sulu Sea.

12. Turris (Turris) cryptorrhaphe (Sowerby, 1825)
67.3mm, 66.0mm; Cebu.
Shell medium sized, lanceolate, the spire being tall and slightly stepped, the siphonal canal short and weakly reflexed; sculptured with a prominently raised shoulder cord, 2 weaker cords above, another below, a weak sinus rib, and numerous weaker lirae on the base; colour greyish brown with fine dark brown lines, aperture violet; reaches 80mm.
DISTRIBUTION: Camotes Sea and Sulu Sea.

13. Ptychobela flavidula (Lamarck, 1822)
71.5mm; Panglao, Bohol.
Shell moderately large, fusiform-elongate, with a tall acute spire and weakly impressed sutures; whorls convex, slightly subangulate, with a smooth, concave subsutural fold, then a series of oblique, arcuate axial ribs, crossed by rather strong spiral cords and weaker interstitial lirae forming weak nodules; outer lip thin, weakly variced and incurved; creamy white in colour with a pale orange band on the spire whorls, 2 or 3 on the body whorl; reaches 75mm.
DISTRIBUTION: Camotes Sea and Sulu Sea.

14. c.f. Brachytoma carmen (Sowerby, 1916)
35.8mm; Bohol.
Shell small to medium sized, fusiform-elongate, with a tall acute spire, convex whorls and adpressed sutures; whorls with a smooth subsutural fold and sculptured with oblique axial folds, noduled from overriding spiral lirae; outer lip thin, arcuate and expanded, with a deep sinus, aperture with a posterior parietal "button" and a slightly glazed columella; pale orange in colour with white nodules and a pinkish white aperture; reaches 40mm.
DISTRIBUTION: Camotes Sea and Sulu Sea.

15. Drillia (Drillia) enna (Dall, 1918)
53.5mm, 46.4mm; Cebu.
Shell medium sized, heavy, with a tall spire, a very short anterior canal and adpressed sutures; sculptured with large, rounded, axially elongate nodules, and crossed by fine spiral striae, sinus deep and "U-shaped"; colour white with brown banding on the lower section of each whorl; reaches 60mm.
DISTRIBUTION: Camotes Sea.

16. Thatcheria mirabilis Angas, 1877
68.7mm; Taiwan.
Recently taken in the Philippines (Lubang Island), this species is characterized by its large size (reaches 110mm), lightweight, and its peculiar pagoda form outline, consisting of sharply keeled, smooth, flat-topped whorls, a wide porcelaneous aperture, a deep sinus, and a dull yellowish surface which is crossed by fine axial and spiral scratches.
DISTRIBUTION: Lubang Island, Mindoro.

17. Gemmula (Gemmula) speciosa (Reeve, 1843)
74.2mm, 71.7mm; Manila Bay.
Shell moderately large and fusiform with a tall spire, a long, reflexed anterior canal and a peripheral keel of yellow, closely-set, spirally bisected nodules; rest of the surface with continuous yellow spiral lines and numerous axial and spiral threads between, sinus deep and "U-shaped"; off-white in colour with yellow spiral lines; reaches 75mm.
DISTRIBUTION: Sporadically found throughout the Philippines.

18. Ptychobela flavidula (Lamarck, 1822) (ecomorph)
62.3mm; Panglao, Bohol.
A form of *P. flavidula* (see also fig. 13), in which the whorls are more convex, the body whorl less angulate than normal, and the axial ribs are more pronounced, especially on the body whorl; the anterior canal is extremely short and the colour is cream with brown mottling and white speckling on the ribs; reaches 75mm.
DISTRIBUTION: Camotes Sea and Sulu Sea.

19. c.f. Inquisitor sterrha (Watson, 1881)
44.0mm; Punta Engaño, Cebu.
Shell small to medium sized, with a tall acute spire and a short, truncated body whorl; sutures adpressed with a solid cord, spire whorls with oblique axial cords crossed by fine, overriding spiral striae, obsolete on the slightly concave presutural slope; outer lip thin and arcuate, sinus deep, siphonal canal short, open and weakly reflexed; off-white in colour with a band of brown blotches between the axial cords of the spire whorls, 2 on the body whorl; reaches 60mm.
DISTRIBUTION: Camotes Sea and Sulu Sea.

20. c.f. Inquisitor kurodai (Habe & Kosuge, 1966)
47.6mm, 50.3mm; Bohol.
Shell medium sized, lightweight, with a tall acute spire and adpressed sutures; the surface is sculptured with numerous arcuate axial folds that are crossed by spiral cords, forming weak nodules on the folds; aperture wide, sinus deep and "U-shaped", outer lip thin and centrally inflected; colour pale brown with dark brown mottling and white coloured nodules. Reaches around 55mm. Although resembling a juvenile form of *P. flavidula* (figs. 13 and 18), it is probably morphologically closer to *I. kurodai*, a species characterized by the yellow marks on the spiral threads, its less convex whorls and the presence of a weak thread below the suture.
DISTRIBUTION: Camotes Sea and Sulu Sea.

21. Gemmula (Gemmula) congener congener (E. A. Smith, 1894)
38.5mm, 41.8mm; Bohol.
Shell medium sized, fusiform-ovate, the subsutural fold enlarged and weakly bicarinate; shoulder area wide, shallow, with fine spiral striae, peripheral carinae nodular, followed on the body whorl by 4

lirae of equal strength, sinus deep; pure white in colour with a yellow subsutural fold; reaches 90mm.
DISTRIBUTION: Sporadically found throughout the Philippines.

22. Xenoturris cingulifera cingulifera (Lamarck, 1822)
51.8mm; Samar.
Shell medium sized, heavy, with a tall acute spire which is slightly stepped, and a short, truncated body whorl; anterior canal short, open and slightly reflexed, sinus deep and "U-shaped", sinus rib thick, bicingulate, white-coloured and decorated with brown spots; remaining sculpture is one of numerous primary and secondary spiral cords, all flecked in brown; reaches 70mm.
DISTRIBUTION: Sporadically found throughout the Philippines in limited quantity.

Plate 77 Turridae

1. Lophiotoma (Lophiotoma) acuta (Perry, 1811)
29.9mm, 40.5mm, 41.9mm; Palawan.
Shell variable, medium sized, fusiform-elongate, spire tall and acute, siphonal canal long, sinus deep and "U-shaped", sutures adpressed; sculptured with a subsutural cord, a centrally bisected sinus rib on the shoulder, a further 3 cords below, and numerous secondary riblets over the whorl surface; off-white in colour with brown spots on the major cords and speckling on the minor cords; reaches 60mm.
DISTRIBUTION: Sporadically found throughout the Philippines.

2. Inquisitor rufovaricosa (Kuroda & Oyama, 1971)
48.0mm; Bohol.
Shell medium sized, thick, fusiform, tall-spired with a short siphonal canal, sutures adpressed with a thick, nodular cord; sculptured with numerous rounded axial cords, irregular varices, and encircled with fine spiral striae, sinus shallow, "U-shaped" and posteriorly thickened; colour off-white with brown-blotched varices and a brown band at the base; reaches around 55mm. This species was figured by Shikama in 1964 under the same name but with Kuroda as the sole author. It is one of several of Kuroda's manuscript names which were widely circulated prior to their validation by Kuroda & Oyama in 1971. It is unlikely that the name *I. rufovaricosa* (Kuroda) is valid, as Shikama only featured a figure and name, with no appended description. Accordingly, we attribute authorship of this species to Kuroda and Oyama (1971) and not to Kuroda (in Shikama, 1964).
DISTRIBUTION: Camotes Sea and Sulu Sea.

3. Bathytoma (Parabathytoma) lühdorfi (Lischke, 1872)
49.5mm, 48.6mm; Cebu.
Shell medium sized, solid, broadly fusiform; sculptured with prominent granular spiral cords, incised linear interspaces, and a prominent, heavily nodulose carinal ridge above mid-whorl height; aperture narrow, columella with a slight central ridge, sinus small and "V-shaped"; pale yellow-brown in colour with reddish brown spiral lirae; reaches 70mm.
DISTRIBUTION: Camotes Sea and Sulu Sea.

4. Spergo (Spheoides) yoshidae (Kuroda & Habe, 1961)
47.4mm; Bohol.
Shell medium sized, lightweight, with a tall spire and a large body whorl comprising about 60% of the total shell length; sculptured with fine axial and spiral threads and a row of nodules below the weakly impressed sutures; aperture elongate-ovate and white-coloured, outer lip thin and arcuate, sinus a shallow fold; yellowish brown in colour with cream axial streaks; reaches 60mm.
DISTRIBUTION: Camotes Sea.

5. Turricula tornata tornata (Dillwyn, 1817)
63.0mm; Sulu Archipelago.
Shell large with a tall, acute spire, a long, slightly twisted canal and adpressed sutures obscured by a granulose cord; surface smooth except for some incised spiral lines at the base of the body whorl; whorls glossy, slightly shouldered below the sutures; white in colour with oblique, reddish brown axial flammules; reaches 95mm.
DISTRIBUTION: Sulu Sea.

6. Clavus (Clavus) flammulatus Montfort, 1810
40.8mm; Punta Engaño, Cebu.
Shell medium sized with a tall, acute spire, a large body whorl, and adpressed sutures; surface smooth, some spiral cords basally; sculptured with a row of echinate nodules on the lower third of the spire whorls and on the upper third of the body whorl; pale brown in colour with brown spots between the nodules and 2 rows of axial flammules on the body whorl; reaches 50mm.
DISTRIBUTION: Cebu and Bohol.

7. Gemmula (Gemmula) concinna (Dunker, 1856) (non Scacchi, 1836)
42.3mm, 35.5mm; Punta Engaño, Cebu.
Shell medium sized, fusiform-elongate, with a tall acute spire, a long, slightly reflexed anterior canal and impressed sutures; sculptured with a tricarinate subsutural fold, followed by a peripheral keel of 2 gemmate cords, and another strong spiral cord; base with 4 strong cords and numerous weaker spiral threads, sinus deep and "U-shaped"; golden yellowish brown in colour with a distinctive purple-coloured siphonal canal; reaches 45mm.
DISTRIBUTION: Camotes Sea and Sulu Sea.

8. c.f. Inquisitor varicosa (Reeve, 1843)
44.8mm, 45.5mm; Punta Engaño, Cebu.
Shell medium sized, with a tall acute spire, a short anterior canal, a weakly granulose cord obscuring the sutures and a smooth, concave subsutural cleft; sculptured with numerous arcuate axial folds, irregular varices, weaker oblique axial striae and encircled with brown-stained spiral grooves; outer lip thin, arcuate and marginally serrate, sinus deep and posteriorly thickened; white in colour with irregular brown blotching, mainly on the varices and a brown band at the base; reaches 50mm.
DISTRIBUTION: Camotes Sea and Sulu Sea.

9. Ptychobela kawamurai (Habe & Kosuge, 1976)
35.5mm; Sulu Archipelago.
Shell medium sized, lightweight, with a tall acute spire, a short, reflexed canal, and a shallow, "U-shaped" sinus; spire whorls sharply angulate with widely spaced, arcuate axial ribs, flattened at the sutures and crossed by spiral cords below the shoulder; ribs and spiral cords pale creamy yellow, interstitial spaces orange-brown; reaches 40mm.
DISTRIBUTION: Sulu Sea, Camotes Sea and Tayabas Bay, Quezon.

10. Turricula nelliae spurius (Hedley, 1922)
28.9mm; Palawan.
Shell small to medium size, fusiform-ovate, tall-spired with a long canal and adpressed sutures; sculptured with a sharp subsutural cord, 3 or 4 spiral lirae on the concave shoulder slope, a large, spirally bisected cord at the shoulder periphery, and 1 or 2 weaker cords below; body whorl with 4 weaker nodulose cords becoming smooth on the canal; pale cream in colour with white nodules and brown mottling; reaches 35mm.
DISTRIBUTION: Camotes Sea and Sulu Sea.

11. Comitas kamakurana (Pilsbry, 1895)
44.8mm, 34.1mm; Cebu.
Shell medium sized, fusiform, with a tall acute spire, a short, recurved siphonal canal, and a shallow "U-shaped" sinus; sculptured with oblique, rounded axial ribs, and crossed with spiral cords forming elongate nodules, sutures adpressed, presutural area finely sculptured; pale cream in colour with brown bands, 2 on the spire whorls, 3 on the body whorl; reaches 50mm.
DISTRIBUTION: Camotes Sea and Sulu Sea.

12. Clavus (Clavus) canicularis (Röding, 1798)
28.8mm, 20.9mm; Punta Engaño, Cebu.
Shell small to medium sized, with a tall acute spire, a truncated body whorl, and virtually no anterior canal, sinus "U-shaped", sutures adpressed; whorls sculptured with a row of vertically open spines on the shoulder and a second row on the base of the body whorl; white in colour with brown sutural banding and a wider band on the body whorl; reaches 30mm.
DISTRIBUTION: Sporadically found throughout the Philippines in limited quantity (deep water).

13. Ptychobela subochracea (E. A. Smith, 1877)
40.4mm, 27.7mm; Bohol.
Shell medium sized, fusiform, with a tall acute spire, adpressed sutures, a deep "U-shaped" sinus, and a small "stromboidal" notch on the anterior margin of the outer lip which is thin and weakly crenulate; sculptured with oblique, arcuate axial ribs, becoming smooth on the sharply angled presutural slope, and crossed by fine spiral threads; pale brown in colour with white granules on the ribs; reaches 45mm; may be a form of *P. flavidula* (Lamarck).
DISTRIBUTION: Camotes Sea and Sulu Sea.

14. Gemmula (Gemmula) graeffei (Weinkauff, 1875)
Syn: *Gemmula (Gemmula) hombroni* (Hedley, 1922)
29.0mm, 28.8mm; Sulu Archipelago.
Shell small, brown-coloured, fusiform-elongate, sculptured with a rounded subsutural cord, followed by a finely striated concave shoulder, a finely nodulose sinus keel, and 1 or 2 secondary cords below this; body whorl with extra fine spiral cords on the canal, outer lip sinuous with a deep, "U-shaped" sinus; reaches 30mm.
DISTRIBUTION: Throughout the Philippines.

15. Inquisitor vulpionis Kuroda & Oyama, 1971
33.0mm; Punta Engaño, Cebu.
Shell medium sized, unusually smooth and glossy, spire tall, acute, body whorl truncate, anterior canal almost obsolete, sutures adpressed, presutural slope smooth; sculptured with oblique, rounded axial ribs, starting at the shoulder and terminating at the sutures on the spire whorls, and mid-line on the body whorl; surface with obsolete spiral striae, more pronounced on the base; white in colour with pale brown bands and blotches, aperture brown; reaches 35mm.
DISTRIBUTION: Camotes Sea.

16. Inquisitor varicosa (Reeve, 1843)
30.8mm, 30.0mm; Palawan.
Shell small to medium sized, with a tall acute spire, a truncated body whorl, a short canal, and sutures that are obliterated by a cord; sculptured with prominent irregular varices, numerous intervarical ribs, and overriding spiral threads; sinus deep and "U-shaped", outer lip margin finely crenulate, with a weak "stromboidal" notch anteriorly; purplish in colour, with

PLATE NO. 77 1. *Lophiotoma acuta* (Perry) 2. *Inquisitor rufovaricosa* (Kuroda & Oyama) 3. *Bathytoma lühdorfi* (Lischke) 4. *Spergo yoshidae* (Kuroda & Habe) 5. *Turricula tornata tornata* (Dillwyn) 6. *Clavus flammulatus* Montfort 7. *Gemmula concinna* (Dunker) (non Scacchi) 8. *c.f. Inquisitor varicosa* (Reeve) 9. *Ptychobela kawamurai* (Habe & Kosuge) 10. *Turricula nelliae spurius* (Hedley) 11. *Comitas kamakurana* (Pilsbry) 12. *Clavus canicularis* (Röding) 13. *Ptychobela subochracea* (E. A. Smith) 14. *Gemmula graeffei* (Weinkauff) 15. *Inquisitor vulpionis* Kuroda & Oyama 16. *Inquisitor varicosa* (Reeve) 17. *Clavus sp.* 18. *Inquisitor c.f. intertincta* (E. A. Smith) 19. *Turridrupa prestoni* Powell 20. *Glyphostoma otohimaea* Kosuge

PLATE 77

white ribs and orange-brown varices; reaches 40mm. DISTRIBUTION: Sporadically found throughout the Philippines in limited quantity (deep water).

17. Clavus sp.
37.3mm; Punta Engaño, Cebu.
Shell medium sized, fusiform, rather ovate, with a tall acute spire, adpressed sutures, a large body whorl, and a short siphonal canal; sculptured with sharp, arcuate axial cords, becoming obsolete on the presutural slope, and crossed by fine spiral threads; aperture ovate, outer lip thin, recurved, sinus shallow; entirely greyish white in colour; reaches 45mm. DISTRIBUTION: Camotes Sea.

18. Inquisitor c.f. intertincta (E.A. Smith, 1877)
37.1mm; Bohol.
Shell medium sized, fusiform-elongate, with an extremely tall, acute spire and adpressed sutures masked by a brown spotted cord; sculptured with 7 large, rounded axial cords, well spaced, and crossed by brown-coloured, slightly nodulose spiral threads; outer lip thin, marginally serrate, sinus deep; greyish white in colour with subsutural brown spots between the ribs, and a row on the lower third of the body whorl; reaches 40mm. Typical *intertincta* has fewer spiral ridges than the illustrated specimen, although this feature may be variable for the species. DISTRIBUTION: Camotes Sea and Sulu Sea.

Type figure of *Inquisitor intertincta* (E. A. Smith, 1877)

19. Turridrupa prestoni Powell, 1967
29.9mm, 29.0mm; Punta Engaño, Cebu.
Shell small to medium sized, with a tall acute spire, a moderately long canal, and adpressed sutures; shell contour weakly stepped by a strong keel on the lower third of each whorl, above which is a row of elongated beads, and below, additional spiral cords and weaker interstitial threads; outer lip thin, arcuate, marginally serrate, sinus deep and "U-shaped", aperture lirate within; honey brown in colour with paler spiral elements and a white aperture and lower canal area. Originally thought to be endemic to the Andaman Islands; reaches 35mm. DISTRIBUTION: Camotes Sea and Sulu Sea.

20. Glyphostoma otohimeae Kosuge, 1981
31.8mm; Bohol.
A medium sized, solid, tall-spired shell, with a large, elongate body whorl; sculptured with oblique axial ribs and overriding spiral threads which become obsolete on the body whorl dorsum, strong spiral threads at the base of the body whorl; aperture narrow, outer lip flaring, arcuate, basally constricted, and dentate within; sinus notch wide and flaring; yellow-brown in colour with white and dark brown bands on the body whorl, and spots between the ribs; reaches 35mm. DISTRIBUTION: Camotes Sea and Sulu Sea.

Turridae — Plate 78

1. Turridrupa cerithina (Anton, 1838)
10.0mm; Punta Engaño, Cebu.
Shell small with a tall, acute spire, a short, truncated body whorl, and adpressed sutures; whorls sculptured with small nodules formed from intersecting axial and spiral cords, the number and size of which varies, some specimens possessing a distinctly coronated appearance, sinus "U-shaped"; deep brown to black in colour with a white apex and nodules, and a blue-white aperture; reaches 25mm. DISTRIBUTION: Throughout the Philippines.

2. Vexitomina regia (Reeve, 1842)
25.0mm; Bohol.
Shell small with a tall acute spire and a truncated body whorl; sculptured with nodular cords at the sutures, followed by 2 spiral lirae, an enlarged nodulose cord on the shoulder and 1 or 2 nodulose rows below; body whorl with 6 nodulose cords below the shoulder, varical remnants visible, sinus deep; white in colour with brown axial streaks; reaches 25mm. DISTRIBUTION: Camotes Sea and Sulu Sea.

3. Clavus (Clavus) unizonalis (Lamarck, 1822)
20.2mm; Punta Engaño, Cebu.
Shell small, lightweight, with a tall spire, a short, truncated body whorl and adpressed sutures; whorls sculptured with a row of fine tubercles just above the sutures and on the upper third of the body whorl; white in colour with a thin brown band below the tubercles, wider on the body whorl; reaches 30mm. DISTRIBUTION: Sporadically found throughout the Philippines.

PLATE NO. 78 1. *Turridrupa cerithina* (Anton) 2. *Vexitomina regia* (Reeve) 3. *Clavus unizonalis* (Lamarck) 4. *Anarithma dorcas* Kuroda and Oyama 5. *Lophiotoma abreviata abreviata* (Reeve) 6. *Asperdaphne peradmirabilis* (Smith) 7. *Turridrupa jubata* (Reeve) 8. *Cochlespira pulchella semiplota* Powell 9. *Cochlespira pulchella fossata* Powell 10. *Eucithara funiculata* (Reeve) 11. *Clavus sp.* 12. *Xenoturris kingae* Powell 13. *Turridrupa astricta consorbrina* Powell 14. *Lienardia roseotincta* (Montrouzier) 15. *Lienardia rubida* (Hinds) 16. *Lienardia mighelsi* Iredale & Tomlin 17. *Clavus pica* (Reeve) 18. *Clavus laetus* (Hinds) 19. *Daphnella ornata* Hinds 20. *Clavus lamberti* (Montrouzier) 21. *Lienardia planilabrum* (Reeve)

PLATE 78

4. Anarithma dorcas Kuroda & Oyama, 1971
22.2mm; Punta Engaño, Cebu.
Shell small, lightweight, fusiform-ovate, with a tall spire and impressed sutures; surface sculptured with fine axial and spiral lines and oblique cords on the base; aperture elongate-ovate, sinus almost obsolete, outer lip thin and smooth, columella with a central ridge; off-white in colour with thin brown axial and spiral threads forming squares all over the surface; reaches 25mm.
DISTRIBUTION: Camotes Sea.

5. Lophiotoma (Lophiotoma) abreviata abreviata (Reeve, 1843)
21.2mm; Samar.
Shell small, solid, fusiform-ovate, with a tall spire and a short canal; sutures obliterated by a thick, rounded, black-and-white spotted cord which is followed by a larger bicingulate sinus cord mottled with black flecks, 1 or 2 secondary cords below this and 3 to 4 on the body whorl with additional lirae on the canal, sinus deep and "U-shaped"; off-white in colour with black spots and flecks; reaches 35mm.
DISTRIBUTION: Sporadically found throughout the Philippines.

6. Asperdaphne peradmirabilis (Smith, 1879)
17.8mm; Punta Engaño, Cebu.
Shell small, fusiform-ovate, with a tall spire, a sharply tapering body whorl and adpressed sutures; whorls sculptured with short oblique axial cords and crossed by fine spiral lirae; aperture ovate and white, sinus short but distinct, outer lip thin, smooth, with an expanded margin peculiarly incurved posteriorly; pure white in colour with pale brown banding on the body whorl; reaches 20mm.
DISTRIBUTION: Camotes Sea and Sulu Sea (deep water).

7. Turridrupa jubata (Reeve, 1843)
21.9mm; Samar.
Shell small, fusiform, with a tall, acute spire and a short siphonal canal; sculptured with elevated spiral cords, strongest at the whorl shoulder and above which is a spiral row of elongated nodules; orange-brown in colour with a white apex, canal and aperture; reaches 25mm.
DISTRIBUTION: Sporadically found throughout the Philippines.

8. Cochlespira pulchella semiplota Powell, 1969
28.2mm; Punta Engaño, Cebu.
Shell resembles a small *Thatcheria mirabilis* and is characterized by its medium size, lightweight and pagoda-form shape which consists of spinose, sharply keeled whorls, a smooth, concave presutural slope and spiral lirae on the lower part of the whorls, sinus deep; colour pale orange with a white aperture and columella; reaches 50mm.
DISTRIBUTION: Camotes Sea and Sulu Sea (deep water).

9. Cochlespira pulchella fossata Powell, 1969
30.7mm; Leyte.
Very similar to fig. 8, but differs in several aspects: the concave shoulder slope is sculptured with nodulose spiral lirae, whereas in *C. semiplota* it is smooth; there is also an additional smaller row of spines adjacent the major spinal row which is missing in *C. semiplota;* very deep water; illustrated specimen from *Xenophora sp.* reaches 33mm.
DISTRIBUTION: Leyte, Mindoro and Tayabas Bay, Quezon.

10. Eucithara funiculata (Reeve, 1846)
17.5mm; Cebu.
Shell small, fusiform-elongate, with a tall spire and a long, tapering body whorl; whorls sharply angled and sculptured with thick, rounded, equally spaced axial cords which become spinose on the shoulder; aperture elongate, outer lip thick; columella and inner lip margin finely dentate, sinus shallow but distinct; black in colour with white ribbing; reaches 20mm.
DISTRIBUTION: Throughout the Philippines.

11. Clavus sp.
19.3mm; Punta Engaño, Cebu.
Shell small, solid, fusiform-elongate, sutures adpressed, spire tall, frequently eroded; sculptured with oblique axial cords, noduled on the shoulder and fine spiral striae; outer lip thick, variced, sinus deep and "U-shaped"; white in colour with brown blotching and fine brown lines forming a reticulate pattern; reaches 25mm.
DISTRIBUTION: Camotes Sea.

12. Xenoturris kingae Powell, 1964
19.1mm; Punta Engaño, Cebu.
A unique species considered endemic to the Hawaiian Islands, but appears to have a wider distribution, several specimens having been found in the Philippines. Shell small, fusiform, with a tall, acute spire; like a small *cingulifera,* but lacks the secondary spiral sculpturing of that species and the brown spots are on the sutural fold, not the sinus rib; reaches 20mm.
DISTRIBUTION: Camotes Sea and Sulu Sea.

13. Turridrupa astricta consorbrina Powell, 1967
8.9mm, 8.2mm; Punta Engaño, Cebu.
Shell small, elongate, with a tall, acute spire; sculptured with narrowly ridged spiral cords, 3 on the spire whorls, 5 on the body whorl, and numerous interstitial spiral threads; yellowish brown in colour, maculated with brown dots on the cords;reaches 20mm.
DISTRIBUTION: Camotes Sea and Sulu Sea.

14. Lienardia (Lienardia) roseotincta (Montrouzier, 1872)
19.3mm; Punta Engaño, Cebu.
Shell small, slender, with a tall, acute spire and rounded whorls, sharply angulate below the sutures which are impressed and constricted; sculptured with fine axial cords and overriding spiral lirae, obsolete in

the interspaces; outer lip thick, angulate and denticulate, columella plicate; pinkish red to orange in colour; reaches 20mm.
DISTRIBUTION: Camotes Sea and Sulu Sea.

15. Lienardia (Lienardia) rubida (Hinds, 1844)
9.7mm, 9.2mm; Punta Engaño, Cebu.
Shell small, elongate-ovate, with convex, rather ventricose whorls; sculptured with broad axial ribs which are crossed by prominent overriding spiral cords that become oblique basally; outer lip thick and denticulate, columella with several folds, sinus deep; pinkish in colour with black sutural bands and another centrally on the body whorl; aperture pink and black-banded; reaches 14mm.
DISTRIBUTION: Camotes Sea.

16. Lienardia (Lienardia) mighelsi Iredale and Tomlin, 1917
Syn: *Pleurotoma rugosa* Mighels, 1845 (non Lea, 1833)
11.0mm; Samar.
Shell small, fusiform-ovate, with a tall spire; spire whorls convex, angled at the shoulder and sculptured with axial cords and overriding spiral lirae; outer lip thin, sinuous, variced and denticulate, columella plicate; greyish white with a brown sutural band, another centrally on the spire whorls and 2 on the body whorl; reaches 12mm. Possibly conspecific with *Lienardia nigrocincta* (Montrouzier, in Montrouzier & Souverbie, 1872).
DISTRIBUTION: Camotes Sea.

17. Clavus (Clavus) pica (Reeve, 1843)
6.5mm; Punta Engaño, Cebu.
Shell small, tall spired with a somewhat truncated body whorl; sculptured with oblique, rounded axial ribs, noduled on the shoulder and obsolete above; outer lip thick, variced and smooth within; white in colour with brown blotches and a brown band on the body whorl dorsum; reaches 25mm.
DISTRIBUTION: Camotes Sea.

18. Clavus (Clavus) laetus (Hinds, 1843)
7.5mm; Punta Engaño, Cebu.
Shell small, with a tall spire and a rather short body whorl; sutures adpressed and wavy; whorls sculptured with oblique, arcuate axial cords, nodulose at the rounded shoulder and becoming obsolete above; outer lip thin, backed by a varix; columella and inner lip margin smooth and a parietal swelling posteriorly, sinus deep; brown in colour with a white band across the nodules; reaches 20mm; Indo-Pacific specimens more colourful with red banding.
DISTRIBUTION: Camotes Sea.

19. Daphnella ornata Hinds, 1844
8.8mm; Punta Engaño, Cebu.
Shell very small, seldom exceeding 10mm; fusiform-elongate in outline with strongly convex whorls; sculptured with arcuate axial threads and intersecting spiral lirae, obsolete on the presutural slope; outer lip thin, finely lirate, sinus moderately deep; white in colour with rows of orange flammules.
DISTRIBUTION: Camotes Sea.

20. Clavus (Clavus) lamberti (Montrouzier, 1860)
11.2mm; Punta Engaño, Cebu.
Shell small and tall spired; spire whorls concavo-convex with adpressed sutures; sculptured with broad, axially oriented nodules and weak interstitial growth striae; aperture ovate, outer lip and columella smooth; orange-brown in colour with a white apex, a brown-lined white band on the nodules and additional bands on the body whorl, aperture banded within; reaches 20mm.
DISTRIBUTION: Camotes Sea.

21. Lienardia (Lienardia) planilabrum (Reeve, 1846)
16.4mm; Punta Engaño, Cebu.
Shell small with a tall spire, a large body whorl and impressed sutures; early spire whorls sculptured with rounded axial cords which become obsolete on the last 2 whorls, and raised spiral threads; outer lip thick, variced and heavily denticulate, sinus small but prominent, columella irregularly corded; shell entirely pinkish brown, lighter within the aperture; reaches 20mm.
DISTRIBUTION: Camotes Sea.

Plate 79 Turridae

1. Tritonoturris cumingii (Powys, 1835)
36.9mm; Punta Engaño, Cebu.
Shell medium sized, thin, with a tall acute spire, well impressed sutures and strongly convex whorls which are angulate at the sutures; sculptured with thin axial cords and overriding spiral lirae intersecting to form weak, elongate nodules; aperture large, ovate, with a thin, convex outer lip and a weak sinus, columella smooth, fasciolar cord ridged; orange brown in colour with darker mottling and white lirae; reaches 40mm.
DISTRIBUTION: Camotes Sea.

2. Gemmula (Unedogemmula) hastula (Reeve, 1843)
38.5mm, 31.2mm; Punta Engaño, Cebu.
Shell medium sized, fusiform-elongate, with a tall spire and a long, straight anterior canal; sculptured with a bicingulate subsutural cord and a large cingulate peripheral cord terminating with a deep "U-shaped" sinus; one major and 2 minor cords between the subsutural and peripheral keels, and a single major cord between the peripheral cord and sutures; numerous major and minor cords on the anterior canal; pale brown in colour with orange-brown mottling; reaches 39mm; described from the Persian Gulf.
DISTRIBUTION: Camotes Sea.

3. Eucithara celebensis (Hinds, 1843)
14.0mm, 12.9mm; Punta Engaño, Cebu.
Shell small, elongate-ovate, whorls convex, somewhat angulate below the moderately channeled sutures;

sculptured with numerous widely-spaced, slightly angulate axial ribs and encircled with very fine spiral striae; aperture narrowly ovate, outer lip thick, variced and denticulate within, sinus shallow but prominent, columella plicate; pale brown in colour with 1 or 2 darker bands or numerous fine, brown-coloured spiral threads; reaches 15mm.
DISTRIBUTION: Throughout the Philippines.

4. Etrema (Etrema) crassilabrum (Reeve, 1846)
10.8mm; Punta Engaño, Cebu.
Shell small, with a fairly tall spire, adpressed sutures and a small, ovate body whorl; whorls convex, slightly concave on the presutural slope; sculptured with rounded axial cords, obsolete below the sutures, and encircled with numerous spiral striae; outer lip thick, variced and marginally denticulate, columella plicate, sinus deep and "U-shaped;" pale brown in colour with a dark band on the body whorl base; reaches 13mm.
DISTRIBUTION: Throughout the Philippines in limited quantity.

5. Lienardia (Lienardia) sp.
10.4mm; Punta Engaño, Cebu.
Shell small, fusiform-ovate, with a tall acute spire, impressed sutures and strongly convex whorls; sculptured with narrow, rounded axial cords and encircled with strong, raised spiral lirae; aperture narrowly ovate, outer lip variced and denticulate within, columella plicate, sinus deep and "U-shaped," canal long and open; deep pink in colour with brown-stained white banding and deep brown spiral cords; reaches 12mm.
DISTRIBUTION: Only known from the Camotes Sea at present.

6. Eucithara marginelloides (Reeve, 1846)
12.1mm; Punta Engaño, Cebu.
Shell small, fusiform-ovate, with a short, acute spire, a large body whorl and impressed sutures; sculptured with slender axial cords, slightly nodulose at the roundly angled shoulder, and encircled with obsolete spiral striae; aperture narrowly elongate, sinus shallow, outer lip denticulate, columella plicate; greyish white in colour with orange spotting on the ribs; reaches 15mm.
DISTRIBUTION: Sporadically found throughout the Philippines in limited quantity.

7. Eucithara questieri (Souverbie, 1872)
Syn: *Mangelia reticulata* Reeve, 1846 (non Risso, 1826)
15.2mm; Punta Engaño, Cebu.
Shell small, fusiform-ovate, spire less than one-half the shell length, body whorl large; sculptured with regularly spaced, arcuate axial cords, slightly nodulose at the roundly angulate shoulder, and reduced in thickness on the presutural slope; interspaces weakly pitted above the shoulder, and below, the shell is encircled with double paired, weakly incised overriding spiral grooves; aperture narrow, outer lip denticulate, columella lirate; colour greyish white with a grey subsutural band on the spire whorls, two on the body whorl, aperture deep brown within; reaches 16mm.
DISTRIBUTION: Sporadically found throughout the Philippines in limited quantity.

8. Philbertia (Pseudodaphnella) philippinensis (Reeve, 1843)
14.2mm (8a), 10.0mm (8b); Punta Engaño, Cebu.
Shell small, fusiform-ovate with a tall acute spire, a large, rounded body whorl and deeply impressed sutures; whorls strongly convex, sculptured with numerous regularly spaced arcuate axial ribs, finely nodulose from overriding spiral threads and making the whole surface strongly cancellate; aperture ovate, outer lip thin, arcuate, marginally serrate and smooth within, columella smooth, fasciolar ridge with 5 to 6 oblique nodulose cords; pure white in colour, though most specimens usually exhibit some darker banding (8b); reaches 15mm.
DISTRIBUTION: Camotes Sea.

9. Drupella angulata (Reeve, 1844) (Family: Muricidae)
8.0mm; Punta Engaño, Cebu.
Shell small but solid, with adpressed sutures and bi-angulate whorls; sculptured with axial ribs, obsolete on the presutural slope, and crossed by fine spiral threads, and 3 nodulose rows on the lower half of the body whorl; aperture elongate-ovate, outer lip denticulate, columella smooth with a few basal plicae, sinus shallow but prominent; pale brownish white in colour with brown spiral lines; reaches 10mm. Incorrectly inserted here as it was initially thought to be a turrid species, but now appears to be an unusual form of the above Drupelline species.
DISTRIBUTION: Throughout the Philippines.

10. Daphnella boholensis (Reeve, 1843)
12.2mm; Punta Engaño, Cebu.
Shell small, fusiform-elongate, with convex whorls and impressed sutures; sculptured with axial and spiral cords, the axial structure becoming obsolete on the last 2 whorls and the spiral sculpture becoming slightly more prominent; outer lip thin, marginally dentate, columella smooth, sinus shallow; pale yellowish white in colour with brown axial streaks and bands; reaches 18mm; close to *D. flammea* but

PLATE NO. 79 1. *Tritonoturris cumingii* (Powys) 2. *Gemmula hastula* (Reeve) 3. *Eucithara celebensis* (Hinds) 4. *Etrema crassilabrum* (Reeve) 5. *Lienardia sp.* 6. *Eucithara marginelloides* (Reeve) 7. *Eucithara questieri* (Souverbie) 8a. *Philbertia philippinensis* (Reeve) 8b. specimen with darker bandings 9. *Drupella angulata* (Reeve) (Muricidae) 10. *Daphnella boholensis* (Reeve) 11. *Daphnella subula* (Reeve) 12. *Lienardia roseotincta* (Montrouzier) 13. *Tritonoturris menecharmes* (Melvill) 14. *Veprecula echinulata* (Thiele) 15. *Lienardia purpurata* (Souverbie) 16. *Mitrolumma sp.* 17. *Mitrolumma sp.* 18. *Paraclathurella gracilenta* (Reeve) 19. *Glyphostoma sp.*

PLATE 79

with much coarser sculpturing.
DISTRIBUTION: Camotes Sea.

11. Daphnella subula (Reeve, 1845)
15.8mm; Punta Engaño, Cebu.
Shell small to medium sized, fusiform-elongate, with a tall acute spire, a short body whorl, and sutures partially obscured by a weak cord; early whorls clathrate with intersecting axial and spiral threads, the axials becoming obsolete on the last 2 whorls and the spiral threads assuming greater prominence; aperture narrowly ovate, outer lip marginally denticulate, columella smooth, sinus deep; cream-coloured with orange-brown mottling, aperture white; reaches 25mm.
DISTRIBUTION: Camotes Sea.

12. Lienardia (Lienardia) roseotincta (Montrouzier, 1872)
8.0mm; Punta Engaño, Cebu.
Shell small, solid, fusiform-ovate, the spire frequently missing; sculptured with rounded axial ribs which become obsolete just before the sutures and on the lower part of the body whorl, and encircled with elevated spiral cords; aperture narrowly elongate, outer lip variced with 4 denticles, sinus deep columella plicate; colour yellowish orange; reaches 20mm. See also Pl. 78, fig. 14 for a more mature specimen.
DISTRIBUTION: Camotes Sea.

13. Tritonoturris menecharmes (Melvill, 1923)
13.2mm; Punta Engaño, Cebu.
Shell small, with sharply angulate whorls conferring a slightly pagoda-form outline; sculptured with narrow, rounded, widely spaced axial cords and overriding elevated spiral ribs forming a slightly reticulate surface and producing weakly scabrous nodules on the cords; aperture ovate, outer lip thick, anteriorly constricted and marginally denticulate, columella smooth, sinus deep; colour pale yellowish brown; reaches 30mm.
DISTRIBUTION: Camotes Sea.

14. Veprecula echinulata (Thiele, 1925)
10.2mm; Punta Engaño, Cebu.
Shell small, elongate-ovate, with a tall spire and impressed sutures; sculptured with strong axial cords which become obsolete on the presutural slope and on the lower half of the body whorl, and encircled with strong spiral lirae, producing strongly echinate nodules on the axial cords and deeply cancellate interspaces; aperture ovate, sinus deep, outer lip and columella smooth, canal long; creamy white in colour; reaches 12mm.
DISTRIBUTION: Camotes Sea.

15. Lienardia (Hemilienardia) purpurata (Souverbie, 1860)
5.6mm; Punta Engaño, Cebu.
Shell extremely small, bright pinkish red in colour with a white band on the body whorl; fusiform-ovate in outline and sculptured with rounded axial ribs and overriding elevated spiral threads, sutures adpressed, spire whorls sharply angulate; aperture narrow, outer lip thick, variced and denticulate within, columella plicate; reaches 7.0mm.
DISTRIBUTION: Camotes Sea.

16. Mitrolumma sp.
9.5mm; Punta Engaño, Cebu.
Shell small, fusiform-ovate, with a moderately tall spire and large body whorl; sculptured with granulose spiral cords and obsolete axial ridges; sutures moderately impressed with a presutural cord and cleft; aperture narrow, outer lip smooth, columella weakly plicate; orange-brown in colour with a white sutural band and white spotting on the brown surface; reaches 10mm.
DISTRIBUTION: Camotes Sea.

17. Mitrolumma sp.
8.3mm; Punta Engaño, Cebu.
Shell small, fusiform-ovate, sutures impressed but obscured by a sutural cord; sculptured with wide spiral cords and interstitial axial striae, the spiral cords becoming wider and flatter towards the base; aperture narrow, outer lip thin, marginally crenulate and weakly lirate internally, columella weakly plicate; greyish brown in colour, aperture dark brown; reaches 10mm.
DISTRIBUTION: Camotes Sea.

18. Paraclathurella gracilenta (Reeve, 1843)
12.2mm; Punta Engaño, Cebu.
Shell small, elongate, with a tall acute spire and impressed sutures; sculptured with oblique axial cords, slightly arcuate on the concave presutural slope, and crossed by spiral threads producing a cancellate surface; outer lip thick, variced and smooth within, columella smooth, sinus deep; colour greyish brown, cords and lirae somewhat paler; reaches 15mm.
DISTRIBUTION: Camotes Sea.

19. Glyphostoma sp.
23.7mm; Lubang Is., Mindoro.
Shell medium sized, fusiform-ovate, with a tall spire and impressed sutures; sculptured with strong axial cords which are nodulose from weaker, overriding spiral threads, whorls slightly angulate, with a concave presutural slope where the axial cords are reduced to thin, arcuate axial striae; aperture triangularly ovate, outer lip thick, variced and denticulate within, columella irregularly plicate, sinus deep and "U-shaped" with a flaring margin; pure white in colour; largest size seen: 26.7mm; not yet found live and presently known only from *Xenophora spp.*
DISTRIBUTION: Mindoro, Tayabas Bay (Quezon) and Leyte.

Turridae Plate 80

1. Turricula javana (Linné, 1767)
46.4mm; Palawan.
Shell medium sized, fusiform-elongate, with a tall

spire and adpressed sutures; spire whorls carinated with a presutural row of nodules below which are 2 or 3 weaker lirae; immediately below the sutural line are 2 strong spiral cords followed by a wide, smooth, concave interspace; below the shoulder row of nodules on the body whorl are numerous close-set spiral cords extending all the way to the base; aperture narrowly elongate, terminating in a moderately long, open anterior canal; outer lip thin, lirate within, sinus deep and "U-shaped"; uniformly pale brown in colour; reaches 75mm.
DISTRIBUTION: Palawan and Sulu Sea.

2. Ptychobela vexillum (Habe & Kosuge, 1966)
35.5mm; Lubang Is., Mindoro.
Shell medium sized, fusiform-ovate with a very tall spire; whorls sculptured with a row of axially elongated, spirally trisected nodules, above which is a smooth, concave, presutural slope, and below a series of 2 or 3 strong spiral cords; body whorl with strong spiral cords extending from the row of nodules all the way to the base; sutures adpressed with a heavy spiral cord; aperture elongate-ovate, smooth within, outer lip thin, reflexed and backed by a varix, sinus deep and "U-shaped", opposite which is a strong parietal denticle; pale brown in colour with whitish cream nodules and spiral elements, aperture creamy white; reaches 40mm. This may be the same as *Ptychobela griffithi* (Reeve, 1843).
DISTRIBUTION: Sporadically found throughout the Philippines in limited quantity.

3. Philbertia (Pseudodaphnella) nexa (Reeve, 1845)
19.3mm; Punta Engaño, Cebu.
Shell small to medium sized, lightweight, with strongly convex whorls, a moderately tall spire and a large, ovate body whorl; sculptured with numerous arcuate axial ribs and overriding spiral threads producing a clathrate surface; aperture widely ovate, outer lip sharply denticulate, columella smooth; colour off-white with reddish brown spiral cords and a dark brown band basally, aperture white within; reaches 20mm.
DISTRIBUTION: Camotes Sea.

4. Gemmula (Gemmula) monilifera (Pease, 1860)
18.8mm; Punta Engaño, Cebu.
Shell small to medium sized, narrowly fusiform, with a tall spire and a long siphonal canal; whorls sculptured with a strong subsutural fold, followed by a deep concavity bearing 3 to 4 spiral threads, then a heavy sinus keel of 2 rows of axially fused gemmules, followed by another spiral cord; last whorl with 4 smooth major cords, followed by numerous weaker cords which become more compact basally; sinus deep and "U-shaped;" shell brown-coloured with a white-banded sinus keel and another white band towards the base; reaches 30mm.
DISTRIBUTION: Camotes Sea and Sulu Sea.

5. Eucithara cylindrica (Reeve, 1846)
11.8mm; Punta Engaño, Cebu.
Shell small, fusiform-elongate, spire and aperture equal in height; sculptured with thin, arcuate axial cords and fine overriding spiral striae; aperture narrowly elongate, outer lip variced and smooth, columella smooth; pale greyish pink in colour with orange-brown bands; reaches 15mm; possibly the same as *E. lyrica* (Reeve, 1846).
DISTRIBUTION: Throughout the Philippines in limited quantity.

6. Lienardia (Acrista) sp.
7.8mm; Punta Engaño, Cebu.
A small, distinctly sculptured species, fusiform-ovate in outline with a tall spire and a large body whorl; sutures adpressed, whorls sharply angulate and sculptured with 7 to 8 axially aligned cords; spirally sculptured with 3 to 4 cords on the spire whorls and 10 to 12 on the body whorl, the interspaces finely clathrate from intersecting axial and spiral threads; aperture ovate-quadrangular, outer lip thin and dentate, columella with 3 folds, a detached peripheral margin and a large parietal denticle bordering a shallow "U-shaped" sinus; brown in colour with white axial and spiral elements; reaches 8mm.
DISTRIBUTION: Cebu.

7. Tritonoturris macandrewi (E. A. Smith, 1882)
19.2mm; Punta Engaño, Cebu.
Shell medium sized with a tall spire and a large, ovate body whorl; sutures impressed, whorls strongly convex and sculptured with numerous thin, arcuate axial ribs which are crossed by numerous spiral threads; aperture elongate-ovate, smooth within, outer lip thin and weakly serrated, columella smooth; colour cream with a subsutural row of brown spots and brown dashes on the body whorl; reaches 20mm.
DISTRIBUTION: Camotes Sea.

Holotype figure of *Pleurotoma (Daphnella) macandrewi* E. A. Smith, 1882

8. Tomopleura coffea (Thiele, 1925)
15.2mm; Punta Engaño, Cebu.
A small to medium sized, fusiform-elongate shell with an extremely tall spire and a relatively small body whorl; sutures adpressed, the entire surface sculptured with raised, concentric, sub-equal carinal ridges which are axially striated in the interspaces; aperture narrowly elongate, outer lip thin and weakly lirate

within, columella smooth with 1 or 2 weak ridges; pure white in colour, occasionally with darker axial streaks; reaches 25mm.
DISTRIBUTION: Camotes Sea.

Type figure of *Tomopleura coffea* (Thiele, 1925)

9. c.f. **Eucithara pulchella** (Reeve, 1845)
14.5mm; Punta Engaño, Cebu.
Shell small, broadly fusiform-elongate with weakly impressed sutures and a slightly angled shoulder; sculptured with numerous thin axial cords and fine overriding spiral striae; aperture narrowly elongate, outer lip thick, weakly variced and weakly lirate within, columella finely plicate; colour off-white with numerous beige spiral bands; reaches 15mm. Close to *E. pulchella* (see insert), but that species has a more angulate outline, a more pronounced presutural slope, and lacks any indication of the fine spiral threads seen in the illustrated specimen.
DISTRIBUTION: Throughout the Philippines in limited quantity.

Type figure of *Eucithara pulchella* (Reeve, 1845) (from Tryon)

10. Mitromorpha (Lovellona) atramentosa (Reeve, 1849)
9.5mm; Punta Engaño, Cebu.
Shell small, coniform, with a short, weakly concave-sided spire, a large body whorl and impressed sutures; early whorls sculptured with nodulose spiral cords which become axial grooves on the penultimate whorl and are crossed by a spiral cord, body whorl with numerous axial threads intersected by finely punctate spiral grooves; aperture narrowly elongate, weakly dentate within, columella non-plicate; colour dark reddish brown (almost black) with a white apex, a row of white spots at the roundly angulate shoulder, a white band at the base and a dark brown aperture; reaches 10mm.
DISTRIBUTION: Camotes Sea.

11. Carinodrillia quadrilirata (E. A. Smith, 1882)
19.0mm; Punta Engaño, Cebu.
Shell medium sized, with a tall acute spire; whorls sculptured with up to 8 thin axial cords, slightly angulate on the shoulder and becoming obsolete on the presutural slope; encircled with fine spiral threads which form weak nodules on the ribs; aperture narrowly elongate, outer lip thin with a weak varix, both outer lip and columella smooth; ground colour pinkish white, with a thin, pinkish brown presutural band and a larger band at the base; reaches 25mm. This is the first Western Pacific record for the genus.
DISTRIBUTION: So far, recorded only from Cebu.

12. Conorbis sp.
14.1mm; Punta Engaño, Cebu.
Shell small, fusiform-ovate, with a moderately tall spire and a large ovate body whorl; sutures weakly impressed, whorls with a concave furrow below the sutures, then strongly convex; sculptured with numerous fine spiral threads crossed by numerous microscopic axial striae producing a weakly granulose surface; aperture narrowly elongate, outer lip and columella smooth; off-white in colour with orange axial streaks in the subsutural furrow, and orange-brown spots and dashes arranged in discontinuous spiral bands; reaches 15mm; may also be an abnormal *Micantaplex* or *Riuguhdrillia sp.*
DISTRIBUTION: Camotes Sea.

13. Clavus (Clavus) fulvus (Hinds, 1843)
15.9mm; Punta Engaño, Cebu.
Shell small, fusiform-ovate, with a tall spire and sculptured with axially oriented nodules, strongest at the shoulder; sutures adpressed, some spiral grooving on the base of the siphonal canal; aperture wide and ovate, sinus weakly "U-shaped", outer lip thin, smooth within, columella smooth; pale brown in colour with a white apex and a white band centrally on each whorl; reaches 45mm.
DISTRIBUTION: Throughout the Philippines in limited quantity.

14. Horaiclavus (Cytharoclavus) filicinctus (Smith, 1882)
15.3mm; Punta Engaño, Cebu.
Shell small, elongate-ovate, with a tall spire and a moderately large body whorl; whorls sharply angulate at the shoulder, presenting a somewhat pagoda-form outline; sculptured with 14 to 16 angulate axial ribs and smooth interspaces; aperture narrowly elongate, outer lip variced, the outer margin sinuous and forming a shallow sinus, columella smooth, anterior

PLATE NO. 80 1. *Turricula javana* (Linné) 2. *Ptychobela vexillum* (Habe & Kosuge) 3. *Philbertia nexa* (Reeve) 4. *Gemmula monilifera* (Pease) 5. *Eucithara cylindrica* (Reeve) 6. *Lienardia sp.* 7. *Tritonoturris macandrewi* (E. A. Smith) 8. *Tomopleura coffea* (Thiele) 9. c.f. *Eucithara pulchella* (Reeve) 10. *Mitromorpha atramentosa* (Reeve) 11. *Carinodrilla quadrilirata* (E. A. Smith) 12. *Conorbis sp.* 13. *Clavus fulvus* (Hinds) 14. *Horaiclavus filicinctus* (Smith) 15. *Clavus opalus* (Reeve) 16. *Eucithara coronata* (Hinds)

PLATE 80

canal slightly tubular and extended; colour creamy, with brown spotting on the presutural slope and on the base, rest of shell encircled with thin brown-coloured threads, strongest on the ribs; reaches 17mm.
DISTRIBUTION: Camotes Sea and Sulu Sea.

15. Clavus (Clavus) opalus (Reeve, 1845)
12.2mm; Punta Engaño, Cebu.
Shell small, shiny and pure white in colour with weakly impressed sutures; sculptured with 5 vertically aligned ribs which become obsolete on the body whorl dorsum and encircled with fine spiral striae, strongest on the body whorl; aperture narrowly elongate, sinus deep and a parietal "button" is present on the upper columellar wall; reaches 20mm.
DISTRIBUTION: Camotes Sea.

16. Eucithara coronata (Hinds, 1843)
16.8mm; Punta Engaño, Cebu.
Shell small with a moderately tall spire and a large body whorl which is greater than half the shell length; sculptured with 10 to 12 roundly angulate ribs which form coronations of varying strengths on the shoulder, some specimens more coronate than others with the ribs frequently extending above the suture line; aperture elongate, outer lip thick, angulate, variced and denticulate, columella plicate; colour off-white with occasional orange lines on the fine interstitial striae; reaches 18mm.
DISTRIBUTION: Throughout the Philippines.

ADDITIONAL NOTES ON THE TURRIDAE

1. Gemmula (Gemmula) graeffei (Weinkauff, 1875)
(Pl. 77, fig. 14)
Examination of the holotype figure of *Pleurotoma (Gemmula) graeffei* Weinkauff (1875: *Jahrb. d. d. M. G. II*, p. 290, t. 9, f. 9. 10) although not very clear, shows the shape and proportions of *Gemmula hombroni* (Hedley, 1922) as depicted in Thiele's type figure of *Pleurotoma padangensis* (1925: *Deutschen Tiefsee-Exped., 17, Gastr. 2,* p. 335, pl. 23, fig. 5) and well illustrated by Powell (1967, p. 262, pl. 201, fig. 8). Powell's concept of *graeffei* (1967, p. 261, pl. 200, fig. 3) depicts a distinctly more ovate shell and appears to be referable to another species entirely. We therefore tentatively propose that *Gemmula hombroni* (Hedley, 1922) be considered a subjective synonym of *Gemmula graeffei* (Weinkauff, 1875), and that the shell commonly illustrated as *graeffei* (i.e., Powell's concept) be relegated to another, as yet undetermined, taxon.

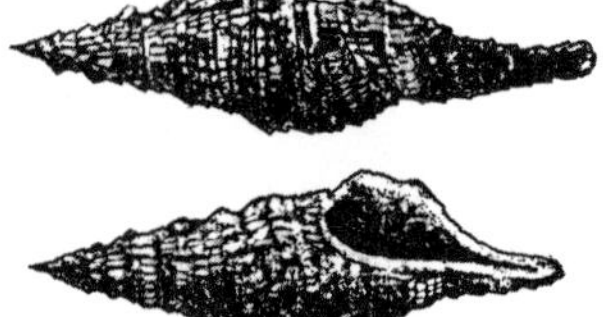

Holotype figure of *Pleurotoma (Gemmula) graeffei* Weinauff, 1875

2. Gemmula (Gemmula) concinna (Dunker, 1856) (non Scacchi, 1836)
(Pl. 77, fig. 7)
This species was originally identified as a form of *graeffei* (fide Orr-Maes) based on material identified by Powell and housed at the Academy of Natural Sciences, Philadelphia. As discussed above, Powell's *"G. graeffei"* seems assignable to another taxon. Dunker's taxon was clearly described (*Proc. Zool. Soc.* p. 356), but curiously enough never figured. His description clearly matches our illustrated specimens, even down to the purple-stained canal, a feature seen in very few Philippine Turridae. The almost identical morphology of our material when compared with Powell's *graeffei* suggests that his shell may in fact represent a shallow water ecomorph of *G. concinna* which is at present known only from deep water and lacks the characteristic intensely purple-coloured canal of its deep water counterpart. The name *concinna* is pre-occupied and there appears to be no substitute currently available. We retain the above name for our material on a conservative basis until more is known on this taxon.

3. Separation of *Inquisitor* Hedley, 1918 and *Ptychobela* Thiele, 1925
In 1966 Powell synonymized these 2 genera on the basis of almost identical protoconch and shell morphology. Subsequent work by Maes (circa 1966 and 1970-unpublished MS) demonstrated the radula differences between *Inquisitor sterrha* (Watson, 1881) (type species of *Inquisitor*) and *Ptychobela crenularis* (Lamarck, 1816) (type species of *Ptychobela*).
The radula of *Ptychobela* is composed of single rolled, occasionally barbed teeth (see fig. 1). The radula of *Inquisitor* has duplex crassispirine teeth (see fig. 2). Apart from anatomical differences, members of *Inquisitor* are generally smaller, have a proportionately taller spire, and a stronger subsutural cord.

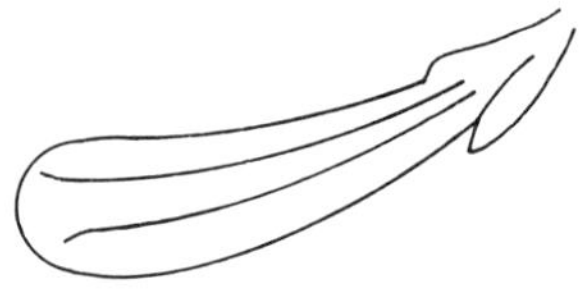

Fig. I
Radula System of *Ptychobela spp.*
Example: *Ptychobela c.f. crenularis* (Lamarck, 1816)
(AFTER POWELL, 1966)

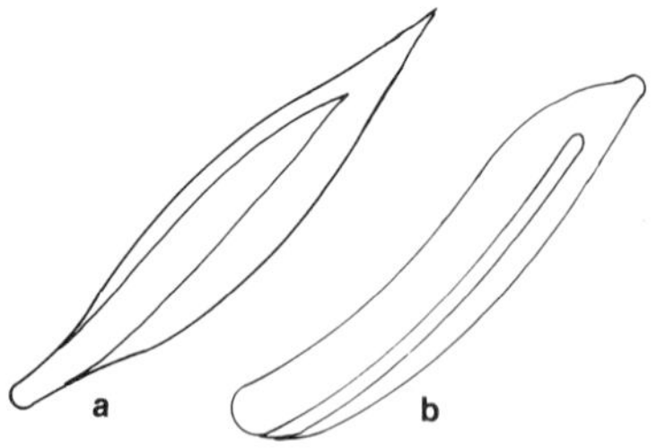

Fig. II
Radula System of *Inquisitor spp.*
Examples:
a. *Inquisitor c.f. aesopus* (Schepman, 1913)
b. *Inquisitor tasconium* (Melvill & Standen, 1901)
(AFTER POWELL, 1966)

4. Gemmula (Gemmula) congener congener (E. A. Smith, 1894)
(Pl. 76, figs. 6 & 21)
The application of this name to our illustrated shells is purely provisional, and is based on the figures and description by Powell (1964: *Indo-Pacific Mollusca,* Vol. I (5):251, pl. 191, figs. 1-4). An analysis of the type figure of *Pleurotoma congener* E. A. Smith (1894: pl. 3, figs. 5, 6) shows it to be different to Powell's interpretation. Of the numerous morphological differences between the two, the "massive" subsutural cord of Powell's *congener* is the most pronounced and is distinctive of *Gemmula (Gemmula) rarimaculata* Kuroda & Oyama, 1971 (*Seashells of Sagami Bay,* p. 222, pl. 57, fig. 7/pl. 111, fig. 6), and may well be that species. Until the holotype of *congener* or topotypic material (from the Indian Ocean) is studied, the true identity of Powell's shell (and our own illustrated material) will remain in doubt.

Class Gastropoda
Subclass Opisthobranchia
Order Entomotaeniata
Superfamily Pyramidellacea
Family PYRAMIDELLIDAE

A large group of small to medium-sized carnivorous gastropods inhabiting sand and coral rubble in shallow intertidal waters. The shells are ovate to cylindrically elongate, have a porcelaneous surface which may be grooved, cancellate or smooth; the aperture is small, the outer lip thin, and the columella bears from 1 to 3 folds (occasionally none); three subfamilies are generally recognized: Pyramidellinae Gray, 1840 (2 to 3 columellar folds), Odostomiinae Pelseneer, 1928 (1 columellar fold), and Turbonillinae Simroth, 1907 (no columellar folds); many small species in the Philippines, but only a few of the larger ones are shown here.

1. Pyramidella acus (Gmelin, 1791)
48.9mm, 43.0mm; Palawan.
Shell medium sized, pyramidally elongate, with impressed sutures and up to 20 smooth convex whorls; aperture semi-lunate, smooth or obsoletely lirate within, columella with 3 folds, posterior one largest and horizontally aligned; bluish white in colour with irregular brown blotching and 3 rows of brown spots on the spire whorls, 4 to 5 on the body whorl, reaches 55mm.
DISTRIBUTION: Sporadically found throughout the Philippines.

2. Pyramidella maculosa Lamarck, 1822
31.5mm; Cebu.
Shell small to medium sized, pyramidally elongate, with slightly impressed sutures and weakly convex whorls which are smooth save for some obsolete axial scratches; aperture elliptically elongate, outer lip thin and weakly lirate within, columella with 3 folds, the posterior one largest and horizontally aligned, the anterior 2 oblique and almost obsolete; colour pale yellow with 3 dark, brown-spotted bands on the spire whorls and 5 on the body whorl; largest seen 35mm; possibly a form *P. acus,* but is smaller and has a different colour pattern and columellar dentition.
DISTRIBUTION: Sporadically found throughout the Philippines.

3. Pyramidella ventricosa (Guerin, 1831)
22.7mm; Palawan.
Shell small, pyramidally ovate, sutures moderately channeled and slightly ledged, whorls smooth and weakly convex with a dull lustre; aperture semi-lunate, smooth within, columella with 3 folds, the uppermost largest and horizontally aligned; pale cream in colour with heavy brown-and-white mottling; reaches 30mm.
DISTRIBUTION: Sporadically found throughout the Philippines.

4. Otopleura auriscati (Holten, 1802)
22.3mm; Cebu.
Shell small, elongate to elongate-ovate; sculptured with numerous heavy axial ribs, constricted subsuturally by a weak groove forming a row of granules, interspaces with weakly incised grooves; aperture semi-lunate, outer lip thin, angulate, and smooth within, columella with 3 folds, the central one smallest; off-white in colour with rows of brown spots and irregular brown blotching; reaches 25mm.
DISTRIBUTION: Sporadically found throughout the Philippines.

5. Otopleura nodicincta (A. Adams, 1855)
17.2mm; Samar.
Similar to the preceding species, but the shell is generally smaller and slightly more ovate; the ribs are more irregular and become weak on the lower part of each whorl, disappearing completely on the body whorl dorsum; the subsutural nodules are much stronger and the whole surface is finely pustulose; the colour is generally lighter than *O. auriscati;* reaches around 20mm.
DISTRIBUTION: Sporadically found throughout the Philippines.

6. Milda cincta (Reeve, 1842)
15.5mm; Cebu.
Shell small, pyramidally ovate with moderately channeled sutures, a weak presutural groove and fine axial scratches on the whorls; aperture elongate-ovate, outer lip thin and smooth within, columella with 3 folds, the upper one strongest and projecting onto the

base of the body whorl forming a strong cord; bluish white in colour with a single presutural brown band, and another on the center of the body whorl; reaches 18mm.
DISTRIBUTION: Sporadically found throughout the Philippines.

7. Pyramidella teres (A. Adams, 1855)
18.7mm; Palawan.
Shell small, pyramidally elongate, whorls strongly convex and smooth with well channeled sutures; body whorl weakly angulate with either a spiral keel or spiral groove; aperture oblique, elliptically elongate and smooth within, columella with 3 folds, posterior one largest; colour greyish white to fawn with darker blotching or spotting, axial lines, or a single dark spiral line on each whorl (as illustrated); reaches 25mm.
DISTRIBUTION: Sporadically found throughout the Philippines.

8. Pyramidella terebellum (Müller, 1774)
15.4mm; Cebu.
Shell small, elongate, with weakly convex to straight-sided whorls and shallow or deeply channeled sutures, whorls smooth with obsolete spiral scratches; aperture elongate-ovate, outer lip thin, arcuate, and may be smooth or dentate within, columella expanded with between 1 and 4 folds (normally 3), the uppermost largest, the lower 2 smaller and closely set; bluish white in colour with brown bands and lines; reaches 30mm.
DISTRIBUTION: Sporadically found throughout the Philippines.

9. Otopleura glans (Reeve, 1843)
12.9mm; Cebu.
Shell small, ovate, with 6 to 8 somewhat telescoped spire whorls which are strongly convex and slightly ledged below the sutures; sculptured with numerous closely-set axial ribs and fine spiral threads in the interspaces; aperture semi-lunate, outer lip smooth, columella with 3 folds, upper one largest; white in colour with a thin brown band at the sutures and a wider band on the body whorl base; reaches 15mm.
DISTRIBUTION: Sporadically found throughout the Philippines.

Class Gastropoda
Subclass Opisthobranchia
Order Cephalaspidea
Superfamily Bullacea
Family BULLIDAE

A small group of carnivorous-herbivorous gastropods living in sand or mud in the shallow intertidal waters of most tropical and temperate seas. The shells are medium sized, thin, globose or elongate-ovate, have a smooth, usually mottled surface, a large flaring aperture, and a thin, convex outer lip which rises above the apex, and is narrowed above and expanded below; columella with a thin, smooth parietal callus; few Philippine species.

10. Bulla ampulla Linné, 1758
52.0mm; Batangas.
Shell medium sized, solid, elongate-ovate in outline with a large, smooth, strongly ventricose body whorl; apical umbilical channel deep, aperture narrow posteriorly, rapidly widening towards the base, outer lip convex and centrally thickened, columellar callus non-expansive and dull white; colour cream, completely obscured by grey and brown streaks and mottling, aperture bluish white; reaches 60mm.
DISTRIBUTION: Throughout the Philippines.

11. Bulla vernicosa Gould, 1859
32.0mm; Samar.
Similar to *B. ampulla*, but usually smaller and more elongate; colour also similar, but the dorsum usually has 3 to 4 dark bands; reaches 50mm.
DISTRIBUTION: Throughout the Philippines.

Class Gastropoda
Subclass Pulmonata
Order Basommatophora
Superfamily Melampidacea
Family MELAMPIDAE

A moderately large group of gastropods inhabiting mangroves and brackish water above the high tide line, living in mud or under rocks; the shells are small to medium sized, ovate to elongate-ovate with a tall spire, smooth surface, and an ovate aperture which may be smooth or lirate within; columella calloused and both apertural margins are heavily dentate; found in most tropical and temperate seas. Moderately well represented in the Philippines, but most are quite small and unlikely to be encountered by the average collector.

12. Ellobium aurisjudae (Linné, 1758)
51.5mm; Palawan.
Shell large, elongate-ovate, with a short, bulbous spire and a large body whorl; sculptured with numerous thin, closely-set axial striae, nodulose from overriding spiral threads; aperture triangularly ovate, outer lip thick and flattened with a single fold, columella heavily glazed with 3 oblique folds, center one largest; creamy white in colour under a brown periostracum; reaches 60mm.
DISTRIBUTION: Sporadically found throughout the Philippines.

13. Cassidula nucleus (Gmelin, 1791)
18.5mm; Samar.
Shell small to medium sized with a short spire, adpressed sutures, a large, strongly convex body whorl and sculptured with fine spiral striae; aperture narrow, outer lip expansive, flattened, sharply angulate posteriorly, with a deep excavation in the posterior margin and a few obsolete denticles anteriorly; columella with 3 folds and an umbilical

shield anteriorly, behind which is a deep umbilical depression; purple in colour with brown banding, a pink-tinted apertural margin and a dark purple apex; reaches 20mm.
DISTRIBUTION: Sporadically found throughout the Philippines.

14. Pythia plicata (Férrusac, 1821)
16.9mm; Palawan.
Shell small, moderately globose, with a short, straight spire, adpressed sutures and a large, rounded body whorl the sides of which are slightly bi-angulate; shell dorso-ventrally compressed and the surface has obsolete axial ridges; aperture narrow, irregularly elongate, with 3 or 4 outer lip denticles on the inner margin and 3 large columellar denticles; deep reddish brown in colour with a cream apertural margin and denticles; reaches 20mm.
DISTRIBUTION: Sporadically found throughout the Philippines.

15. Pythia scabraeus (Linné, 1758)
27.0mm; Palawan.
Shell medium sized, slightly dorso-ventrally compressed, with adpressed sutures and 2 axially aligned varices per whorl; surface with weak axial folds, strongest below the sutures; aperture narrow, outer lip expansive, flattened, marginally reflected with 5 irregularly sized denticles on the inner margin; columella calloused and deeply excavated with 3 teeth; dark brown in colour with paler varices and orange-brown teeth and apertural margin; reaches 50mm.
DISTRIBUTION: Sporadically found throughout the Philippines.

16. Melampus flavus (Gmelin, 1791)
Shell small, elongate-ovate, with a short spire and a large, rounded body whorl; surface smooth except for axial and spiral striae on the earlier whorls; aperture long and narrow, outer lip with a few large internal lirae; columella with 5 to 6 folds, the anterior 2 or 3 largest; variable in colour and pattern; reaches 25mm.
16a. form *nuxcastaneus* Kuroda, 1947: orange-yellow with 1 or 2 dark brown bands; 13mm; Punta Engaño, Cebu.
16b. form *fasciatus* Deshayes, 1830: white with 4 to 5 brown bands; 12.8mm; Punta Engaño, Cebu.
DISTRIBUTION: Sporadically found throughout the Philippines.

17. Auriculastra subula (Quoy and Gaimard, 1832)
10.0mm; Palawan.
Shell small, fusiform-ovate, with a tall spire, weakly channeled sutures and a large, somewhat cylindrical body whorl; surface smooth except for some obsolete axial growth striae; aperture narrowly elongate, outer lip thin with a central swelling, columella with a weak brown-coloured glaze and 3 denticles, the upper strongest, the lower 2 weak; pale greyish white in colour with brown banding; reaches 12mm.
DISTRIBUTION: Sporadically found throughout the Philippines.

18. Auriculastra tornatelliformis (Petit, 1842)
9.5mm; Cebu.
Shell small, fusiform-ovate, glossy, with a tall spire and a large, elongate body whorl; surface smooth except for some obsolete axial scratches; aperture narrow, outer lip thin, 2 to 3 lirae deep within, columella with 5 irregularly sized folds; pale cream or brown in colour or a combination of both; reaches 15mm.
DISTRIBUTION: Sporadically found throughout the Philippines.

Class Gastropoda
Subclass Pulmonata
Order Basommatophora
Superfamily Siphonariacea
Family SIPHONARIIDAE

A small group of herbivorous, air breathing gastropods which live under gravel or rocks near the high-tide mark in most tropical and temperate seas. The shells resemble those of the Patellidae and Acmaeidae in being small, cap-shaped with a tall spire, and sculptured with radial ribs; internally there is a horseshoe-shaped muscle scar impression terminating in a siphonal groove on the right hand side. Only a few Philippine species illustrated here.

19. Siphonaria laciniosa (Linné, 1758)
Syn.: *S. atrata* Quoy & Gaimard, 1833
29.2mm, 20.0mm; Palawan.
Shell medium sized, elongate-ovate, with a slightly eroded, somewhat depressed spire; sculptured with 15 to 16 major ribs which project beyond the periphery, numerous irregularly sized secondary riblets and weakly striated interspaces; the siphonal opening consists of 2 strong, non-fused ribs; external colour deep greyish brown with paler ribbing, internally bluish white with a brownish black muscle scar; reaches 35mm; this species has been incorrectly figured under a variety of names due to faulty interpretation of Linné's type. Linné's reference to Rumphius (Pl. 40, fig. c) is the sole guide to its identity and is illustrated here.
DISTRIBUTION: Throughout the Philippines.

20. Siphonaria sirius Pilsbry, 1894
31.0mm, 24.6mm; Palawan.
Superficially similar to the preceding species but flatter, and the apex always eroded; elongate-ovate in outline, with up to 24 major ribs and 3 to 4 weaker interstitial riblets; siphonal opening prominent and consisting of 2 strong ribs; external colour a dark blackish blue with paler ribs; internally the muscle-scar is a deep chocolate-brown and the margin black and white-stained; reaches around 35mm.
DISTRIBUTION: Throughout the Philippines.

21. Siphonaria javanica (Lamarck, 1819)
15.8mm, 18.1mm; Palawan.
An extremely distinctive and somewhat polypmorphic

species. The typical form (here illustrated) is small, subcircular in outline, with a tall, acute, centrally located apex; sculptured with up to 10 white-coloured, strongly elevated ribs and 2 or 3 black-and-white streaked secondary riblets; internally, the muscle scar is orange to dark brown and the periphery rayed in black and white; reaches around 20mm.
DISTRIBUTION: Throughout the Philippines.

22. Siphonaria javanica (Lamarck, 1819) form **sipho** Sowerby, 1824
26.6mm, 23.0mm; Palawan.
Originally suspected to be a good species, but the occurrence of intermediate forms in a large selection indicates it is no more than an extreme sculptural ecomorph; the shell is larger than the typical form, with up to 40 ribs of varying strengths; the ribs are all white, the interspaces black; internally the muscle scar is a light orange-brown with a wide white margin that is black-spotted at the periphery; can reach 35mm.
DISTRIBUTION: Throughout the Philippines.

Class Scaphopoda
Family DENTALIIDAE

A large group of small to medium sized carnivorous scaphopods inhabiting mud or sand in deep water and having a worldwide distribution. The shells are tubular, straight or curved, tapering at one end, and may be smooth or ribbed. An apical slit may or may not be present. Well represented in the Philippines, but most species are small, difficult to identify and unlikely to be encountered by most collectors.

1. Dentalium elephantinum Linné, 1758
90.0mm; Sulu Archipelago.
Shell moderately large, solid, curved, apex tapered with a notch; sculptured with 10 raised, equi-spaced longitudinal ribs, numerous weaker interstitial riblets and fine spiral scratches; green in colour, white towards the apex; reaches 100mm.
DISTRIBUTION: Sulu Sea.

2. Dentalium aprinum Linné, 1766
73.5mm; Sulu Archipelago.
Shell small to medium sized, slender, rather curved, greenish in colour, white towards the apex; sculptured with numerous slightly raised longitudinal ribs, somewhat variable in number (12 to 15), weaker interstitial riblets and very fine spiral scratches; reaches 80mm.
DISTRIBUTION: Sulu Sea.

3. Dentalium octangulatum Donovan, 1804
80.0mm; Palawan.
Shell medium sized, moderately curved and all white in colour; sculptured with 5 to 8 equi-spaced, elevated longitudinal ribs, and coarse axial scratches in the interspaces; shell tapering rapidly towards the apex, no apical slit; reaches 80mm.
DISTRIBUTION: Palawan and Sulu Sea.

4. Fissidentalium (Fissidentalium) vicdani Kosuge, 1981
111.6mm; Punta Engaño, Cebu.
Shell large, almost straight, some specimens curving; sculptured with numerous (30 to 35), closely-set, irregularly-sized longitudinal cords and overriding spiral threads forming ridges on the tops of the cords, apex with a short, deep groove, aperture ovate with a serrated margin; white in colour with a reddish brown stain towards the anterior aperture; reaches around 115mm.
DISTRIBUTION: Cebu.

5. Fissidentalium (Pictodentalium) formosum (Adams and Reeve, 1850)
77.0mm; Sulu Archipelago.
Shell medium sized, solid and moderately curved; sculptured with around 18 broad, equi-spaced, microscopically scabrous longitudinal cords, shell wall thick with a circular inner lip margin; pale brown in colour with irregular ring-like bands of varying hues of green, purple, and brown; reaches 90mm.
DISTRIBUTION: Palawan and Sulu Sea.

6. Fissidentalium (Pictodentalium) vernedei (Sowerby, 1860)
105.5mm; Palawan.
Shell large, solid and slightly curved; sculptured with numerous closely-spaced longitudinal cords which become almost smooth at the apertural end, and encircled with fine overriding spiral threads; an apical notch is present, and the shell is ovate in cross section; yellowish white in colour with or without brown banding; reaches 120mm.

PLATE NO. 81 1. *Pyramidella acus* (Gmelin) 2. *Pyramidella maculosa* Lamarck 3. *Pyramidella ventricosa* (Guerin) 4. *Otopleura auriscati;* (Holten) 5. *Otopleura nodicincta* (A. Adams) 6. *Milda cincta* (Reeve) 7. *Pyramidella teres* (A. Adams) 8. *Pyramidella terebellum* (Müller) 9. *Otopleura glans* (Reeve) 10. *Bulla ampulla* Linné 11. *Bulla vernicosa* (Gould) 12. *Ellobium aurisjudae* (Linné) 13. *Cassidula nucleus* (Gmelin) 14. *Pythia plicata* (Férrusac) 15. *Pythia scabraeus* (Linné) 16. *Melampus flavus* (Gmelin) form *nuxcastaneus* Kuroda 16b. form *fasciatus* Deshayes 17. *Auriculastra subula* (Quoy & Gaimard) 18. *Auriculastra tornatelliformis* (Petit) 19. *Siphonaria laciniosa* (Linné) 20. *Siphonaria sirius* Pilsbry 21. *Siphonaria javanica* (Lamarck) 22. *Siphonaria javanica* (Lamarck) form *sipho* Sowerby

PLATE 81

DISTRIBUTION: Palawan and Sulu Sea.

7. Antalis weinkauffi (Dunker, 1877)
79.8mm; Samar.
Shell medium sized, lightweight and moderately curved towards the apex; smooth with a waxy lustre and sculptured with weak axial ribs near the apex which has a small slit on the ventral side, aperture circular; shell pale yellowish white in colour; reaches 90mm.
DISTRIBUTION: Samar, Palawan and Sulu Sea.

8. Fustiaria nipponica (Yokoyama, 1922)
54.6mm; Palawan.
Shell small, lightweight, white in colour with a waxy lustre, surface smooth, aperture circular; easily identified by the long fracture-like slit on the ventral side, extending from the apex along almost 25% of the shell length; reaches 55mm.
DISTRIBUTION: Palawan and Sulu Sea.

9. Calliodentalium crocinum (Dall, 1907)
70.0mm; Punta Engaño, Cebu.
Shell medium sized, curved, moderately lightweight, with a smooth glossy surface; the aperture is circular, there is no apical notch, and the surface is encircled with very weak, almost obsolete spiral threads along the entire shell length; colour off-white to brown; reaches 75mm.
DISTRIBUTION: Cebu and Sulu Sea.

10. Antalis tibana (Nomura, 1940)
29.7mm; Sulu Archipelago.
Shell small, very weakly curved, pale brown in colour with a smooth surface, a circular aperture, and a short, rather deep notch on the ventral side of the apex, the body of the shell exhibits distinct spiral growth "bumps"; reaches 50mm.
DISTRIBUTION: Sulu Sea.

Class Gastropoda
Subclass Opisthobranchia
Order Cephalaspidea
Superfamily Acteonacea
Family ACTEONIDAE

A small group of vermivorous opisthobranchs, worldwide in distribution and living on coral or sand at all tidal levels. The shells are small and solid, with short, acute spires and elongate to elongate-ovate body whorls which are usually grooved. The aperture is narrow, rapidly widening anteriorly, and the columella is lightly glazed with 1 or 2 heavy folds. Fairly well represented in the Philippines.

11. Acteon variegatus (Bruguiére, 1789)
24.6mm, 24.8mm; Palawan.
Shell small, elongate-ovate, spire short and bluntly acute, sutures channeled, body whorl large, roundly elongate and sculptured with punctate spiral grooves; aperture narrow, rapidly widening anteriorly, columella weakly glazed, with one basal fold; white in colour with 3 rows of large, brown quadrate spots which occasionally fuse axially to form long flammules; reaches 27mm.
DISTRIBUTION: Sporadically found throughout the Philippines in limited quantity.

12. Pupa solidula (Linné, 1758)
24.2mm, 23.3mm; Bohol.
Shell small, solid, elongate-ovate and short-spired, sutures channeled, body whorl large, sculptured with shallow spiral grooves; aperture narrow and widening anteriorly; columella weakly glazed with a single posterior fold and a bifid anterior fold, white in colour with small red spots on cords between the grooves; reaches 30mm.
DISTRIBUTION: Sporadically found throughout the Philippines in limited quantity.

13. Pupa sulcata (Gmelin, 1791)
19.2mm, 20.5mm, 19.2mm; Samar.
Shell small, solid, similar to *P. solidula,* but is sculptured with shallower grooves; the anterior columellar fold is larger than in *solidula,* and fused, not bifid; greyish white in colour, occasionally yellowish, with black spots and streaks taking on a smudged appearance; reaches 25mm.
DISTRIBUTION: Sporadically found throughout the Philippines.

14. Pupa affinis (A. Adams, 1854)
13.2mm, 12.6mm, 14.2mm; Cebu.
Shell small, narrowly elongate-ovate, with a short acute spire; body whorl large, encircled with shallow spiral grooves, and with a peculiarly oblique left side on the base (prominent in the 3rd specimen); aperture narrow, widening anteriorly, columella lightly glazed with a single posterior and a single bifid anterior fold; pale brown in colour with 3 rows of reddish brown rhomboidal spots; reaches 20mm. Often synonymized with *P. alveola,* which is a more ovate shell, and lacks the peculiar oblique left side contour of *P. affinis.*
DISTRIBUTION: Sporadically found throughout the Philippines in limited quantity.

PLATE NO. 82 1. *Dentalium elephantinum* Linne 2. *Dentalium aprinum* Linné 3. *Dentalium octangulatum* Donovan 4. *Fissidentalium vicdani* Kosuge 5. *Fissidentalium formosum* (Adams & Reeve) 6. *Fissidentalium vernedei* (Sowerby) 7. *Antalis weinkauffi* (Dunker) 8. *Fustiaria nipponica* (Yokoyama) 9. *Calliodentalium crocinum* (Dall) 10. *Antalis tibana* (Nomura) 11. *Acteon variegatus* (Bruguiére) 12. *Pupa solidula* (Linné) 13. *Pupa sulcata* (Gmelin) 14. *Pupa affinis* (A. Adams) 15. *Acteon fabreanus* (Crosse) 16. *Acteon virgatus* (Reeve) 17. *Bullina lineata* (Gray) 18. *Atys naucum* (Linné) 19. *Haminoea fusca* (Pease) 20. *Atys cylindricus* (Helbling) 21. *Atys ooformis* (Habe) 22. *Hydatina albocincta* (Van der Höven) 23. *Hydatina aplustre* (Linné) 24. *Hydatina physis* (Linné)

PLATE 82

15. Acteon fabreanus (Crosse, 1874)
13.0mm; Punta Engaño, Cebu.
Shell small, elongate-ovate, spire moderately tall, sutures channeled, sculptured with punctate spiral grooves, columella with a single basal fold and a narrow umbilical opening; colour pale brown with lavender-coloured axial streaks, frequently broken up into 3 spiral bands; reaches 25mm.
DISTRIBUTION: Sporadically found throughout the Philippines in limited quantity.

16. Acteon virgatus (Reeve, 1842)
22.9mm, 24.5mm; Palawan.
Shell small with a short, acute spire, and a large, elongate-ovate body whorl; sculptured with punctate spiral grooves which become obsolete centrally; aperture narrow, widening anteriorly, columella with a single fold; white in colour with black axial bands; reaches 25mm.
DISTRIBUTION: Sporadically found throughout the Philippines in limited quantity.

17. Bullina lineata (Gray, 1825)
23.3mm, 22.5mm; Sulu Archipelago.
Shell small, elongate-ovate and slightly bulbous, spire short and slightly sunken, sutures well channeled; sculptured with numerous axially striated spiral grooves; aperture narrowly ovate, outer lip thin and arcuate, columella with a weak basal fold; white in colour with pink axial stripes and 2 dark red spiral bands; reaches 25mm.
DISTRIBUTION: Sporadically found throughout the Philippines in limited quantity.

Class Gastropoda
Subclass Opisthobranchia
Order Cephalaspidea
Superfamily Bullacea
Family HAMINOEIDAE

A small group of intertidal, sand-dwelling species, closely allied to the Bullidae. The shells are much thinner, translucent, white in colour and have a concealed, sometimes channeled spire. Only a few Philippine species presently known.

18. Atys (Atys) naucum (Linné, 1758)
38.5mm; Samar.
Shell medium sized, pure white in colour and extremely globose; sculptured with fine axial and spiral lines, spire sunken, umbilicus almost obsolete and obscured by a curving columellar fold, outer lip thin and arcuate, the posterior section detached from the shell and raised above the apex; reaches 40mm.
DISTRIBUTION: Sporadically found throughout the Philippines.

19. Haminoea fusca (Pease, 1863)
21.5mm; Sulu Archipelago.
Shell medium sized, globose and extremely fragile; surface smooth except for weak axial growth striae, spire sunken, outer lip thin and widely arcuate, columella with a thin callus band and curved anteriorly; pale greenish grey in colour with a white columella; reaches 25mm.
DISTRIBUTION: Sporadically found throughout the Philippines in limited quantity.

20. Atys (Aliculastrum) cylindricus (Helbling, 1779)
23.6mm, 24.0mm; Samar.
Shell medium sized, cylindrically ovate in outline with a sunken spire; the surface is essentially smooth, but fine axial scratches adorn the surface and spiral grooves encircle the shell on both the anterior and posterior ends; the outer lip is solid with the posterior margin raised above the apex, the umbilicus obsolete, and the columella has a short, thick fold; off-white in colour and reaching around 20mm in length.
DISTRIBUTION: Sporadically found throughout the Philippines.

21. Atys (Aliculastrum) ooformis (Habe, 1964)
16.2mm, 13.6mm; Cebu.
Shell small, ovate in outline with a sunken spire and a weak, rounded keel on the upper portion of the body whorl; the surface is smooth and encircled with grooves on the anterior and posterior ends; the outer lip is thin and arcuate, with the posterior margin raised above the apex, and the columella has a white fold; off-white in colour and reaching around 20mm, in length.
DISTRIBUTION: Sporadically found throughout the Philippines in limited quantity.

Class Gastropoda
Subclass Opisthobranchia
Order Cephalaspidea
Superfamily Acteonacea
Family APLUSTRIDAE

A small group of carnivorous opisthobranchs inhabiting shallow waters in most tropical and subtropical areas. The shell is medium sized, very thin, lightweight and extremely fragile, with well channeled sutures, a deeply sunken spire, and a large, globose body whorl. The aperture is large, widening anteriorly, and has a lightly calloused columella. Few Philippine species.

22. Hydatina (Hydatina) albocincta (van der Höven, 1839)
46.0mm; Manila Bay.
Shell medium sized, very light, spire sunken, body whorl smooth and extremely globose, aperture wide with a weak, white columellar callus; yellow-white in colour with 4 axially striped spiral bands; reaches 55mm.
DISTRIBUTION: Sporadically found throughout the Philippines in limited quantity.

23. Hydatina (Aplustrum) aplustre (Linné, 1758)
21.0mm, 20.5mm, 21.5mm, 22.3mm; Samar.
Shell small, lightweight, elongate-ovate in outline, sutures well channeled, spire low, almost sunken; aperture narrow, widening anteriorly, columella long, projecting and somewhat twisted anteriorly; shell creamy white with 2 pink bands bordered in black, aperture white; reaches 25mm.
DISTRIBUTION: Sporadically found throughout the Philippines in limited quantity.

24. Hydatina (Hydatina) physis (Linné, 1758)
30.0mm, 33.5mm, 32.7mm, 24.5mm; Palawan.
Shell small to medium sized, spire sunken, sutures deeply channeled, body whorl large and globose, columella moderately glazed and with a strong ridge anteriorly; colour creamy yellow with numerous close-set black or brown spiral lines, aperture white; reaches 40mm.
DISTRIBUTION: Sporadically found throughout the Philippines in limited quantity.

Plate 83 — Nautilidae, Spirulidae

Class Cephalopoda
Subclass Nautiloidea
Family NAUTILIDAE

A small group of cephalopods mostly confined to the Western Pacific regions. Frequently classed as living fossils, this group also includes cuttlefish, squid and octopi. The Nautilidae are the only family within the Cephalopod group to possess an external shell, and this is the main method of differentiating different species. The shells are fairly large, spirally coiled, have an involute spire, weak surface growth striae, an umbilicus which may be open or closed, and a wide aperture. Internally, the shells are gas filled, and by regulating the flow of air in and out of the chambers, the animal can control buoyancy. All members of this family are deep-water dwelling, carnivorous, and the sexes are separate. Overall, the family is well represented in the Philippines.

1. Nautilus repertus Iredale, 1944
215mm; Sulu Archipelago.
Considered by some to be a form of *N. pompilius.* It reaches a much larger size (to 250mm), and the colour bands only extend halfway from the periphery to the brown-coloured umbilicus.
DISTRIBUTION: Mostly from the Sulu Sea.

2. Nautilus pompilius Linné, 1758
This species reaches around 170mm in the mature stage, and is characterized by the lack of an umbilicus, and the orange-brown coloured flammules (numbering 20 to 24) which extend from the periphery to the umbilical area.
2a. typical specimen; 175mm; Bohol.
2b. albinistic specimen; 162mm; Bohol.
2c. dwarf specimen; 104mm; Mindoro.
2d. juvenile; 35mm; Marinduque.
DISTRIBUTION: Sporadically found throughout the Philippines.

3. Nautilus stenomphalus Sowerby, 1849
151mm; Mindoro.
This species has a small umbilical area (less than 15mm), and a very small umbilical channel which is slightly angulate; as in *N. repertus* Iredale, the colour bands only extend halfway across the body whorl; reaches around 150mm. Possibly another form of the extremely variable *N. pompilius* Linné, where the umbilicus has incompletely sealed.
DISTRIBUTION: Mostly from Mindoro and Bohol, always in very limited quantity.

4. Nautilus scrobiculatus Lightfoot, 1786
178mm; Sulu Archipelago.
A distinctive species characterized by a large, deeply channeled umbilicus, subangulate umbilical shoulders, a finely reticulate surface formed from intersecting concentric and radial growth striae (not always present), and the orange-coloured bands which may or may not extend to the umbilicus, and more frequently tend to coalesce on the center of the whorl; reaches 190mm.
DISTRIBUTION: Sulu Sea and Eastern Samar.

Class Cephalopoda
Subclass Coleoidea
Order Sepioidea
Superfamily Spirulacea
Family SPIRULIDAE

This family contains only one member. The shell is small, loosely coiled and has a perforate, chambered interior. It is found embedded in the posterior of a small squid-like animal.

5. Spirula spirula (Linné, 1758)
18.6mm, 19.3mm; Samar.
Shell small, with loose spiral coils that are partitioned, and increase in size towards the last chamber; exterior surface marked with slight concentric constrictions, corresponding to the divisions of the internal septae; reaches 35mm in diameter.
DISTRIBUTION: Sporadically found throughout the Philippines.

Class Cephalopoda
Subclass Coleoidea
Order Octopoda
Superfamily Argonautacea
Family ARGONAUTIDAE

The shells of this family are thin-walled, unchambered and serve as egg receptacles for the female argonautid octopus. Although seemingly closely related to the Nautilidae, the *Argonauta* are free-swimming and not attached to their shell as are the *Nautilus* species; sexes are separate, with the male seldom exceeding 25mm, while the female may reach 170mm. A small family, but well represented in the Philippines.

6. Argonauta argo Linné, 1758
258mm; Masbate.
Shell large, lightweight, rather thin, laterally compressed towards the apertural opening where 2 sharp, spinose projections are formed; the shell has a double keel bearing sharply pointed nodules, and the surface is covered with numerous bifurcate wavy ridges; shell cream in colour, early part of the keel, including the spines, dark brown to black; reaches 270mm.
DISTRIBUTION: Sporadically found throughout the Philippines.

7. Argonauta hians Lightfoot, 1786
This species is much smaller than *A. argo*, somewhat more bulbous, has fewer spinose nodules on the keels which are more widely spaced and have much coarser ridges; spinose projections may or may not be present; brownish in colour, darker on the keel and on the lateral margin; reaches 100mm.
7a. typical specimen; 100mm; Samar.
7b. form with spinose projections; 87mm; Samar.
DISTRIBUTION: Sporadically found throughout the Philippines.

8. Argonauta boettgeri Maltzan, in Kira, 1962
20.5mm; Palawan.
Shell small and extremely fragile, similar to *A. hians*, but the colouration is yellowish white and the tubercles on the keels are extremely prominent; reaches 30mm.
DISTRIBUTION: Palawan.

NOTATUM:
The species *Argonauta nodosa* Lightfoot, 1786, has been reported from the Philippines by many authors. As far as can be ascertained, there are at present no documented cases of the species having been taken in Philippine waters. Its occurrence is suspected, but not yet fully documented.

Arcidae — Plate 84

Class Bivalvia
Order Arcoida
Superfamily Arcacea
Family ARCIDAE

A large family of bivalves inhabiting sand or mud in shallow to deep water, attached to a rocky or coralline substrate by means of a byssus. The shells are small to medium sized, mostly equi-valve, inequilateral, with an inflated posterior surface, a somewhat angulate postero-dorsal surface, and sculptured with strong concentric and radial ridges. The shells are thick and may be triangular, quadrangular or ovate in shape, with broad, incurved umbones, an external ligament, and a porcelaneous, marginally dentate interior lacking a pallial sinus. Well represented in the Philippines, with only some of the more frequently encountered species being illustrated here.

1. Barbatia (Barbatia) amygdalumtosum (Röding, 1798)
32.5mm × 53.7mm; Samar.
Shell medium sized, solid, roundly trapezoidal with a long, somewhat flattened postero-dorsal margin, umbones anteriorly displaced and non-touching, valves moderately inflated and finely granulose from overriding radial and concentric lirae, hinge line long with fine teeth; external colouration a deep chocolate-brown, with 2 to 3 white rays radiating from the umbones, internally bluish white, with a flattened, brown-coloured margin; reaches 60mm in length.
DISTRIBUTION: Sporadically found throughout the Philippines.

2. Arca ventricosa Lamarck, 1819
46.0mm × 70.5mm; Palawan.
Shell medium sized, heavy, unusually distorted but roughly trapezoidal in outline; the umbones are large, inflated, incurved, anteriorly displaced and widely separated, and a strong ridge runs from the umbones to the postero-ventral margin; valves well inflated and sculptured with strong radial ribs and weaker concentric lirae, those ribs on the postero-dorsal slope strongest; the hinge line is long and finely toothed, and a large byssal gape is obvious on the ventral margin; external colouration is dark brown on the posterior slope with brown "zebra-like" stripes

PLATE NO. 83 1. *Nautilus repertus* Iredale 2a. *Nautilus pompilius* Linné 2b. albinistic 2c. dwarf specimen 2d. juvenile specimen 3. *Nautilus stenomphalus* Sowerby 4. *Nautilus scrobiculatus* Lightfoot 5. *Spirula spirula* (Linné) 6. *Argonauta argo* Linné 7a. *Argonauta hians* Lightfoot 7b. form with spinose projections 8. *Argonauta boettgeri* Maltzan, in Kira

PLATE 83

anteriorly, internally white with a large brown blotch; reaches 75mm in length.
DISTRIBUTION: Sporadically found throughout the Philippines.

3. Arca ventricosa Lamarck, 1819 (ecomorph)
33.4mm x 59.8mm; Iloilo.
Markedly different in outline to fig. 2, but appears to have very similar surface sculpturing and colouration; the ventral margin is less sinuous and the ridge from the umbonal area to the postero-ventral margin is weaker; members of this species appear to be extremely variable in shape, a fact which may account for the numerous synonyms applied to it.
DISTRIBUTION: Sporadically found throughout the Philippines.

4. Barbatia (Barbatia) foliata Forskäl, 1775
29.5mm x 42.1mm; Iloilo.
Extremely variable in shape and surface sculpturing; shell medium sized, solid, trapezoidal in outline with rounded ends, umbones large, inflated, non-touching, incurved and anteriorly displaced, a roundly angulate ridge runs from the umbones to the postero-ventral margin, valves moderately inflated, sculptured with numerous fine overriding radial and concentric lirae forming a finely granulose surface, a weak byssal gape on the ventral margin, internal margin minutely crenulate; shell colour pure white inside and out and covered with a dirty greenish brown periostracum; may reach 60mm in length. This is the typical form of the species which is characterized by its more prominent radial sculpturing.
DISTRIBUTION: Sporadically found throughout the Philippines.

5. Barbatia (Barbatia) foliata Forskäl, 1775 form **trapezina** Lamarck, 1819
41.7mm x 60.0mm; Iloilo.
In this very distinct form, the concentric sculpture is much more pronounced, resulting in a coarsely sculptured, roughly granulose surface; the concentric sculpturing can be so coarse that it appears as distinct, somewhat lamellose folds rather than lirae.
DISTRIBUTION: Sporadically found throughout the Philippines.

6. Trisidos semitorta (Lamarck, 1819)
58.7mm x 104.4mm; Palawan.
Shell large, heavy and inflated, sharply twisted on the posterior surface, umbones inflated, off-center and curved anteriorly, hinge line long and simple ligament internal; surface sculptured with concentric growth folds and fine radial riblets; white in colour; reaches 110mm in length.
DISTRIBUTION: Sporadically found throughout the Philippines.

7. Arca navicularis Bruguiére, 1789
24.8mm x 44.7mm; Samar.
Shell medium sized, valves inflated, umbones anteriorly displaced, incurved and widely separated, hinge line straight with numerous fine teeth, ligament external, surface sculptured with numerous primary and secondary radial cords crossed by heavy concentric folds; interior deep purple with blue-white blotches, posterior adductor muscle scar larger than the anterior scar, surface colouration bluish white with brown "zebra-like" stripes; reaches around 45mm in length.
DISTRIBUTION: Sporadically found thoughout the Philippines.

8. Andara maculosa (Reeve, 1844)
57.5mm x 83.2mm; Iloilo.
Shell large, heavy, roundly trapezoidal with a long postero-ventral margin, equi-valve with markedly inflated valves, umbones large, anteriorly displaced and widely separated by a deep, laterally striated V-shaped cleft, postero-dorsal surface somewhat flattened; sculptured with 38 to 40 radial ribs, fluted towards the ventral margin by overriding concentric folds, interior weakly lirate, outer lip margin coarsely crenulate; greyish white in colour, umbones stained black, interior white; reaches 90mm in length.
DISTRIBUTION: Sporadically found throughout the Philippines.

Class Bivalvia
Order Veneroida
Superfamily Solenacea
Family SOLENIDAE

A large group of compressed, cylindrically elongate shells living buried in mud or sand in most geographical areas. Valves gape at both ends which may be truncated or rounded, have small, anteriorly displaced umbones, a large external ligament, and a relatively smooth surface. The shells have a short pallial sinus, cardinal teeth which vary in number from 1 or 2 on the left valve, and 1 on the right valve. The lateral teeth are weak or absent, and the group lacks a byssal membrane. Several species recorded from the Philippines, only one of which is illustrated here.

9. Solen lamarckii Deshayes, 1839
11.3mm x 81.2mm; Manila Bay.
Shell medium sized, light, narrowly rectangular, with

PLATE NO. 84 1. *Barbatia amygdalumtosum* (Röding) 2. *Arca ventricosa* Lamarck 3. *Arca ventricosa* Lamarck — ecomorph 4. *Barbatia foliata* Forskäl 5. *Barbatia foliata* Forskäl form *trapezina* Lamarck 6. *Trisidos semitorta* (Lamarck) 7. *Arca navicularis* Bruguiére 8. *Andara maculosa* (Reeve) 9. *Solen lamarckii* Deshayes 10. *Phaxas cultellus* (Linné) 11. *Tellina rostrata* Linné 12. *Tellina perna* (Spengler) 13. *Tellina virgata* Linné 14. *Tellina foliacea* Linné 15. *Scutarcopagia linguafelis* (Linné) 16. *Scutarcopagia scobinata* (Linné) 17. *Cardites bicolor* (Lamarck) 18. *Cardita variegata* Bruguiére 19. *Solecurtus philippinensis* (Dunker)

PLATE 84

a rounded posterior and a truncated anterior margin, valves inflated and gaping at both ends, sculptured with concentric grown lamellae; purplish in colour with a yellow periostracum; reaches 100mm in length.
DISTRIBUTION: Sporadically found throughout the Philippines.

Class Bivalvia
Order Veneroida
Superfamily Solenacea
Family CULTELLIDAE

Often treated as a Subfamily of Solenidae, but the shells have wider valves, more centrally placed umbones, and the hinge plate usually has 3 cardinal teeth (2 in the left and 1 in the right). The habitat and distribution is similar to the members of the Solenidae. Only a few species illustrated (see also Pl. 86).

10. Phaxas (Ensiculus) cultellus (Linné, 1758)
21.0mm x 73.2mm; Palawan.
Shell small, lightweight, shiny and roundly elongate, dorsal margin straight, the ventral, anterior and posterior margins all curved, valves slightly inflated and gaping anteriorly, surface smooth and encircled with concentric growth lamellae; marbled with white, brown and lavender; reaches 80mm in length.
DISTRIBUTION: Sporadically found throughout the Philippines.

Class Bivalvia
Order Veneroida
Superfamily Tellinacea
Family TELLINIDAE

A large, taxonomically confusing family, worldwide in distribution, and inhabiting tropical and sub-tropical seas. The shells are small to medium sized, ovate to elongate, inequilateral, nearly equivalve, have an external ligament and a flexure or ridge on the posterior margin. The umbones are low and small, the margins smooth, the pallial sinus deep, and the hinge plate narrow, with 2 occasionally bifid cardinal teeth, and an additional 2 lateral teeth on both valves. Fairly well represented in the Philippines. (Note: All members of this family have been illustrated using the right valve, so that the posterior margin appears on the observer's left, not the right as is the usual convention).

11. Tellina (Pharaonella) rostrata Linné, 1758
20.5mm x 54.0mm; Palawan.
Shell small, lightweight, flattened and elongate-ovate in outline; anterior margin rounded, posterior margin produced, forming a long rostration, furrowed on the left valve and channeled on the right forming an interlocking system; surface smooth, umbones weak and posteriorly displaced, ligament external; pink to orange in colour with yellow streaks, internally bright pink; reaches 65mm in length.
DISTRIBUTION: Sporadically found throughout the Philippines in limited quantity.

12. Tellina (Pharaonella) perna (Spengler, 1798)
27.5mm x 49.7mm; Samar.
Shell medium sized, lightweight and quadrangular in outline; anterior margin rounded, posterior margin extended into a short rostration bearing 2 ridges; inequivalve, umbones small, posteriorly displaced and touching; surface smooth except for weak concentric ridges; pink to orange in colour with white umbonal rays; reaches 80mm in length.
DISTRIBUTION: Sporadically found throughout the Philippines.

13. Tellina (Tellinella) virgata Linné, 1758
35.4mm x 55.0mm; Bohol.
Shell medium sized, lightweight, triangularly ovate in outline, with a rounded anterior margin and a slightly projected posterior margin, furrowed on the left valve and channeled on the right forming an interlocking system; sculptured with numerous fine radial cords, umbones small, posteriorly displaced and almost touching; external colour pink with white umbonal rays, interior pink with a yellow umbonal area; reaches 60mm in length.
DISTRIBUTION: Throughout the Philippines.

14. Tellina (Phylloda) foliacea Linné, 1758
38,5mm x 71.7mm; Palawan.
Shell medium sized, lightweight, transparent and ovate-quadrangular in outline; anterior margin long, rounded and slightly gaping and the posterior margin roundly truncated, a sharp ridge running from the umbones to the postero-ventral margin; anterior margin finely dentate, surface sculptured with fine concentric striae; the ligament is external, and the umbones are nearly obsolete; pure yellowish orange in colour; reaches 80mm in length.
DISTRIBUTION: Sporadically found throughout the Philippines in limited quantity.

15. Scutarcopagia linguafelis (Linné, 1758)
36.4mm x 48.5mm; Mindoro.
Shell medium sized, lightweight, somewhat compressed and roundly ovate with a ridge on the postero-dorsal surface; sculptured with fine, closely-set scabrous nodules over the entire surface; umbones incurved and posteriorly displaced, lunule deeply excavated, ligament external; white in colour with a few lavender streaks radiating from the purple stained umbones; reaches 50mm in length.
DISTRIBUTION: Sporadically found throughout the Philippines in limited quantity.

16. Scutarcopagia scobinata (Linné, 1758)
41.2mm x 43.7mm; Iloilo.
Close to the previous species but the shell is rounder, the umbones more centrally placed, the rows of scabrous nodules are more loosely arranged becoming

denser towards the umbonal area, and the lunule is deeply excavated into the right valve; pale brown in colour with brown spotting, interior white with a yellow hue; reaches 65mm in length.
DISTRIBUTION: Sporadically found throughout the Philippines in limited quantity.

Class Bivalvia
Order Veneroida
Superfamily Carditacea
Family **CARDITIDAE**

A large group of shallow to deep water bivalves which inhabit most world seas and attach themselves to the undersides of rocks by means of a byssus. The shells are small to large, thick, solid, and rounded or trapezoidal in shape with strong radial ribs and a crenulate margin. The shells are equivalve, inequilateral, have an external ligament, no pallial sinus, a large posterior adductor muscle scar and a somewhat elevated anterior adductor muscle scar. The hinge line has 1 to 2 cardinal and 1 to 2 lateral teeth. Fairly well represented in the Philippines, with only a few of the more common species being illustrated here.

17. Cardites bicolor (Lamarck, 1819)
31.5mm x 45.0mm; Mindoro.
Shell medium sized, solid, somewhat trapezoidal in outline, sculptured with around 24 flattened, weakly scabrous radial ribs and deep interspaces; interior margin weakly crenulate; shell white in colour with brown banding on the ribs, internally white; reaches 45mm in length.
DISTRIBUTION: Throughout the Philippines.

18. Cardita variegata Bruguiére, 1792
14.5mm x 22.5mm; Samar.
Shell small, rectangularly ovate, umbones off-center and almost above the anterior margin, valves inflated, with a sloping anterior margin; sculptured with scaled radial ribs which are weaker on the anterior slope and encircled with concentric growth striae; white in colour with purplish spotting on the ribs; reaches 30mm in length.
DISTRIBUTION: Sporadically found throughout the Philippines.

Class Bivalvia
Order Veneroida
Superfamily Tellinacea
Family **SOLECURTIDAE**

A small group of "*Solen*-like" shells living in sand and mud in the shallow to deep waters of most world seas. The shells are thin, medium to large sized, and cylindrically elongate with rounded, gaping ends. The valves may be short or long, and the hinge plate is narrow with 3 cardinal teeth (2 on the right valve, 1 on the left) and no laterals. Few Philippine species.

19. Solecurtus philippinensis (Dunker, 1861)
20.8mm x 43.7mm; Palawan.
Shell small, lightweight, rectangular in outline with rounded, gaping anterior and posterior ends, umbones small, anteriorly displaced, and with an external ligament; surface sculptured with incised ribs radiating from the postero-dorsal margin, and crossed by concentric growth folds; shell pinkish in colour with white umbones and 2 white umbonal rays; interior bluish white with an orange margin; reaches 50mm in length.
DISTRIBUTION: Sporadically found throughout the Philippines.

Plate 85 Veneridae

Class Bivalvia
Order Veneroida
Superfamily Veneracea
Family **VENERIDAE**

The largest bivalve family, the members of which are medium sized, inhabit muddy or sandy water just below the surface, and live in shallow to deep water in most geographical provinces. The shells are thick, ovate to heart-shaped, equivalve, and have either a smooth or concentrically sculptured surface. The umbones are displaced and anteriorly curved, the ligament is external, the lunule and escutcheon well developed. There are 3 cardinal teeth, a prominent pallial sinus, and a byssal membrane is lacking. The family is extremely well represented in the Philippines.

1. Katelysia hiantina (Lamarck, 1818)
40.0mm x 53.5mm; Manila Bay.
Shell medium sized, heavy, equivalve, inequilateral, somewhat ovate-rhomboidal in shape, with swollen, touching, anteriorly displaced umbones, escutcheon long and excavated, ligament external; valves inflated and encircled with concentric growth striae, adductor muscle scars equi-sized, pallial sinus deep, a slightly angular ridge extending from the umbones to the postero-ventral margin; off-white in colour with brown streaks and markings, interior white; reaches 55mm in length.
DISTRIBUTION: Throughout the Philippines.

2. Lioconcha castrensis (Linné, 1758)
41.0mm x 44.2mm; Cebu.
Shell small to medium sized, roundly ovate, ventricose, with swollen umbones which are anteriorly displaced and curved; sculptured with fine concentric striae; pale cream in colour with brown "V-shaped" markings, concentrically arranged, interior white; reaches 60mm in length.
DISTRIBUTION: Sporadically found throughout the Philippines.

3. Paphia schnelliana (Dunker, 1866)
37.0mm x 59.3mm; Palawan.
Shell medium sized, lightweight, triangularly ovate in outline and rather flattened; antero-dorsal margin concave, shorter than the postero-dorsal margin which is straight, umbones inflated, smooth, rounded, off-center and anteriorly displaced, sculptured with numerous concentric grooves; light orange in colour with darker lines and blotches, interior white with a pinkish hue; reaches up to 100mm in length.
DISTRIBUTION: Sporadically found throughout the Philippines.

4. Lioconcha fastigiata (Sowerby, 1851)
24.7mm x 29.5mm; Sulu Archipelago.
Shell small, inflated, triangularly ovate with prominent, obliquely incurved, anteriorly displaced umbones, surface with fine concentric growth striae; white in colour with small and large dark triangular blotches, interior white with brown blotching; reaches 35mm in length. Similar to *L. castrensis* but smaller and more triangulate, with coarser concentric grooving and a brown-blotched interior.
DISTRIBUTION: Sporadically found throughout the Philippines in limited quantity.

5. Periglypta reticulata (Linné, 1758)
80.5mm x 85.4mm; Sulu Archipelago.
Shell large, heavy, ventricose, trapezoidal in outline; sculptured with numerous radial ribs and dense overriding concentric cords forming a highly reticulate, sharply nodulose surface; umbones swollen, off-center and anteriorly curved, lunule small but elevated, margin finely crenulate; cream coloured with brownish streaks, interior pale cream, hinge area orange; reaches 95mm in length.
DISTRIBUTION: Sporadically found throughout the Philippines.

6. Tapes sulcarius (Lamarck, 1818)
42.8mm x 71.5mm; Bohol.
Shell medium sized, somewhat flattened, ovately quadrangular in outline, with inflated valves and touching, anteriorly displaced umbones; a ridge runs from the umbones to the postero-ventral margin, and the surface is sculptured with fine concentric striae; orange-brown in colour with darker "V-shaped" markings, interior white with a pale yellow central area; reaches 85mm in length.
DISTRIBUTION: Sporadically found throughout the Philippines.

7. Katelysia ferruginea (Reeve, 1864)
28.9mm x 31.5mm; Iloilo.
Shell medium sized, triangularly ovate in outline with large, anteriorly displaced umbones and a weakly angulate ridge running from the umbones to the postero-ventral margin; valves inflated and sculptured with numerous concentric ribs, adductor scars equal, pallial sinus prominent; surface colouration brown with white maculations, interior bluish white with lavender-coloured lateral teeth, umbones streaked with white; reaches 40mm in length.
DISTRIBUTION: Sporadically found throughout the Philippines.

8. Placamen tiara (Dillwyn, 1817)
22.0mm x 23.0mm; Palawan.
Shell small, trigonally ovate with a roundly ventral margin, a large external ligament and a smooth, deep, ovate lunule, escutcheon smooth and deeply excavated with a sharply angulate shoulder, umbones inflated and anteriorly curved; sculptured with "frilled" concentric lamellae and smooth interspaces; exterior white in colour with darker rays, interior white with a pale lavender adductor scar impression; reaches 20mm in length.
DISTRIBUTION: Sporadically found throughout the Philippines.

9. c.f. Circe scripta (Linné, 1758)
34.8mm x 40.2mm, 31.5mm x 34.4mm; Palawan.
Very close to the typical form (fig. 12), but is more roundly trigonal and lacks the roundly truncated antero-ventral margin of that species; the surface has a weakly divaricate structure, most pronounced on the anterior and posterior umbonal margins; colour white with either brown, concentric zigzag lines, large brown blotches and brown spotting, or lines anterior and posterior to the umbones; close to the species *C. intermedia* Reeve, 1863, but that species has more prominent radial sculpturing; reaches around 40mm in length.
DISTRIBUTION: Sporadically found throughout the Philippines. Mostly from Sulu Sea.

10. Gafrarium aequivocum (Holten, 1803)
27.0mm x 31.8mm; Iloilo.
Shell small to medium sized, trigonally ovate in outline, umbones inflated, touching and anteriorly curved and displaced; surface finely granulose from overriding concentric lirae and radial cords which diverge anteriorly and posteriorly and have their line of divergence nearly central, interior with a finely crenulate margin; surface colour pale bluish white with dark purplish brown radial lines, interior porcelaneous white with some brown spotting below the umbonal area and mauve staining along the hinge line; reaches 35mm in length.
DISTRIBUTION: Sporadically found throughout the Philippines.

PLATE NO. 85 1. *Katelysia hiantina* (Lamarck) 2. *Lioconcha castrensis* (Linné) 3. *Paphia schnelliana* (Dunker) 4. *Lioconcha fastigiata* (Sowerby) 5. *Periglypta reticulata* (Linné) 6. *Tapes sulcarius* (Lamarck) 7. *Katelysia ferruginea* (Reeve) 8. *Placamen tiara* (Dillwyn) 9. *c.f. Circe scripta* (Linné) 10. *Gafrarium aequivocum* (Holten) 11. *Meretrix meretrix* (Linné) 12. *Circe scripta* (Linné) 13. *Asaphis violascens* (Forskål) 14. *Gari elongata* (Lamarck) 15. *Gari amethystus* (Wood) 16. *Glossus vulgaris* (Reeve) 17. *Glycymeris amboinensis* (Gmelin) 18. *Glycymeris reevei* (Mayer) 19. *Mactra mera* Reeve

PLATE 85

11. Meretrix meretrix (Linné, 1758)
41.0mm × 48.5mm; Iloilo.
Shell medium sized, glossy, roundly trigonal, posterior margin longest and somewhat flattened, umbones large, inflated, anteriorly displaced and nearly touching; surface smooth except for weak concentric folds; creamy white in colour with yellowish brown umbones and a purplish coloured posterior slope, shell sometimes with brown surface mottling, interior white; reaches 60mm in length.
DISTRIBUTION: Sporadically found throughout the Philippines in limited quantity.

12. Circe scripta (Linné, 1758)
39.7mm × 42.5mm; Sulu Archipelago.
Shell medium sized, roundly triangulate and rather flattened with a weak ridge from the umbones to the postero-ventral margin, umbones nearly central, flattened and touching; sculptured with numerous concentric ridges, obsolete on the umbones and strongest towards the ventral margin; colour pale yellowish white with brown flecks on the ridges, occasionally radial bands and brown stripes on either side of the umbones, internally white with brown blotching; reaches 45mm in length.
DISTRIBUTION: Sporadically found throughout the Philippines in limited quantity.

Class Bivalvia
Order Veneroida
Superfamily Tellinacea
Family PSAMMOBIIDAE

A moderately large group of mud or sand-dwelling bivalves inhabiting the intertidal zones of most temperate and tropical seas. The shells are medium sized, elongate, roughly rectangular in outline, with moderately inflated valves and gaping ends. The valves are inequilateral, equivalve or inequivalve, with small umbones, a large external ligament, and are sculptured with weak radial and/or concentric growth striae. One to four cardinal teeth with laterals that are weak or absent. Pallial sinus large. Only a few species illustrated here (see also Pl. 86).

13. Asaphis violascens (Forskäl, 1775)
43.0mm × 64.5mm; Palawan.
Shell medium sized, solid, elongate and somewhat ventricose in outline, umbones large and anteriorly displaced, ligament external, escutcheon deep and gaping posteriorly; sculptured with close-set radial cords, overriding concentric grooves and lamellae on the posterior margin, adductor muscle scars prominent; externally cream with radiating purple rays, internally white with a mauve hinge line; reaches 65mm in length.
DISTRIBUTION: Sporadically found throughout the Philippines. Mostly from Sulu Sea.

14. Gari (Gabraeus) elongata (Lamarck, 1818)
23.1mm × 44.9mm; Palawan.
Shell small, light, elongate with small, anteriorly displaced umbones, valves unequal and gaping posteriorly; surface smooth except for some fine concentric growth striae and a slight ridge on the posterior surface; pale bluish white in colour with deep purple rays; reaches 50mm in length.
DISTRIBUTION: Sporadically found throughout the Philippines. Mostly from Sulu Sea.

15. Gari (Gari) amethystus (Wood, 1815)
17.6mm × 36.0mm; Sulu Archipelago.
Shell small, light, flattened, elongate with a rounded anterior margin and a truncated posterior margin, umbones weak and centrally placed, ligament external; sculptured with concentric striae and radial lines radiating from the umbones, stopping suddenly at a ridge running from the umbones to the postero-ventral margin; shell pale pink-lavender in colour with darker radiating bands; reaches 40mm in length.
DISTRIBUTION: Sporadically found throughout the Philippines. Mostly from Sulu Sea.

Class Bivalvia
Order Veneroida
Superfamily Glossacea
Family GLOSSIDAE

A small group of distinctly shaped shells living in sand or mud in fairly deep water. The shells are small to large, globose, inequilateral, equivalve, with large, anteriorly displaced in-rolled umbones, a smooth or concentrically striated surface, an external ligament, a weak or nearly obsolete pallial sinus, and a thin hinge line bearing 2 cardinal teeth per valve and lateral teeth which vary in number and placement. Few Philippine species.

16. Glossus (Meiocardia) vulgaris (Reeve, 1845)
25.0mm × 27.2mm; Tayabas Bay, Quezon.
A uniquely sculptured shell, rather small and inflated, with large, anteriorly incurved umbones and a sharply angled keel from the umbones to the postero-ventral margin, surface sculptured with coarse concentric ribs; colour creamy white with occasional brown blotches; reaches 50mm in length. Similar to *G. moltkiana* (Spengler, 1783) (not illustrated) which is smaller and has fewer, coarser ribs which are serrate at the keel.
DISTRIBUTION: Camotes Sea, Manila Bay and Tayabas Bay, Quezon (deep water).

Class Bivalvia
Order Arcoida
Superfamily Limopsacea
Family GLYCYMERIDIDAE

A large group of shallow water bivalves with a worldwide distribution. The shells are thick, porcelaneous, nearly circular in shape, equivalve, equilateral, smooth or with radial ribs, and have centrally placed, incurved umbones. The hinge plate is broad and strongly arched, with a row of small, transverse or chevron-shaped blunt teeth. Internally, the valves have a crenulate outer margin and 2 adductor muscle scars, the largest anteriorly situated. The family is moderately well represented in the Philippines.

17. Glycymeris (Tucetona) amboinensis (Gmelin, 1791)
51.5mm × 51.3mm; Samar.
Shell medium sized, heavy, rather inflated, umbones nearly central, incurved and non-touching; shell roundly ovate and sculptured with numerous heavy, rather flattened radial ribs which are crossed by concentric growth striae, ligament partially internal, hinge teeth simple, inner margin with "V-shaped" indentations; white in colour with brown concentric zigzag lines, internally white with a brown anterior margin; reaches around 55mm in length. Commonly but erroneously called *Glycymeris (Tucetona) auriflua* (Reeve, 1843), but that species belongs to the subgenus *Axinactis* Mörch, 1861 and has anteriorly curved umbones situated near the end of the ligaments; the illustrated species belongs to the subgenus *Tucetona* Iredale, 1931 which has central umbones as well as the strong radial ribs common to both subgenera.
DISTRIBUTION: Sporadically found throughout the Philippines.

18. Glycymeris (Veletuceta) reevei (Mayer, 1868)
36.8mm × 36.5mm; Iloilo.
Shell medium sized, ovate in outline, valves inflated, umbones central, incurved and non-touching, anterior margin angulate; surface sculptured with flat radial ribs and concentric growth striae, internal margin coarsely serrate; exterior brown in colour with a white anterior margin, interior white with a large brown blotch; reaches 40mm in length.
DISTRIBUTION: Sporadically found throughout the Philippines.

Class Bivalvia
Order Veneroida
Superfamily Mactracea
Family MACTRIDAE

A large group of bivalves inhabiting sandy, shallow water in most world seas. The shells are medium sized to large, thin, elongate or triangularly ovate in outline, equivalve, inequilateral and with moderately inflated porcelaneous valves which may be smooth or concentrically sculptured and have gaping ends. External ligament small or absent, umbones sub-central and anteriorly bent. There is a single cardinal tooth on the left valve, two on the right valve. The laterals are usually present and the shells have a deep pallial sinus and lack a byssus. Moderately well represented in the Philippines.

19. Mactra mera Reeve, 1854
56.0mm × 75.5mm; Batangas.
Shell medium sized, triangular in outline, valves inflated, umbones prominent, non-touching and anteriorly displaced, escutcheon large and prominently lirate; a weak ridge runs from the umbones to the antero-ventral margin, and a second stronger ridge to the postero-ventral margin; surface smooth with concentric growth lamellae; greyish brown in colour with darker concentric bands and white radial streaks, interior brown with a bluish white central area; reaches 90mm in length.
DISTRIBUTION: Sporadically found throughout the Philippines.

Plate 86 Limidae

Class Bivalvia
Order Mytiloida
Superfamily Limacea
Family LIMIDAE

A moderately large family of medium to large sized bivalves living under rocks and coral or in crevices and found in most world seas. The shells are thin, obliquely ovate to elongate and have a long, gaping anterior margin. Equivalve, inequilateral with small equi-sized ears, and sculptured with radial ribs which are smooth or bear spines or erect scales. The ligament is internal with a triangular pit below the umbones, the hinge line is straight but edentulate and there is a single weak, centrally placed adductor muscle scar. Fairly well represented in the Philippines.

1. Acesta rathbuni (Bartsch, 1913)
152mm × 118mm; Bohol.
Shell large, subcircular to elongate-ovate, inflated and semi-transparent, umbones large, inflated and anteriorly displaced, ears unequal, valves slightly gaping on the anterior margin; the surface is smooth with concentric growth lamellae and the ligament is internal; surface colour pale yellow with darker yellow concentric rings, internally cream coloured; reaches 160mm in length.
DISTRIBUTION: Bohol.

2. Ctenoides philippinarum (Masahito & Habe, 1972)
33.0mm × 31.5mm; Panglao, Bohol.
Shell small to medium sized, lightweight, semi-transparent, sub-orbicular in outline with moderately inflated valves, umbones large, inflated, centrally placed and non-touching, ears equal; sculptured with weak radial grooves and partially obsolete concentric lamellae which are clearly visible on the inside surface; shell pure white in colour with a slight bluish tinge; reaches around 40mm in length.
DISTRIBUTION: Bohol.

3. Lima fujitae Oyama, 1943
14.5mm x 12.2mm; Bohol.
Shell small, ovate-quadrangular in outline, valves moderately inflated with a long, straight-sided anterior margin and a curved ventral margin; sculptured with up to 30 strong radial ribs bearing short, sharp spines towards the ventral margin; umbones central and widely separated, posterior ear largest, anterior ear almost obsolete; pure white in colour and reaches around 20mm in length.
DISTRIBUTION: Cebu and Bohol.

4. Ctenoides ales (Finlay, 1927)
68.4mm x 56.0mm; Cebu.
Shell medium sized, solid, elongate-ovate in outline, equivalve, moderately inflated with centrally placed, non-touching umbones, ligamental area deeply excavated, ears unequal, the anterior ones largest and gaping; sculptured with numerous fine, scabrous radial riblets which form a finely serrated margin; colour pale yellow, internally white; reaches 100mm in length.
DISTRIBUTION: Camotes Sea.

5. Limaria orientalis (Adams & Reeve, 1850)
25.9mm x 18.9mm; Cebu.
Shell small, lightweight, semi-transparent, elongate-ovate in outline, umbones moderately inflated, almost smooth, centrally placed but anteriorly curved; sculptured with numerous finely scabrous radial ribs and a few obsolete concentric growth lamellae: shell pure white, internally procelaneous white; reaches around 30mm in length.
DISTRIBUTION: Camotes Sea.

6. Lima lima vulgaris (Link, 1807)
58.5mm x 45.1mm; Nasasa Bay, Zambales.
Shell medium sized, solid, assymetrical with a long, straight-sided anterior slope, a shorter posterior margin and a rounded ventral margin; umbones central, incurved, pointed and separated by a triangular ligamental depression, valves slightly gaping posteriorly; surface sculptured with numerous rounded, lamellate radial ribs, heaviest towards the posterior margin; external colour a dirty creamy white, internally porcelaneous white with a coarsely crenulate margin; reaches 70mm in length.
DISTRIBUTION: Sporadically found throughout the Philippines.

Family ARCIDAE (cont'd.) (see also Pl. 84)

7. Anadara (Tegillarca) granosa (Linné, 1758)
24.0mm x 30.0mm; Iloilo.
Shell medium sized, solid, laterally elongate-ovate with large, inflated, anteriorly displaced umbones; sculptured with up to 26 coarsely granular radial ribs, a finely toothed hinge line and a coarsely serrated inner margin; colour pale brown with darker concentric bands; reaches around 35mm in length.
DISTRIBUTION: Sporadically found throughout the Philippines.

8. Anadara (Potiarca) pilula (Reeve, 1843)
30.2mm x 27.5mm; Iloilo.
Shell medium sized, solid, ovate-quadrangular in outline with large, heavily inflated, strongly incurved umbones; sculptured with around 26 finely granular radial ribs, a finely serrated hinge line and a coarsely crenulate inner margin; external colour off-white with darker blotching, internally white with a pale orange central area; reaches 35mm in length.
DISTRIBUTION: Sporadically found throughout the Philippines.

9. Barbatia (Acar) congenita (E. A. Smith, 1885)
9.1mm x 18.3mm; Cebu.
Shell small, trigonally-ovate, moderately inflated with a truncated posterior margin, a rounded anterior margin and a ridge running from the anteriorly diplaced umbones to the postero-ventral margin; sculptured with intersecting concentric and radial ribs giving the surface a clathrate appearance; colour orange-brown with a white band from the umbones to the ventral margin, internally white; reaches around 20mm in length.
DISTRIBUTION: Sporadically found throughout the Philippines.

Class Bivalvia
Order Veneroida
Superfamily Corbiculacea
Family CORBICULIDAE

A relatively small group of bivalves inhabiting sand or mud in either fresh or brackish water. The shells are moderately large, triangularly ovate in outline, equivalve, inequilateral with large, occasionally eroded umbones which are anteriorly displaced. The valves are moderately inflated and sculptured with concentric ridges. Internally, there are three cardinal teeth and a long anterior and postero-lateral tooth. Few Philippine species.

10. Polymesoda (Geloina) coaxans (Gmelin, 1791)
Syn: *Geloina papua* (Lesson, in Habe, 1964)
51.4mm x 54.9mm; Cavite.
Shell medium sized, trigonally ovate in outline and

PLATE NO. 86 1. *Acesta rathbuni* (Bartsch) 2. *Ctenoides philippinarum* (Masahito & Habe) 3. *Lima fujitai* Oyama 4. *Ctenoides ales* (Finlay) 5. *Limaria orientalis* (Adams & Reeve) 6. *Lima lima vulgaris* (Link) 7. *Anadara granosa* (Linné) 8. *Anadara pilula* (Reeve) 9. *Barbatia congenita* (E. A. Smith) 10. *Polymesoda coaxans* (Gmelin) 11. *Phaxas hanleyi* (Dunker) 12. *Placamen calophylla* (Philippi) 13. *Anomalocardia squamosa* (Linné) 14. *Phapia textile* (Gmelin) 15. *Dosinia troscheli* (Lischke) 16. *Gafrarium tumidum* (Röding) 17. *Pitar philippinarum* (Hanley, in Sowerby) 18. *Lioconcha trimaculata* (Lamarck) 19. *Pitar affinis* (Gmelin) 20. *Periglypta chemnitzii* (Hanley) 21. *Tapes literatus* (Linné) 22. *Pitar sp.* 23. *Timoclea costellifera* (Adams & Reeve) 24. *Dosinia iwakawai* Oyama & Habe 25. *Tellina staurella* (Lamarck) 26. *Gari occidens* (Gmelin)

PLATE 86

sculptured with concentric growth lamellae, umbones swollen, eroded, non-touching and anteriorly displaced pallial sinus, escutcheon and lunule all absent, 3 cardinal and 2 lateral teeth; chalky white with a dirty green periostracum; reaches around 60mm in length.
DISTRIBUTION: Sporadically found throughout the Philippines.

Family CULTELLIDAE (cont'd.) (see also Pl. 84)

11. Phaxas (Phaxas) hanleyi (Dunker, 1861)
31.5mm × 122mm; Manila Bay.
Shell medium sized, solid, elongate with rounded, gaping anterior and posterior ends, umbones small, anteriorly displaced (approximately one-fourth of the way along the dorsal margin) and touching; surface moderately smooth with numerous concentric growth lamellae, interior smooth with irregular calcareous deposits especially towards the posterior margin; external colour a vitreous white, internally porcelaneous white, periostracum a dirty brownish green; may reach 150mm in length.
DISTRIBUTION: Sporadically found throughout the Philippines.

Family VENERIDAE (cont'd.) (see also Pl. 85)

12. Placamen calophylla (Philippi, 1836)
29.8mm × 23.9mm; Iloilo.
Shell medium sized, roundly trigonal in outline with a short, sinuous anterior margin and a long, arcuate posterior margin, umbones moderately inflated, anteriorly displaced and curved; sculptured with up to 10 sharp, raised concentric ridges and numerous weaker intermediate striae; colour off-white; reaches around 30mm in length. The lamellae in the illustrated specimen are somewhat eroded.
DISTRIBUTION: Sporadically found throughout the Philippines.

13. Anomalocardia (Anomalodiscus) squamosa (Linné, 1758)
14.2mm × 18.5mm; Iloilo.
Shell small, trigonally ovate in outline with a short anterior margin and a long, rounded posterior margin; sculptured with numerous radial and concentric cords producing a moderately granulose surface; a ridge extends from the umbones to the postero-ventral margin; shell dirty white externally and greyish white internally; may reach 30mm in length.
DISTRIBUTION: Sporadically found throughout the Philippines.

14. Paphia (Neotapes) textile (Gmelin, 1791)
30.0mm × 52.0mm; Masbate.
Shell medium sized, triangularly elongate-ovate in outline with a moderately smooth, highly glossy surface, umbones weakly inflated, anteriorly displaced and the surface has some obsolete concentric grooves; colour pinkish brown with numerous dark concentric zigzag lines, hinge line with widely spaced red lines, interior white; may reach 80mm in length.
DISTRIBUTION: Sporadically found throughout the Philippines.

15. Dosinia (Phacosoma) troscheli (Lischke, 1873)
16.5mm × 16.7mm; Gasan, Marinduque.
Shell small, circular in outline, valves moderately inflated and sculptured with numerous fine concentric lamellae, umbones inflated, anteriorly displaced and with a prominent lunular plate; colour dirty white with darker radial rays; this species can reach around 50mm in length, but seldom appears to exceed 20mm from this locality.
DISTRIBUTION: Sporadically found throughout the Philippines.

16. Gafrarium tumidum (Röding, 1798)
23.9mm × 26.0mm; Iloilo.
Shell small, solid and irregularly ovate in outline; sculptured with coarsely granulose ribs which become bifurcate towards the ventral margin, the ribs on the anterior slope running obliquely to the main ribs; interior with a deep concavity and a coarsely crenulate margin; colour off-white, occasionally with irregular brownish areas, lunule purple-stained, interior white, the posterior margin often lavender-stained; reaches up to 50mm in length.
DISTRIBUTION: Sporadically found throughout the Philippines.

17. Pitar (Pitarina) philippinarum (Hanley, in Sowerby, 1851)
16.1mm × 19.4mm; Buenavista, Marinduque.
Shell small to medium sized, roundly trigonal in outline with moderately inflated, anteriorly displaced umbones; sculptured with numerous raised concentric ridges; external colouration pale brownish white with darker axial lines and streaks, interior off-white, lavender under the umbones; may reach up to 30mm in length.
DISTRIBUTION: Sporadically found throughout the Philippines.

18. Lioconcha trimaculata (Lamarck, 1818)
17.8mm × 20.4mm; Buenavista, Marinduque.
Shell small, trigonally ovate, with moderately inflated, anteriorly displaced umbones; sculptured with numerous wavy spiral ridges which occasionally split into 2 or 3 additional components; colour off-white with 3 or 4 large, irregularly brown concentric bands and reddish brown banding on the posterior margin, internally white with a purple umbonal area; reaches 25mm in length.
DISTRIBUTION: Sporadically found throughout the Philippines.

19. Pitar (Pitarina) affinis (Gmelin, 1791)
Syn: *Pitar (Pitarina) variegatum* Kuroda & Habe, 1971
24.6mm × 28.1mm; Marinduque.
Shell medium sized, solid, with a short, weakly convex anterior margin and a long, rounded ventral

margin, umbones large, inflated and anteriorly displaced, shell outline roundly trigonal and the surface sculptured with weak concentric growth lines; white in colour with brown rays and markings, interior white; reaches 35mm in length.
DISTRIBUTION: Sporadically found throughout the Philippines.

20. Periglypta (Tigammona) chemnitzii (Hanley, 1844)
17.9mm x 21.3mm; Sta. Cruz, Marinduque.
Shell small to medium sized, ovate-quadrangular in outline with a short anterior margin, a somewhat roundly truncated posterior margin and a long, rounded ventral margin; surface sculptured with numerous concentric lamellae, becoming markedly "frilled"on the posterior slope, umbones inflated and anteriorly displaced, interspaces with numerous fine radial cords; colour off-white with brownish radial streaks and brown spotting on the purple-tipped umbones, interior white; reaches around 30mm in length.
DISTRIBUTION: Sporadically found throughout the Philippines.

21. Tapes literatus (Linné, 1758)
26.8mm x 40.0mm; Palawan.
Shell medium sized, slightly compressed, somewhat ovate-quadrangular in outline with a rounded ridge extending from the umbones to the postero-ventral margin, postero-dorsal margin markedly convex, a feature immediately separating it from the superficially similar *Tapes sulcarius* (Lamarck, 1818) (Pl. 85, fig. 6); umbones inflated, touching and anteriorly displaced; surface sculptured with numerous concentric grooves but smooth towards the umbonal area; colour pale yellowish white with brown spots and lines and light to dark brown streaks radiating from the lemon-tinged umbones, interior white; reaches 50mm in length.
DISTRIBUTION: Sporadically found throughout the Philippines.

22. Pitar (Pitarina) sp.
18.0mm x 22.8mm; Buenavista, Marinduque.
Shell medium sized, trigonally elongate-ovate in outline with moderately inflated, anteriorly displaced umbones; sculptured with weak concentric growth lamellae; pale chalky brown in colour with radial rows of brown spots, interior white; largest seen 25mm in length. Close to *P. australica* (Reeve, 1863).
DISTRIBUTION: Sporadically found throughout the Philippines.

23. Timoclea (Veremolpa) costellifera (Adams & Reeve, 1850)
15.3mm x 18.0mm; Gasan, Marinduque.
Shell small, trigonally ovate in outline with a short anterior margin, a somewhat truncated posterior margin and a rounded ventral margin; a weakly angulate ridge runs from the centrally placed, moderately inflated umbones to the postero-ventral margin; sculptured with numerous radial cords which are granulose from overriding spiral grooves; off-white in colour with dark brown blotching and rays; reaches around 20mm in length.
DISTRIBUTION: Sporadically found throughout the Philippines.

24. Dosinia (Bonartemis) iwakawai Oyama & Habe, 1971
23.5mm x 22.5mm; Marinduque.
Shell small to medium sized, orbicular in outline, valves moderately inflated with the umbones slightly anteriorly displaced; sculptured with numerous concentric lamellae; white in colour with 4 radial rows of brown spots, lunule prominent, strongly impressed and deep brown in colour; reaches around 30mm in length. Possibly a subspecies of *Dosinia (Bonartemis) histrio* (Gmelin, 1791).
DISTRIBUTION: Sporadically found throughout the Philippines.

Family TELLINIDAE (cont'd.) (see also Pl. 84)

25. Tellina (Tellinella) staurella (Lamarck, 1818)
24.0mm x 37.3mm; Palawan.
Shell medium sized, irregularly ovate in outline with a rounded anterior margin and a long, straight posterior margin which is sharply truncated at the edge of the ventral margin; a distinct fold emanates from the umbones to the postero-ventral margin and bends in the direction of the lower valve; umbones central, touching and weakly inflated; surface sculptured with numerous concentric growth folds; white in colour with red rays radiating from the umbonal area; reaches 45mm in length.
DISTRIBUTION: Sporadically found throughout the Philippines.

Family PSAMMOBIIDAE (cont'd.) (see also Pl. 85)

26. Gari (Dysmea) occidens (Gmelin, 1791)
35.0mm x 58.0mm; Palawan.
Shell medium sized, solid, ovate-quadrangular in outline with a rounded anterior margin and a roundly truncated posterior margin, valves gaping at both ends, umbones moderately inflated, touching and centrally placed; sculptured with concentric growth striae, becoming stronger on the angulated posterior slope; pale pinkish lavender in colour with light and dark radial rays, internally pink with a yellow-bordered central area; reaches 60mm in length.
DISTRIBUTION: Sporadically found throughout the Philippines.

Class Bivalvia
Order Veneroidea
Superfamily Cardiacea
Family CARDIIDAE

A large family of medium to large sized shells inhabiting sand or mud in shallow to deep water. The shells may be roundly triangular (heart-shaped) to elongate, inflated, occasionally with gaping valves, and sculptured with prominent radial cords which may be smooth or spinose. Members of this family are often colourful, have large rounded umbones, a small external ligament, and a variable hinge plate, with 2 cardinal and 2 lateral teeth in the left valve and 3 in the right valve. They lack a pallial sinus. Well represented in the Philippines.

1. Trachycardium (Vasticardium) enode (Sowerby, 1841)
87.1mm x 68.5mm; Palawan.
Shell large, solid, inflated and triangularly elongate-ovate in outline, umbones large, inflated, touching and slightly anteriorly displaced; sculptured with up to 41 flattened, weakly scaled ribs and deep, narrow, spirally striated interspaces, outer lip margin strongly crenulate; external colour white with reddish brown maculations, darker towards the ventral margin; internal colour white with a yellowish umbonal area and a dark brown margin; seldom exceeds above specimen size.
DISTRIBUTION: Sporadically found throughout the Philippines.

2. Trachycardium (Vasticardium) mindanense (Reeve, 1844)
31.3mm x 28.0mm; Samar.
Shell medium sized, elongate-ovate in outline, with heavily inflated valves and large, incurved, touching umbones which are centrally placed; surface sculptured with numerous rounded, equi-spaced radial ribs bearing small spines, strongest on the anterior and posterior slopes and towards the ventral margin; a weakly angulate ridge runs from the umbones to the postero-ventral margin; colour creamy white with red blotching and banding, internally pinkish purple, white along the anterior margin; reaches around 35mm in length.
DISTRIBUTION: Sporadically found throughout the Philippines.

Family CARDITIDAE (cont'd.) (see also Pl. 84)

3. c.f. Cardita abyssicola Hinds, 1843
26.9mm x 25.6mm; Samar.
Extremely close in external morphology to certain members of the Cardiidae, but the presence of 2 lamellar cardinal teeth suggest a member of the Cardiitidae; shell medium sized, ovate-quadrangular in outline, heavily inflated with large, incurved, anteriorly displaced umbones; surface sculptured with up to 30 raised, rounded axial ribs bearing numerous fine spatulate spines, ventral margin finely crenulate, ligament external; colour pale brown with wavy, concentric brown bands, internally white; reaches around 30mm in length; normal *C. abyssicola* is uniform white in colour.
DISTRIBUTION: Sporadically found throughout the Philippines.

Family CARDIIDAE (cont'd.)

4. Fragum hemicardium (Linné, 1758)
32.3mm x 25.8mm; Bohol.
Shell small to medium sized, solid, quadrangular in outline, umbones sharply incurved, non-touching and continuing to the postero-ventral margin as a sharply angular ridge, lunule large and prominent; surface anterior to the ridge sculptured with 12 to 14 flat ribs bearing fine scales and weakly grooved interspaces, area posterior to the ridge with somewhat flattened ridges; colour variable, usually white with pink, yellow, or orange markings; may reach 50mm in length.
DISTRIBUTION: Sporadically found throughout the Philippines.

5. Laevicardium attenuatum (Sowerby, 1841)
55.0mm x 34.5mm; Palawan.
Shell medium sized, inflated, triangularly elongate in outline, with rounded anterior and ventral margins and a long, straight-sided posterior margin, umbones centrally located, inflated and touching; surface smooth with obsolete concentric growth lamellae and radial ribs, interior ledged below the umbones, lateral teeth (2) prominent, the posterior one largest, inner posterior margin crenulate, becoming finely serrated towards the ventral margin; pale brownish white in colour with yellow or pinkish blotching, interior white with a pink umbonal area; reaches 60mm in length.
DISTRIBUTION: Sporadically found throughout the Philippines.

6. Laevicardium biradiatum (Bruguiére, 1789)
38.0mm x 30.4mm; Sulu Archipelago.
Shell similar to the preceding species but generally smaller and more ovate; surface sculptured with numerous radial cords, obsolete on the umbones and encircled with numerous fine concentric lirae, radial structure conspicuously absent on the anterior and posterior margins; external colouration bluish white

PLATE NO. 87 1. *Trachycardium enode* (Sowerby) 2. *Trachycardium mindanense* (Reeve) 3. *c.f. Cardita abyssicola* Hinds 4. *Fragum hemicardium* (Linné) 5. *Laevicardium attenuatum* (Sowerby) 6. *Laevicardium biradiatum* (Bruguiére) 7. *Trachycardium elongatum* (Bruguiére) 8. *Lyrocardium lyratum* (Sowerby) 9. *Nemocardium exasperatum* (Sowerby) 10. *Ctenocardia victor* (Angas) 11. *Trachycardium alternatum* (Sowerby) 12. *Trachycardium flavum* (Linné) 13. *Vepricardium multispinosum* (Sowerby) 14. *Fragum unedo* (Linné) 15. *Corculum impressum* (Lightfoot) 16. *Corculum cardissa* (Linné) 17. *c.f. Corculum cardissa* (Linné) 18. *Corculum laevigatum* Bartsch 19. *Corculum monstrosum* (Gmelin) 20. *Corculum kirai* Shikama

PLATE 87

with pinkish brown blotching, internally pale pink with a large orange-brown blotch centrally and a finely crenulate outer margin; reaches 40mm in length.
DISTRIBUTION: Sporadically found throughout the Philippines, usually from deep water.

7. Trachycardium (Vasticardium) elongatum (Bruguiére, 1789)
74.5mm × 55.2mm; Palawan.
Very close to *T. enode* (fig. 1), but slightly more elongate and with fewer radial ribs (around 31), the interstitial grooves are much deeper and ribs on the postero-dorsal surface are strongly spinose; external colouration is light orange-brown, paler on the anterior slope, internally white with a large pinkish orange umbonal blotch, inner lip margin coarsely serrated and golden yellow; reaches around 90mm in length.
DISTRIBUTION: Sporadically found throughout the Philippines in limited quantity.

8. Lyrocardium lyratum (Sowerby, 1841)
36.5mm × 35.5mm; Sulu Archipelago.
Shell medium sized, ovate-quadrangular in outline, valves equally inflated, umbones prominently swollen, incurved, touching and slightly subcentral; sculptured with fine radial riblets, obsolete on the valve posterior and replaced by around 16 oblique sharp ridges, outer lip margin finely serrated; exterior colour cream with a characteristic reddish purple periostracum, internally white suffused with pink and yellow; may reach 50mm in length.
DISTRIBUTION: Sulu Sea.

9. Nemocardium exasperatum (Sowerby, 1841)
24.5mm × 25.3mm; Punta Engaño, Cebu.
Shell small, ovate-quadrangular in outline, with a long, round anterior margin and a short, straight-sided posterior margin, both valves strongly inflated, umbones large and slightly subcentral; sculptured with numerous finely spinose radial ribs and a weakly nodular row in the interspaces, margin finely serrate; colour normally white, Philippine specimen with rows of deep pink banding towards the ventral margin, internally pinkish white with a yellow umbonal area; reaches around 25mm in length.
DISTRIBUTION: Camotes Sea and Sulu Sea.

10. Ctenocardia victor (Angas, 1872)
32.5mm × 28.1mm; Cebu.
Shell medium sized, solid, ovate-quadrangular in outline, umbones moderately large, inflated, incurved, touching and slightly anteriorly displaced, an angulate ridge running from the umbones to the postero-ventral margin; sculptured with numerous close-set radial ribs bearing sharply echinate spines, strongest near the ridge area, weakest on the postero-ventral slope, outer margin crenulate; external colouration off-white with reddish brown blotching and banding, internally yellowish orange with pink suffusion and a white marginal area; may reach 40mm in length.
DISTRIBUTION: Cebu, Bohol and Southern Mindanao.

11. Trachycardium (Vasticardium) alternatum (Sowerby, 1841)
66.6mm × 58.2mm; Samar.
Shell moderately large, elliptically ovate in outline, valves equally inflated with a short, rounded anterior margin and a longer ventral margin, ligament large, prominent and external; sculptured with approximately 32 flat radial cords and narrow interspaces, those on the anterior slope weakly scabrous, those on the posterior slope initially smooth but becoming spinose towards the margin which is coarsely serrated; colour flesh with orange-yellow blotches arranged in bands on the ribs, interior ribbed and white in colour; reaches around 70mm in length.
DISTRIBUTION: Sporadically found throughout the Philippines.

12. Trachycardium (Regozara) flavum (Linné, 1758)
42.9mm × 37.9mm; Samar.
Shell medium sized, elongate-ovate in outline, margins rounded ventrally, somewhat truncated posteriorly, umbones large, inflated, touching and slightly anteriorly displaced; sculptured with around 30 rounded radial ribs and deep, narrow interspaces, the ribs becoming weakly scabrous towards the anterior and posterior margins; white in colour with a yellow tinge, the internal area with a large brownish orange central area; reaches 60mm in length.
DISTRIBUTION: Sporadically found throughout the Philippines.

13. Vepricardium multispinosum (Sowerby, 1841)
50.0mm × 45.9mm; Manila Bay.
Shell medium sized, solid, subcircular in outline, umbones central and touching; sculptured with up to 25 heavy raised ribs bearing small spines, strongest at the margins which are coarsely serrate; external colouration creamy-pink with darker pink margin, internally white with a pale pink streak on the antero-ventral margin; reaches 60mm in length.
DISTRIBUTION: Manila Bay and Tayabas Bay, Quezon.

14. Fragum unedo (Linné, 1758)
43.2mm × 36.7mm; Samar.
Shell medium sized, solid, elongate-quadrangular in outline, umbones sharply incurved, non-touching and continuing to the postero-ventral margin as a sharp angulate ridge; valves sculptured with around 26 strong radial ribs bearing fine scales; colour variable, yellow, pink or white, with red spotting on the rib scales, internal colouration white; reaches around 50mm in length.
DISTRIBUTION: Sporadically found throughout the Philippines.

THE *CORCULUM* SERIES

The following are members of a remarkable group of shells in which the shell morphology has been drastically altered from that of a conventional bivalve shell. The shells have been compressed in an anterior-posterior direction, resulting in the formation of a heart-shaped shell when viewed from the front, and thin, strongly incurved umbones (the left always curved behind the right), from which emanate strong lateral keels running to the mid-line of the ventral margin. The keels may be smooth, undulate or serrate; the diagnosis of the various species appears to depend on the degree of concavity (or convexity) of both anterior and posterior surfaces, the curvature of the lateral keels and the presence or absence of serrations along the keel margins. There appears to be disagreement between various authors as to the interpretation of different species. We have followed Bartsch (1947), but believe that it is most likely that all the species herein described as such represent no more than morphophenotypes of a single taxon (*C. cardissa* Linné, 1758).

Three views of each species are given.
a. Frontal view showing the posterior surface.
b. Dorsal view.
c. Side view showing surface contours and the slope of the lateral keels. Size quoted is for specimen (a.) in each group.
DISTRIBUTION: All specimens figured sporadically found throughout the Philippines, mostly from Cebu.

15. Corculum impressum (Lightfoot, 1786)
45.2mm × 36.6mm; Cebu.
Shell small to medium sized, the anterior surface markedly convex, white coloured and heavily ribbed; posterior surface concave, with a slight central convexity and weaker ribbing across the surface; the lateral keels are slightly anteriorly curved and become thicker and weakly spinose towards the ventral margin; when viewed anteriorly, the outline is boat-like, with the lateral keels flaring outwards and upwards in a wing-like fashion; colour white to cream with pink-spotted keels; reaches 60mm in length.

16. Corculum cardissa (Linné, 1758)
48.2mm × 42.3mm; Cebu.
Shell large, heart-shaped with sharp projections along the lateral keels; both surfaces convex, the anterior surface slightly more so and also more heavily ribbed, the lateral keel is relatively straight but slightly anteriorly curved towards the ventral margin; colour variable, white, yellow or pink, with darker markings; may reach 80mm in length.

17. c.f. Corculum cardissa (Linné, 1758)
64.7mm × 60.3mm; Cebu.
The large size and strongly serrated keels suggest this is *C. cardissa;* however, the rather inflated central area on the posterior surface and the anterior displacement of the keels in the umbonal area [figs. (b) and (c)] suggest something closer to *C. kirai* (fig. 20).

18. Corculum laevigatum Bartsch, 1947
45.2mm × 39.4mm; Cebu.
Shell medium sized, with the anterior surface more convex and more heavily ribbed than the posterior surface which has weaker, more numerous ribs; lateral keels slightly anteriorly curved and smooth; figs. (b) and (c) appear to be intermediate forms between *C. laevigatum* (with the characteristic flattened contours) and *C. kirai* (with the prominently serrated keels); colour usually pale yellow with orange mottling; reaches 50mm in length.

19. Corculum monstrosum (Gmelin, 1791)
32.0mm × 34.5mm; Cebu.
A small, very distinctly sculptured species in which the posterior surface is weakly ribbed and strongly convex; the anterior surface is centrally convex but deeply concave towards the margins of the lateral keels; the keels are strongly curved in an anterior direction and are strongly serrate along their ventral margin; similar to the boat-like outline of *C. impressum* (fig. 15) but in a reverse direction; colour variable, usually cream to yellow with darker mottling; reaches around 35mm in length.

20. Corculum kirai Shikama, 1964
41.5mm × 38.9mm; Cebu.
Shell medium sized, solid, with the anterior surface coarsely ribbed and moderately convex centrally; posterior surface with weaker ribbing but with a more pronounced central convexity; lateral keels strongly curved anteriorly and serrated along their entire length; cream with orange, yellow, lavender or pink mottling or lines; reaches 50mm in length; seems close to *C. monstrosum* (fig. 19).

Plate 88 Mytilidae

Class Bivalvia
Order Mytiloida
Superfamily Mytilacea
Family MYTILIDAE

A large family of colonial, intertidal-dwelling molluscs which prefer cooler waters and may attach themselves to a woody or rocky substrate via a byssal filament, form nests, or burrow into mud or sand. The shells are small to large, equivalve, inequilateral, rather thin, triangularly elongate to elongate-ovate in outline, with a rounded posterior margin and sharp, anteriorly displaced umbones. The ligament is external, and the hinge line long and occasionally dentate, interior nacreous, with unequal adductor muscle scar impressions, the posterior one being largest. Fairly well represented in the Philippines.

1. Modiolus philippinarum (Hanley, 1843)
99.0mm; Leyte.
Shell medium sized, irregularly ovate-quadrangular in outline with a large inflated ridge from the umbones to the postero-ventral margin; surface sculptured with numerous fine concentric ridges; external colouration pale greyish blue on the dorsal slope and off-white on the ventral slope, internal colour similar but more

intense and with a pink suffusion, covered with dark periostracum; reaches 130mm in length.
DISTRIBUTION: Throughout the Philippines.

2. Perna viridis (Linné, 1758)
72.0mm; Manila Bay.
A beautiful species, immediately recognized by its iridescent green periostracum; elongate-ovate in outline with a rounded posterior margin, a sharply tapering umbonal area and 2 small teeth below the umbones; sculptured with numerous concentric ridges; interior a nacreous silver colour with bluish green toning; reaches around 80mm in length; an edible species. Locally known as "tahong"
DISTRIBUTION: Mostly farmed in Cavite along Manila Bay.

3.Septifer excisus (Wiegmann, 1837)
46.2mm; Samar.
Shell small to medium sized, triangularly elongate-ovate in outline, with a long, rounded dorsal margin and a long, almost straight ventral margin, valves moderately inflated with a sharply angulate ridge which in effect forms the upper part of the ventral margin; sculptured with concentric and radial cords forming a finely granulose surface, umbones pointed with a small septum underneath and a finely crenulate inner marginal surface; externally bright pink in colour, internally nacreous silver; reaches 50mm in length.
DISTRIBUTION: Sporadically found throughout the Philippines.

4. Amygdalum politum (Verrill & Smith, 1880)
Syn: *Amygdalum watsoni* (Smith, 1885)
56.3mm; Palawan.
Shell medium sized, lightweight and rather fragile, elongate-ovate in outline, the umbones inflated and rounded, almost obsolete where they merge with the dorsal margin; valves moderately inflated and smooth except for some obsolete concentric lirae; externally of yellow-orange colouration, internally pearly white with a smooth margin; reaches around 60mm in length.
DISTRIBUTION: Sporadically found throughout the Philippines.

5. Lithophaga teres (Philippi, 1846)
37.0mm; Sulu Archipelago.
Shell small to medium sized, elongate in outline, both anterior and posterior margins rounded, valves quite inflated, less strongly so towards the posterior margin, umbones nearly obsolete and merging with the shell contour; sculptured with numerous concentric folds and oblique radial lirae on the ventral slope producing a weakly granulose surface; deep reddish brown in colour, the umbonal area paler, internally pearly silver; seldom exceeds 40mm in length.
DISTRIBUTION: Sporadically found throughout the Philippines.

6. Septifer bilocularis (Linné, 1758)
36.5mm; Manila Bay.
Shell small, ovate-quadrangular in outline, valves moderately inflated with a sharply angulate anterior slope, umbones prominent and anteriorly curved; surface sculptured with strong radial cords and obsolete concentric folds, interior with a finely crenulate margin and a septum below the umbones; dark pinkish purple in colour, umbonal area paler, interior pale bluish white; reaches 40mm in length; close to *S. excisus* but is less elongate and has a more angulate outline.
DISTRIBUTION: Sporadically found throughout the Philippines.

7. Stavelia subdistorta (Récluz, 1852)
169mm; Sulu Archipelago.
Shell large, heavy, elongate-trigonal in outline with a rounded ventral margin, umbones prominent, inflated and anteriorly displaced, anterior slope steeply angulate, the dorsal edge curving under the umbones; sculptured with numerous concentric folds; external colour pale purple with a dark brown periostracum, internally pale yellowish white with a blue margin; may reach 200mm in length.
DISTRIBUTION: Palawan and Sulu Sea.

Class Bivalvia
Order Myoida
Superfamily Pholadacea
Family PHOLADIDAE

A large worldwide family of intertidal-dwelling molluscs which burrow into wood, rocks, coral or mud. The shells are small to large in size, thin and wing-like, with the posterior surface frequently forming a tubular shape. The anterior margin is gaping and frequently filled with a plate or callum. The posterior surface has concentric grooves and the anterior surface fine, rasp-like scales for boring into the substrate. The shells have no ligament, rarely any teeth, but may have between 1-3 plates on the dorsal margin. Some members have a long, internal calcareous spine. Relatively few Philippine species.

8. Barnea (Umitakea) dilatata (Souleyet, 1843)
100.5mm, 60.0mm; Manila Bay.
Shell large, thin, narrowly gaping anteriorly, widely gaping and truncated posteriorly; surface with radial cords and concentric lamellae forming sharp spines on

PLATE NO. 88 1. *Modiolus philippinarum* (Hanley) 2. *Perna viridis* (Linné) 3. *Septifer excisus* (Wiegmann) 4. *Amygdalum politum* (Verrill & Smith) 5. *Lithophaga teres* (Philippi) 6. *Septifer bilocularis* (Linné) 7. *Stavelia subdistorta* (Récluz) 8. *Barnea dilatata* (Souleyet) 9. *Martesia striata* (Linné) 10. *Monothyra orientalis* (Gmelin) 11. *Laternula spengleri* (Gmelin) 12. *Brechites penis* (Linné) 13. *Fimbria fimbriata* (Linné) 14. *Chama brassica* Reeve 15. *Chama limbula* Lamarck 16. *Chama savignyi* Lamy 17. *Chama lazarus* Linné 18. *Codakia tigerina* (Linné)

PLATE 88

the anterior surface; internally a large, thin calcareous spine radiates from each anteriorly displaced umbone; externally, the valves are joined by a large, lanceolate protoplax; colour dirty white with a brownish periostracum; seldom exceeds 100mm in length.
DISTRIBUTION: Sporadically found throughout the Philippines in limited quantity.

9. Martesia striata (Linné, 1758)
26.0mm x 10.5mm, 24.0mm x 9.7mm; Manila Bay.
Shell small, fragile, elongate-ovate in outline, anterior end rounded with a callum, the posterior end drawn out and gaping; valves delineated by a deep sulcus, the anterior portion with numerous sharply granulose ridges, the posterior portion with smoother concentric ridges, a large circular mesoplax above the umbones; dirty white in colour with a yellowish periostracum; may reach 45mm in length.
DISTRIBUTION: Sporadically found throughout the Philippines.

10. Monothyra orientalis (Gmelin, 1791)
69.7mm x 26.5mm; Negros.
Shell medium sized, lightweight, triangularly elongate-ovate in outline, with a widely gaping, sharply attenuate anterior end, and a narrowly gaping, elongated posterior end; surface sculptured with concentric folds and radial striae on the anterior segment forming sharply serrated spines, umbones anteriorly displaced and covered with a lanceolate protoplax; colour dirty white externally, porcelaneous white internally; reaches around 75mm though specimens close to 100mm have been taken in the Philippines.
DISTRIBUTION: Mostly from Negros Is.

Class Bivalvia
Order Pholadomyoida
Superfamily Pandoracea
Family LATERNULIDAE

A small group of mud-dwelling intertidal bivalves, the shells of which are thin, elongate, with rounded margins and a gaping posterior end. The surface is encircled with concentric folds and is generally very finely granulose. The hinge line is edentulate and the umbones fissured, below which is a spoon-shaped chondrophore at the end of an oblique plate.

11. Laternula spengleri (Gmelin, 1791)
47.3mm x 24.0mm; Cebu.
Shell medium sized, thin and extremely fragile, with a rounded anterior end and a gaping, truncated posterior end; valves moderately inflated, the umbones anteriorly displaced, posteriorly curved and have a weak, central fissure; internally there is a small spatulate chondrophore at the end of a thin, oblique plate; surface dirty white in colour, with finely granulose concentric folds which become crowded on the posterior margin; reaches around 60mm in length.
DISTRIBUTION: Sporadically found throughout the Philippines.

Class Bivalvia
Order Pholadomyoida
Superfamily Clavagellacea
Family CLAVAGELLIDAE

The Clavagellids or "Watering Pot" shells are a small group of bivalves which embed themselves in sand or mud in a vertical position. The shell is free in the juvenile stage, but in the adult form, becomes fused at the base of a long calcareous tube which terminates in a perforated fringed disc. The valves are unequal, have an external ligament and a toothless hinge. Only a few Philippine species presently known.

12. Brechites (Penicillus) penis (Linné, 1758)
142mm; Leyte.
The actual shell is extremely small and embedded at the base of a long calcareous tube which is sculptured with concentric growth folds and weak axial striae; the anterior end is ringed with a calcareous "fringe" below which is a sieve-like, perforated convex disc; pure white in colour; the calcareous tube reaches 150mm in length.
DISTRIBUTION: Sporadically found throughout the Philippines in limited quantity.

Class Bivalvia
Order Veneroida
Superfamily Lucinacea
Family FIMBRIIDAE

A small group of intertidal molluscs which burrow into sand or mud and generally prefer tropical waters. The shells are medium sized, equivalve, inequilateral, elongate-ovate in outline and the surface is sculptured with overriding concentric and radial cords forming a granulose surface. The valves are inflated, the umbones broad and rounded, the ligament external and the pallial sinus is quite small. Only a few recorded species in the Philippines.

13. Fimbria fimbriata (Linné, 1758)
62.5mm x 50.2mm; Samar.
Shell medium sized, solid and elongate-ovate in outline, umbones inflated, centrally located, incurved and non-touching; surface sculptured with wide concentric folds which are crossed by numerous radial ribs producing a coarsely granulose surface but is smooth centrally and towards the ventral margin, inner margin finely serrated; external colouration off-white, pink along the margins, internally white, pale yellow below the umbones and pink along the anterior and posterior margins; reaches 100mm in length.
DISTRIBUTION: Sporadically found throughout the Philippines.

Class Bivalvia
Order Hippuritoida
Superfamily Chamacea
Family CHAMIDAE

A relatively small family of bivalves living in shallow to deep water, mainly tropical in distribution and attached to the substrate by a cement-like substance. The shells are small to medium sized, thick, irregularly ovate, and inequivalve. The ligament is external, the hinge thick, bearing 1 or 2 cardinal teeth, and the laterals are weak or obsolete. There are two adductor muscle scar impressions but no pallial sinus. Fairly well represented in the Philippines with only a few species illustrated here.

14. Chama brassica Reeve, 1846
57.5mm x 55.8mm, 49.5mm x 57.0mm; Sulu Archipelago.
Shell medium sized, heavy and inequivalve, the left valve being moderately inflated with a weakly developed umbone, the right valve large and more inflated with a larger umbone, ligament partially external; surface sculptured with concentric rows of extremely foliated growth lamellae; external colouration variable, being pink or yellow, occasionally with darker coloured rays, internally white; reaches around 90mm in length.
DISTRIBUTION: Sporadically found throughout the Philippines. Mostly from Palawan and Sulu Sea.

15. Chama limbula Lamarck, 1819
Syn: *Chama iostoma* Conrad, 1837
73.0mm x 57.0mm; Palawan.
Shell medium sized, solid, inequivalve, the lower valve moderately concave with a flat base and a somewhat truncated ventral margin; the upper valve is slightly smaller and more inflated with a ridge running from the umbones to the postero-ventral margin; lower valve with an obsolete umbone, that of the upper valve large and curved to the right; surface sculptured with concentric "frilly" lamellae which are rather spinose on the anterior slope; external colour lavender, internally white with a purplish margin; reaches 80mm in length.
DISTRIBUTION: Sporadically found throughout the Philippines. Mostly from Palawan and Sulu Sea.

16. Chama savignyi Lamy, 1921
Syn: *Chama imbricata* Broderip, 1835 (non Lamarck, 1801)
70.0mm x 50.5mm x 41.0mm; Palawan.
Shell medium sized, solid, irregularly ovate in outline, lower valve deeply concave with a large cardinal tooth, inner margin finely crenulate, the outer posterior margin forming a series of coarse lamellae or ridges that are blotched in purple; the upper valve slightly smaller than the lower, rather flat, but with a deep central concavity, muscle scar impressions prominent on both the upper and lower valves; surface of the upper valve with concentric lamellae, more pronounced towards the postero-ventral margin; external colouraton variable, white, brown or pinkish, internally off-white with a mauve-spotted margin; may reach 65mm in length.
DISTRIBUTION: Sporadically found throughout the Philippines. Mostly from Sulu Sea and Palawan.

17. Chama lazarus Linné, 1758
94.5mm x 93.2mm, 78.0mm x 71.0mm, 54.0mm x 69mm; Palawan.
The most beautiful of all the *Chama* species; shell medium sized, irregularly ovate, the lower valve more concave than the upper and the umbones posteriorly curved; surface with numerous foliated spines and lamellae, strongest towards the ventral margin, internal margin finely crenulate; colour variable, usually off-white with pink, orange, yellow or lavender suffusion, internally white; reaches 100mm in length.
DISTRIBUTION: Sporadically found throughout the Philippines. Mostly from Palawan and Sulu Sea.

Class Bivalvia
Order Veneroida
Superfamily Lucinacea
Family LUCINIDAE

A moderately large group of mainly intertidal-dwelling bivalves living in coral sand or mud in most tropical and subtropical provinces. The shells are small to large, usually ovate in outline and the surface may be smooth or crossed with concentric and radial ribs forming a granulose surface. The umbones are centrally located and anteriorly curved, the ligament external or partially internal, the hinge line may be smooth or with weak cardinal and lateral teeth, and all lack a pallial sinus. The inner margin may be smooth *(Codakia spp.)* or crenulated *(Lucina spp.)*. Fairly well represented in the Philippines.

18. Codakia tigerina (Linné, 1758)
94.0mm x 101mm; Palawan.
Shell large, ovate in outline, rather heavy, with weakly inflated valves and moderately enlarged, anteriorly displaced umbones below which is a small, heart-shaped lunule; surface sculptured with radial and concentric lirae producing a coarsely granulose surface; internal sculpture consists of a long anterior adductor muscle scar, a rounded posterior one and a weak groove around the inner ventral margin which is finely crenulate; external colour pale brown, internally pale yellow with a white margin; reaches 100mm in length.
DISTRIBUTION; Sporadically found throughout the Philippines.

Class Bivalvia
Order Mytiloida
Superfamily Pteriacea
Family ISOGNOMONIDAE

A relatively small family of bivalves inhabiting intertidal waters under stones, in mangrove roots or around coral boulders. The shells are moderately large, solid, variable in shape, usually elongate to elongate-ovate and the surface is frequently heavily encrusted. The shells are flat and have a straight hinge line consisting of numerous vertical ligamental pits. Few Philippine species.

1. Isognomon (Isogonum) isognomum (Linné, 1758)
151.0mm, 68.0mm, 51.0mm, 48.0mm; Iloilo.
Shell moderately large, extremely solid, variable in shape, usually elongate to elongate-ovate, the external surface deep blackish purple with irregular concentric lamellae; interior bright nacreous silver, with a straight hinge line bearing numerous ligamental pits and a well impressed byssal depression; seldom exceeds 150mm in length.
DISTRIBUTION: Throughout the Philippines.

2. Crenatula modiolaris Lamarck, 1819
15.5mm × 19.9mm; Iloilo.
Shell small, thin, roundly triangulate in outline, with a short anterior ear and a nearly obsolete posterior ear, upper valve with a moderately inflated umbone, the lower valve nearly flat; sculptured with fine concentric striae and radiating riblets, hinge line straight, simple, with a few widely-spaced ligamental grooves; colour brown with white rays, internally silvery pearl; may reach around 20mm in length.
DISTRIBUTION: Throughout the Philippines.

Class Bivalvia
Order Mytiloida
Superfamily Pteriacea
Family MALLEIDAE

A comparatively small group of bivalves, often termed "hammer oysters" (*Malleus spp.*) or "sponge-fingers" (*Vulsella spp.*). The shells are medium sized to large, usually elongate, but in the hammer oysters the ears are drawn out in a "wing-like" fashion. The margins are frequently gaping, the hinge line is simple, edentulate, and there is a large external ligament in a triangular ligamental pit. The interior is nacreous, there is a single large adductor muscle scar, and the exterior may be smooth or irregular and heavily encrusted. Members inhabit intertidal reefs (*Malleus spp.*) or sponges (*Vulsella spp.*).

3. Malleus malleus (Linné, 1758)
171mm × 167mm; Samar.
Shell large, solid, irregularly elongate with 2 wings on the dorsal margin, the anterior "wing" usually smaller than the posterior; external surface heavily sculptured with irregular growth folds, hinge line straight and simple, with a large, centrally located ligamental pit; internally there is a deep muscle scar impression; external colour black purplish brown, internally vitreous purple-brown, the muscle scar pit a deep nacreous purple; may reach 200mm in length.
DISTRIBUTION: Throughout the Philippines.

4. Malleus albus Lamarck, 1819
150mm × 175mm; Palawan.
Shell very similar to *M. malleus* (Linné), but is more regular in outline, the surface is not as rugosely sculptured, the wings are frequently more equi-sized, and the ligamental pit is more ovate rather than triangular; colour characteristically cream externally and internally, with a nacreous pale-blue cavity; may reach 200mm in length but usually smaller.
DISTRIBUTION: Throughout the Philippines.

5. Malleus anatinus (Gmelin, 1791)
92.0mm × 51.0mm; Iloilo.
Shell medium sized, rather lightweight for the genus, irregularly elongate in outline, the margins with regular wavy folds and a single "wing-like" extension on the posterior edge, the ligamental pit is small, obliquely aligned, and the body cavity is shallow; externally pale cream in colour with irregular blackish blotches, internally dull yellowish cream with a dark purple cavity, and nacreous silver just under the ligamental area; reaches around 100mm in length.
DISTRIBUTION: Sporadically found throughout the Philippines in limited quantity.

6. Vulsella vulsella (Linné, 1758)
53.4mm; Marinduque.
This is the so called "sponge-finger"; shell medium sized, obliquely elongate-ovate, and the surface sculptured with numerous concentric striae, valves moderately inflated, hinge line short with a large, central ligamental pit; external colouration pale brown with darker rays, internal colouration pearly silver, margins light brown; may reach 100mm in length.
DISTRIBUTION: Throughout the Philippines.

Class Bivalvia
Order Mytiloida
Superfamily Ostreacea
Family OSTREIDAE

A large family of bivalves inhabiting shallow waters in the swamps and mangroves of most world seas. The shells are small to large, inequilateral, irregularly

PLATE NO. 89 1. *Isognomon isognomum* (Linné) 2. *Crenatula modiolaris* Lamarck 3. *Malleus malleus* (Linné) 4. *Malleus albus* Lamarck 5. *Malleus anatinus* (Gmelin) 6. *Vulsella vulsella* (Linné) 7. *Lopha cristagalli* (Linné) 8. *Lopha cristagalli* (Linné) typical form 9. *Lopha cristagalli* (Linné) *form folium* (Linné) 10a. to 10b. *Pinctada margaritifera* (Linné) 11. *Pteria penguin* (Röding) 12. *Pteria avicular* (Holten)

PLATE 89

elongate-ovate in outline and ornamented with irregular radial ribbing on a thick lamellar surface. Inequivalve, the lower or right valve thicker and more convex than the upper or left valve which is almost flat. Umbones prominent, margins folded or denticulate, the ligament is external in a triangular pit and the hinge line is straight and edentulate. The family is fairly well represented in the Philippines. The following 3 specimens, although bearing no morphological resemblance to each other, are now believed to represent extreme ecomorphs of a single, variable taxon, namely *Lopha cristagalli* (Linné, 1758).

7. Lopha cristagalli (Linné, 1758) (ecomorph)
64.0mm; Palawan.
Shell medium to large sized, solid, upper valve slightly inflated, the surface with nodulose, lamellate ridges and the margin with numerous "V-shaped" ribs or folds, strongest on the ventral margin, lower valve flat with curiously shaped spinose projections on the base; surface colouration brownish purple, interior nacreous silver-white with orange and brown suffusion; reaches 80mm in length.
DISTRIBUTION: Sporadically found throughout the Philippines.

8. Lopha cristagalli (Linné, 1758) (typical)
59.0mm; Palawan.
Shell medium sized, solid, sculptured with 4 to 6 raised, sharply angulate radial ribs forming large, "V-shaped" folds on the ventral margin; short rounded spinose projections adorn the ribs, becoming prominent towards the umbonal area; entire exterior surface dark purple in colour and finely pustulose, interior orange-purple with a finely granulose peripheral margin; usually occurs in large clusters and reaches around 80mm in length.
DISTRIBUTION: Throughout the Philippines.

9. Lopha cristagalli (Linné, 1758) form **folium** (Linné, 1758)
64.0mm; Sulu Archipelago.
Shell medium sized, solid and "leaf-shaped", upper valve with a rounded central ridge from which radiate several sharply angulate, obliquely aligned ribs forming "V-shaped" folds on the margin, these folds interlocking with those of the lower valve which is similarly sculptured, but has a central groove bordered by spinose projections; brownish purple in colour; reaches 100mm in length.
DISTRIBUTION: Mostly from Palawan and the Sulu Sea.

Class Bivalvia
Order Mytiloida
Superfamily Pteriacea
Family PTERIIDAE

These shells are popularly known as "pearl oysters" and include a moderately large group which vary in size from small to large, are aviculoid or suborbicular in outline and have a pearly interior. The shells have a small anterior ear, a deep byssal notch on the anterior margin of the lower valve, an external ligament, and a hinge line which may be short (*Pinctada spp.*) or produced (*Pteria spp.*) in the form of a "wing-like" extension. *Pinctada spp.* inhabit coral rubble, while the *Pteria spp.* are found in gorgonoid formations. Few Philippine species.

10. Pinctada margaritifera (Linné, 1758)
Shell large, solid, slightly compressed, ovate-quadrangular in outline; external surface sculptured with numerous scaly concentric lamellae, forming projecting spines at the margin, anterior ear short with a deep byssal gape below; external colour pale greenish grey to black with white radiating bands, internally nacreous silver-pearl with a yellowish blue tint and black at the margins; may reach 250mm in length. Popularly known as the "black lip pearl shell".
10a. typical form (interior); 156mm; Samar.
10b. young specimen clearly showing external colour and pattern; 76mm; Samar.
DISTRIBUTION: Throughout the Philippines.

11. Pteria (Pteria) penguin (Röding, 1798)
190mm; Palawan.
Shell large, solid, elliptically ovate in outline, the upper valve more inflated with a rounded keel; surface of both valves sculptured with irregular concentric growth lamellae, "wings" variable, sometimes equi-sized, in other cases the posterior "wing" is produced, hinge line long and straight with 2 denticles; external colour dark purple to black, internally nacreous silver, marginally purple-black; reaches 240mm in length.
DISTRIBUTION: Throughout the Philippines.

12. Pteria (Austropteria) avicular (Holten, 1802)
Syn: *Pteria cypsellus* (Dunker, 1872)
66.0mm, 70.0mm, 108mm; Samar.
An extremely variable species; shell medium sized, aviculoid in outline with a narrowly elongate to rectangularly ovate body, a short anterior "wing", and a longer, thinner posterior "wing", upper valve usually inflated with a rounded ridge; external colour dark purple to light yellowish brown with faint, yellow-coloured radial lines, internal colour nacreous silver-blue, with a darker margin; may reach 130mm in length.
DISTRIBUTION: Throughout the Philippines.

Pteriidae Plate 90

1. Pinctada maxima (Jameson, 1901)
170mm × 185mm; Sulu Archipelago.
This is known as the "golden-lip pearl oyster", and is similar to *P. margaritifera* (Linné), 1758 in shape. However, it grows much bigger in size and differs mainly in its golden brown exterior which lacks the radiating white rays and the pearly silver interior

which has a greenish gold inner margin and a golden-brown outer margin; can reach 300mm in length.
DISTRIBUTION: Throughout the Philippines, mainly from the Sulu Sea.

Class Bivalvia
Order Mytiloida
Superfamily Anomiacea
Family ANOMIIDAE

A small group of shallow-water gastropods inhabiting mainly tropical regions occasionally attached to the substrate via a byssus. The shells are thin, translucent and inequilateral. The surface is generally smooth, occasionally with some granular lirae. Members of the genus *Placuna* (here illustrated), are rounded or saddle-shaped and have 2 "V-shaped" teeth in the upper umbonal area.

2. Placuna ephippium (Retzius, 1788)
Syn: *P. sella* (Gmelin, 1791)
120mm x 165mm; Palawan.
Shell large, solid, irregularly ovate-pentagonal in outline, the 2 valves curved giving the shell a characteristic saddle shape; surface scaly, with irregular concentric growth lamellae and numerous fine radial striae; interior purplish silver with a small circular, centrally placed muscle scar impression and 2 "V-shaped" teeth in the upper area; external colour silvery white suffused with dark purple; may reach 200mm in length. Specimen featured has a *Spondylus* attached to it.
DISTRIBUTION: Throughout the Philippines.

3. Placuna placenta (Linné, 1758)
105mm x 112mm; Capiz.
Shell moderately large, subcircular in outline with translucent valves; sculptured with fine concentric ridges and numerous faint radial striae, cardinal teeth prominent and "V-shaped," adductor muscle scar small, circular and centrally placed, valves almost flat, some specimens with a slight convexity; reaches around 180mm in length. Locally known as "kapis".
DISTRIBUTION: Throughout the Philippines.

Family OSTREIDAE (cont'd.) (see also Pl. 89)

4. Crassostrea gigas (Thunberg, 1793)
180mm; Palawan.
This species reaches an extremely large size in some Western Pacific localities (over 400mm); shell irregular in outline, somewhat triangularly elongate, the surface with coarse concentric lamellar folds, inequivalve, inequilateral, the lower valve usually larger and somewhat more concave than the upper which is nearly flat; hinge line simple and edentulate with a thick ligament, inner surface with a finely crenulate outer margin and a prominent, centrally located adductor muscle scar impression; colour pale silvery white with purple suffusion externally, internally vitreous silver-white with a brown-and-purple stained margin.
DISTRIBUTION: Throughout the Philippines.

5. Crassostrea iredalei (Faustino, 1928)
65.0mm; Manila Bay.
Shell medium sized but can reach a fairly large size (around 150mm); the shell is irregularly elongate-ovate in outline with sinuous, folded margins; the surface is sculptured with mainly irregular, concentric, somewhat scaly lamellar folds; internally the surface is irregular, somewhat pock-marked and has dark brown, semi-circular adductor muscle scar impression; external colour off-white with purple, brown and yellow suffusion, internally a chalky white colour.
DISTRIBUTION: Throughout the Philippines.

Class Bivalvia
Order Mytiloida
Superfamily Pinnacea
Family PINNIDAE

The Pinnidae, or "Pen Shells" as they are more frequently called, are a relatively small group of bivalves with a worldwide distribution. They inhabit shallow to deep water in mud, sand or gravel, and attach to the substrate by means of a byssus, orientating the shell in a vertical position. Shells of this family are large, thin, transparent and wedge-shaped with a gaping posterior margin. The surface may be smooth or scaly, the umbones are near the anterior margin, the hinge teeth are obsolete, and the valves are united by a long, non-elastic primary ligament on the antero-dorsal margin and by a so-called secondary ligament of fused shell nacre along the postero-dorsal margin. Internally, the shells have a long nacreous layer, a large posterior adductor muscle scar and an almost obsolete anterior adductor muscle scar. Well represented in the Philippines.

6. Atrina (Servatrina) pectinata pectinata (Linné, 1767)
form **kinoshitai** Habe, 1953
165mm; Sulu Archipelago.
An exquisitely sculptured form of this species in which the entire surface is covered with regular radial rows of short, sharp, open, semi-circular spines, strongest towards the gaping posterior margin and absent on the umbonal area.
DISTRIBUTION: Sulu Sea.

7. Atrina (Servatrina) pectinata pectinata (Linné, 1767)
135mm; Bohol.
Shell medium sized to large, triangularly wedge-shaped with moderately inflated valves and a gaping posterior margin; surface with up to 30 radial ribs which may be smooth or bear numerous, somewhat spatulate spines, dorsal margin relatively straight, ventral margin weakly curved, most prominently so near the posterior margin, internal nacreous layer occupying the anterior two-thirds of the shell, posterior adductor muscle scar large, subcircular and within the nacreous layer; external colouration translucent brown to tan, internally iridescent dark

brown to black; can reach 370mm and is extremely variable in surface sculpturing.
DISTRIBUTION: Sporadically found throughout the Philippines.

8. Streptopinna saccata (Linné, 1758)
210mm, 170mm; Cebu.
Shell moderately large, solid, extremely variable in shape; anterior portion typically pinnoid, but after that the shell may develop a variety of environmentally controlled shapes ranging from roundly triangular to broadly elongate; outer surface dull with 6 to 12 radiating ribs on the posterior slope, none on the anterior slope, internal nacreous layer small, limited to the antero-dorsal margin and enclosing the posterior adductor muscle scar; external colour grey to dark brown, occasionally darker on the posterior margin, internally iridescent with a smokey white nacreous layer; reaches 235mm in length.
DISTRIBUTION: Sporadically found throughout the Philippines in limited quantity.

9. Pinna muricata Linné, 1758
255mm; Samar.
Shell large, triangularly attenuate with a sharply to roundly truncated posterior margin, valves thin with between 12 and 24 occasionally spinose radiating ribs, dorsal margin straight to concave, ventral margin concave posteriorly then convex anteriorly; longitudinal sulcus narrow, bringing the dorsal and ventral nacreous lobes into close proximity, posterior adductor muscle scar sub-terminal on the dorsal nacreous lobe; shell light brown in colour with occasional darker reddish brown banding, internally light green, posterior area purple-black, nacreous layer silvery white; can reach over 300mm in length.
DISTRIBUTION: Sporadically found throughout the Philippines.

10. Atrina (Atrina) vexillum (Born, 1778)
250mm; Palawan.
Shell large, solid, somewhat triangularly ovate in outline, dorsal margin relatively straight, ventral margin lobate; valves inflated, heavy and sculptured with 10 to 17 radiating, occasionally spinose ribs, weaker interstitial riblets and concentric growth striae, internal nacreous layer iridescent, posterior adductor muscle scar large, prominent and subcentral, nacreous layer not divided by a sulcus and occupies the anterior two thirds of the shell interior, umbones with concentric growth lines; shell colour reddish brown to black, slightly translucent, interior dark brown to black with lighter greenish mottling; reaches 420mm.
DISTRIBUTION: Throughout the Philippines.

11. Pinna bicolor Gmelin, 1791
390mm; Palawan.
Shell large, solid, varying in outline from triangularly ovate to triangularly elongate, posterior margin sharply truncate to strongly convex, valves moderately inflated with a strong keel on the anterior surface; sculptured with between 8 and 16 radial ribs occasionally becoming lamellose towards the posterior margin, and numerous coarse concentric folds; internally the nacreous layer occupies the anterior half of the shell, the dorsal and ventral nacreous lobes separated by a well defined sulcus, and the dorsal lobe encloses a large circular, sub-terminal posterior muscle scar; external colour purplish brown with light and dark coloured concentric bands, internally light to dark brown, the nacreous lobes a silver-white colour; reaches 500mm.
DISTRIBUTION: Throughout the Philippines.

Tridacnidae Plate 91

Class Bivalvia
Order Veneroidea
Superfamily Tridacnacea
Family TRIDACNIDAE

A small group of tropical bivalves often referred to as "Giant Clam" shells. Members of this family are large, ponderous, and hermaphroditic in their reproductive cycle. The adult shell is non-locomotory and may be confined within a coralline burrow or firmly attached by a byssus to the substrate. The Tridacnidae feed not only by extraction of particulate matter from the surrounding seawater, but have also evolved a unique supplementary food source, utilizing symbiotic zooxanthella algae present in the mantle tissue. These algae manufacture nutrients via photosynthesis and are utilized as food when the need arises. The shell of the Tridacnidae is sculptured with strong radial ribs which may be fluted or ribbed and have scalloped margins which interlock the 2 valve sections. The morphological anatomy of the Tridacnidae is somewhat different from that of normal bivalves due to the curious pattern of evolution they appear to have followed. The umbones are on the ventral margin, the byssal margin is internally plicate, the ligament is on the same side as the hinge teeth, and both the posterior adductor and pedal retractor scars tend to merge centrally on the pallial line. Well represented in the Philippines with most species coming from the Sulu Sea. Besides being edible all members of this family are used extensively for commercial purposes.

PLATE NO. 90 1. *Pinctada maxima* (Jameson) 2. *Placuna ephippium* (Retzius) 3. *Placuna placenta* (Linné) 4. *Crassostrea gigas* (Thunberg) 5. *Crassostrea iredelei* (Faustino) 6. *Atrina pectinata pectinata* (Linné) form *kinoshitai* Habe 7. *Atrina pectinata pectinata* (Linné) 8. *Streptopinna saccata* (Linné) 9. *Pinna muricata* Linné 10. *Atrina vexillum* (Born) 11. *Pinna bicolor* Gmelin

PLATE 90

MORPHOLOGICAL FEATURES OF THE VALVES OF TRIDACNA SPECIES

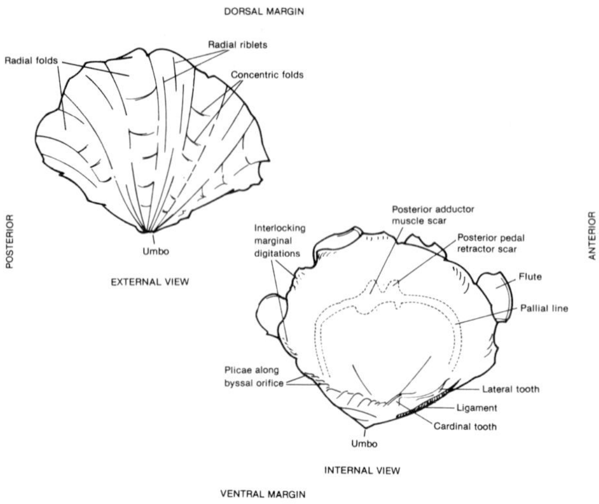

Three views of each species are given and the measurements quoted are length X breadth X thickness (co-joined valves) and refer to specimen displaying internal and external sculpturing.

1. Tridacna (Chametrachea) maxima (Röding, 1798)
Syn.: *T. elongata* Lamarck, 1819.
168mm x 89.0mm x 82.0mm; Jolo, Sulu Archipelago.
Shell large, heavy, elongate to triangularly ovate in outline, equivalve, inequilateral, hinge line short, umbones off-center and postero-medially directed; valves moderately inflated, sculptured with 8 to 12 broad, convex radial folds, weaker radial riblets on (10 to 20 in number) and between (3 to 7 in number) these folds and crossed by concentric growth folds, producing crowded scales on the primary ribs; byssal orifice large (approximately half the ventral margin length) and marginally plicate; colour white to grey, occasionally tinged with yellow or orange; may reach 330mm in length.
DISTRIBUTION: Sporadically found throughout the Philippines, mostly from Sulu Sea.

2. Hippopus hippopus (Linné, 1758)
150mm x 100mm x 87.0mm; Palawan.
Shell large, heavy, triangularly ovate in outline, with heavily inflated valves; sculptured with 11 to 12 irregularly sized, convex radial folds bearing short, prickly, semi-tubular spines, numerous weaker radial riblets on and between the major folds and all crossed by concentric growth striae; umbones nearly central, postero-medially directed, hinge-line long, byssal orifice nearly closed, the margin with up to 12 interdigitating processes, the size of which increases posteriorly; externally off-white in colour with yellowish orange suffusion, lavender to purple bands on the primary radial ribs and a pale orange byssal margin, internally white with orange-yellow colouration on the ligamental and byssal margins; may reach 400mm in length.
DISTRIBUTION: Throughout the Philippines.

3. Tridacna (Chametrachea) crocea Lamarck, 1819
111mm x 86.0mm x 58.0mm; Samar.
Shell medium sized, heavy, triangularly ovate in outline, valves markedly inflated, byssal orifice extremely large, hinge line short; externally sculptured with 6 to 10 broad, flattened radial folds, the central 4 or 5 very strong, weaker radial riblets (5 to 6 on the folds, 3 to 4 between the folds) and crossed by undulate concentric growth lines forming low scales on the primary folds near the dorsal margin, umbones off-center, postero-medially directed; external colour pale brown with orange-yellow suffusion, interior porcelaneous white with a yellow or pinkish margin; reaches 150mm in length.
DISTRIBUTION: Sporadically found throughout the Philippines.

4. Tridacna (Chametrachea) squamosa Lamarck, 1819
220mm x 150mm x 150mm; Siasi, Sulu Archipelago.
Shell large, heavy, triangularly ovate in outline, valves inflated with a small to medium sized, marginally plicate byssal orifice; sculptured with 4 to 8 convex radial folds, the central 5 or 6 very strong, and weaker interstitial radial riblets which are crossed by concentric growth striae producing pronounced, erect, scale-like flutes on the major folds; umbones centrally placed, hinge line half the shell length, internally a single cardinal tooth, 2 postero-lateral teeth in the right valve, 1 in the left valve; colour very variable, comes in different hues of orange, yellow, pink and lavender with fluted scales of darker shade or different colour, internally porcelaneous white; reaches 400mm in length.
DISTRIBUTION: Sporadically found throughout the Philippines, colorful specimens mainly from the Sulu Sea.

5. Tridacna (Persikima) derasa (Röding, 1798)
102mm x 66.0mm x 39.0mm; Bohol.
Shell large, semi-circular in outline and may be weakly or strongly inflated, with a narrow, marginally plicate byssal orifice; sculptured with between 8 and 12 weak, nearly obsolete radial folds, numerous weaker radial riblets and overriding concentric growth lamellae, dorsal margin of the valves undulate and interlocking, the internal margin weakly crenulate, umbonal area large, subcentral in juvenile forms, but posteriorly displaced in adults, hinge line longer than half the shell length; external colour dirty white, internally porcelaneous white with a yellow hinge line; may reach over 500mm in length; of infrequent occurrence, it is distinguished from *T. gigas*, which it most closely resembles, by the nearly obsolete radial folds, the long hinge line, and the large posteriorly displaced umbones.

PLATE NO. 91 1. *Tridacna maxima* (Röding) 2. *Hippopus hippopus* (Linné) 3. *Tridacna crocea* Lamarck 4. *Tridacna squamosa* Lamarck 5. *Tridacna derasa* (Röding) 6. *Hippopus porcellanus* Rosewater

PLATE 91

DISTRIBUTION: Sporadically found throughout the Philippines, mainly from the Sulu Sea.

6. Hippopus porcellanus Rosewater, 1982
350mm x 245mm x 90.0mm (large), 117mm x 84.0mm x 57.0mm (small); Jolo, Sulu Archipelago.
Shell large, triangularly ovate in outline, with heavily inflated valves, and a long, rounded, strongly undulate dorsal margin; the hinge line is long, the anterior margin rather drawn out, the posterior margin sharply truncated and delineated by a keel which becomes concave beneath the umbones; surface sculptured with up to 9 strong, wide, convex primary ribs which are somewhat radially striate, and 2 or 3 weaker interstitial radial riblets, posterior slope with 4 weak ribs followed by up to 10 weaker riblets which terminate as short, sharp denticles on the margin of the slightly open byssal orifice; external colour off-white with pale purple bands on the radial ribs and a yellowish tinge along the byssal margin, internally porcelaneous white, some purplish colouration apparent on the dorsal margin in younger specimens; may reach up to 380mm in length; similar to *H. hippopus* (Linné) but has a rounded, not triangulate dorsal margin, smoother and flatter ribs and shallower interstitial spaces; the surface colouration is slightly different; one of the most commercially important species, known locally as "China Clam". Curiously enough, it has only just been recently described.
DISTRIBUTION: Mainly found in Sulu Sea.

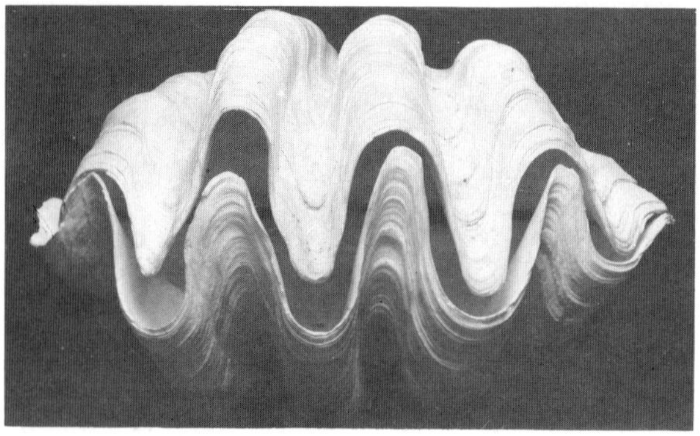

Tridacna (Tridacna) gigas (Linne, 1758)
457.2mm x 279.4mm x 254mm; Palawan

Tridacna (Tridacna) gigas (Linné, 1758)
457.2mm x 279.4mm x 254mm
Shell extremely large, heavy, sub-ovate to fan shaped, valves moderately inflated, byssal orifice slightly open or closed; sculptured with 5 to 6 strong, convex radial folds, the middle 4 being strongest, weaker radial riblets on and between the folds, and overriding concentric growth lamellae forming weak scales or nodules; umbones nearly central, hinge line long, byssal margin with 7 or 8 obsolete plicae, a single cardinal tooth in each valve; colour dull white externally, surface often covered with coralline encrustations, interior porcelaneous white; reaches over 1300mm in length and may weigh over 550 lbs.
DISTRIBUTION: Mainly from Palawan and Sulu Sea.

Spondylidae — Plate 92

Class Bivalvia
Order Mytiloida
Superfamily Pectinacea
Family SPONDYLIDAE

Shells medium to large sized, irregularly ovate to elongate-ovate in shape and generally having a convex upper valve which is sculptured with radial ribs bearing spinose projections, and a lower valve which is normally, but not always, flat and which may also bear spines. The shells are generally quite colourful with a white interior and a crenulate margin. The lower valve bears 2 cardinal teeth and a large adductor muscle scar impression. Mainly tropical in distribution, they inhabit shallow to deep water, lack a byssus and attach directly to the substrate via the lower valve. The extreme variation observed within this family often makes identification difficult. Measurement quoted are exclusive of spines. Well represented in the Philippines with many species at present unnamed.

1. Spondylus (Eleutherospondylus) regius Linné, 1758
92.0mm x 81.0mm, 96.0mm x 83.0mm, 106mm x 84.0mm; Palawan
Shell large, heavy, roundly ovate in outline; sculptured with 6 large, widely spaced primary radial ridges bearing 4 to 6 long, thick, partially closed spines whose size increases towards the ventral margin; between the major ribs are numerous weaker ribs and riblets all bearing short, sharp spines, upper and lower valves equally inflated; attached to the substrate via the lower umbonal area which is larger and more inflated than that of the upper valve; colour variable: purple, pink, reddish brown and occasionally orange, interior white, the margin finely crenulate and coloured the same as the exterior; grows to 120mm in length. Younger specimens (**1a.** 43mm × 46mm) differ from adult forms in the reduction of secondary spination; the primary spination is sometimes one of flattened, marginally sinuous spines.
DISTRIBUTION: Sporadically found throughout the Philippines, mostly from Palawan and Sulu Sea.

PLATE NO. 92 1. *Spondylus regius* Linné 1a. juvenile 2. *Spondylus versicolor* Schreibers 3. *Spondylus plurispinosus* Reeve 4. *Spondylus squamosus* Schreibers 5. *Spondylus barbatus* Reeve 6. *Spondylus candidus* Lamarck 7. *Spondylus sanguineus* Dunker 8. *Spondylus butleri* Reeve 9. *Spondylus anacanthus* Mawe 10. *Spondylus imbutus* Reeve 11. *Spondylus wrightianus* Crosse 12. *Spondylus imperialis* Chenu 13. *Spondylus multimuricatus* Reeve 14. *Spondylus setiger* Reeve 15. *Spondylus sinensis* Schreibers 16. *Spondylus varians* Sowerby

PLATE 92

2. Spondylus (Spondylus) versicolor Schreibers, 1793
Syn: *S. aurantius* Lamarck, 1819
64.0mm x 54.0mm; Palawan.
Shell medium sized, solid, roundly ovate, both valves equally inflated, but the umbonal area of lower valve is larger and deeper than the upper; sculptured with numerous radial rows of ribs each bearing long, curved, somewhat spatulate spines, strongest at the periphery, interstitial sculpturing weaker; colour deep yellowish orange, with pale umbones, a white interior and a finely crenulate, orange-coloured inner margin; reaches 80mm in length.
DISTRIBUTION: Sporadically found throughout the Philippines.

3. Spondylus (Spondylus) plurispinosus Reeve, 1856
81.0mm x 72.0mm; Palawan.
Shell large, solid, variable in outline from trigonally ovate to elongate-ovate, upper valve more inflated than the lower, but the umbonal area of the lower valve is more deeply excavated; surface sculptured with up to 26 radial ribs, each bearing numerous close-set spines which become flattened towards the margin; externally, ribs and spines white, brown in the interspaces, interior white, the margin coarsely crenulate and stained yellowish brown, the umbones white and stained with blackish brown; reaches 75mm in length. Close to *S. nicobaricus* Schreibers, 1793, but that species has much finer spination and is red or white with rays of dark umbonal spots.
DISTRIBUTION: Sporadically found throughout the Philippines, mostly from Palawan and Sulu Sea.

4. Spondylus (Spondylus) squamosus Schreibers, 1793
Syn: *S. ducalis* Röding, 1798
85.0mm x 70.0mm; Palawan.
Shell moderately large, solid, elongate-ovate in outline, both upper and lower valves equally inflated; scupltured with low primary radial ridges bearing flattened spines which enlarge towards the margins; between the primary ridges are numerous close-set interstitial radial striae; shell blackish in colour with white primary ridges and spines, internally white, brown at the hinge line and black spotted around the finely crenulate inner margin; reaches 100mm in length. (Specimen here with lower valve uppermost to illustrate colour and structure).
DISTRIBUTION: Sporadically found throughout the Philippines.

5. Spondylus (Spondylus) barbatus Reeve, 1856
Syn: *Spondylus sinensis* Sowerby, 1847 (non Schreibers, 1793)
63.0mm x 50.0mm; Jolo, Sulu Archipelago.
Shell medium sized, solid, ovate in outline, the upper valve more inflated than the lower, but the lower umbonal area quite deep; sculptured with numerous radial rows of spines, short and sharply spinose initially, but becoming larger, spatulate and more densely packed towards the ventral margin; colour variable, usually pinkish purple with paler spotted umbones and white spines, interior white with a finely crenulate, gold-coloured margin; reaches 80mm, in length.
DISTRIBUTION: Sporadically found throughout the Philippines, mostly from Sulu Sea and Palawan.

6. Spondylus (Eltopera) candidus Lamarck, 1819
94.0mm x 82.0mm; Cebu.
Shell large, solid, irregularly subcircular in outline, the upper valve slightly more inflated than the lower; sculptured with numerous radial ribs which bear very short, sharp spines; surface colouration pinkish purple but generally eroded to a whitish shade, with lavender spines; upper valve umbone small and streaked with black lines, internally creamy white with some yellowish orange suffusion and a brownish purple, coarsely crenulate inner margin; reaches 100mm in length.
DISTRIBUTION: Sporadically found throughout the Philippines.

7. Spondylus (Eltopera) sanguineus Dunker, 1852
52.0mm x 50.0mm; Punta Engaño, Cebu.
Shell medium sized, outline irregularly triangular-ovate, with a short anterior margin and a long posterior margin; umbones centrally placed, the lower one sharply truncated, ears unequal, valves moderately inflated, the lower one frequently box-like and attached to the substrate; sculptured with strong, weakly nodulose radial cords and weaker intermediate riblets; external colouration deep pinkish purple to red, internally white with a finely crenulate, purple-coloured margin; reaches 55mm in length.
DISTRIBUTION: Camotes Sea.

8. Spondylus (Spondylus) butleri Reeve, 1856
89.0mm x 62.0mm (yellow), 72.0mm x 92.0mm (red); Palawan.
Shell medium sized, variable in outline (ovate to elongate-ovate), the upper valve moderately flat, the lower valve deeply concave especially in the umbonal area; sculptured with numerous radial ribs bearing short, sharp spines which increase in length and become flattened towards the ventral margin; variable in colour from light to dark hues of orange, yellow, brown and purple, the spines being orange, yellow or lavender, umbones off-white, spotted and frequently eroded, internal colouration white with a finely crenulate margin the same colour as the external spination; reaches 100mm in length.
Erroneously sold as *Spondylus coccineus.*
DISTRIBUTION: Sporadically found throughout the Philippines, mostly from Palawan and Sulu Sea.

9. Spondylus (Eltopera) anacanthus Mawe, 1823
43.0mm x 41.0mm; Cebu.
Shell small, rather lightweight and irregularly ovate in outline, upper valve moderately inflated, lower valve deeply concave especially at the umbonal area; sculptured with numerous radial ribs, bearing fine lavender-coloured spines which increase in size towards the margin, especially the postero-ventral margin, base relatively spine-free except for a few large marginal ones; white in colour with pinkish

umbonal rays and pink spines, interior, creamy white with a finely crenulate margin; reaches 60mm in length.
DISTRIBUTION: Camotes Sea, Sulu Sea and Palawan.

10. Spondylus (Spondylus) imbutus Reeve, 1856
45.0mm x 46.0mm; Davao.
An exquisitely beautiful species immediately recognized by the fine, delicate leaf-shaped spinose projections which develop towards the margin; the shell is subcircular in outline, the upper valve weakly inflated, the lower deeply concave; sculptured with 8 to 10 major radial ribs bearing short, sharp spines which become leaf-like towards the margin, a single secondary and numerous tertiary radial riblets, all of which are finely granulose and have very fine spines; light pinkish red in colour, the leaf-like projections paler, internally white with a finely crenulate, pinkish yellow margin; few specimen so far seen.
DISTRIBUTION: Southern Mindanao.

11. Spondylus (Eleutherospondylus) wrightianus Crosse, 1872
42.0mm x 42.0mm; Bulan, Sorsogon.
Shell small, solid, ovate in outline, both valves equally inflated; sculptured with numerous sharply angulate radial ribs bearing spines; there are 6 to 7 major ribs upon which the spinal development is more pronounced, especially towards the margins where the spines become extremely long; the remaining ribs bear spines of a weaker strength, usually a strong central row and two weaker rows on either side; shell pure white internally and externally, with pale yellow umbones; reaches around 50mm in length.
DISTRIBUTION: Sorsogon, Camotes Sea and Sulu Sea.

12. Spondylus (Eleutherospondylus) imperialis Chenu, 1843
Syn: *Spondylus (Spondylus) ciliatus* Sowerby, 1847
61.0mm x 54.0mm; Palawan.
Shell medium sized to large, heavy, triangularly ovate in outline, umbones centrally placed, valves equally inflated; sculptured with 6 widely spaced, primary radial ribs bearing between 4 and 6 thick, semi-tubular spines; between the major ribs are 3 secondary riblets which may in themselves be further bisected into tertiary riblets, and all bearing smaller, more numerous spines; shell pure white in colour, occasionally with some darker suffusion, internally white, with a crenulate margin and a yellowish tinge near the teeth; reaches around 80mm in length.
DISTRIBUTION: Sorsogon, Camotes Sea and Sulu Sea.

13. Spondylus multimuricatus Reeve, 1856
65.0mm x 57.0mm; Palawan.
Shell medium sized, solid, elongate-ovate in outline with equally inflated valves; sculptured with numerous primary and secondary scabrous radial ribs, bearing short, sharp spines which are slightly more pronounced on the anterior and posterior margins; externally, the shell is bright yellow, internally white with a finely crenulate, yellow-coloured margin; appears to reach around 75mm in length.
DISTRIBUTION: Palawan and Sulu Sea.

14. Spondylus (Spondylus) setiger Reeve, 1856
65.0mm x 86.0mm; Davao.
Shell medium sized, solid, subcircular in outline, upper valve moderately inflated, lower valve deeply concave especially at the umbonal area; surface sculptured with numerous primary and secondary radial cords, the primary cords with large, solid spines, increasing in size and length peripherally and those on the secondary cords much finer in structure; pinkish lavender in colour, the spines paler, internally white with a finely crenulate, yellowish mauve margin; reaches around 75mm in length.
DISTRIBUTION: Southern Mindanao.

15. Spondylus (Eleutherospondylus) sinensis Schreibers, 1793
Syn: *S. pesaninus* Röding, 1798
73.0mm x 65.0mm (yellow), 65.0mm x 60.0mm (red), 86.0mm x 81.0mm (reddish brown); Palawan.
Shell medium sized, subcircular in outline, lower valve more inflated than the upper especially at the umbonal area; sculptured with 6 primary radial ribs, somewhat elevated and bearing between 5 and 10 spines which increase in size and become more spatulate towards the margin; interstitial spaces with finely scabrous secondary and tertiary riblets, some of which bear a few weak spines; colour variable, white, pink, orange or purple, frequently more intense towards the umbones; spines pale lavender, internally white with a finely crenulate margin coloured the same as the external surface; reaches 100mm in length.
DISTRIBUTION: Palawan and Sulu Sea.

16. Spondylus (Spondylus) varians Sowerby, 1829
95.0mm x 77.0mm (yellow), 74.0mm x 72.0mm (red), 140mm x 110mm; Palawan.
Shell large, solid, heavy, ovate to elongate-ovate in outline, immediately recognized by its pure white colour and the presence of a highly colored umbonal area; upper valve almost flat, the lower deeply concave and the surface sculptured with fine concentric growth folds and weak radial ribs bearing short, bead-like spines that become longer and sharper towards the margin; the umbonal area of the upper valve is small, while that of the lower valve is extremely large; colour pure white with highly coloured umbones (red, yellow or orange), internally white with a finely crenulate, occasionally brown-stained margin; specimens are known to approach 180mm in length; often called the "water spondylus" on account of a large, crater-like depression in the upper and lower valves which is covered with a thin, calcareous membrane, entrapping pockets of water within the shell structure. (See arrows).
DISTRIBUTION: Camotes Sea, Palawan and Sulu Sea.

Class Bivalvia
Order Mytiloida
Superfamily Pectinacea
Family PECTINIDAE

A large group of bivalves living in shallow to deep water in most temperate and tropical seas. The shells range from small to very large, are auriculate, ovate in outline and inequivalve, the lower valve usually more inflated than the upper. Externally sculptured with radial ribs and occasionally concentric lamellae. Internally, the shells have a single adductor muscle scar impression, a byssal notch below the right anterior ear, and weak, symmetrical, lamelliform hinge teeth. The taxonomy of this group is still under revision. Measurements represent the maximum shell length. Well represented in the Philippines.

1. Amusium pleuronectes pleuronectes (Linné, 1758)
73.2mm; Manila Bay.
Shell medium sized, thin, circular in outline, with small equal ears, centrally located umbones and moderately inflated valves; interior of the upper valve with around 30 to 35 ribs which are obsolete on the anterior and posterior margins, lower valve similarly sculptured, and the outer surface of both valves is smooth; external colouration white on the lower valve, the upper valve pinkish grey with darker concentric striae and pale purple radial lines, internal colouration white on the lower valve, upper valve white with central yellow blotching and lavender margin; reaches around 80mm.
DISTRIBUTION: Sporadically found throughout the Philippines.

2. Pecten (Minnivola) pyxidatus (Born, 1778)
36.7mm; Tayabas Bay, Quezon.
Shell small, ovate, umbones central, ears nearly equal; upper valve flat to slightly concave with around 26 rounded, equi-spaced radial ribs which are also present on the inner surface, lower valve deeply concave with up to 30 radial ribs on the inner and outer surface; upper valve pale pink to dark brown with darker radial bands and white blotching near the ventral margin, lower valve white with some brown umbonal spotting, interior of both valves white; reaches 40mm.
DISTRIBUTION: Sporadically found throughout the Philippines in limited quantity; always from deep water.

3. Serratovola tricarinata (Anton, 1839)
27.9mm; Tayabas Bay, Quezon.
Shell small, ovate, umbones central, ears nearly equal; upper valve flat with up to 20 rounded, equi-spaced radial ribs, lower valve moderately convex, with 18 to 20 trisected radial ribs producing a serrated margin; upper valve light pinkish brown in colour with darker concentric streaks in the interspaces, lower valve white with a brown-tinted, yellow-coloured umbonal area, internally white with a brown umbonal stain; seldom exceeds 30mm.
DISTRIBUTION: Sporadically found throughout the Philippines in limited quantity; always from deep water.

4. Serratovola aspera (Sowerby, 1842)
37.0mm; Palawan.
Shell medium sized, ovate, umbones central, ears nearly equal; upper valve flat and sculptured with 18 to 20 finely scabrous, equi-spaced radial ribs, lower valve moderately convex with a similar surface sculpture; colour off-white with pinkish brown staining between the ribs and fine concentric striae on the ribs, internally white with a lavender-coloured umbonal blotch; reaches 50mm.
DISTRIBUTION: Sporadically found throughout the Philippines in limited quantity; always in deep water.

5. Chlamys (Mimachlamys) lentiginosa (Reeve, 1865)
50.3mm, 48.0mm; Siasi, Sulu Archipelago.
Shell medium sized, slightly elongate-ovate, umbones central, ears unequal, those on the lower valve more coarsely sculptured; surface with up to 30 radial ribs bearing small fluted spines and 1 or 2 radial threads in the interstitial spaces; colour variable, black, reddish brown, and purplish, maculated with white streaks and bands; interior colour reflecting external colouration, somewhat paler at the umbones; reaches 50mm.
DISTRIBUTION: Sulu Sea.

6. Gloripallium speciosum (Reeve, 1853)
45.8mm; Mactan, Cebu.
Shell medium sized, ovate-triangular, umbones central, anterior ear larger than the posterior, posterior byssal notch prominent; entire surface sculptured with wide, rounded, narrowly spaced, fluted radial cords, 4 to 5 on the ears and up to 14 on the valves; colour creamy white with light and dark reddish maculations, somewhat darker on the upper valve which has additional brown spotting, interior white; reaches 50mm.
DISTRIBUTION: Camotes Sea, deep water.

7. Annachlamys macassarensis (Chenu, 1845)
43.5mm; Tayabas Bay, Quezon.
Shell medium sized, ovate, height and width almost equal, ears equal, gaping at the margins, umbones central and somewhat flattened, upper valve less

PLATE NO. 93 1. *Amusium pleuronectes pleuronectes* (Linné) 2. *Pecten pyxidatus* (Born) 3. *Serratovola tricarinata* (Anton) 4. *Serratovola aspera* (Sowerby) 5. *Chlamys lentiginosa* (Reeve) 6. *Gloripallium speciosum* (Reeve) 7. *Annachlamys macassarensis* (Chenu) 8. *Anguipecten superbus* (Sowerby) 9. *Gloripallium pallium* (Linné) 10. *Semipallium tigris* (Lamarck) 11. *Chlamys squamata* (Gmelin) 12. *Bractechlamys vexillum* (Reeve) 13. *Decatopecten striatus* (Schumacher) 14. *Comptopallium radula* (Linné) 15. *Semipallium luculentum* (Reeve) 16. *Chlamys squamosa* (Gmelin) 17. *Chlamys mirifica* (Reeve) 18. *Annachlamys reevei* (Adams & Reeve) 19. *Chlamys senatoria* (Gmelin) 20. *Chlamys rastellum* (Lamarck) 21. *Chlamys thaanumi* (Dall, Bartsch & Rehder)

PLATE 93

inflated than the lower; sculptured with between 15 and 20 rounded radial ribs and undulate, overriding concentric grooves; pinkish white in colour with red lines and blotches on the ribs, lower valve paler with fewer markings, interior white, suffused with yellow; reaches 50mm.
DISTRIBUTION: Manila Bay and Tayabas Bay, Quezon.

8. Anguipecten superbus (Sowerby, 1842)
58.0mm; Cebu.
Shell medium sized, triangularly elongate-ovate in outline, the posterior margin somewhat more produced than the anterior, giving a slightly lobed appearance to the shell outline; umbones central, ears unequal, no visible byssal notch; valves moderately flat with up to 22 rounded ribs; colour pale pinkish orange with darker blotching lined in creamy white, lower valve paler in colouration, interior white with a bright pink suffusion; reaches 65mm.
DISTRIBUTION: Camotes Sea, deep water.

9. Gloripallium pallium (Linné, 1758)
57.0mm; Cebu.
Shell medium sized, ovate, height and width almost the same, both valves equally inflated, umbones central, anterior ear largest with a prominent, orange-stained byssal notch; sculptured with 13 to 14 large, rounded, widely spaced radial ribs, each bearing 2 or 3 highly scabrous radial riblets, interspaces with 2 or 3 scabrous radial cords; external colour in varying shades of deep reddish purple with white concentric banding, umbonal area white, interiorly white, the margin the same colour as the exterior; reaches 90mm.
DISTRIBUTION: Camotes Sea and Sulu Sea.

10. Semipallium tigris (Lamarck, 1819)
57.8mm; Palawan.
Shell medium sized, ovate, with a long, rounded ventral margin, and slightly lobate on the postero-ventral margin, umbones central, anterior ear larger than the posterior with a pink coloured byssal notch; surface with around 10 rounded radial folds and numerous thin radial riblets on and between the folds; colour creamy white with black blotching, paler on the lower valve, umbones yellow, interior white and suffused with bright yellow, hinge line brown; reaches 60mm.
DISTRIBUTION: Palawan and Sulu Sea.

11. Chlamys (Azumapecten) squamata (Gmelin, 1791)
29.5mm, 37.3mm, 45.0mm; Zamboanga.
Shell small to medium sized, lightweight, length and width almost equal, umbones central, anterior ear largest, the one on the lower valve marginally serrate; upper valve sculptured with 10 elevated radial ribs bearing fluted spines and numerous interstitial radial riblets, lower valve with more numerous weakly spinose ribs; colour variable, off-white to dark red, yellow and black, oftentimes suffused with large lavender areas and thin white radial lines, interior white and suffused with pink or lavender; reaches 50mm.
DISTRIBUTION: Sulu Sea.

12. Bractechlamys vexillum (Reeve, 1853)
40.0mm, 41.5mm; Sulu Archipelago.
Shell medium sized, solid, length and width almost equal, umbones central, ears equal, byssal notch weak; upper valve less inflated than the lower, both sculptured with 11 to 12 rounded radial ribs and numerous radial riblets on and between the major ribs; colour variable, yellow, orange or brown, umbones paler and blotched with white, interior white and suffused with the exterior colouration, hinge line brown; reaches 50mm.
DISTRIBUTION: Sporadically found throughout the Philippines, mostly from Sulu Sea.

13. Decatopecten striatus (Schumacher, 1817)
30.8mm, 39.8mm; Manila Bay.
Shell medium sized, solid, triangularly ovate in outline, umbones central, anterior ear slightly larger than the posterior; both valves equally inflated, sculptured with 5 large, elevated, well rounded radial cords and numerous radial riblets on and between these cords, internal margin finely crenulate; colour variable, off-white with dark brown or orange-coloured concentric banding, lower valve usually devoid of markings, internally white suffused with brown; reaches 50mm.
DISTRIBUTION: Manila Bay and Tayabas Bay, Quezon.

14. Comptopallium radula (Linné, 1758)
67.8mm; Cebu.
Shell large, solid, triangularly ovate, left margin shorter than the right, upper valve with a flattened umbonal area and less inflated than the lower; both valves sculptured with 12 raised, rounded radial ribs on and between which are smaller scabrous radial riblets, ears equal, byssal notch obsolete; cream-coloured with concentric brown spots and bands on the upper valve, absent on the lower, interior white and suffused with yellow, hinge line brown, umbones yellowish; reaches 75mm.
DISTRIBUTION: Sporadically found throughout the Philippines.

15. Semipallium luculentum (Reeve, 1853)
32.2mm, 35.3mm, 44.1mm; Siasi, Sulu Archipelago.
Shell small to medium sized, length greater than the width, anterior margin shorter than the posterior, umbones central, anterior ear largest with a distinct byssal notch; valves weakly inflated, almost flat, sculptured with 7 to 8 rounded, widely spaced radial folds and numerous finely scabrous radial riblets on and between these folds; colour variable, purple, yellow, orange or red, with white lines or spots, lower valve usully paler; reaches 50mm.
DISTRIBUTION: Sulu Sea.

16. Chlamys (Chlamys) squamosa (Gmelin, 1791)
43.5mm, 45.3mm, 38.7mm; Siasi, Sulu Archipelago.
Shell small to medium sized, triangularly ovate, umbones central, the upper one somewhat flattened, anterior ear largest with a prominent byssal notch; valves equally inflated but almost flat, sculptured with scabrous radial ribs, weaker radial riblets and fine concentric striae; colour variable, yellow, red, orange or purple often with darker markings, internal colouration the same as the exterior; reaches 50mm.
DISTRIBUTION: Sulu Sea.

17. Chlamys (Mirapecten) mirifica (Reeve, 1853)
29.3mm; Cebu.
Shell small, roundly triangular in outline, umbones central, anterior ear slightly larger than the posterior, byssal notch distinct; upper valve less inflated than the lower, and both are sculptured with 7 or 8 large, rounded, irregularly spaced radial ribs bearing wide fluted scales on the upper valve and rounded nodulose spines in the lower valve; fluted scales also occur on the ears of both valves; external colour variable, bright crimson suffused with yellow or purple, interior highly porcelaneous and the same colour as the exterior; reaches 30mm.
DISTRIBUTION: Sulu Sea.

18. Annachlamys reevei (Adams & Reeve, 1850)
34.3mm; Siasi, Sulu Archipelago.
Shell medium sized, triangularly ovate in outline, umbones central, anterior ear larger than the posterior, byssal notch small but distinct; valves equally inflated, sculptured with 20 to 22 rounded radial ribs which are crossed by numerous closely packed, deeply incised concentric grooves producing a finely scabrous surface; colour usually white with red blotches and bands, interior white with a pink suffusion; reaches 40mm.
DISTRIBUTION: Sulu Sea.

19. Chlamys (Mimachlamys) senatoria (Gmelin, 1791)
65.5mm, 59.6mm, 48.0mm; Sulu Archipelago.
Shell medium to large sized, triangularly ovate, length and width approximately equal, umbones central, anterior ear much larger than the posterior, those of the lower valve with coarsely crenulate margins; valves equally inflated, sculptured with numerous rounded radial ribs, radial grooves and overriding concentric lamellae; colour variable, red, yellow, orange or purple with white markings, or yellow with darker markings, internally white with a marginal colouration the same as the exterior; reaches 75mm.
DISTRIBUTION: Sulu Sea and Bantayan Island, Northern Cebu.

20. Chlamys (Mirapecten) rastellum (Lamarck, 1819)
36.5mm; Punta Engaño, Cebu.
Shell small to medium sized, triangularly ovate, umbones central, anterior ear slightly larger than the posterior, those on the lower valve with coarse ribs and a fluted margin, byssal notch distinct; upper valve less inflated than the lower, sculptured with 8 to 10 elevated, rounded radial ribs bearing numerous upright, widely rectangular fluted scales, more numerous on the lower valve but weaker in strength; the ventral margin has a characteristic "W-shaped" profile; off-white in colour with pinkish purple bands; reaches 50mm but rarely exceeding 35mm.
DISTRIBUTION: Cebu and Bohol.

21. Chlamys (Mirapecten) thaanumi (Dall, Bartsch & Rehder, 1939)
41.7mm; Punta Engaño, Cebu.
Often synonimized with *C. rastellum* but appears to differ in several fairly constant features: the valves are normally less inflated, the top one being flat, not weakly inflated as in *rastellum;* the ventral margin is undulate, not sharply angulate ("W-shaped") and the spines are tall and semi-circular; the colour pattern is also quite variable, usually pale yellowish white with dark purple suffusion and white zig-zag markings, as compared with the fairly constant colour and pattern of *rastellum;* reaches 45mm; now believed to be a form of *C. mirifica* (fig. 17).
DISTRIBUTION: Cebu and Bohol.

Gastropod Species — Plate 94

This plate, together with the following two (Pls. 95 & 96), illustrates an assortment of gastropod species from various family groups already covered earlier within this guide. They represent species which are either extremely rare and difficult to obtain, or species which, although not considered rare, have proved to be extremely elusive.

Family CYMATIIDAE (cont'd.) (see also Pls. 30-31)

1. Cymatium (Ranularia) pyrulum (A. Adams & Reeve, 1850)
60.9mm; Samar.
Best described as being like *C. caudatum* (Gmelin), but lacking the deep sutural groove of that species. Shell medium sized, solid, spire moderately tall with 4 to 5 whorls, the earliest of which are finely clathrate from intersecting axial and spiral threads, later whorls slightly angulate with 2 axially filleted spiral cords at and below the shoulder and weaker threads on the presutural slope; undulate axial folds on the last 2 whorls, intersecting the spiral cords to produce flattened tubercles; body whorl with 6 major cords, a single interstitial thread and two varices which are characteristically deeply channeled behind; canal moderately long, open and slightly curved, with up to 16 cords along its length; aperture elongate-ovate, outer lip with 7 paired denticles, columella smooth with a moderately thickened, sharply delineated callus; colour pale orange-brown, varices brown and white-banded, aperture white; reaches 65mm.
DISTRIBUTION: Sporadically found throughout the Philippines in limited quantity.

2. Cymatium (Ranularia) c.f. moniliferum (A. Adams & Reeve, 1850)
58.0mm; Bohol.
Nearly always mixed with *C. gutturnium* but differs in the following ways: it is generally smaller in size, has a pink aperture (bright orange-brown in *gutturnium*), a darker external colouration (pale to medium tan vs. creamy white), a lower varix which is flattened and pushed back from the aperture (varix evenly rounded in typical *gutturnium*), a curious channel at the top of the aperture between the lip, varix and penultimate whorl (only seen occasionally in *gutturnium*), and 4 or 5 prominent, raised cords on the siphonal canal; reaches around 60mm.

Triton monilifer A. Adams & Reeve, 1850 has been the subject of much controversy since its original description. The somewhat brief, poorly detailed description, the lack of any type material, and the nebulous type locality (Eastern Seas) have all compounded to make the identification of this species difficult. Currently available evidence has established that true *moniliferum* is endemic to East Africa, with material from that area almost identically matching the type figure. Through the courtesy of Dr. R. N. Kilburn (Natal Museum, South Africa), the authors have been able to examine a specimen from East Africa and to compare it with the Philippine shells. The East African taxon was found to exhibit the following differences: the size is smaller with a white-coloured aperture, the columellar dentition is strongest on the parietal wall, reducing rapidly anteriorly, the whorls are more angulate at the shoulder, the shoulder tubercles on the dorsum are more echinate and more closely crowded together, the outer lip varix is more rounded, the anterior canal is less reflexed and the 4 or 5 raised cords on the canal are very poorly defined.

The relationship between the two taxa is very close, but whether the Philippine shells are conspecific with true *moniliferum* or an entirely distinct taxon is unknown at this stage. It is most likely that subspeciation may be occurring, but until the distribution characteristics of the 2 shells are more accurately known, we provisionally annotate Philippine shells as *c.f. moniliferum*.
DISTRIBUTION: Sporadically found throughout the Philippines in limited quantity.

3. Cymatium (Ranularia) oblitum (Lewis & Beu, 1976)
(Paratype: New Zealand Geological Survey no. 11895)
38.4mm; 1½ miles off Vlaming Head Lighthouse, N.W. Cape, W. Australia.
Species small for the group, with a rounded body, low spire, and a long, narrowly open, distally recurved siphonal canal; early whorls with spiral lirae which are later crossed by axial folds forming nodules, a single shoulder row on the penultimate whorl, 2 on the body whorl; aperture narrowly elongate-ovate, outer lip thick with a flattened varix, dentate within, the upper two denticles bifid, columella calloused and strongly plicate. This species is recorded from the Philippines (Punta Engaño, Cebu) and is known to us from 2 specimens (H. Lee collection and V. Dan collection).

Family CYPRAEIDAE (cont'd.) (see also Pls. 20-24)

4. Cypraea katsuae (Kuroda, 1950)
21.5mm, 19.7mm; Panglao, Bohol.
Shell medium sized, roundly elongate and moderately inflated, tips moderately produced; dorsum pinkish grey in colour with numerous small brown dots (frequently absent) and 1 to 3 dark brown transverse bands, sides orange-yellow with large, dark marginal blotches which are absent from the base; columellar teeth numerous, fine and restricted to the aperture, labral teeth less numerous, brown-stained and produced across the base; reaches around 23mm.
DISTRIBUTION: Bohol, Cebu and Southern Mindanao.

5. Cypraea sakuraii (Habe, 1970)
48.6mm; Balut Is., Southern Mindanao.
Shell medium sized, elongate-ovate and fairly inflated with moderately produced tips; aperture narrow, slightly curved, labral teeth numerous and coarse, columellar teeth finer and more numerous, both groups of teeth restricted to the aperture; base white, sides tan coloured, extending onto the dorsal surface where it is replaced with a finely reticulate, dark brown blotch; reaches 50mm.
DISTRIBUTION: Camotes Sea and Southern Mindanao.

Family TEREBRIDAE (cont'd.) (see also Pls. 73-75)

6. Terebra vicdani Kosuge, 1981
79.8mm; Bohol.
Shell large, moderately glossy, with a tall spire of around 21 straight-sided whorls, weakly incised sutures below which is a deep presutural groove; surface sculptured with numerous axial threads and crossed by spiral lirae, 3 above the presutural groove, 5 below, producing a finely latticed surface; aperture narrowly elongate, columella calloused with a single basal fold; spire white, rest of the shell pale yellow with 3 spiral rows of brown dots on all but the last whorl which has 4; reaches 80mm.
DISTRIBUTION: Bohol.

Family CORALLIOPHILIDAE (cont'd.) (see also Pls. 43, 44 & 95)

7. Latiaxis (Babelomurex) filiaregis Kurohara, 1959
48.2mm; Bohol.
Shell medium sized, elongate-ovate, somewhat

PLATE NO. 94 1. *Cymatium pyrulum* (A. Adams & Reeve) 2. *Cymatium c.f. moniliferum* (A. Adams & Reeve) 3. *Cymatium oblitum* (Lewis & Beu) 4. *Cypraea katsuae* (Kuroda) 5. *Cypraea sakuraii* (Habe) 6. *Terebra vicdani* Kosuge 7. *Latiaxis filiaregis* Kurohara 8. *Coralliobia fimbriata* (A. Adams) 9. *Latiaxis vicdani* Kosuge 10. *Trophon sp.* or *Latiaxis sp.* 11. *Oliva vicdani* da Motta 12. *"Fasciolaria" walleri* Ladd (teratological)

PLATE 94

fusiform in outline, with a tall, stepped spire and a large body whorl which is sharply attenuated towards the base; sculptured with numerous axial cords that are crossed by highly scabrous spiral lirae producing short, sharp spines, strongest at the angled shoulder; interstitial spaces with weaker spiral threads and numerous finely scabrous axial lamellae; aperture elongate-ovate, outer lip finely serrate and lirate within, columella smooth, canal long, fasciolar ridge finely corded; pure white throughout; reaches 50mm.
DISTRIBUTION: Cebu and Bohol.

8. Coralliobia fimbriata (A. Adams, 1854)
14.0mm x 13.5mm, 16.0mm x 16.5mm; Mactan Is., Cebu.
Shell small, dome-shaped, somewhat flattened, with a characteristically well expanded outer lip which is finely lirate on the inner margin and finely fimbriate on the outer margin; spire somewhat submerged, surface sculptured with sharp axial lamellae and crossed by spiral cords forming a coarsely reticulate surface; aperture sub-circular, outer margin ledged, columella smooth; off-white in colour; may reach 20mm in length.
DISTRIBUTION: Cebu and Bohol.

9. Latiaxis (Latiaxis) vicdani Kosuge, 1981 (paratype)
33.0mm; Mactan Is., Cebu.
Shell medium sized and delicately sculptured, fusiform in shape and pagoda-form in outline due to a thick, elevated, weakly spinose carinal ridge on the shoulder of each whorl; surface sculptured with numerous highly scaled spiral cords and weak axial lamellae in the wide interspaces, body whorl with around 9 spiral cords above the shoulder, the third one strongest, and up to 25 below; aperture ovate, outer lip finely crenulate with a narrow, deep sinus posteriorly, columella smooth, canal long, fasciolar cord densely lamellate; pure white in colour; reaches 35mm.
DISTRIBUTION: Cebu.

10. Trophon sp. or **Latiaxis (Lamellatiaxis) sp.**
31.8mm; Mactan Is., Cebu.
Shell medium sized, lightweight, elongate-ovate in outline, spire moderately tall and somewhat pagoda-form, body whorl large; sculptured with numerous weakly fimbriate blade-like varices, weakly channeled on the leading edge and forming short, sharp semi-circular spinose projections on the sharply angled shoulder; whorls with several spiral cords in the interspaces but absent on the presutural slope; aperture elongate-ovate, outer lip thin and flattened into a varical blade which has weak sinus posteriorly and several lirae on the internal margin; columella smooth with a weakly expansive, posteriorly adherent callus, siphonal canal long, open and recurved, fasciolar cord coarsely lamellate; shell colour pale pinkish orange with a porcelaneous deep pink aperture; figured specimen the only one seen. Until anatomical studies are done, the exact taxonomic status of this shell (even at a family level) will remain in doubt. Possibly *L. (L.) marumai* Habe & Kosuge
DISTRIBUTION: Cebu.

Family OLIVIDAE (cont'd.) (see also Pls. 50-52)

11. Oliva (Strephona) vicdani da Motta, 1982 (paratype)
31.5mm; Sulu Sea (near Zamboanga).
Shell medium sized, glossy, with a moderately tall spire, channeled sutures and a long cylindrical, uniformly wide body whorl; aperture narrow, widening anteriorly, columella calloused and lirate along almost the entire length, outer lip thick and smooth; nuclear whorls white, rest of the shell with a characteristic rosy vermillion tint, slightly more intense on the spire whorls; the body whorl is superimposed with numerous white tents of varying sizes, most prominent on the upper area, aperture and columella white; reaches 32mm. Named after V. Dan
DISTRIBUTION: Sulu Sea.

Family FASCIOLARIIDAE (cont'd.) (see also Pls. 47, 48 & 95)

12. "Fasciolaria" walleri Ladd, 1976 (teratological)
53.1mm; Panglao, Bohol.
Shell medium sized, fusiform-elongate in outline, with a tall acute spire, deeply incised sutures and a long, narrowly open, weakly recurved siphonal canal; 1½ to 2 apical whorls, followed by 7 to 8 post-natal whorls which are sculptured with narrow axial folds, weakly tuberculate at the sharply angulate shoulders, and encircled with numerous fine spiral lirae; below the sutures are 2 rows of small, axially fused granules; aperture elongate-ovate and lirate within, outer lip thick, columella smooth with a weak callus; shell pale brown in colour with diffuse brown banding, brown streaks on the axial folds, white shoulder tubercles, and a pale, bluish white aperture; largest seen 57.4mm. Apart from the weakly tuberculate, sharply angled shoulder and 2 rows of subsutural granules, the shell is almost morphologically indistinguishable from the typical form (Pl. 47, fig. 17). A fair number of specimens have been examined, all identical to the above form, suggesting that it may represent a good taxon rather than a freakish, teratological form of *walleri*.
DISTRIBUTION: Presently known only from Bohol.

Gastropod Species — Plate 95

Family NASSARIIDAE (cont'd.) (see also Pls. 41-42)

1. Nassarius accuminatus (Marrat, 1880)
30.4mm; Palawan.
Shell medium sized, solid and glossy, whorls 7 to 8 in number, weakly convex with ledged, deeply channeled sutures, and sculptured with numerous axial furrows, restricted to the sutures of the body whorl as a row of sutural nodules, a few spiral grooves on the body whorl base; aperture elongate-ovate, outer lip lirate within, with a small notch posteriorly and coarsely serrated basally, columella moderately calloused and

plicate; pale greyish brown in colour with 1 or 2 obsolete brown bands on the body whorl dorsum, aperture white, greyish brown within; reaches 35mm.
DISTRIBUTION: Sporadically found throughout the Philippines in limited quantity.

Family CORALLIOPHILIDAE (cont'd.) (see also Pls. 43, 44 & 94)

2. Latiaxis (Echinolatiaxis) tuberosus Kosuge, 1980
29.0mm; Panglao, Bohol.
Shell medium sized, solid, with a tall spire and a somewhat inflated body whorl; early whorls with 7 to 8 axial cords and 3 to 4 weaker spiral threads, later whorls with lamellated folds on the axial cords and small spinose projections where crossed by spiral lirae; shoulder spines strongest, those above and below weaker, on the body whorl the spines are occasionally stunted, upturned and narrowly open, interspaces between axial and spiral elements deeply pitted and brown-coloured; aperture ovate and lirate within, outer lip thin and coarsely serrated, columella smooth, fasciolar cord coarsely lamellate; pale whitish yellow in colour, aperture white; reaches 35mm.
DISTRIBUTION: Camotes Sea and Bohol.

3. Coralliophila (Lepadomurex) solutistoma (Kuroda & Shikama, 1966)
25.5mm; Panglao, Bohol.
Shell medium sized, lightweight, with a moderately tall spire and a somewhat inflated body whorl; sculptured with numerous rounded axial folds which are crossed by densely packed, scabrous spiral lirae; aperture widely elongate-ovate, smooth within, outer lip finely serrate, columella smooth and somewhat lamelliform, umbilicus slightly open, fasciolar cord finely scabrous; external colour a deep orange-brown, aperture bluish white; close to *C. elvirae* (Pl. 43, fig. 17), but that species has a finer surface sculpture and much narrower contours; reaches around 25mm; appears to be a fairly polymorphic taxon.
DISTRIBUTION: Camotes Sea and Bohol.

Family FASCIOLARIIDAE (cont'd.) (see also Pls. 47, 48 & 94)

4. Peristernia angulata (Sowerby, 1888)
22.7mm; Panglao, Bohol.
Shell small, roughly fusiform-ovate in outline, with a tall spire, a moderately inflated body whorl and a short, slightly curved siphonal canal; spire whorls strongly convex, sculptured with rounded axial folds which become quite large on the body whorl and number between 6 and 7, entire surface encircled by finely scabrous lirae, strongest on the folds and occasionally forming weak spines; aperture ovate, weakly lirate within, outer lip thin and finely serrate, columella smooth, fasciolar cord lamellate; apical whorls white, rest of shell dark orange-brown, aperture bluish white; reaches 25mm.
DISTRIBUTION: Camotes Sea, Bohol and Southern Mindanao.

Family CORALLIOPHILIDAE (cont'd.) (see also Pls. 43, 44 & 94)

5. Latiaxis (Echinolatiaxis) problematicus Kosuge, 1980
24.5mm; Panglao, Bohol.
Shell medium sized, fusiform-ovate in outline, with a moderately tall spire, a large body whorl and a short, distally recurved siphonal canal; sculptured with 8 rounded axial folds, each bearing a single short, triangular, upturned spine which is narrowly open towards the aperture, encircled with numerous, densely lamellated spiral cords which bear numerous fine spines, most pronounced where they cross the axial folds; aperture ovate, lirate within, outer lip thin and coarsely serrate, columella smooth, fasciolar cord with 4 to 5 spinose lamellae; pale orange-brown in colour, the shoulder spines slightly pinkish, aperture pale lavender; reaches 25mm.
DISTRIBUTION: Camotes Sea and Bohol.

6. Latiaxis (Babelomurex) pagodus A. Adams, 1853 form **multispinosus** Shikama, 1966
22.4mm; Panglao, Bohol.
One of the many forms of *L. pagodus* (Pl. 43, fig. 1), and is characterized by its smaller size and longer, more numerous spinose projections.
DISTRIBUTION: Camotes Sea and Bohol.

Family MURICIDAE (cont'd.) (see also Pls. 35-41)

7. Favartia salmonea (Melvill & Standen, 1899)
15.9mm; Mactan Is., Cebu.
Shell small, with a tall spire of 5 to 6 convex whorls and a broadly fusoid body whorl; sculptured with 6 widely spaced varices on the body whorl, strongly fimbriate on the leading edge, and encircled with 5 to 6 spiral lirae, strongest on the varices where they form small, open spinose processes; aperture ovate, outer lip thin, raised, weakly crenulate and weakly lirate within, columella smooth; colour pale pink, apex light orange, aperture light pinkish lavender; reaches 23mm.
DISTRIBUTION: Camotes Sea, Bohol and Southern Zamboanga.

8. Pterynotus aparii D'Attilio & Bertsch, 1980
16.4mm; Bohol.
Shell small, broadly fusiform, with a tall spire of 8 to 9 subangulate whorls, undulate, weakly impressed sutures and a broad body whorl; sculptured with 3 varices which are covered with irregularly contoured "wing-like" blades, axially sculptured with numerous lamellate striae; earlier whorls with numerous non-webbed varices, body whorl with a flattened, upturned spine on the shoulder of each varix; aperture ovate, outer lip finely crenulate and weakly dentate within, columella smooth, canal long, open and recurved; colour pale orange; reaches 22mm.
DISTRIBUTION: Camotes Sea and Sulu Sea.

9. Phyllocoma convoluta (Broderip, 1833)
21.8mm; Mactan Is., Cebu.
Shell small to medium sized with a tall spire of 9 to 10 strongly convex whorls, impressed sutures and a large, rounded body whorl; sculptured with 2 thin varices per whorl (every 270°), deeply grooved on the trailing edge and encircled with finely scabrous spiral cords, some axial threads present in the interspaces of the earlier whorls; aperture ovate, columella smooth, flaring its entire length, outer lip variced and marginally crenulate from the spiral sculpture; external colour dirty white, aperture porcelaneous white; reaches 35mm.
DISTRIBUTION: Camotes Sea, Sulu Sea and Marinduque.

Family CANCELLARIIDAE (cont'd.) (see also Pl. 18)

10. Scalptia nodosivaricosa (Petuch, 1979)
13.0mm; Bohol.
Shell small, inflated, with deeply impressed sutures and a moderately tall, somewhat scalariform spire; body whorl with 9 to 12 rounded varices, finely nodulose from overriding spiral cords; aperture elongate-ovate, attenuated towards the base, columella with 3 folds, umbilicus weakly open, outer lip thin and lirate within; colour pale brown with 3 darker brown bands on the body whorl, aperture off-white; reaches around 13mm.
DISTRIBUTION: So far only recorded from Bohol.

Family CYMATIIDAE (cont'd.) (see also Pls. 30-32)

11. Distorsio (Personella) pusilla Pease, 1861
10.0mm; Punta Engaño, Cebu.
Shell small, moderately distorted, with a tall spire, and a somewhat ovate body whorl; sculptured with 5 to 6 varices, strongest on the body whorl, the surface finely granulose from intersecting axial and spiral cords, and the intespaces have fine, microscopic spiral threads; the aperture is irregularly quadrangular in outline, and its features are rather unique and quite distinctive among modern *Distorsio spp.;* the outer lip margin is thick with 5 to 6 denticles, the posterior most one being largest (usually the 2nd or 3rd in other *Distorsio spp.),* the columellar edge to the anterior canal is thin and simple (lacking the anteriorly decreasing rows of nodules typically seen in *Distorsio),* and there is a short row of 4 to 5 elliptical, obliquely orientated, anteriorly decreasing nodules on the thickened innermost edge of the columellar callus; colour pale creamy brown; reaches 18mm, though Philippine specimens seldom exceed 12mm; often considered a juvenile of some other species, but the proportions of aperture, canal and spiral cords suggest a fully mature, valid taxon even if of unusually small size.
DISTRIBUTION: So far only recorded from the Camotes Sea.

Family EPITONIIDAE (cont'd.) (see also Pl. 18)

12. Amaea (Scalina) sericogazea (Masahito, Kuroda & Habe, 1971)
32.8mm; Bohol.
Shell medium sized, with a tall spire of some 14 strongly convex whorls and deeply channeled sutures; surface finely clathrate from the intersection of numerous spiral cords and finely fimbriate axial threads, body whorl with a weak keel, below which spiral sculpturing is absent; aperture ovate, outer lip thin and weakly serrate, columella smooth and arcuate with a weak basal fold; shell white in colour, brown below the body whorl keel; reaches around 35mm.
DISTRIBUTION: Visayan Sea, Camotes Sea and Sulu Sea.

Gastropod Species **Plate 96**

Family CYPRAEIDAE (cont'd.) (see also Pls. 20-24, 94)

1. Cypraea leucodon Broderip, 1828
87.8 × 52.7mm; Punta Engaño, Cebu.
Shell large, solid, ovate in outline, with a heavily inflated dorsum and short, prominent terminals; aperture uniformly narrow with both the labral and columellar teeth thick, heavy, and uniform in strength; dorsum dark orange brown with numerous white spots and a pale mantle line, base light tan, teeth white; reaches around 83mm in length. This species is extremely rare and several specimens are reported as having been taken "ex pisce". Philippine shells have been appended the name *leucodon angioyna* Massilia, 1979, and are said to differ from the nominate subspecies (which occurs in the Indian Ocean) in having a lighter coloured base and in having fewer, less compacted columellar teeth, the central ones being shorter than the rest.
DISTRIBUTION: Cebu, Bohol and Sulu Sea.
(Photo: M. Blöcher)

2. Cypraea musumea (Kuroda & Habe, 1961)
20.5 × 11.2mm; South China Sea.
Shell small, pyriform, with a moderately inflated dorsum and slightly extended terminals; the base is smooth and convex and both the labral and columellar teeth are small and restricted to the narrow, posteriorly curved aperture; dorsal colouration pale yellow-tan with numerous small, light brown-coloured dots, and larger blotches on the

PLATE NO. 95 1. *Nassarius accuminatus* (Marrat) 2. *Latiaxis tuberosus* Kosuge 3. *Coralliophila solutistoma* Kuroda & Shikama 4. *Peristernia angulata* (Sowerby) 5. *Latiaxis problematicus* Kosuge 6. *Latiaxis pagodus* A. Adams form *multispinosus* Shikama 7. *Favartia salmonea* (Melvill & Standen) 8. *Pterynotus aparrii* D' Attilio & Bertsch 9. *Phyllocoma convoluta* (Broderip) 10. *Scalptia nodosivaricosa* (Petuch) 11. *Distorsio pusilla* Pease 12. *Amaea sericogazea* (Masahito, Kuroda & Habe)

PLATE 95

margins, base becoming white; reaches around 25mm in length.
DISTRIBUTION: Cebu, Bohol and Southern Mindanao. (Photo: M. Blöcher)

3. Cypraea teramachii (Kuroda, 1938)
56.8mm x 34.1mm; South China Sea.
Shell large, solid and pyriform, with a well-humped dorsum and well developed terminal processes; aperture narrow, strongly curved posteriorly, labral teeth strong, confined to the aperture but extending slightly anteriorly, columellar teeth extremely weak, almost obsolete centrally; dorsal colouration yellowish cream to light buffy tan, lightly spotted with darker orange-tan and crossed by 3 faint, purplish brown bands just under the surface nacre; margins a light orange-brown with diffuse brown spotting and a reddish brown spot on the posterior end, base light orange tan peripherally, becoming white towards the aperture; reaches around 75mm.
DISTRIBUTION: Bohol. (Photo: M. Blöcher)

Class Gastropoda
Subclass Prosobranchia
Order Neogastropoda
Suborder Rachiglossa
Superfamily Volutacea
Family MARGINELLIDAE

A moderately large family of carnivorous gastropods inhabiting sand or coral in the intertidal regions of most geographic localities. The shells are mainly small, ovate to elongate in outline, with a glossy surface. The spire may be tall and ribbed, or flattened and calloused. The aperture is narrow, the outer lip thick, arcuate, occasionally denticulate on the inner margin, and the columella is plicate. There are only 5 moderately sized species of note in the Philippines, although numerous smaller species (under 5mm) are known but their identity is still under review. (Photo: P. Clover)

4a. Bullata (Cryptospira) tricincta (Hinds, 1844)
Shell moderately large for the family with adult specimens reaching around 30mm; elongate-ovate in outline, with somewhat ventricose sides, a smooth, arcuate, thickened outer lip, the posterior margin of which projects above a flattened spire; columella with 6 folds, the posterior 3 thin and widely spaced, the last 3 thicker and crowded together; colour grey, occasionally with 3 darker bands on the dorsum. Known to us from deep water off Bohol.
4b. Bullata (Closia) quadrilineata (Gaskoin, 1849)
Shell similar in size to *B. tricincta* but more cylindrical in outline; the outer lip is fairly straight and there are 4 to 5 folds set very low in the aperture; the surface colouration is grey and the dorsum is encircled with 4 regularly spaced, thin red bands. Trawled in 80 meters off Southern Palawan.
4c. Granulina guttula (Sowerby, 1837)
Shell small, elongate-ovate in outline, with a very low spire and 3 small folds set low in a narrow aperture; the shell is burnt orange all over and usually reaches around 8 to 10mm; known from intertidal reef areas around Samar.
4d. Hyalina (Volvarina) philippinarum (Redfield, 1848)
Shell medium sized, cylindrically elongate-ovate with a moderately tall spire and a smooth, thickened, straight-sided outer lip; the aperture is narrow posteriorly, widening anteriorly, and the columella has 4 low-set folds; shell colour pale yellowish brown with 3 wide, dark orange bands on the dorsum; reaches around 15mm in length and is found all over the Philippines in shallow water down to 40 meters. Mainly from Palawan.
4e. Marginella fischeri Bavay, 1902
Shell small to medium sized, elongate-ovate in outline with a flattened spire and a thickened outer lip; the aperture is uniformly narrow and the columella has 6 folds; surface colouration is grey with 2 or 3 thin red bands; reaches around 15mm in length and is known from Southern Palawan (trawled in 80 meters).

Family CONIDAE (cont'd.) (see also Pls. 61-72)

5. Conus artroptus Sowerby I, in Sowerby II, 1833
37.8mm x 13.1mm; Sulu Archipelago.
Shell medium sized, cylindrically elongate, with a moderately tall, weakly concave-sided spire and a long, straight-sided body whorl which is attenuated towards the base; spire whorls weakly stepped and convex and sculptured with 5 or 6 strong spiral ridges and weak axial threads; body whorl with a roundly angulate shoulder and encircled with numerous closely spaced, somewhat roughened ridges which become fairly weak basally; aperture narrow posteriorly, widening anteriorly, with a thin outer lip; shell colour cream, occasionally with a violet tinge, and encircled with 3 irregular bands of pale, orange-tan coloured blotches, entire body whorl surface also covered with numerous discontinuous bands of small, brown-coloured quadrate spots; aperture white, occasionally with a pinkish or purplish tinge; reaches around 50mm in length.
DISTRIBUTION: Sulu Sea. (Photo: R. Martin)

6. Conus pergrandis (Iredale, 1937)
125mm; S. W. of Taiwan.
Shell large, biconic, with a low gloss; spire moderately tall, weakly concave-sided with a pointed apex; spire

PLATE NO. 96 1a. to 1b. *Cypraea leucodon* Broderip 2a. to 2b. *Cypraea musumea* (Kuroda & Habe) 3a. to 3b. *Cypraea teramachii* (Kureda) 4a. *Bullata tricincta* (Hinds) 4b. *Bullata quadrilineata* (Gaskoin) 4c. *Granulina guttula* (Sowerby) 4d. *Hyalina philippinarum* (Redfield) 4e. *Marginella fischeri* Bavay 5. *Conus artroptus* Sowerby I, in Sowerby II 6. *Conus pergrandis* (Iredale) 7. *Conus rogmartini* da Motta 8. *Conus kimioi pseudokimioi* da Motta & Martin 9. *Conus lani* Crandall 10. *Cancilla cloveri* Cernohorsky

PLATE 96
1a
1b
2a
2b
3a
3b
4a
4b
4c
4d
4e
5
6
7
8
9
10

whorls concave, the early ones weakly nodular then becoming undulate and sculptured with spiral grooves and axial threads; body whorl with a rounded shoulder, widest below the shoulder and tapering to a narrow base; encircled with deep spiral grooves, crowded anteriorly but becoming more widely separated posteriorly and with a few spiral threads in the interspaces; shell colour off-white, the spire whorls with small brown blotches, and the body whorl with brown axial flammules on the shoulder area, an irregular, brown-coloured central band, and a smaller, somewhat discontinuous brown band at the base; aperture uniformly narrow, white within; reaches around 170mm and known in the Philippines only from Bohol. The shell is difficult to get in gem condition and is nearly always marked with growth flaws.
DISTRIBUTION: Bohol. (Photo: T. C. Lan)

7. Conus rogmartini da Motta, 1982
31.2mm × 17.4mm; Balut Is., Southern Mindanao. (Holotype).
Original description: "Shell turbinated, consisting of 10 whitish, moderately stepped spiral whorls, the first 2 forming a globose protoconch, immediately followed by 4 flat-surfaced whorls; the next 4 being deeply excavated and sculptured with arcuate striae along closely-coiled wavy sutures, which are symetrically coronated with white nodules, increasing progressively in size with each volution, and having brown quadrangular spots in the intervening spaces between these protruberances, culminating in an acutely angulated shoulder, and having sides that are flat and tapering. The body colour is rosy-pink, including the penultimate whorl, and ornamented with a solid band of vermillion tint towards the basal area. The last whorl is sculptured with twenty-six spirally raised ridges, seven of which, in irregular intervals from the shoulder to the base, have rows of the same white nodules seen on the shoulder, except that these knobby protruberances do not occur in any particular arrangement, and the intervening brown spots are instead aligned in irregular, interrupted rows of brown dots on the crest of all encircling ridges. The intervening areas between the ridges on the entire surface are sculptured with perpendicular striae. The aperture is moderately wide with a sharp undulating outer lip, its interior a porcelaneous light rose shade." Appears to reach around 35mm in length. Taken in 120 fathoms. An exquisitely sculptured species which has occasionally been compared to a granulated form of *Conus otohimeae* Kuroda & Ito, 1961. In late 1982, a granulated form of the latter species was discovered at Balut Island, and although superficially resembling *C. rogmartini* to some extent, is a completely unrelated species.
DISTRIBUTION: So far only known from the type locality. (Photo: R. Martin)

8. Conus kimioi pseudokimioi da Motta & Martin, 1982 (Holotype)
20.0mm × 11.0mm; Siquijor Is., Mindanao Sea (100 fathoms).
Type description: "Shell subcylindrically turbinated, small, smooth and glossy, with a somewhat depressed spire consisting of 8 slightly-stepped spiral whorls, sculptured with fine threads overlapped by short curving striae; shoulder angulate and broad, the sides slightly inflated at its upper section, gradually tapering down the length of its attenuated body whorl. The base colour is lavender; spire is decorated with brown radial maculations, which overflow beyond the shoulder line into the border of the body surface, creating a necklace of evenly-spaced brown dots; the entire body whorl is marked with parallel spiral lines of chestnut-brown, overlaid with flame-like longitudinal brown streaks and patches flowing irregularly down its length. The aperture is lavender with parallel sides, and a thin, trenchant outer lip." A distinct subspecies known only from the type locality. It differs from the nominate subspecies in its broader shoulder and characteristic surface pattern; appears to reach around 25mm.
DISTRIBUTION: At present, known only from the type locality.
(Photo: R. Martin)

9. Conus lani Crandall, 1979
Shell medium sized, biconic, with a tall, straight spire and a long tapering body whorl; spire whorls 8 in number, each encircled with a few spiral grooves, body whorl with a well-rounded, sub-angulate shoulder and encircled with around 50 to 55 weakly incised, evenly spaced spiral grooves; aperture narrow, white within; shell a dark, glossy chestnut-brown colour with a white protoconch and 3 white-dotted bands on the body whorl, one at the shoulder, one at mid-body whorl, and the other basally.
Left: Holotype specimen; 53mm; N. E. Taiwan.
Right: Philippine specimen; 42mm; Bohol.
DISTRIBUTION: Presently known only from a single specimen from Bohol.
(Photo: T. C. Lan)

Family MITRIDAE (cont'd.) (see also Pls. 54-57)

10. Cancilla (Ziba) cloveri Cernohorsky, 1971
25.6mm, 21.0mm; N.E. off Taiwan.
Shell medium sized, elongate ovate in outline, with a tall spire and well incised sutures; spire whorls convex and encircled with punctate spiral grooves; body whorl long, more than half the shell length, sculptured with a few punctate spiral grooves on the weakly rounded shoulder, becoming weaker towards the base; aperture narrowly elongate with 5 plaits; shell colour pale yellowish white, ornamented with reddish brown spiral lines that are crossed by irregular axial lines forming a reticulated network; the surface is fairly glossy and the shell reaches around 25mm in length; known in the Philippines from Balut Island, Southern Mindanao.
DISTRIBUTION: Presently known only from Balut Island, S. Mindanao
(Photo: T. C. Lan)

INTRODUCTION TO LAND SNAILS

The Philippine archipelago contains one of the largest assortments of land snail species to be found anywhere in the world. Its vast tropical rain forests and mountains have produced some of the most colorful land snails in the world that can vie in beauty and variety with those of New Guinea, Florida, Hawaii, Cuba, and other countries. Unfortunately, very few shell collectors in the Philippines are seriously collecting this group despite the abundance of material. To the average Filipino, "susô" (local name for land snails) brings to mind the varieties coming from the rivers and rice paddies, which when cooked in coconut milk and hot peppers, make an exotic dish which is popular all over the islands.

Among the very few Filipino shell collectors, the late Pedro de Mesa is well remembered as having been one of the earliest and most active shell collectors of both marine and land snails in the Philippines. During the span of more than 50 years, he donated a great number of specimens to various museums all over the world and corresponded with some of the world's great malacologists, to name a few, Dr. W. Clench (Harvard Museum), Dr. H. Rehder (Smithsonian Institute), Dr. T. Kuroda of Japan, Dr. T. Abbott, and many others. Because of Mr. de Mesa's contributions, the genus *Mesanella* and several new species were named in his honour especially by Dr. Clench. The Philippine government, being aware of the importance and value of his collection, granted him reparations when his collection was destroyed during the last war. Among the local collectors who were influenced by Mr. de Mesa, Mr. Fernando Dayrit stands out as the most knowledgeable collector of land snails in the Philippines at present.

Many early conchologists such as Sowerby, Möllendorff, Pfeiffer, Lea, and others made attempts to catalogue Philippine land snails, but the more recent works of Bartsch, Webb, and Solem are probably the most significant over the last 50 years. Regrettably, many earlier works on this complex group are now extremely difficult to acquire and those that are available, in many cases, lack detailed drawings or photographs to complement the descriptions. This lack of existing or dated literature on land snails is the primary reason why these colorful land molluscs have not gained as much popularity among shell collectors as their marine counterparts have. The following section is a serious attempt to illustrate a broad cross-section of many of the more commonly encountered species.

The following 4 plates are devoted to a broad selection of land snails found at various localities within the Philippine archipelago. They represent a specialized group of Gastropods belonging to the sub-class *Pulmonata**. In contrast to their marine counterparts, these shelled, air-breathing molluscs have evolved on land and have taken on an incredible diversity of shapes and colours. The natural barriers of land, including mountain ranges, rivers, lakes, and even seemingly unimportant barriers like river channels and hill slopes, have given rise to an endless assortment of species, subspecies, and forms. The apparent successful adaptation to the harsher environment of land in comparison with a marine environment may be attributable to the fact that most land snail species are hermaphroditic. This remarkable evolutionary development has no doubt arisen to compensate for the relative loss of mobility in comparison with their marine counterparts.

All land snails prefer a moist, shaded environment and nearly all are herbivorous, feeding on decaying matter or small fungi. Many species are morphologically very similar, and it is sometimes extremely difficult to ascertain the exact status of a particular specimen, even at a family level. Land snails are well known for their diversity of colour and surface pattern, and it is these two characteristics which often provide the means of ascertaining the taxonomic position of a particular specimen. Almost all land snails possess a periostracal covering and, in many cases, this is augmented by the appearance of white material, usually referred to as cuticle. Within a particular species at any one locality, members appear to possess fairly uniform morphology and periostracal/cuticle characteristics. Members of the same species situated in a close but geographically isolated area, although possessing nearly identical morphology, often exhibit differences in periostracal and cuticular pattern. Whether such differences can be construed as to representing a different form, a subspecies, or even a completely different species, is today, the subject of much controversy. Only a detailed examination of the habitat and population characteristics will eventually solve this perplexing problem.

The taxonomy of land snails is still very much in a state of flux. There is a bewildering array of genera and subgenera, and even at a family level, there is still considerable disagreement among experts. The generic and family groupings in this section are used only in the broadest sense and many will no doubt change when more detailed research is done. In the main, we have followed L. A. Faustino in his monumental work, "Summary of Philippine Land Snails", (*Philippine Journal of Science* 1930, Vol. 42:85-198), which probably represents the most complete compilation of Philippine land snails. The names appearing in this section are *not* consistent with normal taxonomical procedure as regards classification. In all cases, the name appearing after

* Some species appearing in this section under the broad heading of Land Snails are not true pulmonates, but terrestrial prosobranchs (see classification section). A small selection of fresh water snails is also presented on Plate 100.

the species' name is intended to represent a form only, though in many cases, the subspecies' rank may be valid. The authors have made no attempt to resolve this problem, and have merely presented the currently accepted names.

N.B.
A single representative of the family STENOGYRIDAE is illustrated on Pl. 100 (fig. 24). This is a South African Group and will not be discussed here. This single representative, belonging to the genus *Homorus* Albers, 1850, is an introduced species, and not an endemic one.

The following list represents the various families covered within this section (including 2 fresh water groups), together with a diagnosis of the family and the genera employed.

Class: GASTROPODA

I. Subclass Prosobranchia
Order Mesogastropoda
Suborder Taenioglossa

A. Superfamily Cyclophoracea
Family CYCLOPHORIDAE

DIAGNOSIS: A large group of shells varying in size from small to large, with a short or tall spire. The whorls are strongly convex, the sutures deeply channeled and the aperture is rounded with a thickened margin. Members have a calcareous or horny multispiral operculum which is centrally concave.
Genera employed: *Cyclotus* Guilding, in Swainson, 1840; *Leptopoma* Pfeiffer, 1847; *Eulota* Hartman, 1843; *Chloraea* Albers, 1850; *Cyclophorus* Montfort, 1810.

B. Superfamily Viviparacea
Family AMPULLARIIDAE

DIAGNOSIS: A small group of shells, large to moderately large in size and mostly globular to ovoid in outline. Operculum horny or calcareous, with an eccentric nucleus. Sexes are separate and they generally inhabit stagnant or fresh running water wherever vegetation is abundant.
Genera employed: *Pila* Röding, 1798.

C. Superfamily Cerithiacea
Family THIARIDAE

DIAGNOSIS: A moderately large group of fresh water gastropods containing small to medium-sized shells which are many-whorled and turreted. The surface may be smooth or with a combination of axial and spiral ridges, nodules or spines. The aperture is oval and a siphonal canal is lacking. The animals are parthenogenetic (i.e. no males occur) and widely distributed. They occur in most continents and inhabit stagnant or fresh running water.
Genera employed: *Melanoides* Olivier, 1804; *Faunus* Montfort, 1810; *Thiara* Röding, 1798.

D. Superfamily Littorinacea
Family POMATIASIDAE

DIAGNOSIS: A large group of land and fresh water dwellers. The shells are small, solid, elongate-ovate in outline, with numerous convex whorls, the surface of which may be smooth or weakly striated. The aperture is ovate, the peristome entire and thickened. Usually covered with a thin, brown periostracum.
Genera employed: *Schistoloma* Kobelt, 1901.

II. Subclass: Pulmonata
Order: Stylommatophora

A. Superfamily Ariophantacea
Family TROCHOMORPHIDAE

DIAGNOSIS: A small family group of relatively small sized shells which are discoidal in outline, have a low to moderately tall spire, a simple aperture with a non-thickened outer lip margin, and a narrowly open to closed umbilicus. Normally brown to yellowish brown in colour.
Genera employed: *Ryssota* Albers, 1850; *Hemiglypta* Möllendorff, 1893.

B. Superfamily Helicacea
1. Family CAMAENIDAE

DIAGNOSIS: A small group of ovate to elongate-ovate shells, medium sized, being almost indistinguishable from the members of the family BRADYBAENIDAE, separation being based on an anatomical rather than morphological basis. Some members of this family illustrated here are distinguished by their surface sculpturing, which consists of numerous, oblique axial ridges.
Genera employed: *Camaena* Albers, 1850; *Amphidromus* Albers, 1850.

2. Family BRADYBAENIDAE

DIAGNOSIS: A large group of pulmonates containing a large assemblage of genera. The shells are small to large, have an ovate to elongate outline, and around 7 to 8 whorls. The aperture is generally ovate to elongate-ovate, smooth within and has a flattened, reflexed outer lip margin. The umbilicus may be partially open or closed. The surface has a smooth or hairy periostracum, and is ornamented with numerous coloured bands.
Genera employed: *Helicostyla* Férussac, 1821; *Obba* Beck, 1837; *Hemitrichiella* Zilch, 1956 (= *Hemitrichia* Möllendorff, 1889 (non Rostafinski, 1873)); *Eulota* Hartmann, 1843 ((?) = *Bradybaena* Beck, 1837)

Family **BRADYBAENIDAE**

1. Helicostyla rollei Möllendorff, 1890
74.7mm × 42.1mm; Mindoro.
Shell medium sized, trigonally elongate-ovate in outline, with 6 convex whorls, well impressed sutures, and a partially open umbilicus; shell dark brown with pale greyish white axial streaks, earlier whorls light brown; aperture elongate-ovate, pale blue within, outer margin flattened and brown in colour; reaches 80mm in height. Commonly known as *Helicostyla mindoroensis rollei* (Möllendorff). Because of the 2-year priority of Möllendorff's name over that of Broderip, it is expedient to consider this as a distinct species to avoid taxonomic confusion.

2. Helicostyla mindoroensis Broderip, 1892
66.0mm × 32.2mm, 61.0mm × 30.5mm; Mindoro.
This is the typical form of the species and is generally more elongate than *H. rollei;* the shell may be light brown or almost pure white, with a darker apex; axial streaks adorn the body; the aperture is white (sometimes light blue) and the outer lip margin may be brown or lavender; reaches 70mm in height.

3. Helicostyla chrysalidiformis Sowerby, 1892
59.0mm × 30.3mm; Mindoro.
Very much like a more elongated version of *mindoroensis,* but the white axial streaks are strongest just below the sutures of the spire whorls and on the upper half of the body whorl; reaches 65mm in height.

4. Helicostyla filaris (Valenciennes, 1845)
Syn. *H. nympha* (Pfeiffer, 1849)
57.0mm × 30.0mm, Polilio Is., Quezon.
Shell medium sized, triangularly elongate-ovate, with 6 convex whorls, well impressed sutures, and a nearly obsolete keel on the lower body whorl; aperture elongate-ovate; light brown with some brownish streaks on the body whorl and a thin white band centrally; aperture white within, outer margin brownish orange, columella white with a brown peri-umbilical band; reaches 60mm in height.

5. Helicostyla subcarinata moellendorfi Hidalgo, 1890
44.5mm × 22.2mm; Marinduque.
Shell small to medium sized, rather lightweight, with 6 to 7 convex whorls, well incised sutures and a closed umbilicus; shell dark brown, earlier whorls paler, sutures with a white band, and the surface with pale brown streaking; aperture pale blue within, outer lip thin, reflexed, dark brown, columella white with a purplish peri-umbilical band; reaches 45mm.

6. Helicostyla partuloides Broderip, 1892
50.0mm × 24.0mm, 43.5mm × 21.3mm; Mindoro.
Shell medium sized, elongate to elongate-ovate in outline, with 5 to 6 weakly convex whorls; aperture ovate, columella and outer lip thick, non-umbilicate; usually pale cream to brown with white axial streaks and occasionally some brown banding, aperture outer lip and columella white; reaches 55mm in height.

7. Helicostyla calamianica Quadras & Möllendorff, 1894
48.4mm × 24.0mm, 44.8mm × 23.0mm, 38.0mm × 18.2mm; Mindoro.
Shell medium sized, elongate to elongate-ovate in outline, with 6 convex whorls, weakly impressed sutures, and an elongate aperture bearing a flattened outer margin; variable throughout its range, may be pale greenish yellow or white, with or without brown banding, white axial blotches and streaks, and a white aperture; reaches 50mm in height.

8. Helicostyla satyrus (Broderip, 1840)
41.5mm × 24.4mm; Balabac Is., Palawan.
A small elongate-ovate shell with 5 to 6 convex whorls and a somewhat flattened apex; body whorl large in relation to total shell size, the aperture is elongate-ovate, with a slightly flattened outer margin, and the umbilicus is obsolete; entire surface with fine oblique axial scratches; early whorls off-white with a brown sutural band, body whorl dark brown with some white cuticle, aperture and columella bluish white with a brown margin; reaches 45mm in height.

9. Helicostyla leai Pfeiffer, 1892
36.0mm × 26.5mm, 33.8mm × 27.9mm; Batanes Islands.
Shell medium sized, elongate-ovate, almost globose in some specimens, with 5 to 6 rounded whorls, a flattened spire, and a large body whorl; umbilicus almost obsolete, aperture roundly quadrangular with a slightly flattened outer lip margin; most specimens either white, orange, pink or yellow; reaches 40mm in height.

10. Helicostyla boholensis Broderip, 1892
42.0mm × 18.8mm; Bohol.
Shell medium sized, lightweight, pyramidally elongate, with 6 to 7 convex whorls which terminate in a slightly mammillate apex; aperture roundly-quadrangular, outer lip thin, slightly flattened umbilicus obsolete; shell pale brownish orange with or without white cuticle; aperture bluish white, outer margin brown, columella white with a purplish peri-umbilical band; may reach 50mm in height.

11. Helicostyla fischeri Hidalgo, 1892
39.0mm × 22.0mm, 40.5mm × 24.0mm, 41.0mm × 25.0mm; Burias Is., Masbate.
Shell medium sized, trigonally ovate in outline, with 5 to 6 rounded whorls and occasionally a weak keel on the lower body whorl; entire surface with fine, oblique axial scratches; pale to dark brownish lavender, with irregular brownish blotching and white cuticle bands; aperture ovate-quadrangular, bluish white within, outer lip margin thin, lavender-coloured, columella white with a lavender peri-umbilical band; reaches 45mm in height.

12. Helicostyla effusa Pfeiffer, 1892
40.0mm × 27.0mm; Sibuyan Is., Romblon.
Shell medium sized, lightweight, with 5 strongly convex whorls and a somewhat flattened spire; non-umbilicate, aperture ovate with a thin, slightly flaring outer lip; pure white with a brown band at the sutures, one centrally on the spire whorls, 2 on the body whorl, and another bordering the columella; reaches 50mm in height.

13. Helicostyla subcarinata Pfeiffer, 1890
45.6mm × 25.0mm; Burias Is., Masbate.
This is the typical form of fig. 5 and is pale greenish white in colour, with a brown outer lip margin and columellar patch; seldom exceeds 45mm in height.

14. Helicostyla bicolorata Lea, 1891
57.0mm × 36.7mm; Marinduque.
Shell medium sized, triangularly ovate in outline, whorls 6 in number, slightly squashed and strongly convex in outline; entire surface with fine, oblique axial growth striae; typically dark brown, spire whorls paler with a green and white cuticle on the body whorl; aperture ovate-quadrangular, bluish white within and with a brown margin; reaches 65mm in height.

15. Helicostyla camelopardalis Broderip, 1892
32.0mm × 18.5mm; Mindoro.
A small trigonally ovate shell, with 4 to 5 convex whorls, well channeled sutures, a flattened spire, an elongate-ovate aperture, and a thin, slightly reflected outer lip margin; dark brown with white axial cuticle occasionally interrupted on the body whorl by 2 brown bands; aperture and apertural margin bluish white; may be a form of *H. boholensis* Broderip (fig. 10); reaches around 35mm in height.

16. Helicostyla cincinna (Sowerby, 1840)
44.0mm × 25.0mm; Burias Is., Masbate.
Shell medium sized, trigonally elongate-ovate in outline, with 6 to 7 strongly convex whorls, a slightly flattened spire, impressed sutures and the surface covered with numerous fine, oblique axial scratches; aperture ovate-quadrangular, outer lip thin, weakly reflected, umbilicus closed; greenish white with a brown band at the sutures, mid-body whorl, and around the white columella; aperture white with a brown margin; reaches 50mm in height.

17. Helicostyla ovoidea (Bruguiére, 1792)
42.0mm × 23.0mm, 40.2mm × 23.0mm, 40.3mm × 25.5mm; Burias Is., Masbate.
Shell medium sized, elongate-ovate in outline, some specimens more squat than others; whorls convex, sutures impressed, spire somewhat flattened; aperture ovate-quadrangular, umbilicus nearly obsolete, outer lip thick and slightly reflected; pure white with varying numbers of brown bands; aperture white with a greenish parietal glaze; variable; reaches 45mm in length.

18. Helicostyla fulgetrum Broderip, 1892
52.0mm × 29.5mm; Burias Is., Masbate.
Shell medium size to large, triangularly-elongate, whorls strongly convex, somewhat globose, 5 to 6 in number; aperture elongate-ovate, umbilicus obsolete, outer lip thick, reflected; colour pinkish white, with a brown band at the sutures, mid-body whorl and around the umbilical area; covered with a greenish brown periostracum and white cuticle especially on the last 2 whorls; variable; reaches 55mm in height.

19. Helicostyla virgata Jay, 1892
50.0mm × 23.0mm; Baco, Mindoro.
Shell medium sized, triangularly-elongate, with 5 to 6 convex whorls, a somewhat mamillate apex, moderately impressed sutures, and a large, elongate, white-coloured aperture with a thickened, markedly reflected outer lip margin; shell pinkish white, covered with a greenish brown periostracum, a white sutural band, a brown subsutural band, and a brown-stained peri-umbilical area; reaches 50mm in height.

20. Helicostyla fictilis Broderip, 1892
47.0mm × 24.0mm; Baco, Mindoro.
Shell medium sized, rather lightweight, triangularly elongate in outline with 5 to 6 convex whorls, a somewhat mamillate apex, no umbilicus, and a large bluish white, elongate-ovate aperture with a thin, reflected outer lip margin; spire whorls almost black, the rest of the shell covered with a dark periostracum and axial flammules of white cuticle; reaches 50mm in length.

21. Helicostyla stabilis (Sowerby, 1840)
31.0mm × 22.0mm; Masbate.
Shell medium sized, elongate-ovate, somewhat triangular in outline, the whorls short, weakly convex, with weakly impressed sutures; aperture irregularly ovate, with a thickened, slightly reflected outer lip margin; shell white with a presutural brown band which enlarges to cover almost all of the body whorl; aperture bluish white; reaches 35mm in height.

22. Helicostyla cincinniformis (Sowerby, 1840)
33.0mm × 24.0mm; Lubang Is., Mindoro.
Shell medium sized, lightweight, triangularly ovate in

PLATE NO. 97 1. *Helicostyla rollei* Möllendorff 2. *Helicostyla mindoroensis* Broderip 3. *Helicostyla chrysalidiformis* Sowerby 4. *Helicostyla filaris* (Valenciennes) 5. *Helicostyla subcarinata moellendorfi* Hidalgo 6. *Helicostyla partuloides* Broderip 7. *Helicostyla calamianica* Quadras & Möllendorff 8. *Helicostyla satyrus* (Broderip) 9. *Helicostyla leai* Pfeiffer 10. *Helicostyla boholensis* Broderip 11. *Helicostyla fischeri* Hidalgo 12. *Helicostyla effusa* Pfeiffer 13. *Helicostyla subcarinata* Pfeiffer 14. *Helicostyla bicolorata* Lea 15. *Helicostyla camelopardalis* Broderip 16. *Helicostyla cincinna* (Sowerby) 17. *Helicostyla ovoidea* (Bruguiére) 18. *Helicostyla fulgetrum* Broderip 19. *Helicostyla virgata* Jay 20. *Helicostyla fictilis* Broderip 21. *Helicostyla stabilis* (Sowerby) 22. *Helicostyla cincinniformis* (Sowerby) 23. *Helicostyla ventricosa decorata* Férussac 24. *Helicostyla dryas* Broderip 25. *Helicostyla carinata rugata* Hidalgo

PLATE 97

outline, almost squat with 5 convex whorls, a flattened apex, an almost obsolete umbilicus and an elongate-ovate aperture with a thin outer lip margin; yellowish white with a brown sutural band and another two, centrally and basally on the body whorl; aperture and columella bluish white; reaches 35mm in height.

23. Helicostyla ventricosa decorata Férussac, 1898
40.0mm × 20.5mm; Guimaras Is., Iloilo.
Shell medium sized, elongate, with 6 convex whorls, a flattened, weakly mamillate apex, and an elongate-ovate outer lip with a thickened, slightly reflected outer lip margin; shell pure white with a brown sutural band and another mid-body whorl; aperture and columella white, with a greenish white parietal glaze; reaches 40mm in height.

24. Helicostyla dryas Broderip, 1892
51.0mm × 21.0mm; Mindoro.
Shell medium sized, elongate, with 6 to 7 weakly convex whorls and a flattened spire; aperture elongate-ovate, outer lip margin thick and markedly reflexed; colour whitish brown, earlier whorls paler, with a white, brown-lined sutural band, and a bluish white aperture; reaches 50-55mm in height.

25. Helicostyla carinata rugata Hidalgo, 1890
54.0mm × 29.0mm, 54.5mm × 29.0mm, 56.0mm × 29.0mm; Polilio Is., Quezon.
Shell medium sized, elongate, somewhat triangular in outline, with 6 to 7 strongly convex whorls, a weakly mamillate apex, and an almost obsolete keel on the lower body whorl; aperture ovate, non-umbilicate, with a thin, reflexed outer lip margin; white to yellowish brown, with or without brown sutural banding, a large brown band on the lower body whorl, a bluish white aperture and a brown outer lip margin; quite variable; reaches 60mm in height.

Family BRADYBAENIDAE (cont'd.)

1. Helicostyla portei (Pfeiffer, 1861)
65.0mm × 52.0mm; Polilio Is., Quezon.
Shell moderately large with 4 strongly convex spire whorls and a large inflated body whorl; aperture large, laterally elongate-ovate, with an open umbilicus and a thickened, reflexed outer lip margin; base colour pinkish brown with a rich green periostracum and bands of white cuticle (not always obvious), aperture bluish white with a brown margin; reaches 80mm in height.

2. Helicostyla pithogaster Férussac, 1891
Syn: (?) *H. philippiensis* Pfeiffer
48.0mm × 43.5mm; Masbate.
Shell medium sized, triangularly ovate, with 4 strongly convex spire whorls, a flattened apical area, and a large globose body whorl; aperture large, ovate, umbilicus open, outer lip slightly thickened and reflexed; pale brown with a greenish brown periostracum and axial streaks of white cuticle; earlier whorls sometimes paler with a brown sutural band; aperture bluish white, columella pinkish, outer lip brown; reaches 55mm in height.

3. Helicostyla rufogaster Lesson, 1891
63.0mm × 50.0mm; Marinduque.
Shell large, triangularly ovate, with 4 strongly convex spire whorls, a flattened apex, and a large ovate body whorl with a nearly obsolete spiral keel on the lower section; aperture wide, ovate, umbilicus open; early whorls pinkish brown, later whorls dark brown, with closely packed bands of white cuticle on the last two; aperture bluish white, columella pink, outer lip margin brownish purple; reaches 70mm in height.

Family HELICARIONIDAE

4. Ryssota lamarckiana (Lea, 1853)
48.0mm × 85.0mm; Marinduque.
Shell large, discoidal, with a low spire of 4 strongly convex whorls, flattened at the apex and a large, sharply-keeled body whorl; body whorl weakly convex above the keel, markedly inflated below; nuclear whorls axially wrinkled, later whorls with axially and spirally incised lines; aperture large, laterally ovate-quadrangular, outer lip non-thickened and angulate; colour dark pinkish brown, greenish below the body whorl keel, aperture bluish white within, with a partially open umbilicus; reaches 95mm in diameter.

PLATE NO. 98 1. *Helicostyla portei* (Pfeiffer) 2. *Helicostyla pithogaster* Férussac 3. *Helicostyla rufogaster* Lesson 4. *Ryssota lamarckiana* (Lea) 5. *Ryssota ovum* (Valenciennes) 6. *Helicostyla marinduquensis* Hidalgo 7. *Helicostyla mirabilis trichroa* Pilsbry 8. *Mesanella palawanica* (Pfeiffer) 9. *Ryssota mülleri* (Pfeiffer) 10. *Helicostyla norrissi* (Sowerby) 11. *Helicostyla submirabilis* (Möllendorff) 12. *Helicostyla turbo* (Pfeiffer) 13. *Helicostyla melanocheila* (Valenciennes) 14. *Camena trailli* (Pfeiffer) 15. *Helicostyla codonensis* Hidalgo 16. *Helicostyla intorta* (Sowerby) 17. *Ryssota sagittifera* (Pfeiffer) 18. *Ryssota sagittifera carinata* (Möllendorff) 19. *Helicostyla samarensis* Semper 20. *Helicostyla polillensis hectori* (Pfeiffer) 21. *Helicostyla hidalgoi* Möllendorff 22. *Helicostyla cryptica* (Broderip) 23. *Helicostyla luzonica* (Sowerby) 24. *Helicostyla sphaerica* (Sowerby) 25. *Ryssota quadrasi* Hidalgo

PLATE 98

5. Ryssota ovum (Valenciennes, 1854)
53.0mm × 71.0mm; Luzon.
Shell large and heavy, with a moderately tall spire; spire whorls convex, slightly ledged at the deeply incised sutures, body whorl large with a weak central keel; earlier whorls with axial wrinkles, the later whorls crossed by weak spiral folds; aperture moderately large, ovate-quadrangular with a slightly open umbilicus and a thick, angulate outer lip margin; spire whorls purple, later whorls greenish brown with a white band on the body whorl keel, aperture bluish white within; reaches 85mm in diameter. Said to be a s.s. of *Ryssota otaheitana* Férussac.

Family BRADYBAENIDAE (cont'd.)

6. Helicostyla marinduquensis Hidalgo, 1891
69.0mm × 48.0mm, 61.0mm × 44.0mm; Marinduque.
Shell large, conoidal, with a tall spire of approximately 5 swollen, strongly convex whorls, flattened apically, and a large rounded body whorl; aperture large, subcircular, with a thick, markedly reflexed outer lip margin, umbilicus partially open; early whorls white, later whorls with brown banding, some specimens are almost white; aperture bluish white within with a brown-lined, pinkish outer lip margin and columella; reaches 75mm in height.

7. Helicostyla mirabilis trichroa Pilsbry, 1891
45.0mm × 40.0mm; Luzon.
Shell medium sized, somewhat globular, with a low spire of 4 strongly convex whorls, flattened apically, and a large ovate body whorl; aperture ovate, umbilicus slightly open, outer lip margin thick and reflexed; colour pale greenish brown with a weak brown spiral thread on the center of the spire whorls and on the upper part of the body whorl, a thicker band at the sutures and mid-body whorl; aperture and columella porcelaneous white; reaches 45mm in height.

8. Mesanella palawanica (Pfeiffer, 1898)
31.5mm × 41.0mm, 33.0mm × 41.0mm; Palawan.
Shell medium sized, slightly globose with a low, flattened spire and a large inflated body whorl; aperture laterally elongate-ovate, with a thin, weakly angulate, reflexed outer lip margin, umbilicus slightly open; pale yellowish white with brown banding and white cuticle especially on the body whorl; seldom exceeds 40mm in diameter.

Family HELICARIONIDAE (cont'd.)

9. Ryssota mülleri (Pfeiffer, 1854)
27.0mm × 48.0mm, 23.0mm × 48.5mm; Mindoro.
Shell medium sized, depressed helicoid in outline, with a low spire, the first whorls flat, the remaining weakly convex; body whorl large with a strongly angulate keel, sculptured with numerous closely-set, obliquely inclined axial grooves on the earlier whorls and crossed by weak spiral folds on the upper section of the body whorl; area below the body whorl keel very finely granular; upper area of shell pale chestnut brown, base horn coloured, with a white band around the peripheral keel; aperture narrowly elongate-ovate, with a thick, non-reflexed outer lip margin, bluish white within, umbilicus slightly open; reaches 55mm in diameter.

Family BRADYBAENIDAE (cont'd.)

10. Helicostyla norrissi (Sowerby, 1842)
38.0mm × 52.0mm, 37.5mm × 47.0mm; Sibuyan Is., Romblon.
Shell medium sized, globose, with a low, flat spire and a large, rather swollen body whorl; aperture ovate-quadrangular, with a slightly open umbilicus and a thin, reflexed outer lip margin; pale yellowish brown with darker brown banding and white cuticular banding especially on the last whorl; aperture and columella bluish white, margin brown; reaches 55mm in diameter.

11. Helicostyla submirabilis (Möllendorff, 1897)
25.0mm × 28.0mm, 22.0mm × 25.0mm; Tonquil Is., Sulu Archipelago.
Shell small to medium sized, globular, with a low, flat-topped spire and a large inflated body whorl; aperture irregularly elongate-ovate, umbilicus slightly open, outer lip thin and reflexed; yellowish brown with brown banding at the sutures and mid-body whorl; aperture and columella bluish white with a lavender margin; seldom exceeds 30mm in diameter.

12. Helicostyla turbo (Pfeiffer, 1845)
49.0mm × 43.0mm; Burias Is., Masbate.
Shell medium sized, spire moderately tall, somewhat stepped with 4 to 5 moderately convex whorls; surface with numerous crowded, fine, oblique axial grooves; aperture elongate-ovate, umbilicus slightly open, partially occluded by a columellar callus, outer lip margin thick and reflexed; golden brown in colour with darker brown-and-white banding on the body whorl, aperture bluish white with a light brown margin; reaches 60mm in height.

13. Helicostyla melanocheila (Valenciennes, 1854)
37.0mm × 44.0mm; 30.0mm × 40.0mm; Sibuyan Is., Romblon.
Shell medium sized, somewhat globose, with a low, apically flattened spire, a slightly open umbilicus, a large ovate aperture and a weakly thickened, reflexed lip; ground colour pale brown with a greenish periostracum and brown banding mainly on the body whorl; some specimens entirely brown; aperture and columella bluish white with a brown margin; reaches 45mm in diameter.

Family CAMAENIDAE

14. Camaena trailli (Pfeiffer, 1855)
36.0mm × 48.3mm; Palawan.
Shell medium sized, solid, sub-globose in outline, the entire surface with fine, oblique axial ridges; aperture large, elongate-ovate, with a thick, reflected outer lip and a slightly open umbilicus, partially occluded by a columellar "pad"; ground colour of two shades of

yellowish brown, a dark brown band at the sutures and mid-body whorl, a bluish white aperture and a dark brown margin; reaches 55mm in diameter.

Family BRADYBAENIDAE (cont'd.)

15. Helicostyla codonensis Hidalgo, 1890
34.0mm x 45.0mm; Sibuyan Is., Romblon.
Similar in shape to the brown form of *H. melanocheila* (fig. 13), but the umbilicus is closed, the columella has a pronounced thickening and the outer lip margin is white, not brown; seldom exceeds 45mm in diameter.

16. Helicostyla intorta (Sowerby, 1840)
34.5mm x 43.3mm; Sibuyan Is., Romblon.
Shell medium sized, lightweight, globose, with a pink coloured, somewhat flattened spire and a large green-coloured body whorl; a brown band at the sutures and occasionally on the body whorl; aperture ovate, bluish white in colour, outer lip thin, reflexed, with a brown margin; columella thick, bluish white with a brown band, umbilicus slightly open; reaches 45mm in diameter.

Family HELICARIONIDAE (cont'd.)

17. Ryssota sagittifera (Pfeiffer, 1842)
29.0mm x 52.0mm, 26.5mm x 47.0mm; Northwest to Central Luzon.
Shell medium size to large, spire whorls flat with finely incised axial and spiral lines, body whorl large, imperceptively keeled, coarsely sculptured above and markedly inflated below, with fine axial and spiral scratches; umbilicus open or closed, aperture laterally ovate-quadrangular, with a thin or thick outer margin; whorls marked with arrow-shaped dashes of light brown, base of body whorl darker with a peripheral band, aperture off-white, periostracum dark brown; may exceed 60mm in diameter.

18. Ryssota sagittifera carinata (Möllendorff, 1892)
23.0mm x 43.0mm, 24.0mm x 43.0mm; Marinduque.
A beautiful form which has a pronounced body whorl keel. The surface is finely granular from overriding axial and spiral threads; base of the body whorl inflated, smooth, umbilicus partially open, aperture laterally elongate-ovate; whorls pale to dark brown, base greenish brown and a white and brown band on the keel, aperture white; reaches 60mm in diameter.

Family BRADYBAENIDAE (cont'd.)

19. Helicostyla samarensis Semper 1891
28.0mm x 39.0mm; Samar and Sibuyan Is., Romblon
Shell medium sized, sub-globose in outline with a low spire and a large body whorl which is keeled on the lower third; aperture laterally elongate with a thickened, reflexed, angulate outer margin, umbilicus closed; spire whorls purplish, body whorl dark brown with white cuticle, aperture bluish white with a pale pink margin; reaches 40mm in diameter.

20. Helicostyla polillensis hectori (Pfeiffer, 1861)
35.0mm x 35.0mm; Polilio Is., Quezon.
Shell medium sized, triangularly ovate in outline; whorls strongly convex, spire slightly flattened; aperture ovate, umbilicus slightly open, outer lip thin and weakly reflexed; whitish in colour with a greenish periostracum, some white cuticle and up to 4 brown bands on the body whorl; reaches 35mm in diameter.

21. Helicostyla hidalgoi Möllendorff, 1890
33.0mm x 39.0mm; Sibuyan Is., Romblon.
Very close to *H. samarensis* (fig. 19), but with a taller spire and the colouration is generally paler; reaches 35mm in height.

22. Helicostyla cryptica (Broderip, 1840)
49.0mm x 62.0mm; Suluan Is., E. Samar.
Shell large, heavy, triangularly ovate in outline, with a low spire and a large, weakly keeled body whorl; aperture large and ovate, umbilicus closed, outer lip thick and reflexed; shell surface with a light to dark brown periostracum and white cuticle, aperture white; reaches 60 to 65mm in diameter.

23. Helicostyla luzonica (Sowerby, 1842)
38.5mm x 39.0mm; Tablas Is., Romblon.
Shell triangularly ovate, with strongly convex whorls, a slightly flattened spire, a closed umbilicus, and a large, laterally elongate-ovate, bluish white aperture; spire whorls pale brown, later whorls dark brown with a white sutural band and some white cuticle; reaches 45mm in height.

24. Helicostyla sphaerica (Sowerby, 1841)
22.0mm x 24.0mm; Ilocos Norte, N. Luzon.
A small globose shell with a low spire and a large inflated body whorl; aperture ovate, umbilicus slightly open; colour greenish yellow with a brown band at the sutures and around the umbilical opening, aperture white; reaches 25mm in diameter.

Family HELICARIONIDAE (cont'd.)

25. Ryssota quadrasi (Hidalgo, 1890)
29.0mm x 51.0mm, 28.0mm x 48.0mm; Polilio Is., Quezon
Shell medium size to large, helicoid, spire depressed, body whorl large and rounded, sutures deeply channeled; early whorls with finely incised, arcuate axial grooves, becoming coarser and crossed by spiral folds on the upper body whorl, lower body whorl finely scratched; aperture wide, ovate-quadrangular in outline, umbilicus slightly open; upper whorls pale greenish brown, lower body whorl darker with a brown band at the periphery, aperture white; reaches 65mm in diameter.

Family BRADYBAENIDAE

1. Helicostyla lignaria Pfeiffer, 1891
46.0mm x 57.0mm; Luzon.
Shell large, heavy, triangularly ovate in outline with 6 whorls having deep, weakly ledged sutures, convex sides and a large, roundly angulate body whorl; umbilicus closed, heavily glazed, aperture ovate-quadrangular, centrally angulate, outer lip moderately thick and slightly reflexed; early whorls pale yellow, later ones light brown with dark brown periostracum, aperture bluish white, with a pink margin; reaches 60mm in height.

2. Helicostyla polychroa (Sowerby, 1840)
36.8mm; 32.1mm; Romblon.
Shell medium sized, lightweight, triangularly ovate in outline, whorls 4 to 5 in number, slightly globose with a flattened apex; aperture ovate, umbilicus slightly open, columella moderately thick; shell white in ground colour, early whorls with a presutural brown band, body whorl white below the sutures, rest of the whorl with a green band of axial streaks crossed by 2 brown bands; aperture and columella white with a brown band around the umbilical area; reaches 42mm in height.

3. Helicostyla sarcinosa (Férussac, 1840)
45.0mm x 39.0mm; San Antonia, N. Samar.
Shell medium sized, lightweight, triangularly ovate, with 5 to 6 well rounded whorls and a flattened apex; aperture ovate, outer lip thick, non-reflexed, umbilicus slightly open; pale pinkish white with a greenish brown periostracum and some white cuticle, aperture white, pinkish around the umbilicus; reaches 55mm in height.

4. Helicostyla florida (Sowerby, 1840)
40.3mm x 30.5mm; Mindoro.
An extremely variable shell, typically triangularly elongate-ovate, with 4 to 5 well rounded whorls, a flat apex, a slightly open umbilicus, and an ovate aperture with a thin, reflexed outer lip margin; early whorls white, later whorls with a greenish periostracum and a subsutural white band; aperture white with a characteristic yellowish brown parietal glaze; reaches 45mm in height.

5. Helicostyla dimera (Jonas, 1854)
30.0mm x 26.8mm; Sibuyan Is., Romblon.
Shell medium sized, triangularly ovate, with 4 to 5 convex whorls and a flat apex; aperture ovate-quadrangular, umbilicus slightly open; spire purple, rest of the shell off-white, covered with a pale greenish yellow periostracum and diffused white cuticle, a brown band at the sutures, another at the body whorl base and around the umbilical area; aperture bluish white with a thin, brown stained, weakly reflexed outer lip margin; reaches 40mm in height.

6. Helicostyla montana Semper, 1891
25.8mm x 24.6mm; Basilan Is., Zamboanga.
Shell small, lightweight, globose, with 3 to 4 whorls and a flattened apex; aperture ovate, umbilicus slightly open, outer lip thin and weakly reflexed; off-white with a light brown periostracum, a brown band below the sutures and behind the outer lip margin, aperture and columella bluish white; reaches 28mm in height.

7. Helicostyla dimera (Jonas, 1854)
33.4mm x 39.5mm; Sibuyan Is., Romblon.
A more squat form of fig. 5 but the colouration is more intense and the brown band on the body whorl extends over the entire lower area.

8. Helicostyla annulata (Sowerby, 1840)
20.2mm x 17.5mm, 19.5mm x 17.5mm, 23.4mm x 18.7mm, 23.1mm x 20.0mm; N. Luzon.
Shell small, distinctly globose, with 4 to 5 whorls and a flattened apex; aperture ovate, umbilicus slightly open, outer lip thick, reflexed; colour variable, usually combinations of yellow, brown and white in the form of bands, aperture white; reaches 25mm in height.

9. Helicostyla pudibunda Semper, 1891
22.5mm x 33.2mm; N. Luzon.
Shell medium sized, lightweight, triangularly ovate, almost discoidal in outline; whorls weakly convex, body whorl large with a weak, roundly angulate shoulder; aperture widely ovate, outer lip thin, umbilicus a slight indentation; a pure white shell, occasionally with some brown banding on the spire whorls; reaches 35mm in diameter.

10. Helicostyla effusa Pfeiffer, 1892
41.0mm x 31.8mm; Sibuyan Is., Romblon.
Shell medium sized, triangularly elongate-ovate, with a moderately tall spire of 4 to 5 strongly convex whorls, the last 2 flattened at the apex; aperture ovate, umbilicus closed, outer lip thick and reflexed; early whorls pinkish brown, last 2 whorls with dark green periostracum, and a white and brown band below the sutures, aperture bluish white, outer lip margin light purple; reaches 50mm in height.

PLATE NO. 99 1. *Helicostyla lignaria* Pfeiffer 2. *Helicostyla polychroa* (Sowerby) 3. *Helicostyla sarcinosa* (Férussac) 4. *Helicostyla florida* (Sowerby) 5. *Helicostyla dimera* (Jonas) 6. *Helicostyla montana* Semper 7. *Helicostyla dimera* (Jonas) 8. *Helicostyla annulata* Sowerby 9. *Helicostyla pudibunda* Semper 10. *Helicostyla effusa* Pfeiffer 11. *Helicostyla metaformis* (Férussac) 12. *Helicostyla viridostriata* Lea 13. *Camaena brachyodon* (Sowerby) 14. *Camaena arata* (Sowerby) 15. *Camaena campanula* (Pfeiffer) 16. *Amphidromus versicolor* Fulton 17. *Amphidromus maculiferus* Sowerby 18. *Amphidromus quadrasi* Hidalgo 19. *Obba livesayi* (Pfeiffer) 20. *Obba listeri* (Gray) 21. *Obba planulata* (Lamarck) 22. *Obba listeri* (Gray) 23. *Obba rota* (Broderip) 24. *Obba horizontalis* (Pfeiffer) 25. *Obba subhorizontalis* (Möllendorff) 26. *Obba columbaria* (Sowerby) 27. *Obba planulata marmorata* Möllendorff 28. *Obba bigonia* (Férussac)

PLATE 99

11. Helicostyla metaformis (Férrusac, 1840)
38.5mm x 30.0mm; Mindoro.
Close to *H. florida* in shape, the whorls somewhat more convex in outline; shell is white, with or without a brown subsutural band, and the last 2 whorls are covered with a dark green periostracum; aperture ovate, bluish white within, outer lip thin, reflexed, and dark purplish brown in colour; reaches 45mm in height.

12. Helicostyla viridostriata Lea, 1891
35.0mm x 31.0mm; Sibuyan Is., Romblon.
Shell medium sized, triangularly ovate in outline, spire whorls strongly convex, body whorl roundly angulate towards the base; aperture ovate-quadrangular, outer lip thin, reflexed, umbilicus slightly open; off-white in ground color, the last 2 whorls with a brownish green periostracum and occassionally a subsutural white band; aperture and columella white; reaches 35mm in height. Said to be a form of *H. polychroa* (fig. 2).

Family CAMAENIDAE

13. Camaena brachyodon (Sowerby, 1840)
55.3mm x 39.3mm; Mindoro.
Shell moderately large, solid, triangularly elongate in outline, with 5 to 6 strongly convex whorls, all sculptured with fine, oblique axial riblets and well impressed sutures; aperture large, sub-circular, umbilicus closed, outer lip thick and reflexed; early whorls pale brown, body whorl dark brown with a pale white band centrally; aperture bluish white, with a light brown margin; reaches 60mm in height.

14. Camaena arata (Sowerby, 1840)
41.1mm x 30.5mm; Mindoro.
A characteristic shell, medium sized, solid, triangularly ovate, with 5 to 6 strongly convex whorls, the first 2 or 3 smooth, the rest with numerous arcuate axial ridges all over the surface; aperture ovate, umbilicus slightly open, outer lip thin and reflexed; early whorls greyish brown, last whorl dark brown, all with a white, brown-lined presutural band; aperture bluish white, outer margin deep brown; reaches 45mm in height.

15. Camaena campanula (Pfeiffer, 1890)
26.5mm x 24.3mm; Calamian, Palawan.
Shell small, globose, almost circular in outline, with 4 to 5 rounded whorls, the sutures weakly impressed and the surface with numerous fine, arcuate axial scratches; aperture ovate, umbilicus almost occluded, outer lip thick and weakly reflexed; light to dark brown with a white band at the sutures and centrally on the body whorl; aperture bluish white with a pale brown margin; reaches 30mm in height.

16. Amphidromus versicolor Fulton, 1896
37.5mm x 17.5mm; Balabac, Palawan.
Shell small, sinistral, triangularly elongate in outline with around 8 smooth, weakly convex whorls; aperture semi-lunate, non-umbilicate, outer lip thick and weakly reflexed; colour pale greyish white with brown axial maculations on the spire whorls, absent on the body whorl, aperture pale brown; reaches 45mm in height; considered a form of *quadrasi* (fig. 18).

17. Amphidromus maculiferus (Sowerby, 1838)
53.2m x 26.5mm; Tonquil Is., Sulu Archipelago.
Shell medium size to large, rather heavy, sinistral, with 8 smooth, strongly convex whorls, a large semi-lunate aperture, a thick, weakly reflexed outer lip margin and a slightly open umbilicus; colour pale greyish white with dark brown axial maculations and thin, dark purplish brown axial lines at irregular intervals, especially on the last 2 whorls, aperture white; reaches 60mm in height. Extremely variable, with no less than 18 forms having been described.

18. Amphidromus quadrasi Hidalgo, 1887
40.5mm x 21.5mm, 52.1mm x 22.3mm; Tonquil Is., Sulu Archipelago.
Shell medium sized, solid, triangularly elongate to elongate-ovate in outline, with 7 to 8 smooth convex whorls and deeply channeled sutures; aperture elongate-ovate, umbilicus closed or slightly open, outer lip thick and reflexed; shell pure yellow with a white subsutural band, aperture white; reaches 55mm in height.

19. Obba livesayi (Pfeiffer, 1860)
12.5mm x 28.6mm, 12.0mm x 26.8mm; Mindoro.
Shell small, discoidal, with 5 to 6 flat whorls and a keeled, roundly angulate body whorl shoulder, convex above and more inflated below; umbilicus deep and open, aperture widely elongate; outer lip thick and reflexed with a weak denticle on the lower margin, the upper margin curved downwards; pale greyish white with darker brown blotches, a discontinuous brown spiral band on the spire whorls and 2 on the body whorl, aperture pale brownish white; reaches 30mm in diameter.

20. Obba listeri Gray, 1840
9.3mm x 23.3mm, 11.1mm x 25.0mm; Batanes Is., N. Luzon.
This appears to be a dwarf form of the species found in other parts of Luzon. See fig. 22.

21. Obba planulata (Lamarck, 1822)
11.2mm x 29.1mm, 10.5mm x 29.5mm; Quezon Province.
Shell medium sized, discoidal, whorls flat, body whorl sharply keeled, somewhat inflated below; umbilicus deeply open, aperture elongate, outer lip thick, reflexed, with a basal denticle and the upper margin projecting downwards; colour whitish brown with a spiral band on the spire whorls and 2 on the body whorl, aperture white; reaches 40mm in diameter.

22. Obba listeri Gray, 1840
14.5mm x 33.0mm, 13.3mm x 31.9mm; Luzon.
Shell medium sized, discoidal, whorls flat to weakly convex, body whorl with a sharp keel; aperture laterally elongate, outer lip thick and reflexed, with a basal denticle, umbilicus wide and deep; colour pale

greyish white with light brown suffusion, some white cuticle and a discontinuous brown band on the spire whorls, 2 on the body whorl, aperture white; reaches 40mm in diameter.

23. Obba rota Broderip, 1841
10.2mm x 24.6mm, 10.3mm x 22.6mm; Masbate.
Shell small to medium sized, discoidal, the whorls very weakly stepped and sculptured with very fine axial threads which sometimes become ridge-like; body whorl rounded, weakly carinate, umbilicus open and partially occluded, aperture laterally elongate, outer lip thin, reflexed; colour off-white with dark brown maculations and a brown spiral band, base pale brown with a single brown band, aperture pinkish brown; reaches 25mm in diameter.

24. Obba horizontalis (Pfeiffer, 1845)
15.0mm x 37.0mm, 16.0mm x 36.8mm; Sibuyan Is., Romblon.
Shell large, heavy, discoidal, spire whorls weakly convex, body whorl rounded to sharply carinate; umbilicus deep and open, aperture laterally elongate, outer lip thick, reflexed, with no basal denticle and the upper margin perpendicular to the apertural opening; light brown with white and dark brown bands, aperture pale brown, margin pinkish; reaches 40mm in diameter.

25. Obba subhorizontalis (Möllendorff, 1894)
16.0mm x 38.0mm, 16.0mm x 38.1mm; Palawan.
Similar to *O. planulata* (fig. 21), but more inflated, with the body whorl only weakly carinate and with a slight subsutural ledge; aperture laterally elongate, thick and reflexed basally, but thinner and non-reflexed on the upper margin; colour pale yellowish white with dark brown mottling and banding, the inner part of the base devoid of any markings; reaches 40mm in diameter.

26. Obba columbaria (Sowerby, 1841)
17.5mm x 28.0mm, 18.2mm x 28.9mm; Polilio Is., Quezon.
Shell medium sized, solid, triangularly ovate in outline, with 4 to 5 strongly convex whorls sculptured with numerous fine axial ridges; aperture laterally ovate, outer lip thick and reflexed with a basal denticle, umbilicus open and deep, partially occluded; coloured with light and dark brown banding, and a subsutural row of white quadrate spots.

27. Obba planulata marmorata Möllendorff, 1892
13.3mm x 32.1mm, 12.9mm x 29.0mm; Mindoro.
A form of fig. 21, in which the shell is a pale greyish white, mottled with pale brown; the base is devoid of any markings except for a thin brown band; this form reaches around 30mm in diameter.

28. Obba bigonia (Férrusac, 1854)
11.5mm x 24.2mm, 13.0mm x 27.0mm; Leyte.
Shell small, discoidal, but with a slightly conical apex of 5 to 6 almost flat whorls; aperture laterally ovate, outer lip thick and reflexed basally, thin and down-curved on the upper margin; umbilicus open, deep and partially occluded; pure white with brown banding; reaches 30mm in diameter.

Plate 100 — Bradybaenidae, Helicarionidae

Family BRADYBAENIDAE (cont'd.)

1. Helicostyla roissyana (Férussac, 1840)
32.5mm × 32.0mm, 28.0mm × 30.5mm; Mindoro.
Shell medium sized, globose, with 4 to 5 whorls flattened at the apex, an ovate aperture, a slightly open umbilicus and a thin, weakly reflexed outer lip margin; shell white with varying degrees of dark purple to black banding, a white aperture and columella, and a blackish outer margin; a characteristic species; reaches 35mm in diameter.

Family HELICARIONIDAE (cont'd.)

2. Hemiglypta semperi (Möllendorff, 1893)
24.3mm × 48.5mm; N. Luzon.
Shell medium sized, solid, triangularly ovate in outline, with a low, flat-sided spire, and a strongly carinate body whorl which is flat above and convex below; surface with numerous close-set radial striae, becoming finely granulose on the last 2 whorls from overriding spiral threads, base relatively smooth; aperture laterally elongate, with a thick, sharply angulate margin, umbilicus occluded by a small callus pad; light greenish brown in color, the base shining and a dark brown colour, aperture pale bluish white; reaches 55mm in diameter.

3. Hemiglypta blainvilleana Lea, 1898
23.0mm × 35.0mm, 23.0mm × 40.0mm; Lubang Is., Mindoro.
Shell medium sized, circular in outline, with a low spire of 5 to 6 strongly convex whorls, and a large body whorl with a roundly angulate shoulder; whorls finely granulose from overriding axial and spiral lirae, base smooth and shiny; aperture elongate-ovate, outer lip thick, slightly angulate, umbilicus occluded by a callus pad; whorls yellowish brown, a brown band on the body whorl shoulder, and iridescent greenish brown below, aperture white; reaches 45mm in diameter.

Family BRADYBAENIDAE (cont'd.)

4. Hemitrichiella setigera (Sowerby, 1898)
17.0mm × 25.2mm, 16.8mm × 25.8mm; Sibuyan Is., Romblon.
Shell medium sized, very light, weakly conical, with 6 to 7 weakly convex whorls and a large inflated body whorl; whorls finely granulose from overriding axial and spiral threads, body whorl base smooth and shiny; aperture irregularly ovate, outer lip thin,

umbilicus slightly open, whorls covered with a hairy periostracum; light greenish white in colour, with a brownish red presutural band, aperture and columellar area white; reaches 25mm in diameter.

Family CYCLOPHORIDAE

5. Cyclophorus appendiculatus Pfeiffer, 1886
18.0mm x 29.5mm, 19.5mm x 30.0mm; Mindoro.
Shell medium sized, with a low spire of strongly convex whorls and deeply channeled sutures; surface with fine spiral threads and weaker axial scratches, base of body whorl nearly smooth, umbilicus deep, outer lip circular and thick; pale brown, with a row of white rectangular subsutural spots, and a white band on the body whorl periphery and around the umbilicus; reaches 30mm in diameter.

6. Cyclophorūs picturatus Pfeiffer, 1862
19.0mm x 25.2mm, 17.7mm x 23.5mm; Burias Is., Masbate.
Shell medium sized, solid, the spire somewhat taller than fig. 5, and the whorls more inflated; the sutures are deeply grooved, not channeled, and the wide, open umbilicus is partially occluded by a callus extending from the inner lip; light brown in colour, becoming darker on the body whorl, white axial streaks initially confined to the sutures, then extending to the white band on the body whorl periphery, aperture and peri-umbilical area white; reaches 25mm in diameter.

7. Cyclophorus linguiferus Sowerby, 1886
14.0mm x 22.2mm, 13.6mm x 21.9mm; Burias Is., Masbate.
Shell small to medium sized, somewhat trochiform in shape, and with 5 to 6 rounded whorls; sculptured with fine spiral lirae, deeply gooved sutures, and a relatively smooth base, umbilicus deep and open, partially occluded by a callus, aperture circular with a thick margin; pale reddish brown with white axial zigzag lines and a white aperture and peri-umbilical area; reaches 25mm in diameter.

8. Cyclophorus woodianus Lea, 1862
17.2mm x 30.8mm, 20.5mm x 32.9mm; Marinduque.
Shell medium sized, with a low spire of 4 to 5 strongly convex whorls and deeply channeled sutures; sculptured with coarse spiral threads, almost smooth on the base; umbilicus deep and open, aperture circular, outer lip thick; pale yellowish brown with darker brown maculations, a row of brown and white quardate spots at the sutures, and a thin brown and white spiral band on the body whorl periphery; reaches 35mm in diameter.

9. Cyclophorus daraganicus Hidalgo, 1890
15.3mm x 29.5mm, 18.5mm x 27.9mm; Quezon.
Shell medium sized with a low spire, rounded whorls and deeply channeled sutures; surface with major and minor spiral lirae, base nearly smooth; umbilicus wide and deep, aperture circular, outer margin thick; a fairly pale shell with pale brown axial streaks terminating at the body whorl periphery where there is a thin, circular brown band, base white; reaches 30mm in diameter.

10. Cyclophorus reevei Hidalgo, 1890
16.0mm x 19.5mm, 16.8mm x 20.5mm; Coron, Palawan.
Shell small to medium sized, trochiform, sutures deeply impressed, spire whorls strongly convex, body whorl rounded with a weak peripheral keel; sculptured with major and minor spiral threads, base fairly smooth; umbilicus partially open and deep, aperture rounded, outer margin thin and weakly reflexed; early whorls dark, later ones pale yellow with dark brown streaks, base with concentric spiral bands, aperture and outer lip margin white; reaches 22mm in height.

11. Cyclophorus alabatensis Kobelt, 1886
14.0mm x 15.0mm, 14.2mm x 15.5mm; Sibuyan Is., Romblon.
Shell small, trochiform, with well impressed sutures, 5 to 6 strongly convex whorls, and sculptured with major and minor spiral cords, producing a weakly bicarinate body whorl; umbilicus deep, partially closed, aperture circular with a thin, slightly reflexed outer lip; lightweight, semi-transparent and dark brown in colour with pale yellow stripes and blotches, the aperture reflecting the external colouration, outer lip margin white; reaches 15mm in height.

12. Cyclophorus fernandezi Hidalgo, 1890
13.5mm x 14.0mm, 14.6mm x 15.2mm; Sibuyan Is., Romblon.
Very similar to *C. alabatensis* Kobelt (fig. 11) in outline and sculpturing, but the whorls are more rounded and the colour is pale yellowish white with a subsutural row of brown and white quadrate spots, and white banding on the spiral lirae; aperture and outer lip margin bluish white; reaches 15mm, in height.

PLATE NO. 100 1. *Helicostyla roissyana* (Férussac) 2. *Hemiglypta semperi* (Möllendorff) 3. *Hemiglypta blainvilleana* Lea 4. *Hemitrichiella setigera* (Sowerby) 5. *Cyclophorus appendiculatus* Pfeiffer 6. *Cyclophorus picturatus* Pfeiffer 7. *Cyclophorus linguiferus* Sowerby 8. *Cyclophorus woodianus* Lea 9. *Cyclophorus daraganicus* Hidalgo 10. *Cyclophorus reevei* Hidalgo 11. *Cyclophorus alabatensis* Kobelt 12. *Cyclophorus fernandezi* Hidalgo 13. *Cyclophorus sericinus* Quadras & Möllendorff 14. *Cyclotus caroli* Kobelt 15. *Leptopoma calamanensis* Bartsch 16. *Cyclophorus validus* Sowerby 17. *Leptopoma perlucidum* Grateloup 18. *Trochomorpha schmakerii* Möllendorff 19. *Trochomorpha metcalfei* (Pfeiffer) 20. *Trochomorpha sibuyanica* Hidalgo 21. *Eulota mighelsiana* (Pfeiffer) 22. *Chloraea dryope* (Broderip) 23. *Schistoloma alta sibuyanica* Bartsch 24. *Homorus burnessi* M. Connolpy 25. *Melanoides torulosa* (Bruguiére) 26. *Melanoides maculata* (Bruguiére) 27. *Faunus ater* (Linné) 28. *Melanoides granifera* (Lamarck) 29. *Melanoides tuberculata* (Müller) 30. *Thiara winteri* (Von Dem Busch) 31. *Thiara cancellata* (Röding) 32. *Pila ampullacea* (Linné)

PLATE 100

1
2
3
4
5
6
7
8
9
10
11
12
13
14
15
16
17
18
19
20
21
22
23
24
25
26
27
28
29
30
31
32

13. Cyclophorus sericinus Quadras & Möllendorff, 1894
16.9mm x 21.0mm, 17.0mm x 20.2mm; Coron Is., Palawan.
Shell medium sized, lightweight, somewhat trochiform in outline, with 4 to 5 strongly convex whorls and deeply channeled sutures below which are weak spiral lirae which become stronger down the whorls and smooth on the body whorl base; umbilicus deep, partially occluded, aperture circular, outer lip thin and reflexed; deep chocolate brown in colour, with strong, white axial zigzag flammules, aperture and outer lip margin white; reaches 18mm in height.

14. Cyclotus caroli Kobelt, 1886
7.7mm x 14.3mm, 6.8mm x 13.3mm, 8.0mm x 15.9mm; Palawan.
Shell small, lightweight, discoidal in outline, with a low, almost sunken spire; whorls rounded and smooth, umbilicus wide and deep, aperture circular, outer lip thin and slightly reflexed; white with thick brown axial streaks, nuclear whorls dark purple; reaches 18mm in diameter.

15. Leptopoma calamanensis Bartsch
9.7mm x 9.5mm, 9.4mm x 9.9mm, 10.0mm x 10.7mm; Batanes Islands.
Shell small, conoidal, with 5 to 6 strongly convex whorls, nuclear whorls smooth, the rest with spiral lirae; umbilicus narrow, deep, partially occluded, aperture circular, outer lip thin and weakly reflexed; a dark greenish brown apex, the rest of the shell pale yellowish white with a subsutural row of brown spots, numerous brown-dotted axial streaks, and a darker band on the white base, aperture white; reaches 10mm in height.

16. Cyclophorus validus Sowerby, 1862
19.0mm x 20.5mm, 20.2mm x 19.6mm; Leyte.
Shell medium sized, solid, trochiform with 5 to 6 whorls, the first 3 or 4 strongly convex with weak, arcuate axial threads, the rest with strong spiral lirae and weaker spiral threads producing slightly angulate whorls; umbilicus deep, narrow, partially occluded, aperture eccentrically elongate-ovate, outer lip margin thick; dark chocolate brown in colour with yellowish axial flammules; reaches 22mm in height.

17. Leptopoma perlucidum Grateloup, 1886
14.0mm x 13.3mm, 14.0mm x 14.0mm; Palawan.
Shell small, light, semi-translucent with a moderately tall spire and a large, somewhat globose body whorl; whorls strongly convex and smooth, umbilicus narrow, deep and partially occluded; aperture large and circular, with a thin, reflexed outer lip margin; pale greyish white in colour with brown spiral threads; reaches 15mm in height.

Family TROCHOMORPHIDAE

18. Trochomorpha schmakerii Möllendorff, 1894
5.5mm x 13.5mm, 6.3mm x 123.8mm; Sibuyan Is., Romblon.
Shell small, discoidal, with a low, flat-sided spire, a sharply keeled body whorl, a strongly convex base, a narrow, deep umbilicus and a narrow, ovate aperture with a thin outer lip; shell yellowish brown in colour with a brown presutural band and a greenish brown base; reaches 15mm in diameter.

19. Trochomorpha metcalfei (Pfeiffer, 1845)
8.0mm x 10.5mm; Cebu.
Shell medium sized, discoidal, spire low, spire whorls almost flat, base slightly inflated below a sharply angulate peripheral keel; umbilicus deep, aperture ovate with a thin outer lip; shell pale yellowish brown, with a dark brown presutural band and 2 on the base; reaches 24mm in diameter.

20. Trochomorpha sibuyanica Hidalgo, 1890
7.2mm x 23.5mm; Sibuyan Is., Romblon.
Shell medium sized, discoidal, spire almost flat, body whorl with a sharply angled peripheral keel, concave above, inflated and convex below; umbilicus wide and deep, aperture ovate-quadrangular, whorls with fine arcuate radial scratches; colour entirely brownish green; reaches 25mm in diameter.

Family BRADYBAENIDAE (cont'd.)

21. Eulota mighelsiana (Pfeiffer, 1846)
15.2mm x 16.1mm; Batanes Islands.
Shell small, distinctly globose in outline with weakly channeled sutures and the entire surface sculptured with numerous fine, close-set axial ridges; aperture ovate with a thin outer lip, umbilicus partially open; pinkish brown, spire whorls darker; reaches 20mm in height.

Family CYCLOPHORIDAE (cont'd.)

22. Chloraea dryope (Broderip, 1840)
11.5mm x 26.2mm, 10.5mm x 20.9mm; Sibuyan Is., Romblon.
Shell medium sized, circular in outline, with a flat to moderately tall spire of 4 to 5 weakly convex whorls; body whorl with a sharp peripheral keel, concave above and inflated below; umbilicus closed and brown coloured, aperture laterally elongate with a thin, weakly reflexed outer lip; nuclear whorls black, early whorls off-white, later whorls light green, aperture white; reaches 30mm in diameter.

Family POMATIASIDAE (cont'd.)

23. Schistoloma alta sibuyanica Bartsch, 1915
23.3mm; Sibuyan Is., Romblon.
Shell small, elongate, with 7 to 8 strongly convex whorls, reducing rapidly in size towards the apex; sutures impressed and the whorls have fine axial scratches; aperture ovate, peristome entire, the outer lip margin slightly reflexed, umbilicus almost entirely obliterated; colour deep reddish brown, with a paler outer lip margin; reaches 28mm.

Family STENOGYRIDAE

24. Homorus burnessi M. Connolly
41.5mm; Sibuyan Is., Romblon.
A South African species apparently introduced to the Philippines; shell medium sized, lightweight, elongate, with 9 to 10 weakly convex whorls and moderately impressed sutures; surface with fine axial scratches; aperture elongate-ovate with a thin, weakly arcuate outer lip; off-white with a dirty green periostracum; reaches 45mm.

Family THIARIDAE

25. Melanoides torulosa (Bruguiére, 1789)
43.5mm, 40.2mm, 34.0mm; Bukid River, Leyte.
Shell medium sized, highly turreted, the spire frequently truncated; spire whorls flat sided, body whorl large and strongly convex; sculptured with axial and spiral ridges producing a granulose surface, the subsutural row of granules being strongest and somewhat more elongate than the rest; axial ridges become obsolete towards the body whorl base, with only the spiral ridges persisting; aperture pyriform, pointed above, rounded below, outer lip thin and crenulate, columella twisted and weakly glazed; colour pale brown to white with a thick greenish black periostracum; reaches 80mm.

26. Melanoides maculata (Bruguiére, 1789)
39.5mm; Bukid River, Leyte.
Shell medium sized, lightweight, highly turreted, extremely attenuated with up to 12 strongly convex whorls, weakly incised sutures, and sculptured with spiral grooves and weak axial striae; aperture pyriform, pointed above, rounded below, outer lip thin; colour light olivaceous brown with red axial streaks, restricted to the suture on the body whorl; reaches 70mm.

27. Faunus ater (Linné, 1758)
67.0mm; Palawan.
Shell medium sized, solid, very highly turreted, the apex frequently eroded, whorls flat sided, sutures weakly impressed; sculptured with axial threads and weak spiral grooves; aperture elongate-ovate with an anterior and posterior sinus; outer lip thick and curved, columella arcuate, narrow but heavily glazed; generally covered by an almost black periostracum; reaches 85mm.

28. Melanoides granifera (Lamarck, 1822)
18.0mm; Cavite, Luzon.
Shell small, elongate to elongate-ovate in outline, with a tall spire and a large body whorl; whorls granular from intersecting spiral and axial ridges with only the spiral elements persisting towards the body whorl base; aperture ovate, pointed above and rounded below, outer lip thin and arcuate, columella weakly glazed; spire whorls almost black, last whorl pale green, aperture white; reaches 40mm.

29. Melanoides tuberculata (Müller, 1774)
28.5mm; Palawan.
Shell small to medium sized, with a tall spire and a large body whorl; early whorls strongly convex with axial and spiral striae, later whorls somewhat flatter, with coarse, undulating axial ribs becoming the predominant feature, but disappearing on the body whorl base; aperture vertically ovate, outer lip thin, slightly angulate basally; early whorls dark brown, later whorls with yellowish axial ribs, brown interspaces, and a solid brown band on the body whorl base; reaches 35mm.

30. Thiara winteri (Von Dem Busch, 1842)
37.3mm; Palawan.
Shell large, solid, with a tall spire and a large body whorl; whorls 6 to 8 in number and somewhat stepped; sculptured with weak axial folds which form a crown of short spines on the sharply angled whorl shoulder, some flat spiral ridges towards the body whorl base; aperture vertically ovate, pointed above and rounded below, the outer lip thin, lirate within and the columella heavily ridged; greenish yellow with a white aperture; reaches 50mm.

31. Thiara cancellata (Röding, 1798)
27.8mm; Eastern Quezon.
Shell medium sized, ovately biconic, with a low, occasionally truncated spire, the whorls of which descend in a step-like fashion; whorls sharply angulate, a row of spines on the shoulder, slightly concave above and encircled with numerous fine spiral threads, almost vertical below and crossed by 2 or 3 granular cords; body whorl strongly convex below the shoulder spines and encircled with up to 20 flat ridges; aperture elongate-ovate, with a shiny callus on the lower columellar area; early whorls black, later whorls pale yellowish white with brown spots, spines tinged in brown; reaches 45mm.

Family AMPULLARIIDAE (Fresh Water)

32. Pila ampullacea (Linné, 1758)
36.8mm x 30.0mm; Cavite, Luzon.
Shell large, roundly pyriform with a low spire and a large, globose body whorl; sutures impressed, surface sculpturing obsolete; aperture semi-lunate, obliquely aligned, outer lip thin and arcuate, columella weakly calloused and reflexed towards the basal area, umbilicus closed or very weakly open, operculum calcareous, semi-lunate, with concentric ridges around an eccentric nucleus; yellowish green with brown spiral bands, frequently obscured by a shiny, dark green periostracum, aperture deep purple within, outer area with a yellowish orange tinge; reaches 100mm in height.

Appendix

Glossary of Malacological Terminology

ABUTTING. Adjacent, or next to.
ACUMINATE. Pointed; tapering to a point.
ACUTE. Sharply angled.
ADAPERTURAL. View of shell with aperture facing observer.
ADDUCTOR SCARS. Muscle scar impressions within a bivalve shell.
ADHERENT. Attached.
ADPRESSED. Whorls overlapping; no distinct suture line.
ALATE. Winged.
ANAL CANAL/ANAL SIPHON. An opening in the shell where waste material is excreted; found on the posterior edge of aperture.
ANAL SULCUS. A posteriorly-located groove in the aperture, which contains the ANAL SIPHON.
ANGULATE. Referable to whorls meeting at an angle; also applicable to two surfaces.
ANNULATE. Bearing concentric rings (i.e. as on opercula).
ANTERIOR/ANTERIOR MARGIN. Portion of shell nearest the head of the mollusc when in motion. In gastropods, it is opposite the apex; in bivalves, it is opposite the ligament.
ANTERIOR SLOPE. Referable to bivalves; the slope of the surface from umbo to antero-ventral margin.
ANTERO-DORSAL SURFACE. Referable to bivalves; dorsal surface anterior to umbones.
ANTERO-VENTRAL SURFACE. Referable to bivalves; ventral surface anterior to umbones.
APERTURAL. Pertaining to the aperture, or on the same side as the aperture.
APERTURAL LIP. The outer margin of the aperture.
APERTURE. Pertaining to the opening into the shell.
APEX. The first-formed whorls of the gastropod spire (protoconch); usually pointed.
APICAL. Referring to the apex or tip of the spire.
APOPHYSIS. Referable to a projecting area within some bivalves, below the umbones, to which certain muscles are attached; also refers to a projecting structure on the inside of some gastropods.
ARCUATE. Arched or curved.
ASYMMETRICAL. Unequal.
AURICAL. Ear-shaped appendage (as in the Pectinidae).
AXIAL. Longitudinal; parallel to or following the shell axis (usually applied to gastropod shells).
AXIAL STRUCTURE/AXIAL SCULPTURE. Sculpturing on a gastropod shell running parallel to the shell axis.
AXIS. An imaginary line through the apex, around which the whorls of univalve shells are laid down.
BASAL. Pertaining to the base.
BASE. The last-formed part of a gastropod shell; area opposite the apex, below the periphery of the body whorl (not including aperture).
BATHYMORPH. A form of a species, the morphology of which appears to be dependent on the depth where it is found.
BEAD. A small, rounded knob, usually on a rib.
BEADED. Resembling a string of beads.
BEAK. A term applying to the umbones in a bivalve shell.
BIANGULATE. Having 2 angles; applies to shell surfaces which meet at different angles on different areas of the shell.
BICONIC. Descriptive term applied to shells whose structure resembles two cones placed base-to-base; diamond-shaped.
BIFID. Split in two, separated by a cleft.
BIFURCATE. Divided into two parts; double-pronged; resembling a fork.
BISECTED. Divided into two parts.
BINOMIAL. Consisting of two names.
BIVALVE. Shells with two halves joined by a hinge (N.B. The bivalve Gastropod Family Juliidae is excluded from this interpretation).
BLADE. A thin, erect structural element originating from the shell surface (i.e. seen in certain Murex species e.g. *S. alabaster*).
BODY. Refers to the body whorl, excluding the siphonal canal.
BODY WHORL. The last formed whorl of any spiral shell.
BUCCAL. Refers to the mouth area.
BUCCAL MASS. A mass of tissue in the mouth area involved in the initial stages of food digestion. Usually supports the radula teeth.
BYSSAL ORIFICE/BYSSAL GAPE. Marginal opening of bivalve shells through which the byssal strands pass.
BYSSAL NOTCH. A structural indentation below the anterior ear of the right valve in Pectinidae, for passage of the byssus.
BYSSUS. Fine, silky, fibrous strands, by which some bivalves attach themselves to the substrate.
BULBOUS. Bulging or globular.
CALLUS/CALLOUSED. A thickened area of calcareous matter or enamel, usually found around the aperture or on the parietal wall.
CANAL. A narrow, tube-like extension of the aperture which allows for the protrusion of the siphon.
CANALICULATE. Channeled or grooved; usually applies to suture.
CANCELLATE. The resultant surface structure formed from spiral and vertical cords intersecting at right angles.
CANTED. Refers to a non-perpendicular structural element.
CARDINAL TEETH. The main teeth of bivalves; situated centrally near the umbones.
CARINA.Keel; sharply-edged ridge.
CARINATE. Having a keel; frequently applied to the body whorl of some gastropod shells.
CHANNEL. Groove.
CHINK. A small cleft, usually applied to the area around the umbilicus.
CINGULATE. With a girdle or cingulum.
CINGULUM. A band of colour or elevated sculpturing.
CLATHRATE. A latticed surface element formed from bisecting axial and spiral sculpture.
CLUB-SHAPED. Having a rounded body whorl and a long siphonal canal (e.g. *Haustellum spp.*).
COLUMELLA. A pillar or wall surrounding shell axis, opposite the outer apertural lip.
COLUMELLAR FOLD. A raised fold on the columella, which projects into the apertural opening.
COLUMELLAR WALL. The wall of the columella.
COMPLANATE. Refers to a flattened, unsculptured surface.
COMPRESSED. Flattened or reduced in thickness.
CONCAVE. Hollowed or excavated.
CONCENTRIC. Concentric striae following the direction of growth lines; particularly applicable to bivalves.
CONFLUENT. Merging or coalescing.
CONICAL. Cone-shaped.
CONSPECIFIC. A term applied to 2 specimens which have been shown to be the same species.
CONSTRICTED. Pinched; an abrupt narrowing.
CONTRACTED. Having shrunken or been reduced in size.
CONVEX. Outwardly curved; rounded; opposite to concave.
CONVOLUTE. Last whorl completely obscuring earlier whorls.
CORD. Thickened, elevated axial or spiral sculpturing.
CORNEOUS. Horny.
CORONA. Refers to a crown-like structural element.
CORONATE. Possessing a corona.
CORRUGATE. Structured with heavy ridges or folds.
COSTA (AE). An axial ridge, usually more prominent than a cord.
COTYPE. A shell from a selection of type material, from which no holotype was ever selected. (= Syntype).

CRASSATE. Coarse.
CRENATE. Descriptive of a scalloped outer-lip margin.
CRENULATE. Similar to crenate, but usually implies a finer wrinkling with smaller subdivisions.
CRISPATE. Fine, irregularly folded curves.
DECUSSATE. Surface formed from intersecting axial and spiral elements, but usually at acute angles. (c.f. Cancellate).
DENTATE. With teeth or short ridges.
DENTICLE. Small, tooth-like protruberance.
DENTICULATE. Sculptured with small denticles.
DEPRESSED. Low in relation to diameter.
DEXTRAL. Right-handed; aperture on observer's right when viewed from an adapertural position, with the apex pointed upwards.
DIGITATE. Having finger-like projections.
DILACERATE. Interrupted.
DIMORPHIC. Having two forms.
DISCOIDAL. Disc-shaped, more or less compressed.
DISTAL. Situated away from the point of attachment.
DIVARICATE. Forked or branched. Usually applied to ridges or grooves in bivalves which deviate from concentric or radial growth patterns in a fairly non-specific way.
DORSAL. The back, or upper surface of a shell; in gastropods, opposite the aperture.
DORSAL MARGIN. Usually applied to the hinge area in Bivalves.
DORSUM. The dorsal surface of a shell.
DULL. Lacking sheen or gloss.
DWARF. Refers to forms fully developed, but much smaller than normal.
EARS. Elongated appendages of the dorsal areas of bivalves, situated on either side of the umbones (i.e. as in Pectinidae).
ECCENTRIC. Off centre; in gastropods refers to structural features not following the normal growth patterns.
ECOLOGY. The relationship between organism and environment.
EDENTULATE. Lacking teeth.
ELEVATED. Raised; high in relation to diameter; opposite to depressed.
ELONGATE. Lengthened along one axis (i.e. axially or laterally elongate).
EMARGINATE. Margin with fine indentations or notches.
EMBRYONIC WHORL. The protoconch.
ENDEMIC. Confined to a particular area or geographical region.
ENTIRE. Continuous, without any visible break (i.e. the apertural margin).
EQUILATERAL. Usually applied to valves of bivalves which are the same size and shape.
EQUIVALVE. See EQUILATERAL.
ERECT. Elevated or upright.
ESCUTCHEON. A depressed, elongated area behind the umbonal area of bivalves which usually includes the ligamental area.
EVANESCENT. Applied to describe structural features which are barely visible and which appear and disappear on the shell surface.
EXCAVATED. Strongly depressed or hollowed out.
EXPANDED. Enlarged; increased in size.
FALSE UMBILICUS. A depression at the base of some gastropod shells which is only slightly penetrating (i.e. Trochidae).
FASCIOLE. A groove or raised ridge at the base of some gastropod shells; usually consisting of a series of growth lines.
FENESTRATE. Have deep, pit-like surface depressions, much akin to windows.
FIMBRIATE. Bordered by thin, frilly, sculptural elements.
FISSURE. Cleft or slit.
FLANGE. A rib-like structure which re-enforces the juncture of 2 other structural elements.
FLARING. Opening or spreading outwards; usually applied to the outer lip of certain gastropods.
FLATTENED. Level.
FLUTED. Sculptured with grooves or channels; in bivalves usually refers to the erect, somewhat arched scale-like structures.
FOLD. A ridge on the columellar wall (see also COLUMELLAR FOLD).
FOLIACEOUS. Shaped like, or resembling foliage.
FOLIATE. Bearing foliage-shaped structures; branched with somewhat crimped outer ends (e.g. applied to some Muricid species).
FORM. Refers to 1 or several morphophenotypes of a single species. The use of the term "form", although having some practical advantages, has no taxonomical validity.
FOSSULA. A shallow depression on the inner lip of many cowry species.
FRILLED. Refers to structures that have crimpled or fluted edges
FROND. A sculptural element with a leaf-like extension.
FRONDOSE. Resembling a leaf with numerous divisions (i.e. a palm leaf).
FURROW. A weak, narrow channel or groove.
FUSED. Bound together as a single unit of surface sculpturing.
FUSIFORM. Spindle-shaped; biconic.
GAPE. The opening between two valve halves which are unable to completely close (see also BYSSAL GAPE).
GEMMATE. Having a fine, bead-like structure.
GENUS. A taxonomic status applied to individuals sharing a common phylogeny.
GERONTIC. Applying to specimens of an old age.
GLOBOSE. Rounded, spherical.
GRANULOSE/GRANULATE. Covered with fine pustules or granules.
GROOVE. A uniform channel or depression.
GROWTH LINE. A series of elevated, concentric lines or ridges, indicating earlier growth margins.
GROWTH MARGIN. The edge of the shell surface from which new growth appears to originate; in gastropods, it is considered to be the outer apertural edge.
GROWTH STAGES. Generally considered to be the area between two growth lines, which represents an area of active growth.
HEIGHT. In gastropods, this parameter refers to the distance between the protoconch and the base of the siphonal canal; in bivalves, this distance refers to the greatest distance from the umbones to the ventral margin along a perpendicular line.
HINGE. The thickened internal area in the dorsal region of a bivalve which contains the interlocking teeth arrangement and ligament.
HINGE LINE. Dorsal side of hinge.
HINGE PLATE. The section of hinge which bears the teeth.
HINGE TOOTH. A small, projecting structure from the hinge plate which interlocks with a socket-like arrangement on the opposite valve (i.e. the ball, in a ball and socket arrangement).
HOLOTYPE. The single specimen on which the characteristics of a given species are based; the salient features of the species are based on the features of this one particular shell.
HOMONYM. A taxonomical situation, where 2 different species bear the same name.
HYPOTYPE. A specimen known from published illustrations and descriptions, but not in the original type series.
HYBRID. The offspring from two biologically distinct species; usually sterile.
IMBRICATE. Usually applied to scales or lamellae which overlap (similar to tiles of a roof).
IMPERFORATE. Not a good term. Usually refers to a closed umbilicus. An open umbilicus, which would, conversely, be said to be perforate is in fact, not a perforation.
IMPRESSED. Indented. Usually applies to sutures.
INCISED. Deeply or sharply grooved.
INCURVED. Applies to a structure which curves in upon itself, as do the umbones in some bivalves.
INDENTATION. A notch or excavation on any structural margin.
INDENTED. Pressed in.
INDUCTURA. A large expansion of the inner apertural lip of gastropod shells (usually erect).
INEQUILATERAL. (Bivalves) — Said of valves on either side of the umbones which are unequal in size or shape.
INEQUIVALVE. (Bivalves) — One valve larger than the other.
INFLATED. Swollen, ventricose; often infers strong whorl convexity.
INFLECTED. Turning inwards; same as INCURVED.
INNER LIP. Apertural margin opposite outer lip, extending from columellar base to apertural apex.
INHALENT SIPHON. Tubular appendage of molluscs which facilitates water intake.
INTERCALATE. Interspersed.
INTERGRADE. Applied to specimens which possess characters intermediate between two morphologically distinct population groups. Often seen on the edge of certain population areas, where there is apparent merging of one form with another.
INTERMEDIATE RIBS. Ribs between the major ribs, which are of a smaller and weaker nature.
INTERSPACE. Intermediate spaces between structural features on a shell surface; usually between ribs, cords or teeth (linear surface ornamentation).
INTERSTICE. Space between structures; can be considered same as INTERSPACE.
INTERSTITIAL. Pertaining to interstices.

INTERVAL. The gap between two patterns or structural elements on a shell surface.
INTERVARICAL. Between 2 varices.
INTERVARICAL SPACE. The space between 2 varices, which represents area of active growth.
INTRITACALX. A fine, chalky layer on the surface of some gastropod shells which is often of help in species identification (i.e. certain Muricid species).
JUVENILE. Immature; not fully developed.
KEEL. A sharp, raised sculptural element, usually arranged in a spiral fashion; a ridge or carina.
KEELED. Ridged; carinate.
LABIAL. Refers to the inner lip.
LABIAL TEETH. A somewhat indecisive term which, in Cowry terminology, is used when referring to the columellar teeth.
LABIUM. Inner lip of the aperture.
LABRAL. Refers to the outer lip.
LABRAL TEETH. A tooth or spur-like extension from the outer lip margin of some gastropod shells.
LABRUM. The outer lip of the aperture.
LAMELLA. A thin, normally erect, plate-like ridge or scale.
LAMELLAR. Plate-like; scale-like.
LAMELLATE. Covered with scales or plates.
LAMELLOSE. Same as LAMELLATE.
LAMINA. A plate or scale.
LAMINATE. Covered with laminae; i.e. surface covered with plates or scales. Basically the same as LAMELLOSE or LAMELLATE.
LAMINOSE. Having many laminae; distinction from laminate is fine, but term usually applied when only part of surface covered with laminae.
LANCEOLATE. Extending to a point at both ends; "like a lance".
LATERAL. Arising from the side or the margin of a structure.
LATERAL TOOTH. (Bivalves) — Tooth on the hinge plate, relatively distant to the umbones; (Radula) — An element of the radula structure lying midway between the central and marginal teeth structures.
LATTICED. The resulting structural pattern from intersecting axial and spiral elements.
LEADING EDGE. The area closest the line of active shell growth.
LECTOTYPE. A specimen selected from the type material to act as the holotype; done when an original holotype was never selected, or when the selected holotype has been lost or destroyed.
LEFT VALVE. (Bivalves) — Valve on observer's left, when bivalve is orientated with hinge line uppermost and anterior end pointing away from him.
LENTICULAR. Lens-shaped; narrowly double-convex in form.
LIGAMENT. A horny, elastic structure joining the two bivalve halves, either internally or externally, on the dorsal surface, along the hinge line; facilitates valve opening when adductor muscles are relaxed.
LIP. Refers to either the inner or outer-lip margin.
LIRA (AE). Fine, linear, ridge-like elevations on shell surface; usually applies to the inner surface of the outer lip; can also be applied to surface structural elements.
LIRATE. Sculptured with lirae.
LITHODESMA. Calcareous structure integrated with the internal ligament of some bivalves.
LOBATE. Having prominent rounded projections.
LUMEN. Cavity or hollow center.
LUNULE. (Bivalve) — A broad or narrow, heart-shaped depression anterior to the umbones in certain bivalve species.
MACROSPIC. Visible to the naked eye.
MACROSCULPTURE. Surface sculpture clearly visible to the naked eye.
MACULATION. Blotches or spots regularly or irregularly arranged.
MALLEATE. Surface with a hammered appearance.
MAMMILATE. 'Nipple-like' structural elements.
MANTLE. The fleshy outer layer of the molluscan body which secretes the shell.
MANTLE EDGE. The outer or peripheral edge of the mantle which contains the shell-secreting cells.
MARGIN. Shell edge or border; in gastropods the edge of the aperture; in bivalves, the extreme valve edge.
MARGINAL. Located along the margin.
MEDIAL. Refers to the middle.
MESOPLAX. An accessory plate lying across the umbones of the Pholalidae.
MICROSCULPTURE. Surface structure not readily visible to the naked eye.
MOLLUSCA. Highly developed, soft-bodied animals, without a backbone and developing an internal or external shell.
MONOTYPE. Type species of a genus, designated by virtue of it being the only species within the designated genus.
MORPH. Form.
MORPHOPHENOTYPE. A form of a species which appears to be environmentally controlled.
MULTISPIRAL. With numerous spiral whorls; usually used in reference to opercular sculpture.
NACREOUS. Lustrous in appearance; like mother-of-pearl.
NEBULOUS. Cloudy; indistinct.
NODE. A protruberance or knob.
NODOSE. Surface sculptured with nodes.
NODULE. A small knob or protruberance.
NODULOSE. Sculptured with nodules.
NOMEN DUBIUM. A name of dubious taxonomic validity.
NOMEN INQUIRENDUM. A name which requires further taxonomical clarification before it can be correctly applied to a particular species. Not necessarily invalid.
NOMEN NOVUM. A new name or species.
NOMEN NUDUM. A name which has been shown to be taxonomically invalid.
NOMEN OBLITUM. A valid name which is considered taxonomically "extinct", usually by virtue of the fact that it has never been used or because, for various reasons, its continuous use is taxonomically undesirable. A difficult situation to validate.
NOMINATE. Usually used in reference to a subspecies bearing a particular species name.
NOTCH. Indentation in any structural element.
NUCLEAR. Refers to the protoconch or larval whorls; constitutes the apex in gastropod shells.
NUCLEAR WHORL. Protoconch.
OBESE. Fat; corpulent.
OBLIQUE. Slanting; inclined at an angle to either the vertical or horizontal.
OBLITERATED. Worn away.
OBSOLETE. Very weakly marked; almost indiscernible.
OBTUSE. Rounded; not pointed.
ODONTOPHORE. Refers to the gastropod radular ribbon.
OPERCULUM. An oval, lid-like structure, horny or calcareous in composition, situated on the foot of many gastropods, which seals the apertural entrance when the animal withdraws into the shell.
ORIFICE. An opening or cavity.
ORNAMENTATION. Sculptural elements of the shell surface.
OUTER LIP. The outer apertural margin, extending from the base of the columella to the suture.
OVATE. Egg-shaped.
OVERLAPPING. Refers to the partial covering of one structural element by another; i.e. as in overlapping lamellae.
PALATAL. Belonging to the outer lip.
PALLIAL. Refers to the mantle or gill regions.
PALLIAL LINE. An impressed line, on the inside of each valve of a bivalve shell near the outer valve margins, produced by the animal's mantle edges. Usually seen as the line joining the anterior and posterior adductor-muscle scars.
PALLIAL SINUS. An embayment or recess in the pallial line, indicative of the former line of attachment of the siphonal retractor muscle.
PALMATE. Shaped like a palm leaf.
PAPILLA (AE). A small bump or projection.
PAPILLATE. Nipple-shaped. See also MAMMILATE.
PARALECTOTYPE. Any one of the remaining shells in a syntypic series after a lectotype has been chosen.
PARATYPE. One of the specimens left in a type series after the holotype has been selected.
PARIETAL. The region of a gastropod shell extending from the posterior region of the columella and following the shell contour.
PARIETAL WALL. The region above the columella, extending slightly inwards and outwards from the aperture.
PATELLIFORM. Limpet-shaped; resembling members of the genus Patella.
PATULOUS. Expanded or distended.
PAUCISPIRAL. Having few whorls; normally used in reference to the spiral lines on opercula.
PEDAL. Referring to the foot.
PEDAL GLAND. A secretory gland on the foot of gastropods, which provides a lubricant to facilitate locomotion.

PELAGIC. Free swimming; used frequently in reference to the veliger stages of many molluscan species.
PELLUCID. Transparent.
PELECYPODS. The bivalves; not used much. The term Bivalvia is more common.
PENULTIMATE. Next to last.
PERIOSTRACUM. The outer fibrous covering of many molluscs; may be smooth or hairy; epidermal covering.
PERIPHERY. Outermost area of any whorl on a gastropod shell.
PERISTOME. Margin of the aperture; may be continuous as in many species of land snail, or discontinuous, as in most gastropod species where the continuity is interrupted by the parietal wall.
PILLAR. Columella.
PIT. A minute surface depression.
PLAIT. Columellar fold.
PLICA (AE). A wrinkle, fold or plait; applied to columellar folds but strictly speaking, the term is used when describing a structural feature which is less pronounced; plait and plicae are normally interchangeable.
PLICATE. Bearing plicae.
PLICATION. The structural arrangement of plicae.
POLYPORPHIC. Literally means 'many-forms'; applied to a species which can exhibit a variation in structure and ornamentation; this phenomenon is frequently seen in populations inhabiting different clinal areas, but strictly speaking, the term is generally applied to the species variation within a specific population group.
PORCELANEOUS. Polished with the texture of porcelain.
POSTERIOR. The rear end.
POSTERIOR RIDGE. A ridge, originating at or near the umbones, and running diagonally to the postero-ventral margin of a bivalve shell.
POSTERIOR SLOPE. The surface shape of the area of a bivalve from the umbones to the postero-ventral margin.
POSTNUCLEAR. Refers to any growth stage, other than that of the protoconch, in a gastropod shell.
POSTNUCLEAR WHORL. Any whorl other than the embryonic or protoconch whorls.
PREOCCUPIED NAME. A name used by an earlier author to describe a species within a genus, making the name unavailable for any other species within the same genus.
PRESUTURAL. Literally 'just before the suture'; usually refers to a band or groove, just above or just below the sutural line.
PROBISCUS. A tube-like organ in gastropods, located anteriorly, between the tentacles, and extending outwards, in a 'snout-like' manner from the body surface; contains the mouthparts at the tip.
PRODUCED. Drawn out or extended.
PROTOCONCH. The larval shell formed from the embryonic whorls.
PROTOPLAX. A descriptive term used in connection with the Bivalve family PHOLADIDAE; refers to a long, narrow accessory plate, single or longitudinally divided, situated at the anterior end of the dorsal margin.
PROXIMAL. The point nearest.
PUNCTATE. Surface with many pits or pin-prick indentations; frequently in lines.
PUSTULE. A very small surface protruberance; smaller than a tubercle; a small pimple-like swelling.
PUSTULOSE. Surface covered with pustules.
PYRAMIDAL. Shaped like a pyramid.
PYRIFORM. Pear-shaped.
QUADRANGULAR. Irregularly four-sided; the term OVATELY QUADRANGULAR is used to describe features where one or more of the sides are convexly curved, rather than straight.
QUADRATE. Square; nearly square.
RACHIDIAN TOOTH. The largest, central tooth of the three typical stenoglossan radula teeth.
RACHIGLOSSATE. With radula teeth arranged like backbone vertebrae.
RADIAL. Radiating in a ray-like manner from a common centre; following a path along the radius of a particular structural feature.
RADULA. The teeth of molluscs, which are a series of chitonous structures, arranged in a ribbon-like fashion in the oral cavity; not present in clams and some toothless molluscs.
RADULAR RIBBON. The organ formed from the arrangement of the radular teeth in molluscs.
RAMUS. A branching process on the end of a spine.
RECESSED. Located within a cavity.
RECURVED. Curved or bent away from a central axis.
REFLECTED. Turned outwards and backwards from the margin; usually refers to outer lip margin of some gastropods.
REFLEXED. Turned outwards or in the opposite direction; see REFLECTED.
RETICULATE. Marked with a netted pattern.
RETICULATION. The pattern formed from oblique intersecting ridges or striae.
RHOMBOIDAL. Diamond-shaped; having 4, approximately equal sides and no right angles.
RIB. A broad, raised, elongate, elevated structure usually axially orientated.
RIBLET. A small rib.
RIDGE. A sharp-edged, elevated surface structure; usually a single line.
RIGHT VALVE. The valve on the right when a bivalve is held with hinge line uppermost and the anterior end pointing away.
RIPPLED. Finely ridged.
ROSTRATE. Having a long, sharp, beak-like projection.
ROTUND. Rounded, somewhat globular.
RUGOSE. Strongly wrinkled.
SCABROUS. Surface roughened with minute denticles or scales.
SCABROUS LAMELLAE. Lamellae bearing small scale-like projections where they intersect with spiral sculpture.
SCALE. A weak, thin, erect, plate-like structure; usually small and located on a rib.
SCALLOPED. A margin with rounded or fluted projections.
SCULPTURE. The morphological features and contours of a shell; the ornamentation.
SEALED. Closed.
SEAM. The line of fusion or joining of two surfaces.
SEPTUM. A transverse, plate-like structure which seals off the chambers in some shells; e.g. *Nautilus* and *Spirula*.
SERRATE. A margin which is finely notched like the blades of a saw.
SERRATION. Refers to a single tooth in a serrate edge.
SHELF. A flattened surface; i.e. on the top of a columellar plait.
SHELL. The calcareous structure secreted by molluscs.
SHOULDER. The area of a whorl lying below the suture and extending to the whorl periphery.
SHOULDER MARGIN. The peripheral margin of a whorl.
SINISTRAL. Left-handed; with the aperture on the observer's left when the shell is held with the apex pointed upwards.
SINUOUS. Undulating or wavy.
SINUS. An embayment, notch or slit; in gastropods the so-called anal sinus is located on the posterior area of the outer lip margin; for bivalves see PALLIAL SINUS.
SIPHON. A tube-like extension of gastropods, formed from a folding of the mantle surface near the head region; a similar structural feature seen in Bivalves; for inhalent currents.
SIPHONAL CANAL. A tubular shell extension which may be opened or closed, and through which the anterior siphon can extend; frequently termed the anterior canal.
SIPHONAL NOTCH. An embayment at the base of the columella on the apertural margin where the anterior siphon protrudes.
SIPHONAL FASCIOLE. A ridge adjacent the siphonal canal on the columellar side; normally has the remnants of earlier siphonal canals.
SPINATION. The arrangement of the spinal structure of a shell.
SPINDLE-SHAPED. Fusiform.
SPINE. A slender, elongate projection from the shell surface; may be flat, rounded, solid or tubular, and can taper, or spread distally into fronds or foliations.
SPINELET. A small spine.
SPINOSE. Bearing many spines.
SPIRAL. Following a continuous line, roughly parallel to the suture.
SPIRE. Refers to all the whorls excluding the body whorl.
SQUAMOSE. Scaly.
STEPPED. Whorls taking on the appearance of steps when viewed in profile.
STOUT. Solid; bulky.
STRIA (AE). Finely incised or raised lines, or furrows.
STRIATION. The pattern formed by a series of striae.
STROMBOID NOTCH. Refers specifically to the family STROMBIDAE: An embayment on the apertural margin, above the siphonal notch, through which the left eye stalk protrudes.
SUBACUTE. Moderately acute.
SUBANGULATE. Obtuse.
SUBSUTURAL. Under the suture.
SUFFUSED. Flushed with a particular colour, tint or hue.

SULCATE. Possessing a slit or fissure.
SULCUS. A slit, furrow or fissure.
SUPRASUTURAL. Above the suture.
SUTURE. The line of 2 whorls in a gastropod shell; shells are generally considered to have several sutures, but in reality the suture is a single continuous line.
SYNONYM. One of two or more names given to the same species.
SYNTYPE. Refers to one of a number of specimens examined by the author of a species, but from which no holotype was ever selected or designated.
TABULATE. Sharply angulate.
TAPERING. A gradual decrease in circumference size.
TAUTONOMY. A curious situation where the type species of a newly erected genus has the same species name as the genus i.e. *Haustellum haustellum*.
TEETH. A rather indecisive term, usually used in reference to structures seen on the outer lip or the parietal wall of gastropod shells and also to the structural elements of the hinge line in bivalves; also used in reference to the radula structural components.
TELEOCONCH. The total gastropod shell excluding the protoconch; i.e. the opposite to protoconch.
TERMINAL. At the end of a shell; the outermost peripheral point.
THREAD. A very fine sculptural element; a very fine lira.
TRAILING EDGE. The edge to the rear of a structure; behind the leading edge.
TRANSVERSE. Crossing the shell surface in a spiral direction in gastropod shells and in a concentric fashion in bivalves.
TRIGONAL. Having three sides; roughly triangulate in shape.
TRIVARICATE. Having three varices on every whorl (i.e. 3 varices within one 360° rotation of the shell sculpture).
TRUNCATE. Finishing abruptly; cut off suddenly; in bivalves, term often applied to square-ended appearance of some species; in gastropods, it usually refers to the abrupt termination of the columella.
TUBERCULATE. Possessing tubercles.
TUMID. Swollen, enlarged; distended.
TURBINATE. Resembling members of the gastropod genus Turbo; turban-shaped, with a broad, conical spire and convex base.
TURRETED. Having a tall, generally stepped spire.
TURRICULATE. Shaped in the form of a tower; having an acutely pointed, elongate, flat-sided spire of many whorls as in the TEREBRIDAE.
ULTIMATE. Final; last.
UMBILICAL. Referring to the umbilicus.
UMBILICAL PIT. A depression in the umbilicus.
UMBILICATE. Possessing an umbilicus.
UMBILICUS. The hole or the imaginary axis around which the inner surface of a gastropod is coiled.
UMBO (ONES). The earliest formed part of a bivalve shell; often referred to as the beak (s).
UMBONAL CAVITY. Space within the umbones.
UNDULATE. Wavy.
UNGUICULATE. Claw-shaped.
UNIVALVE. Mollusc having only a single shell piece.
VALVE. One of the halves of a bivalve shell.
VARICAL. With reference to a varix.
VARICAL FLANGE. An elevated, wing-like structure growing from the top portion of a varix.
VARICAL MARGIN. The peripheral, normally sculptured edge of a varix.
VARICAL SPINE. A spinose projection arising from the varix.
VARICAL WING. A thin, blade-like varical extension.
VARICOSE. Bearing one of more varices.
VARIETY. Taxonomically a bad term, the use of which should be restricted to 'domesticated strains'. Usually has the same connotation as form.
VARIX (VARICES). Thickened axial cords, usually quite prominent, which represents earlier outer lip margins; deposited during resting stages or interrupted growth stages.
VAULTED SCALE. A curved scale, orientated in the direction of shell growth.
VELIGER. The larval stage of molluscan development; usually free swimming.
VENTRAL. In gastropods, the surface on the same side as the apertural opening; in bivalves, the area opposite the ligament where the valves open widest.
VENTRAL MARGIN. The border opposite the hinge line and umbones in a bivalve shell.
VENTRICOSE. Swollen; inflated.
WEBBING. Lamellate surface structure, re-inforced by other sculpturing.
WHORL. One complete 360° turn of a gastropod shell about the imaginary central axis.
WRINKLED. Having ridges or furrows.

Bibliography

Abbott, R. T.
1959 The Family Vasidae in the Indo-Pacific. *Indo-Pacific Mollusca* 1(1):15-32, pls. 1-10
1960 The Genus *Strombus* in the Indo-Pacific. *Indo-Pacific Mollusca* 1(2):33-146, pls. 11-117
1961 The Genus *Lambis* in the Indo-Pacific. *Indo-Pacific Mollusca* 1(3):147-174, pls. 118-134
1962 *Sea Shells of the World.* Golden Press, Western Pub. Co. Inc., N.Y., 160 pp., 790 col. figs.
1968 The Helmet Shells of the World (Cassidae). Part I. *Indo-Pacific Mollusca* 2(9):7-202, pls. 1-187
1974 *American Seashells,* Ed. II. Van Nostrand Reinhold Co. (U.S.), 663 pp., 25 col. pts., textfigs. and diag.

Adams, A., Reeve, L. et. al.
1850 *The Zoology of the Voyage of H.M.S. Samarang.* Edited by A. Adams. Publish. Reeve & Benham, London.

Allan, J.
1956 *Cowry Shells of the World Seas.* Georgian House, Melbourne. 170 pp., 15 pls.
1959 *Australian Shells.* Georgian House, Melbourne, 487 pp., 44 pls., textfigs.

Bartsch, P.
1918 New Marine Molluscs from the Philippine Islands. *Proc. Biol. Soc. Wash.* Vol. 31, pp. 181-188
1919(a) Critical Remarks on Philippine Landshells with Descriptions of New Forms. *Proc. Biol. Soc. Wash.* Vol. 32, pp. 15-20
1919(b) Critical Remarks on Philippine Island Land Shells. *Proc. Biol. Soc. Wash.* Vol. 32, pp. 177-184
1938 A Synopsis of the Philippine Land Molluscs of the Genus *Ryssota. Proc. Biol. Soc. Wash.* Vol. 51, pp. 101-120
1942 A Synopsis of the Philippine Land Molluscs of the Genus *Hemitrichia. Proc. Biol. Soc. Wash.* Vol 55, pp. 27-44
1947 The Little Hearts *(Corculum)* of the Pacific and Indian Oceans. *Pacific Science* Vol. 1, No. 4, pp. 221-226, 2 pls.

Bertsch, H. & A. D'Attilio
1980 New Species of Muricidae (Gastropoda) from the Indian Ocean, the Philippines and Brazil. *Venus* (Jap. Jour. Malac.). Vol. 39, No. 3:131-138

Beu, A. G.
1970(a) The Mollusca of the Genus *Charonia* (Family Cymatiidae). *Trans. R. Soc. N. Z. Biol. Sciences,* 11(16):205-223, 5 pls., 1 textfig.
1970(b) The Mollusca of the Subgenus *Monoplex* (Family Cymatiidae). *Trans. R. Soc. N. Z. Biol. Sciences,* 11(17):225-237, 5 pls., 1 textfig.
1971 Some Synonymies in the Molluscan Family Cymatiidae. *J. Malac. Soc. Aust.,* 2(2):101-111, 2 pls.
1976(a) An Unnamed Species of *Ranularia* (Gastropoda: Cymatiidae) from Australia. *J. Malac. Soc. Aust.,* 3:233-238, 14 textfigs.
1976(b) Revision of the Southwest Pacific Species of *Morum* (Oniscidia) (Gastropoda:Cassidae). *J. Malac. Soc. Aust.,* 3(3-4):223-231, 15 textfigs.

1978(a) New Records and Species of Cymatiidae (Gastropoda:Prosobranchia) from the Kermadec Islands, Norfolk Ridge, and New Zealand. *J. Malac, Soc, Aust.*, 4(1-2):29-42, 30 textfigs.

1978(b) A New Species of *Bursa* from Deep Water off the Philippine Islands. *J. Malac. Soc. Aust.*, 4(1-2):23-27, 11 textfigs.

1979 A Revision of the Species of *Bolma* Risso, 1826 (Gastropoda:Turbinidae). *Rec. Aust. Mus.*, 32(1):1-68, figs. 1-19

1980 Australian Gastropods of the Family Bursidae, Part I. The Families of Tonnacea, the Genera of Bursidae, and Revision of Species previously assigned to *Tutufa* Jousseaume, 1881. *Rec. Aust. Mus.*, 33(5):248-324, figs. 1-22

Bratcher, T.

1977 Deshayes' Terebrid Types in Ecole Des Mines, Paris. *Nautilus*, 91(2), pp. 39-42

1979 A New Indo-Pacific Terebrid. *Veliger* 22(1), pp. 65-66

1981 Four Previously Undescribed Indo-Pacific Terebrids (Mollusca:Gastropoda). *Veliger*, 23(4), pp. 329-332.

Bratcher, T. & W. O. Cernohorsky

1982 Six New Species of Indo-Pacific Terebridae (Gastropoda). *Nautilus:* 96(2), pp. 62-66

Burgess, C. M.

1970 *The Living Cowries.* A. S. Barnes & Co. Inc., Cranbury, New Jersey, 389 pp., 44 pls.

Cate, C. N.

1967 The Cowries of the Ryukyu Islands. *Veliger* 10(1):13-41

1973 A Systematic Revision of the Recent Cypraeid Family Ovulidae (Mollusca:Gastropoda). *Veliger*, Vol. 15 (supplement), 116 pp.

1979 A Review of the Triviidae (Mollusca:Gastropoda). *San Diego Society of Natural History, Memoir 10.* 126 pp.

Cate, J. M.

1962 Discussion of *Vexillum regina* (Sowerby, 1825) and Related Species, With Descriptions of a New Subspecies. *Veliger:* 4(2), pp. 76-83, pls. 18 & 19.

Cernohorsky, W. O.

1967(a) The Muricidae of Fiji (Mollusca:Gastropoda). Part I. Subfamilies Muricinae and Tritonaliinae. *Veliger* 10(2):111-132, pls. 14-15, textfigs.

1967(b) *Marine Shells of the Pacific.* Pacific Publications, Australia, 249 pp., 60 b. & w. pls.

1969 The Muricidae of Fiji. Part 2. Subfamily Thaidinae. *Veliger* 11(4):293-315, pls. 47-49, textfigs.

1970 Systematics of the Families Mitridae and Volutomitridae (Mollusca:Gastropoda). *Bull. Auckland Inst. Mus.* 8:1-190, pls. 1-18, 222 textfigs.

1971(a) Indo-Pacific Pisaniinae (Mollusca:Gastropoda) and Related Buccinid Genera. *Bull. Auckland Inst. Mus.* 8:137-167.

1971(b) The Family Naticidae (Mollusca:Gastropoda) in the Fiji Islands. *Rec. Auckland Inst. Mus.* 8:169-208.

1972 *Marine Shells of the Pacific.* Vol. II., Pacific Publications, Sydney, Australia, 411 pp., 68 pls.

1974(a) The Taxonomy of Some Indo-Pacific Mollusca, with Description of a New Species. Part 2. *Rec. Auckland Inst. Mus.* 11:121-142. 38 textfigs.

1974(b) Type Specimens of Mollusca in the University Zoological Museum, Copenhagen. *Rec. Auckland Inst. Mus.* 11:143-192, 67 textfigs.

1975(a) Supplementary Notes on the Taxonomy of Buccinid Species of the Subfamily Pisaniinae (Mollusca:Gastropoda). *Rec. Auckland Inst. Mus.* 12:175-211, 75 textfigs.

1975(b) The Taxonomy of some Indo-Pacific Mollusca. Part 3. With Descriptions of New Taxa and Remarks on an Ecuadorian Fossil Species of Turridae. *Rec. Auckland Inst. Mus.* 12:213-234, 60 textfigs.

1976(a) The Taxonomy of some Indo-Pacific Mollusca. Part 4. With Descriptions of New Taxa, and Remarks on *Nassarius coppingeri* (Smith). *Rec. Auckland Inst. Mus.* 13:111-129, 43 textfigs.

1976(b) The Mitridae of the World. Part I. The Subfamily Mitrinae. *Indo-Pacific Mollusca* 3(17):273-528

1977 The Taxonomy of some Indo-Pacific Mollusca. Part 5. With Descriptions of New Taxa, and Remarks on *Nassarius albus* (Say). *Rec. Auckland Inst. Mus.* 14:121-132, 26 textfigs.

1978(a) The Taxonomy of some Indo-Pacific Mollusca. Part 6. *Rec. Auckland Inst. Mus.* 15:67-86, 40 textfigs.

1978(b) *Tropical Pacific Marine Shells.* Pacific Publications (Aust.) Pty. Ltd., 352 pp., 68 pls.

1979 The Taxonomy of some Indo-Pacific Mollusca. Part 7 *Rec. Auckland Inst. Mus.* 16:171-187, 34 textfigs.

1980 Systematics of some West Pacific Lyria (Mollusca: Gastropoda) with Description of a New Species. *Rec. Auckland Inst. Mus.* 17:127-134

1981(a) The Taxonomy of some Indo-Pacific Mollusca. Part 9. *Rec. Auckland Inst. Mus.* 18:193-202, 19 textfigs.

1981(b) The Family Buccinidae. Part I. The Genera *Nassaria, Trajana & Neoteron. Monographs Marine Mollusca* No. 2, 52 pp., 42 pls.

1981 On a Collection of Buccinacean and Mitracean Gastropods (Mollusca, Neogastropoda) from the Mozambique Channel and New Caledonia. *Bull. Mus. natn. Hist. Nat., Paris,* sér., **3**, section A, No. 4:985-1009

1982 The Taxonomic Status of *Cronia fiscella* (Gmelin, 1791) and *C. margariticola* (Broderip, 1933) (Gastropoda: Muricidae) *Rec. Auckland Inst. Mus.* 19:113-124

1982 The Taxonomy of some Indo-Pacific Mollusca. Part 10. *Rec. Auckland Inst. Mus.* 19:125-127

Cernohorsky, W. O. & M. Azuma

1974 Description of a New Species of Vexillidae (Mollusca: Gastropoda) from Japan. *Venus (Jap. Jour. Malac.)* Vol. 33, No. 1:7-10

Cernohorsky, W. O. & T. Bratcher

1976 Notes on the Taxonomy of Indo-Pacific Terebridae (Mollusca:Gastropoda) with Description of a New Species. *Rec. Auckland Inst. Mus.*, 13:131-140, 27 textfigs.

Coleman, N.

1975 *What Shell is That.* Paul Hamlyn Pty. Ltd., N.S.W., Australia, 308 pp., 811 col. textfigs.

Coomans, H. E. & R. G. Moolenbeek

1982 Studies on Conidae (Mollusca:Gastropoda). 1. *Conus papuensis* and *C. kintoki.* Two New Species from Deeper Water in the Western Pacific. *Bulletin Zoologisch Museum, Universiteit Van Amsterdam.* Vol. 8, No. 15.

Crosse, H.

1872 Diagnoses molluscorum Novae Caledoniae. *Jour. de Conchyl.*, v. 20, pp. 74, 218, pl. 13, fig. 2

da Motta, A. J. & P. Lenavat

1979 *Cone Shells of Thailand.* da Motta and Lenavat, Thailand, 20 pp., 8 col. pls.

da Motta, A. J.

1983 Two New Species for the Genus Conus (GASTROPODA: CONIDAE) *Publ. Ocas. Soc. Port. Malac. No. 2,* pp. 1-8 (1 Plate).

1985 Conus asiaticus n. sp. *La Conchiglia,* Year XVII, 192-193:25

Dance, S. P.

1969 *Rare Shells.* Fabre & Fabre, London, 128 pp., 24 pls.

1974 *The Collectors Encyclopedia of Shells.* McGraw-Hill Publishing Co., England, 288 pp., textfigs.

Eisenberg, J. M.

1981 *A Collectors Guide to Shells of the World.* 239 pp. 158 pls. (col.) McGraw-Hill Book Company (New York)

Emerson, W. K. & W. O. Cernohorsky

1973 The Genus *Drupa* in the Indo-Pacific. *Indo-Pacific Mollusca* 3(13):801-864

1977 Notes on some Indo-Pacific Species of *Morum* (Gastropoda:Tonnacea). *Nautilus* 95(3):81-86, figs. A-J

1981 Two New Species of *Morum* (Gastropoda:Tonnacea). *Nautilus* 95(3):101-105, figs, 1-5

Emerson, W. K. & A. D'Attilio

1965 Remarks on *Colubraria soverbii* (Reeve, 1844) and Related Species (Mollusca:Gastropoda). *Veliger,* 8(3), pp. 173-177.

1979 Six New Living Species of Muricacean Gastropods. *Nautilus* 93(1):1-10, figs. 1-21

1981 Remarks on Muricodrupa Iredale, 1918 (Muricidae:Thaidinae), with the Descriptions of a New Species. *Nautilus* 95(2):77-82, figs. 1-9

Emerson, W. K. & E. L. Puffer
1953 A Catalogue of the Molluscan Genus *Distorsio* (Gastropoda:Cymatiidae). *Proc. Biol. Soc. Wash.*, 66:93-108

Fair, R. H.
1976 *The Murex Book.* Ruth H. Fair, Honolulu, 138 pp., 23 pls.

Fujioka, Y.
1982 On the Secondary Sexual Characters Found in the Dimorphic Radula of *Drupella* (Gastropoda:Muricidae) with Reference to its Taxonomic Revision. *Venus (Jap. Jour. Malac.).* Vol. 40, No. 4:203-223

Garrard, T. A.
1977 A Revision of the Australian Architectonicidae (Gastropoda:Mollusca). *Rec. Aust. Mus.*, 31(13):506-584, pls. 1-12

Habe, T.
1964 *Shells of the Western Pacific in Colour.* Vol. II. Hoikusha Publishing Co. Ltd., Osaka, 233 pp., 66 pls., textfigs.

Habe, T. & K. Ito
1965 *Shells of the World in Colour. Vol. I The Northern Pacific.* Hoikusha Publishing Co. Ltd., Osaka, 176 pp., 56 pls. (Japanese)

Habe, T. & S. Kosuge
1966 *Shells of the World in Colour. Vol. II. The Tropical Pacific.* Osaka, 193 pp., 68 pls. (Japanese)
1967 *The Standard Book of Japanese Shells in Colour.* Hoikusha Publishing Co. Ltd., Osaka, 223 pp., 64 pls., textfigs.

Habe, T. & T. Okutani
1980 Two New Turbinid Gastropods from the Philippines. *Venus (Jap. Jour. Malac.).* Vol. 39, No. 2:77-81
1981 Two New Gastropods from the Philippines. *Venus (Jap. Jour. Malac.).* Vol. 39, No. 4:193-196

Hinton, A.
1972 *Shells of New Guinea and the Central Indo-Pacific.* Robert Brown and Associates Pty. Ltd./Jacaranda Press, 94 pp., 44 pls.
1975 *Guide to Shells of Papua New Guinea.* Robert Brown & Associates Pty. Ltd. (Port Moresby), 74 pp., 68 pls.
1978 *Guide to Australian Shells.* Robert Brown & Associates Pty. Ltd. (Port Moresby), 82 pp., 77 pls.

Houbrick, R. S.
1978 The Family Cerithiidae in the Indo-Pacific. Part I. The Genera *Rhinoclavis, Pseudovertagus,* and *Clavocerithium. Monographs of Marine Mollusca,* No. I. American Malacologists (Delaware, U.S.A.)
1980 Review of the Deep Sea Genus *Argyropeza* (Gastropoda:Prosobranchia:Cerithiidae). *Smithsonian Contributions to Zoology,* No. 321, 30 pp., 12 textfigs.

Kay, E. A.
1979 Hawaiian Marine Shells. Reef and Shore Fauna of Hawaii, Section 4 Mollusca. *Bernice P. Bishop Museum Special Publications 64 (4).* Bishop Museum Press, Honolulu, Hawaii, 653 pp., pls. (col/b & w), textfigs.

Kira, T.
1962 *Shells of the Western Pacific in Colour.* Vol. I. Hoikusha Publishing Co. Ltd., Osaka, 224 pp., 72 pls, textfigs.
1971 *Coloured Illustrations of the Shells of Japan.* Hoikusha Publishing Co. Ltd., Osaka [Reprint of Enlarged & Revised Edition (1959)]. 240 pp., 71 pls.

Kosuge, S.
1979(a) Descriptions of Two New Species of the Family Muricidae (Gastropoda:Mollusca). *Bull. Inst. Malac. Tokyo.* 1(1):1-2, pl. 1
1979(b) Descriptions of Two New Subgenus and Seven New Species of the Genus *Latiaxis* (Gastropoda:Mollusca). *Bull. Inst. Malac. Tokyo.* 1(1):3-8, pls. 2-3
1979(c) Description of a New Species of the Genus *Calliostoma* from the Central Pacific (Trochidae: Gastropoda). *Bull. Inst. Malac. Tokyo.* 1(2):19, pl. 4
1979(d) Description of a New Species of the Genus *Guildfordia* (Turbinidae:Gastropoda). *Bull. Inst. Malac. Tokyo.* 1(2):20, pl. 4, figs. 4-7
1979(e) Descriptions of New and Rare Cones from the Western Pacific (Conidae:Gastropoda). *Bull. Inst. Malac. Tokyo.* 1(2):21-22, pl. 4, figs. 8-10
1980(a) Study on the Collection of Mr. Victor Dan (1). Descriptions of New Species of the Genera *Perotrochus, Angaria* and *Latiaxis. Bull. Inst. Malac. Tokyo.* 1(3):37-38, pls. 7-9
1980(b) Descriptions of Six New Species of the Genus *Latiaxis* from the Philippine Sea. *Bull. Inst. Malac. Tokyo.* 1(3):41-45, pls. 9-10.
1980(c) Description of New Species of the Genus *Mitrella. Bull. Inst. Malac. Tokyo.* 1(3):46, figs. 10-11.
1980(d) Descriptions of Three New Species of the Family Muricidae (Gastropoda:Muricacea). *Bull. Inst. Malac. Tokyo.* 1(4):53-56, pls. 14-15
1980(e) Studies on the Collection of Mr. Victor Dan (2). Description of a New Species of the Genus *Haustellum* (Gastropoda:Muricacea). *Bull. Inst. Malac. Tokyo.* 1(4):57-58, pl. 17, figs. 2-4
1980(f) Descriptions of Three New Species of the Family Coralliophilidae (Gastropoda:Muricacea). *Bull. Inst. Malac. Tokyo.* 1(4):59-61, pls. 15-16
1980(g) Descriptions of Two New Species of the Genus *Conus* (Gastropoda:Conacea). *Bull. Inst. Malac. Tokyo.* 1(4):62-64, pl. 18
1980(h) Studies on the collection of Mr. Victor Dan (3). Note on the Local Variations of *Lyria taiwanica* (Gastropoda:Volutacea). *Bull. Inst. Malac. Tokyo.* 1(4):65-66, pl. 17
1981(a) Descriptions of Three New Species of the Genus *Latiaxis* from Japan and Philippines (Gastropoda: Muricacea). *Bull. Inst. Malac. Tokyo.* 1(6):88-90, pl. 29
1981(b) Description of a New Species of the Family Fasciolariidae from Philippines (Gastropoda:Buccinacea). *Bull. Inst. Malac. Tokyo.* 1(6):91-92, pl. 31
1981(c) Notes of Newly Recorded Species of the Superfamily Conacea from Philippines with Descriptions of New Species of the Genera *Terebra, Conus* and *Glyphostoma. Bull. Inst. Malac. Tokyo.* 1(6):93-96, pls. 31-32
1981(d) Notes on the Family Triphoridae from the Philippines. *Bull. Inst. Malac. Tokyo.* 1(6):97, pl. 31
1981(e) Note on *Conus sulcocastaneus* Kosuge Nom. Nov. *Bull. Inst. Malac. Tokyo.* 1(6):99
1981(f) Description of Two New Species of the Genus *Morum* with Remarks on the Recent Species from the Philippines. *Bull. Inst. Malac. Tokyo.* 1(7):101-104, pl. 33
1981(g) Studies on the Collection of Mr. Victor Dan (4). Descriptions of New Species of the Genera *Lyria, Conus* and *Fissidentalium. Bull. Inst. Malac. Tokyo.* 1(7):113-115, pl. 39

Kuroda, T.
1964 A New Muricid Species from Japan. *Venus:* 23(3), pp. 129-130

Kuroda, T., Habe, T., & K. Oyama
1971 *The Sea Shells of Sagami Bay.* Maruzen Co. Ltd., Tokyo, 741 pp. (Japanese), 489 pp. (English), 120 pls.

Ladd, H. S.
1976 New Pleistocene Neogastropoda from the New Hebrides. *Nautilus* 90(4):127-133, figs. 12-15

Linder, G.
1977 *Seashells of the World.* Australian & New Zealand Book Co. Pty. Ltd., Brookvale, N.S.W., Australia, 271 pp., 64 col. pls., textfigs.

Marsh, J. A. & O. H. Rippingdale
1964 *Cone Shells of the World.* Jacaranda Press Pty. Ltd., Brisbane, Australia 166 pp., 21 col. pls.

Melvin, A. G.
1966 *Seashells of the World with Values.* 167 pp. 54 pls. (B/W) C. E. Tuttle Company, Inc. (Tokyo, Japan)

Mühlhäusser, von Heinrich
1981 *Strombus kleckhamae boholensis* n. subsp. (Gastropoda:Strombidae). *Spinaxia* 4(3):319-324, 4 pls.
1982 Beltráge zue Kenntnis del Arten *Chicoreus aculeatus* (Lamarck, 1822) und *Chicoreus rossiteri* (Crosse, 1872). *Spinaxia*, 5(1):35-45, 8 figs.

Okutani, T.
1979 A New Volutid Gastropod, *Lyria habei* n. sp. Collected from a Submarine Bank, Hirase, in the South of Kyushu, Japan. *Venus (Jap. Jour. Malac.).* Vol. 38, No. 1:5-8

Oliver, A. P. H.
1975 *The Hamlyn Guide to Shells of the World.* Hamlyn Publishing Group Ltd., England, 320 pp., textfigs.
Pechar, P., Prior, C., and B. Parkinson
1981 *Mitre Shells from the Pacific and Indian Oceans.* Robert Brown & Associates Pty. Ltd. (Aust.); 56 pls.
Petuch, E. J.
1979 Twelve New Indo-Pacific Gastropods. *Nemouria* 23:1-21, 35 textfigs.
Ponder, W. F.
1967 The Classification of the Rissoidae and Orbitestallidae with Descriptions of Some New Taxa. *Trans Royal Soc. New Zealand, Zoology* 9(17):193-224.
1983 A revision of the Recent Xenophoridae of the world and of the Australian fossil species (Mollusca:Gastropoda with appendix by Ponder, W. F. & J. Cooper). *Aust. Mus. Mem.* 17:1-126
Powell, A. W. B.
1942 The New Zealand Recent and Fossil Mollusca of the Family Turridae. *Bull. Auckland Inst. Mus.* No. 2, 188 pp., 10 pls.
1964 The Family Turridae in the Indo-Pacific. Part I. The Subfamily Turrinae. *Indo-Pacific Mollusca,* 1(5):227-340, 172-262
1966 The Molluscan Families Speightiidae and Turridae. *Bull. Auckland Inst. Mus.* No. 5, 184 pp., 23 pls.
1967 The Family Turridae in the Indo-Pacific. Part 1(a). The Subfamily Turrinae concluded. *Indo-Pacific Mollusca* 1(7):409-436, pls. 298-312
1969 The Family Turridae in the Indo-Pacific. Part 2. The Subfamily Turriculinae. *Indo-Pacific Mollusca* 2(10): 215-416, pls. 188-324
1973 The Patellid Limpets of the World (Patellidae). *Indo-Pacific Mollusca.* 3(15):75-206, pls. 60-181
Radwin, G. E. & A. D'Attilio
1976 *Murex Shells of the World.* Stanford University Press, Stanford, U.S.A. 284 pp., 32 pls., textfigs.
Richardson, L., Tucker Abbott, R. and G. M. Davis
1979 Early References to the Figures in the Conchylien Cabinet of Martini and Chemnitz: Volumes I-XIII. *Tryoma* 2(1):1-427. 432 pls., 70 Vigns.
Rippingdale, O. H. and D. F. Mc Micheal
1961 *Queensland and Great Barrier Reef Shells.* Jacaranda Press Pty. Ltd., 210 pp., 29 pls.
Rockel, D. & T. C. Lan
1981 Description of a New Species of Conidae from Taiwan. *Bull. Malacol. R.O.C.*:8, pp. 15-17
Rockel, D. & A. J. da Motta
1983 Descriptions of Two New Species of the Genus *Conus Bull. Inst. Malac. Tokyo.* Vol. 1, No. 8:117-118, pl. 40, figs. 3, 4, 5 & 6.
Rosewater, J.
1970 The Family Littorinidae in the Indo-Pacific. Part I. The Subfamily Littorininae. *Indo-Pacific Mollusca,* 2(11):417-506, pls. 325-387.
Salvat, B. and C. Rives
1975 *Coquillages de Polynesie.* Publications — Elysees, France, 391 pp.
Shikama, T.
1963 *Selected Shells of the World Illustrated in Colours.* Vol. I. Hokuryu-Kan Publishing Co. Ltd., Tokyo, 154 pp., 70 pls., textfigs. (Japanese)
1964 *Selected Shells of the World Illustrated in Colours.* Vol. II. Hokuryu-Kan Publishing Co. Ltd., Tokyo, 212 pp., 70 pls., textfigs.
1977 Description of New and Noteworthy Gastropoda. *Science Report of Yokahama Nat. University,* Sect. II, No. 24, p. 14, pl. 2, figs. 9a, b.
Smith, B. J.
1976 Revision of the Recent Species of the Family Clavagellidae (Mollusca, Bivalvia) *J. Malac. Soc. Aust.* 3(3-4):187-209
Taylor, J. and J. G. Walls
1975 *Cowries.* T. F. H. Publications Inc., Neptune City, N.J., U.S.A. 286 pp., numerous pls., textfigs.
Tinker, S. W.
1958 *Pacific Seashells.* Charles E. Tuttle Co. Inc., Japan, 240 pp., 109 pls. (B/W)
Tryon, G. N.
1881 *Man. Conch. III.* Pl. 7. figs. 41-44
Wagner, R. J. L. and R. Tucker Abbott
1978 *Wagner and Abbott's Standard Catalogue of Shells.* American Malacologists, Inc., Greenville, Delaware, U.S.A. (Ed. III with supplements)
Walls, J. G.
1978 *Cone Shells. A Synopsis of the Living Conidae.* T. F. H. Publications Inc., Neptune City, N.J., U.S.A., 1011 pp., numerous col./b & w pls., textfigs.
1980 *Conchs. Tibias and Harps.* T. F. H. Publications Inc., Neptune City, N.J., U.S.A., 191 pp., numerous col. pls., textfigs,
Weaver, C. S. and J. E. du Pont
1970 *Living Volutes, A Monograph of the Recent Volutidae of the World.* Monograph Series No. 1. Delaware, Museum of Natural History, Greenville, Delaware, 375 pp., 75 pls.
Weinkauff, H. C.
1875 *Jahrb. deut. malak. Ges.* 2, p. 290, pl. 9, figs. 9, 10.
Wilson, B. and K. Gillett
1971 *Australian Shells.* A. H. & A. W. Reed, Sydney, Australia, 168 pp., 106 pls.
Zeigler, R. F. and H. C. Porreca
1969 *Olive Shells of the World.* Zeigler & Porreca, W. Henrietta, N.Y., 96 pp., 13 pls., textfigs.

Additional References

In addition to the detailed references given, xerox figures and descriptions were taken from the following Conchological Works . . .

Chemnitz, J. H.
1780-1795 *Neues systematisches Conchylien — Cabinet* : Pts. 4-11. Nurnberg.
Chenu, J. C.
1842-2964 *Illustrations conchyliologiques ou descriptions et figures de toutes les coquilles connues, vivants et fossiles.* Vols. 1-4. Paris.
Deshayes, G. P.
1835-1845 *Histoire naturelle des animaux sans vertèbres,* (2nd ed.) 11 vols. Paris.
Perry, G.
1811 *Conchology: Or the Natural History of Shells.* London.
Pilsbry, H. A.
1888-1898 *Manual of Conchology* (in G. W. Tryon), Vols. 10-17
Reeve, L.
1843-1878 *Conchologia Icon.* 20 vols.
Sowerby, G. B. (II)
1842-1887 *Thesaurus Conchyliorum.* 5 Vols. London

Miscellaneous Publications

Carfel Philippine Shell News: Private publication of the Carfel Seashell Museum (Manila, Philippines)
The Festivus: Publication of the San Diego Shell Club (California, U.S.A.)
Hawaiian Shell News: Publication of the Hawaiian Malacological Society (Honolulu, Hawaii)
La Conchiglia: Italian Publication edited by K. Nicolay.
Keppel Bay Tidings: Publication of the Keppel Bay Shell Club (Queensland, Australia)

A List of Genera and Subgenera

The following section contains a list of genera and subgenera used within the text of this book, together with family and subfamily (where applicable) placements. To facilitate easy reference, the families are arranged in alphabetical order, together with their appropriate plate numbers. Within each family, the subfamily, generic and subgeneric placements are also arranged alphabetically. The Land snails and fresh water Gastropoda (Pls. 97-100) are ommitted from this list, and are to be found at the start of their appropriate section.

The splitting of certain families into a number of subfamily groups is, at present, still under debate by many malacologists, and their employment here is provisional and based on currently accepted ideas, though this may change as more research is done. Similarly, some subgeneric groups may be worthy of elevation to a full generic status, and are so annotated within the list.

Family **Acmaeidae** Carpenter, 1857 (Pl. 3)
Patelloida Quoy & Gaimard, 1834
Family **Acteonidae** Orbigny 1842 (Pl. 82)
Acteon Montfort, 1810
Bullina Ferussac, 1822
Pupa Röding, 1798
Family **Angariidae** Thiele, 1924 (Often considered a subfamily group within the Trochidae) (Pl. 6)
Angaria Röding, 1798
Family **Anomiidae** Rafinesque, 1815 (Pl. 90)
Placuna Lightfoot, 1786
Family **Aplustridae** Gray, 1847 (Pl. 82)
Hydatina (Aplustrum) Schumacher, 1817
Hydatina (Hydatina) Schumacher, 1817
Family **Aplysiidae** Rafinesque, 1815 (Pl. 12)
Subfamily **Dolabriferinae** Pilsbry, 1895
Dolabella Lamarck, 1801
Family **Architectonicidae** Gray, 1850 (Pl. 2)
Subfamily **Architectonicinae** Gray, 1850
Architectonica (Architectonica) Röding, 1798
Discotectonica Marwick, 1931
Philippia (Philippia) Gray, 1847
Philippia (Psilaxis) Woodring, 1828
Subfamily **Heliacinae** Troschel, 1867
Heliacus (Heliacus) Orbigny, 1842
Heliacus (Torinista) Iredale, 1936
Family **Arcidae** Lamarck, 1809 (Pls. 84/86)
Subfamily **Anadarinae** Reinhart, 1935
Anadara (Anadara) Gray, 1847
Anadara (Potiarca) Iredale, 1939 (possibly a full genus)
Anadara (Tegillarca) Iredale, 1939 (possibly a full genus)
Subfamily **Arcinae** Lamarck, 1809
Arca Linné, 1758
Barbatia (Barbatia) Gray, 1842
Bentharca Verrill & Bush, 1898
Trisidos Röding, 1798
Family **Buccininae** Rafinesque, 1815 (Pls. 44/45)

This family contains around 120 genera which are dispersed in some 11 subfamily groups recognized by various authors. The lack of detailed anatomical studies makes the use of so many subfamily groups rather tenuous. Certain groups and their associated genera have been well established and are here employed, but the remaining genera are provisionally placed in the subfamily Buccininae until more detailed work establishes their true taxonomic position.

Subfamily **Babyloniinae**
Babylonia Schlüter, 1838
Subfamily **Buciininae** Rafinesque, 1815
Siphonofusus Kuroda & Habe, 1962
Subfamily **Photinae** Gray, 1857
Nassaria (Microfusus) Dall, 1916
Nassaria (Nassaria) Link, 1807
Phos (Phos) Montfort, 1810
Phos (Strongylocera) Mörch, 1852
Subfamily **Pisaniinae** Gray, 1857
Cantharus (Cantharus) Röding, 1798
Cantharus (Clivipollia) Iredale, 1929
Cantharus (Pollia) Gray in Sowerby, 1834
Cantharus (Prodotia) Dall, 1924
Engina Gray, 1839
Pisania (Caducifer) Dall, 1904
Pisania (Ecmanis) Gistel, 1848
Pisania (Pisania) Bivona, 1832
Family **Bullidae** Rafinesque, 1815 (Pl. 81)
Bulla Linné, 1758
Family **Bursidae** Thiele, 1925 (Pls. 33/34)
Bufonaria (Bufonaria) Schumacher, 1817
Bursa (Bursa) Röding, 1798
Bursa (Colubrellina) Fischer, 1884
Tutufa (Tutufa) Jousseaume, 1881
Tutufa (Tutufella) Beu, 1980
Family **Cancellariidae** Forbes & Hanley, 1853 (Pls. 18/95)
Subfamily **Cancellariinae** Forbes & Hanley, 1853
Cancellaria (Cancellaria) Lamarck, 1799
Cancellaria (Merica) H. & A. Adams, 1854
Subfamily **Trigonostomoninae** Cossmann, 1899
Agatrix (Agatrix) Petit, 1967
Agatrix (Olssonella) Petit, 1970
Trigonostoma Blainville, 1827
Subfamily uncertain
Pisanella von Koenen, 1865 (syn. *Plesiotriton* Fischer, 1884)
Scalptia Jousseaume, 1887
Family **Cardiidae** Lamarck, 1809 (Pl. 87)
Subfamily **Cardiinae** Lamarck, 1809
Vepricardium Iredale, 1929
Subfamily **Fraginae** Stewart, 1930
Corculum Röding, 1798
Ctenocardia H. & A. Adams, 1857
Fragum Röding, 1798
Subfamily **Laevicardiinae** Keen, 1936
Laevicardium Swainson, 1840
Subfamily **Protocardiinae** Keen, 1951
Lyrocardium Meek, 1876
Nemocardium Meek, 1876
Subfamily **Trachycardiinae** Stewart, 1930
Trachycardium (Trachycardium) Mörch, 1853
Trachycardium (Regozara) Iredale, 1938
Trachycardium (Vasticardium) Iredale, 1927
Family **Carditidae** Fleming, 1820 (Pl. 84)
Subfamily **Carditinae** Fleming, 1820
Cardita Bruguiére, 1792
Cardites Link, 1807
Family **Cassidae** Swainson, 1832 (Pl. 27)
Casmaria H. & A. Adams, 1853
Cassis (Cassis) Scopoli, 1777
Cassis (Cypraecassis) Stutchburry, 1837
Galeodea Link, 1807
Morum (Morum) Röding, 1798
Morum (Oniscidia) Swainson, 1840
Phalium (Echinophoria) Sacco, 1890
Phalium (Phalium) Link, 1807
Phalium (Semicassis) Mörch, 1852
Family **Cerithiidae** Fleming, 1822 (Pls. 13/14)
Subfamily **Cerithiinae** Fleming, 1822
Cerithium (Cerithium) Bruguiére, 1789
Clypeomorus Jousseaume, 1888
Pseudovertagus (Pseudovertagus) Vignal, 1904
Rhinoclavis (Proclava) Thiele, 1929
Rhinoclavis (Longicerithium) Houbrick, 1978
Rhinoclavis (Rhinoclavis) Swainson, 1840
Family **Chamidae** Lamarck, 1809 (Pl. 88)
Chama Linné, 1758
Family **Clavagellidae** Orbigny, 1844 (Pl. 88)
Brechites (Brechites) Guettard, 1770
Brechites (Penicillus) Bruguiére, 1789
Family **Columbellidae** Swainson, 1840 (Pl. 46)
Aesopus Gould, 1860
Columbella Lamarck, 1799
Mitrella (Mitrella) Risso, 1826
Mitrella (Sundamitrella) Oostingh, 1952
Pictocolumbella Habe, 1962
Pyrene Röding, 1798

Family Colubrariidae Dall, 1909 (Often considered a subfamily of **Buccinidae**) (Pl. 34)
Antemetula (Antemetula) Rehder, 1943
Antemetula (Colubrarina) Kuroda & Habe, 1971
Colubraria (Colubraria) Schumacher, 1817
Colubraria (Nivitriton) Iredale, 1929

Family Conidae Rafinesque, 1815 (Pls. 61-71 incl.)
This family contains a large assemblage of subgeneric groups, and at the present time, their application is inadvisable until more work has been done on this family. For this reason, we have restricted the classification of all species to a single genus, *Conus* Linné, 1758.

Family Coralliophilidae Chenu, 1859 (Pls. 43/44)
Coralliobia H. & A. Adams, 1853
Coralliophila (Coralliophila) H. & A. Adams, 1853
Coralliophila (Latimurex) Coen, 1922
Coralliophila (Lepadomurex) Coen, 1922
Latiaxis (Babelomurex) Coen, 1922
Latiaxis (Echinolatiaxis) Kosuge, 1979
Latiaxis (Laevilatiaxis) Kosuge, 1979
Latiaxis (Lamellatiaxis) Habe & Kosuge, 1970
Latiaxis (Latiaxis) Swainson, 1840
Latiaxis (Pseudomurex) Monterosato, 1872
Latiaxis (Tolema) Iredale, 1929
Magilus Montfort, 1810
Rapa Röding, 1798

Family Corbiculidae Gray, 1847 (Pl. 86)
Subfamily **Polymesodinae** Habe, 1977
Polymesoda (Geloina) Gray, 1842
Polymesoda (Polymesoda) Rafinesque, 1820

Family Costellariidae Macdonald, 1860 (Pls. 57-60 incl.)
Thala H. & A. Adams, 1853
Vexillum (Costellaria) Swainson, 1840
Vexillum (Pusia) Swainson, 1840
Vexillum (Vexillum) Röding, 1798
Zierliana Gray, 1847

Family Crepidulidae Fleming, 1822 (Pl. 12)
Subfamily **Calyptraeinae** Blainville, 1824
Calyptraea (Calyptraea) Lamarck, 1799
Calyptraea (Desmaulus) Rehder, 1943
Cheilea Modeer, 1793
Crucibulum Schumacher, 1817
Subfamily **Crepidulinae** Fleming, 1822
Crepidula (Crepidula) Lamarck, 1799
Crepidula (Siphopatella) Lesson, 1831

Family Cultellidae Davies, 1935 (Pls. 84/86)
Phaxas (Ensiculus) H. Adams, 1860
Phaxas (Phaxas) Gray, 1852
Pharella Gray, 1854 (often placed in its own family **Pharellidae** Tryon, 1884)

Family Cyclostrematidae Fischer, 1885 (Pl. 3)
Liotina Fischer, 1885

Family Cymatiidae Iredale, 1913 (Pls. 30-32 incl.)
Subfamily **Cymatiinae** Iredale, 1913
Charonia Gistel, [1847]
Cymatium (Cymatium) Röding, 1798
Cymatium (Gutturnium) Mörch, 1852
Cymatium (Lotoria) Emerson & Old, 1963
Cymatium (Ranularia) Schumacher, 1817
Cymatium (Septa) Perry, 1810
Cymatium (Turritriton) Dall, 1904
Linatella (Gelagna) Schaufuss, 1869
Linatella (Linatella) Gray, 1857
Sassia Bellardi, 1873
Subfamily **Distorsioninae** Kuroda, Habe & Oyama, 1971
Distorsio (Distorsio) Röding, 1798
Distorsio (Personella) Conrad, 1865
Subfamily **Ranellinae** Powell, 1933
Gyrineum (Biplex) Perry, 1811
Gyrineum (Gyrineum) Link, 1807

Family Cypraeidae Rafinesque, 1815 (Pls. 20-24 incl.)
This family is, at present, in a rather disordered state as regards the correct taxonomic status of its various taxa. There is a profusion of genera and subgenera being used, and a critical revision is clearly indicated. We have restricted our classification of the various species to a single genus, *Cypraea* Linné, 1758.

Family Dentaliidae Gray, 1834 (Pl. 82)
Subfamily **Antalinae** Stoliczka, 1868
Antalis H. & A. Adams, 1854
Subfamily **Dentaliinae** Gray, 1834
Dentalium Linné, 1758
Fissidentalium (Fissidentalium) Fischer, 1885
Fissidentalium (Pictodentalium) Kira in Habe, 1963
Subfamily **Laevidentaliinae** Palmer, 1974 (considered a full family by some authors)
Calliodentalium Habe, 1964
Fustiaria Stoliczka, 1868

Family Epitoniidae S. S. Berry, 1910 (Pl. 18)
Amaea (Amaea) H. & A. Adams, 1853
Amaea (Elegantiscala) Boury, 1911
Amaea (Scalina) Conrad, 1865
Cirsotrema Mörch, 1852
Epitonium (Epitonium) Röding, 1798
Epitonium (Gyroscala) Boury, 1887
Epitonium (Hirtoscala) Monterosato, 1890
Epitonium (Laniscala) Lan, 1976
Epitonium (Papyriscala) Boury, 1909
Epitonium (Spiniscala) Boury, 1909

Family Fasciolariidae Gray, 1853 (Pls. 47/48)
Subfamily **Fasciolariinae** Gray, 1853
Fasciolaria (Fasciolaria) Lamarck, 1799
Fasciolaria (Pleuroploca) Fischer, 1884
Subfamily **Fusininae** Swainson, 1840
Fusinus (Fusinus) Rafinesque, 1815
Fusinus (Simplicifusus) Kira, 1962
Granulifusus Kuroda & Habe, 1952
Subfamily **Peristerniinae** Tryon, 1880
Dolicholatirus Bellardi, 1884
Fusolatirus McGinty, 1955
Latirolagena G. F. Harris, 1897 (? s.g. of *Leucozonia* Gray, 1847)
Latirus (Latirulus) Cossmann, 1901
Latirus (Latirus) Montfort, 1810
Peristernia Mörch, 1852
Pseudolathyrus Bellardi, 1884

Family Ficidae Conrad, 1867 (Pl. 29)
Ficus Röding, 1798

Family Fimbriidae Nicol, 1950 (Pl. 88)
Fimbria Megerle v Mühlfeld, 1811

Family Fissurellidae Fleming, 1822 (Pl. 3)
Subfamily **Emarginulinae** Gray, 1834
Hemitoma (Hemitoma) Swainson, 1840
Hemitoma (Montfortia) Reclúz, 1843
Scutus Montfort, 1810
Subfamily **Diodorinae** Odhner, 1932
Diodora Gray, 1821

Family Glossidae Gray, 1847 (Pl. 85)
Glossus (Glossus) Poli, 1795
Glossus (Meiocardia) H. & A. Adams, 1857 (possibly a full genus)

Family Glycymerididae Newton, 1922 (Pl. 85)
Subfamily **Glycymeridinae** Newton, 1922
Glycymeris (Axinactis) Mörch, 1861
Glycymeris (Glycymeris) da Costa, 1778
Glycymeris (Tucetona) Iredale, 1931 (possibly a full genus)
Glycymeris (Veletuceta) Iredale, 1931 (may = *Glycymeris* s.s.)

Family Haliotidae Rafinesque, 1815 (Pl. 2)
Haliotis (Haliotis) Linné, 1758
Haliotis (Ovinotis) Cotton, 1943
Haliotis (Sanhaliotis) Iredale, 1929
Haliotis (Sulculus) H. & A. Adams, 1854

Family Haminoeidae Pilsbry, 1895 (Pl. 82)
Atys (Aliculastrum) Pilsbry, 1896
Atys (Atys) Montfort, 1810
Haminoea Turton & Kingston, in Carrington, 1830

Family Harpidae Bronn, 1849 (Pl. 28)
Harpa Röding, 1798

Family Hipponicidae Troschel, 1861 (Pl. 12)
Hipponix (Hipponix) Defranc, 1819
Hipponix (Antisabia) Iredale, 1937
Sabia Gray, 1847

Family Isognomonidae Woodring, 1925 (Pl. 89)
Crenatula Lamarck, 1803
Isognomon (Isognomon) Solander in Lightfoot, 1786
Isognomon (Isogonum) Röding, 1798

Family Laternulidae Hedley, 1918 (Pl. 88)
Laternula Röding, 1798

Family Limidae Rafinesque, 1815 (Pl. 86)
Acesta H. & A. Adams, 1858
Ctenoides Mörch, 1853
Lima Bruguiére, 1797
Limaria (Limaria) Link, 1807

Family Littorinidae Gray, 1840 (Pl. 3)
Subfamily **Echinininae** Rosewater, 1973
Echininus (Echininus) Clench & Abbott, 1942
Subfamily **Littorininae** Gray, 1840
Littorina (Littoraria) Gray, 1834
Littorina (Littorina) Ferussac, 1822
Littorina (Littorinopsis) Mörch, 1876
Subfamily **Tectariinae** Rosewater, 1973
Tectarius (Tectarius) Valenciennes, 1832

Family Lucinidae Fleming, 1828 (Pl. 88)
Subfamily **Lucininae** Fleming, 1828
Codakia Scopoli, 1777

Family Mactridae Lamarck, 1809 (Pl. 85)
Subfamily **Mactrinae** Lamarck, 1809
Mactra (Mactra) Linné, 1767

Family Malleidae Lamarck, 1819 (Pl. 89)
Malleus Lamarck, 1799
Vulsella Röding, 1798 (often placed in its own family **Vulsellidae** H. & A. Adams, 1857)

Family Marginellidae Fleming, 1828 (Pl. 96)
Subfamily **Cystiscinae** Coan, 1965
Granulina Jousseaume, 1888
Subfamily **Marginellinae** Fleming, 1828
Bullata (Bullata) Jousseaume, 1875
Bullata (Closia) Gray, 1857
Bullata (Cryptospira) Hinds, 1844
Hyalina (Hyalina) Schumacher, 1817
Hyalina (Volvarina) Hinds, 1840 (possibly a full genus)
Marginella Lamarck, 1799

Family Mathildidae Dall, 1889 (Pl. 12)
Mathilda Semper, 1865

Family Melampidae Stimpson, 1851 (Pl. 81)
Auriculastra von Martens, 1880
Cassidula Férusac, 1821
Ellobium Röding, 1798
Melampus Montfort, 1810
Pythia Röding, 1798

Family Melanellidae Bartsch, 1917 (Pl. 12)
(For convenience, we have incorporated 2 additional families into this group: **Stiliferidae** H. & A. Adams, 1853 and **Aclididae** G. O. Sars, 1878)
Melanella (Balcis) Leach in Gray, 1847
Melanella (Melanella) Bowdich, 1822
Mucronalia A. Adams, 1862

Family Melongenidae Gill, 1867 (Pl. 29)
Subfamily **Melongeninae** Gill, 1867
Volema (Hemifusus) Swainson, 1840
Volema (Melongena) Schumacher, 1817
Volema (Pugilina) Schumacher, 1817
Volema (Volema) Röding, 1798

Family Mitridae Swainson, 1831 (Pls. 53-57 incl.)
Subfamily **Cylindromitrinae** Cossmann, 1899
Pterygia Röding, 1798
Subfamily **Imbricariinae** Troschel, 1867
Cancilla (Cancilla) Swainson, 1840
Cancilla (Domiporta) Cernohorsky, 1970
Cancilla (Ziba) H. & A. Adams, 1853
Imbricaria Schumacher, 1817
Neocancilla Cernohorsky, 1966
Scabricola (Scabricola) Swainson, 1840
Scabricola (Swainsonia) H. & A. Adams, 1853
Subcancilla Olsson & Harbison, 1853
Subfamily **Mitrinae** Swainson, 1831
Mitra (Dibaphus) Philippi, 1847
Mitra (Mitra) Lamarck, 1798
Mitra (Nebularia) Swainson, 1840
Mitra (Strigatella) Swainson, 1840

Family Modulidae Fischer, 1884 (Pl. 12)
Modulus Gray, 1842

Family Muricidae da Costa, 1776 (Pls. 35-41 incl. 95/96)

This family contains a large and diverse assortment of genera and subgenera, the exact placement of which is often difficult to ascertain. A number of subfamily groups have been introduced to contain these genera, but in many cases, the boundaries are often ill-defined, and there is some degree of overlap between one group and another. The result has been that many genera are often placed in more than one subfamily grouping by different authors. This apparent inconsistency merely reflects the need for more detailed research into this complex family. The classification given below is based on a number of references, and agrees for the most part, with the currently accepted positioning of the genera concerned. Where a genus has been assigned to more than one subfamily grouping, this is annotated, with the provision that future research may possibly alter its position.

Subfamily **Aspellinae** Keen, 1971
Attiliosa Emerson, 1968 (? **Muricinae** da Costa, 1778)
Favartia (Caribiella) Perrilliat Monteya, 1871 (? **Muricopsinae** Radwin & D'Attilio, 1971)
Favartia (Favartia) Jousseaume, 1880 (? **Muricopsinae** Radwin & D'Attilio, 1971)
Phyllocoma Tapparone-Canefri, 1881
Subfamily **Muricinae** da Costa, 1776
Bedeva Iredale, 1924 (? **Trophoninae** Marwick, 1924)
Chicoreus (Chicomurex) Arakawa, 1964 (? = *Phyllonotus* Swainson, 1833)
Chicoreus (Chicoreus) Montfort, 1810
Chicoreus (Naquetia) Jousseaume, 1880
Cytharomorula Kuroda, 1953
Dermomurex Monterosato, 1890
Haustellum Schumacher, 1817
Hexaplex Perry, 1810
Homalocantha Mörch, 1852
Lataxiena Jousseaume, 1883
Murex Linné, 1758
Murexiella Clench & Farfante, 1945 (?**Muricopsinae** Radwin & D'Attilio, 1971)
Muricopsis Bucquoy & Dautzenberg, 1882 (? **Muricopsinae** Radwin & D'Attilio, 1971)
Pterynotus Swainson, 1833
Siratus Jousseaume, 1880

Subfamily **Rapaninae** Gray, 1853
Rapana Schumacher, 1817
Subfamily **Thaidinae** Jousseaume, 1888
Cronia (Cronia) H. & A. Adams, 1853
Cronia (Ergalatax) Iredale, 1931
Cronia (Muricodrupa) Iredale, 1918
Cronia (Pascula) Dall, 1908 (possibly a full genus)
Drupa (Drupa) Röding, 1798
Drupa (Drupina) Dall, 1923
Drupa (Ricinella) Schumacher, 1817
Drupella Thiele, 1925
Maculotriton Dall, 1904
Morula Schumacher, 1817
Nassa Röding, 1798
Pinaxia H. & A. Adams, 1853
Spinidrupa Habe & Kosuge, 1966
Thais (Mancinella) Link, 1807
Thais (Thais) Röding, 1798
Vexilla Swainson, 1840
Vitularia Swainson, 1840 (? **Muricopsinae** Radwin & D'Attilio, 1971)
Subfamily **Trophoninae** Cossmann, 1903
Boreotrophon Fischer, 1884
Subfamily **Typhinae** Cossmann, 1903

Talityphis Jousseaume, 1882
Typhinellus Jousseaume, 1880

Family Mytilidae Rafinesque, 1815 (Pl. 88)
Subfamily **Lithophaginae** H. & A. Adams, 1857
Lithophaga Röding, 1798
Subfamily **Modiolinae** Keen, 1958
Modiolus Lamarck, 1799
Stavelia Gray, 1858
Subfamily **Musculinae** Iredale, 1939
Amygdalum Megerle v Mühlfeld, 1811
Subfamily **Mytilinae** Rafinesque, 1815
Perna Philipsson, 1788
Septifer Reclüz, 1848

Family Nassariidae Iredale, 1916 (Pls. 41-42 incl.)
Cyllene Gray, in Griffith & Pidgeon, 1834
Nassarius (Alectrion) Montfort, 1810
Nassarius (Hebra) H. & A. Adams, 1853
Nassarius (Hima) Leach in Gray, 1852
Nassarius (Nassarius) Dumeril, 1806
Nassarius (Niotha) H. & A. Adams, 1853
Nassarius (Plicarcularia) Thiele, 1929
Nassarius (Telasco) H. & A. Adams, 1853
Nassarius (Zeuxis) H. & A. Adams, 1853

Family Naticidae Gray, 1840 (Pls. 9-10 incl.)
Subfamily **Ampullospirinae** Cox, 1930 (=**Globulariinae** Wenz, 1941)
Globularia Swainson, 1840
Subfamily **Naticinae** Gray, 1840
Natica (Natica) Scopoli, 1777
Natica (Naticarius) Dumeril, 1806
Tanea Marwick, 1931
Subfamily **Polinicinae** Gray, 1847
Polinices (Mammilla) Schumacher, 1817
Polinices (Neverita) Risso, 1826
Polinices (Polinices) Montfort, 1810
Subfamily **Sininae** Woodring, 1928
Eunaticina Fischer, 1885

Family Neritidae Rafinesque, 1815 (Pls. 10-11 incl.)
Subfamily **Neritinae** Rafinesque, 1815
Clithon Montfort, 1810
Nerita (Amphinerita) von Martens, 1887
Nerita (Nerita) Linné, 1758
Nerita (Ritena) Gray, 1858
Nerita (Theliostyla) Mörch, 1852
Neritina (Neritina) Lamarck, 1816
Neritina (Provittoida) Baker, 1923
Neritina (Vittoida) Baker, 1923
Neritodryas Martens, 1869
Septaria (Septaria) Férussac, 1807
Theodoxus (Clithon) Montfort, 1810
Theodoxus (Theodoxus) Montfort, 1810
Subfamily **Smaragdiinae** H. B. Baker, 1923
Smaragdia Issel, 1869

Family Neritopsidae Gray, 1847 (Pl. 11)
Neritopsis Grateloup, 1832

Family Olividae Latreille, 1825 (Pls. 50-52 incl.)
Subfamily **Olivinae** Swainson, 1840
Oliva (Carmione) Gray, 1858
Oliva (Galeola) Gray, 1858
Oliva (Neocylindrus) Fischer, 1883
Oliva (Oliva) Bruguiére, 1789
Oliva (Omogymna) von Martens, 1897
Oliva (Parvoliva) Thiele, 1929
Oliva (Strephona) Mörch, 1852

Family Ostreidae Rafinesque, 1815 (Pls. 89/90)
Subfamily **Lophinae** Vyalov, 1936
Lopha Röding, 1798
Subfamily **Ostreinae** Rafinesque, 1815
Crassostrea Sacco, 1897

Family Ovulidae Fleming, 1822 (Pls. 24-25 incl.)
Subfamily **Ovulinae** Fleming, 1822
Calpurnus (Calpurnus) Montfort, 1810
Calpurnus (Procalpurnus) Thiele, 1939
Crenavolva (Crenavolva) Cate, 1973
Delonovolva Cate, 1973
Dentiovula Habe, 1961
Ovula Bruguiére, 1789
Prionovula Iredale, 1930
Pseudosimnia (Diminovula) Iredale, 1930
Pseudosimnia (Inflatovula) Cate, 1973
Pseudosimnia (Pseudosimnia) Schilder, 1927
Testudovula Cate, 1973
Subfamily **Simniinae** Schilder, 1927
Hiata Cate, 1973
Phenacovolva (Calcaria) Cate, 1973
Phenacovolva (Phenacovolva) Iredale, 1939
Phenacovolva (Turbovula) Cate, 1973
Prosimnia Schilder, 1927
Sandalia Cate, 1973
Volva Röding, 1798

Family Patellidae Rafinesque, 1815 (Pl. 3)
Subfamily **Nacellinae** Thiele, 1929
Cellana H. Adams, 1869
Subfamily **Patellinae** Rafinesque, 1815
Patella (Patella) Linné, 1758
Patella (Scutellastra) H. & A. Adams, 1854

Family Pectinidae Rafinesque, 1815 (Pl. 93)
Subfamily **Amusiinae** Ridewood, 1903
Amusium Röding, 1798
Subfamily **Chlamydinae** Korobkov, 1960
Anachlamys Iredale, 1939
Anguipecten Dall, Bartsch & Rehder, 1938
Bractaechlamys Iredale, 1939
Chlamys (Azumapecten) Habe, 1977
Chlamys (Chlamys) Röding, 1798
Chlamys (Mimachlamys) Iredale, 1929
Chlamys (Mirapecten) Dall, Bartsch & Rehder, 1938
Comptopallium Iredale, 1939
Decatopecten Sowerby, 1838
Gloripallium Iredale, 1938
Semipallium Jousseaume in Lamy, 1928
Subfamily **Pectininae** Rafinesque, 1815
Pecten (Minnivola) Iredale, 1939
Pecten (Pecten) Müller, 1776
Serratovola Habe, 1951

Family Phasianellidae Swainson, 1840 (Pl. 5)
Phasianella Lamarck, 1804

Family Pholadidae Lamarck, 1809 (Pl. 88)
Subfamily **Pholadinae** Lamarck, 1809
Barnea (Barnea) Risso, 1826
Barnea (Umitakea) Habe, 1952
Monothyra Tryon, 1862

Family Pinnidae Leach, 1819 (Pl. 90)
Atrina (Atrina) Gray, 1847
Atrina (Servatrina) Iredale, 1939
Pinna Linné, 1758
Streptopinna Martens, 1880

Family Placunidae — see **Anomiidae**

Family Planaxidae Gray, 1850 (Pl. 12)
Planaxis Lamarck, 1822

Family Pleurotomariidae Swainson, 1840 (Pl. 1)
The genus *Pleurotomaria* Defranc, 1826, is based on fossil species, and should not be used for living taxa.
Perotrochus (Entemnotrochus) Fischer, 1885
Perotrochus (Mikadotrochus) Lindholm, 1927
Perotrochus (Perotrochus) Fischer, 1885

Family Potamididae H. & A. Adams, 1854 (Pl. 14)
Subfamily **Batillariinae** Thiele, 1929
Batillaria Benson, 1842
Subfamily **Potomidinae** H. & A. Adams, 1854
Cerithidea Swainson, 1840
Telescopium Montfort, 1810
Terebralia Swainson, 1840

Family Psammobiidae Fleming, 1828 (Pls. 85/86)
Subfamily **Psammobiinae** Fleming, 1828
Asaphis Modeer, 1793
Gari (Dysmea) Dall, Bartsch & Rehder, 1939
Gari (Gabraeus) Brown, 1844 (possibly a full genus)
Gari (Gari) Schumacher, 1817

Family Pteriidae Gray, 1847 (Pls. 89/90)
Pinctada Röding, 1798
Pteria (Austropteria) Iredale, 1931 (? = *Pteria* s.s.)
Pteria (Pteria) Scopoli, 1777

Family Pyramidellidae Gray, 1840 (Pl. 81)
Subfamily **Pyramidellinae** Gray, 1840
Milda Dall & Bartsch, 1904
Otopleura Fischer, 1885
Pyramidella Lamarck, 1799

Family Rissoidae Gray, 1847 (Pl. 12)
Subfamily **Rissoinidae** Stimpson, 1865
Rissoina (Rissoina) Orbigny, 1840
Rissoina (Zebinella) Mörch, 1876
Zebina (Morchiella) G. Nevill, 1885
Zebina (Zebina) H. & A. Adams, 1854

Family Siliquariidae Anton, 1839 (Pl. 8)
Siliquaria Bruguiére, 1789

Family Siphonariidae Gray, 1840 (Pl. 81)
Siphonaria Sowerby, 1824

Family Solecurtidae d'Orbigny, 1846 (Pl. 84)
Solecurtus Blainville, 1824

Family Solenidae Lamarck, 1809
Subfamily **Soleninae** Lamarck, 1809
Solen Linné, 1758

Family Spondylidae Gray, 1826 (Pl. 92)
Spondylus (Eleutherospondylos) Dunker, 1882
Spondylus (Eltoperna) Iredale, 1939
Spondylus (Spondylus) Linné, 1758

Family Strombidae Rafinesque, 1815 (Pls. 15-17 incl.)
Lambis (Harpago) Mörch, 1852
Lambis (Lambis) Röding, 1798
Lambis (Millipes) Mörch, 1852
Strombus (Canarium) Schumacher, 1817
Strombus (Conomurex) P. Fischer, 1884
Strombus (Dolomena) Iredale, 1831
Strombus (Doxander) Iredale, 1931
Strombus (Euprotomus) Gill, 1870
Strombus (Gibberulus) Jousseaume, 1888
Strombus (Labiostrombus) Coostingh, 1925
Strombus (Laevistrombus) Kira, 1955
Strombus (Lentigo) Jousseaume, 1886
Strombus (Strombus) Linné, 1758
Terebellum (Terebellum) Röding, 1798
Tibia (Rimella) Agassiz, 1841
Tibia (Tibia) Röding, 1798

Family Tellinidae Blainville, 1814 (Pls. 84/86)
Subfamily **Tellininae** Blainville, 1814
Scutarcopagia Pilsbry, 1918
Tellina (Pharaonella) Lamy, 1918
Tellina (Phylloda) Schumacher, 1817
Tellina (Tellina) Linné, 1758
Tellina (Tellinella) Mörch, 1853 (possibly a full genus)

Family Terebridae H. & A. Adams, 1854 (Pls. 73-75 incl.)
This family contains a myriad of genera and subgenera. Due to the uncertainty of the taxonomic position of many species, we have restricted ourselves to the use of only 4 genera.
Duplicaria Dall, 1908
Hastula H. & A. Adams, 1853
Impages Smith, 1873
Terebra Bruguiére, 1789

Family Tonnidae Peile, 1926 (Pl. 26)
Malea Valenciennes, 1833
Tonna Brünnich, 1772

Family Trichotropidae Gray, 1850 (Pl. 12)
Amathina Gray, 1842
Trichotropis (Trichotropis) Broderip & Sowerby, 1829
Trichotropis (Turritopsis) Habe, 1961

Family Triphoridae Gray, 1847 (Pl. 46)
Subfamily **Mastoniinae** Kosuge, 1966
Inella (Inella) Dayle, 1878
Inella (Cautotriphora) Laws, 1940
Mastonia Hinds, 1843
Tetraphora Laseron, 1958
Viriola Jousseaume, 1884

Family Triviidae Troschel, 1863 (Pl. 25)
Trivirostra Jousseaume, 1884

Family Trochidae Rafinesque, 1815 (Pls. 4/5)
Subfamily **Calliostomatinae** Thiele, 1921
Astele (Astele) Swainson, 1855
Astele (Coralastele) Iredale, 1930
Calliostoma Swainson, 1840 (= *Tristichotrochus* Ikebe, 1942)
Subfamily **Gibbulinae** Stoliczka, 1868
Gibbula (Cantharidella) Pilsbry, 1889
Gibula (Gibbula) Risso, 1826
Subfamily **Margaritinae**, Stoliczka, 1868
Euchelus (Euchelus) Philippi, 1847
Euchelus (Herpetopoma) Pilsbry, 1889
Lischkeia Fischer, 1879
Subfamily **Monodontinae** Cossmann, 1916
Cantharidus (Cantharidus) Montfort, 1810
Cantharidus (Iwakawatrochus) Kuroda & Habe, 1954
Monodonta (Monodonta) Lamarck, 1799
Subfamily **Stomatellinae** Gray, 1840
Pseudostomatella (Pseudostomatella) Theile, 1924
Stomatella (Stomatella) Lamarck, 1816
Stomatia (Stomatia) Helbling, 1799
Subfamily **Trochinae** Rafinesque, 1815
Clanculus (Clanculus) Montfort, 1810
Clanculus (Mesoclanculus) Cotton & Godfrey, 1934
Tectus (Tectus) Montfort, 1810
Trochus (Trochus) Linné, 1758
Subfamily **Umboniinae** H. & A. Adams, 1854
Ethminolia (Ethminolia) Iredale, 1924
Monilea (Monilea) Swainson, 1840
Umbonium (Suchium) Makiyama, 1925
Umbonium (Umbonium) Link, 1807

Family Turbinellidae Swainson, 1840 (Pls. 28/29)
Subfamily **Ptychatractinae** Stimpson, 1865
Teramachia Kuroda, 1931
Subfamily **Vasinae** H. & A. Adams, 1854
Vasum (Vasum) Röding, 1798

Family Turbinidae Rafinesque, 1815 (Pls. 7/8)
Subfamily **Astraeinae** Davies, 1933
Astraea Röding, 1798
Astralium Link, 1807 (? s.g. of *Astraea)*
Bolma (Bolma) Risso, 1826
Bolma (Galeoastraea) Kuroda & Habe, 1958
Guildfordia Gray, 1850
Subfamily **Turbininae** Rafinesque, 1815
Turbo (Lunella) Röding, 1798
Turbo (Marmarostoma) Swainson, 1829
Turbo (Turbo) Linné, 1758

Family Turridae Swainson, 1840 (Pls. 76-80 incl.)
Subfamily **Borsoniinae** Bellardi, 1875
Anarithma Iredale, 1916
Subfamily **Clathurellinae** H. & A. Adams, 1858 (= **Mangeliinae** Tryon, 1884)
Etrema (Etrema) Hedley, 1918
Eucithara Fischer, 1883
Glyphostoma Gabb, 1872
Horaiclavus (Cytharoclavus) Kuroda & Oyama, 1971
Horaiclavus (Horaiclavus) Oyama, 1954
Lienardia (Acrista) Hedley, 1922
Lienardia (Hemilienardia) Boettger, 1895
Lienardia (Lienardia) Jousseaume, 1834
Mangelia Risso, 1826
Paraclathurella Boettger, 1895
Subfamily **Clavinae** Powell, 1942
Brachytoma Swainson, 1840
Carinodrillia Dall, 1919
Clavus (Clavus) Montfort, 1810
Drillia (Drillia) Gray, 1838
Inguisitor Hedley, 1918
Ptychobela Thiele, 1925
Tomopleura Casey, 1904
Subfamily **Cochlespirinae** Powell, 1942 (= *Torriculinae* Powell, 1942)
Comitas Finlay, 1926
Vexitomina Powell, 1942

Subfamily **Mitromorphinae** Casey, 1904
Lovellona Iredale, 1917
Mitrolumna Bucquoy, Dautzenberg & Dollfus, 1883
Mitromorpha Carpenter, 1865
Mitromorpha (Lovellona) Iredale, 1917
Mitromorpha (Mitromorpha) Carpenter, 1865
Subfamily **Raphitominae** Bellardi, 1875 (= *Daphnellinae* Casey, 1904)
Asperdaphne (Asperdaphne) Hedley, 1922
Daphnella (Daphnella) Hinds, 1844
Philbertia (Philbertia) Monterosato, 1884
Philbertia (Pseudodaphnella) Boettger, 1895
Spergo (Spergo) Dall, 1895
Spergo (Speoides) Kuroda & Habe, 1961
Tritonoturris Dall, 1924
Veprecula Melvill, 1917
Subfamily **Thatcheriinae** Powell, 1942
Thatcheria Angas, 1877
Subfamily **Turrinae** Swainson, 1840
Bathytoma (Bathytoma) Harrison & Burrows, 1891
Bathytoma (Micantaplex) Iredale, 1936
Bathytoma (Parabathytoma) Shuto, 1961
Bathytoma (Riuguhdrillia) Oyama, 1951
Gemmula (Gemmula) Weinkauff, 1875
Gemmula (Unedogemmula) Mac Neil, 1960
Lophiotoma (Lophiotoma) Casey, 1904
Lophiotoma (Lophioturris) Powell, 1964
Pinguigemmula Mac Neil, 1960
Turridrupa Hedley, 1922
Turris (Turris) Röding, 1798
Xenoturris Iredale, 1929

Family Turritellidae Clarke, 1851 (Pl. 12)
Turritella Lamarck, 1799

Family Vanikoroidae Gray, 1845 (Pl. 12)
Vanikoro Quoy & Gaimard, 1832

Family Veneridae Rafinesque, 1815 (Pls. 85/86)
Subfamily **Chioninae** Frizzell, 1936
Anomalocardia (Anomalocardia) Schumacher, 1817
Anomalocardia (Anomalodiscus) Dall, 1902
Placemen Iredale, 1925
Timoclea (Timoclea) Brown, 1827
Timoclea (Veremolpa) Iredale, 1930 (possibly a full genus)
Subfamily **Circinae** Dall, 1898
Circe Schumacher, 1817
Gafrarium Röding, 1798
Subfamily **Dosiniinae** Deshayes, 1853
Dosinia (Bonartemis) Iredale, 1929 (possibly a full genus)
Dosinia (Dosinia) Scopoli, 1777
Dosinia (Phacosoma) Jukes-Brown, 1912 (possibly a full genus)
Subfamily **Lioconchinae** Iredale & Mc Michael, 1962
Lioconcha (Lioconcha) Mörch, 1853
Subfamily **Meretricinae** Fischer, 1887
Meretrix Lamarck, 1799
Subfamily **Pitarinae** Stewart, 1930
Pitar (Pitar) Römer, 1857
Pitar (Pitarina) Jukes-Browne, 1913
Subfamily **Tapetinae** H. & A. Adams, 1857
Katelysia Römer, 1857
Paphia (Neotapes) Kuroda & Habe, 1971
Paphia (Paphia) Röding, 1798
Tapes Megerle von Muhlfeld, 1811
Subfamily **Venerinae** Rafinesque, 1815
Periglypta (Periglypta) Jukes-Browne, 1914
Periglypta (Tigammona) Iredale, 1930
Venus (Antigonia) Schumacher, 1817 (possibly a full genus)
Venus (Venus) Linné, 1758

Family Vermetiidae Rafinesque, 1815 (Pl. 8)
Serpulorbis Sassi, 1827
Vermetus, Daudin, 1800

Family Volutidae Rafinesque, 1815 (Pl. 49)
Subfamily **Cymbiinae** H. & A. Adams, 1853
Cymbiola (Aulica) Gray, 1847
Cymbiola (Aulicina) Rovereto, 1899
Cymbiola (Cymbiola) Swainson, 1831
Melo (Melo) Broderip in Sowerby I, 1826
Melo (Melocorona) Pilsbry & Olsson, 1954
Subfamily **Lyriinae** Pilsbry & Olson, 1954
Lyria (Lyria) Gray, 1847

Family Xenophoridae Philippi, 1853 (Pl. 19)
Xenophora (Onustus) Swainson, 1840
Xenophora (Stellaria) (Schmidth MS) Möller, 1832
Xenophora (Xenophora) Fischer von Waldheim, 1807

Index

A

B

C

D

E

F

G

H

Q

R

S